EQUIVARIANT COHOMOLOGY IN ALGEBRAIC GEOMETRY

Equivariant cohomology has become an indispensable tool in algebraic geometry and in related areas including representation theory, combinatorial and enumerative geometry, and algebraic combinatorics. This text introduces the main ideas of the subject for first- or second-year graduate students in mathematics, as well as researchers working in algebraic geometry or combinatorics. The first six chapters cover the basics: definitions via finite-dimensional approximation spaces, computations in projective space, and the localization theorem. The rest of the text focuses on examples – toric varieties, Grassmannians, and homogeneous spaces – along with applications to Schubert calculus and degeneracy loci. Prerequisites are kept to a minimum, so that one-semester graduate-level courses in algebraic geometry and topology should be sufficient preparation. Featuring numerous exercises, examples, and material that has not previously appeared in textbook form, this book will be a must-have reference and resource for both students and researchers for years to come.

David Anderson is Associate Professor at The Ohio State University. He works in combinatorial algebraic geometry and has written over three dozen papers on topics including Schubert calculus, Newton–Okounkov bodies, and equivariant K-theory. In 2020, he received a CAREER Award from the National Science Foundation.

William Fulton is Oscar Zariski Distinguished University Professor Emeritus at the University of Michigan. He is an algebraic geometer, and author or co-author of approximately five dozen papers and a dozen books, including *Intersection Theory*, which won a Steele Prize from the American Mathematical Society. Fulton is a member of the National Academy of Sciences, and a foreign member of the Royal Swedish Academy of Sciences.

CAMBRIDGE STUDIES IN ADVANCED MATHEMATICS

All the titles listed below can be obtained from good booksellers or from Cambridge University Press. For a complete series listing, visit www.cambridge.org/mathematics.

Equivariant Cohomology
in Algebraic Geometry

DAVID ANDERSON
Ohio State University

WILLIAM FULTON
University of Michigan

Shaftesbury Road, Cambridge CB2 8EA, United Kingdom

One Liberty Plaza, 20th Floor, New York, NY 10006, USA

477 Williamstown Road, Port Melbourne, VIC 3207, Australia

314–321, 3rd Floor, Plot 3, Splendor Forum, Jasola District Centre,
New Delhi – 110025, India

103 Penang Road, #05–06/07, Visioncrest Commercial, Singapore 238467

Cambridge University Press is part of Cambridge University Press & Assessment,
a department of the University of Cambridge.

We share the University's mission to contribute to society through the pursuit of
education, learning and research at the highest international levels of excellence.

www.cambridge.org
Information on this title: www.cambridge.org/9781009349987
DOI: 10.1017/9781009349994

First published 2024

Printed in the United Kingdom by CPI Group Ltd, Croydon CR0 4YY

A catalogue record for this publication is available from the British Library

*A Cataloging-in-Publication data record for this book is available from the Library
of Congress*

ISBN 978-1-009-34998-7 Hardback

Contents

Preface

Given a Lie group G acting on a space X, the *equivariant cohomology ring H_G^*X* packages information about the interaction between the topology of X and the representation theory of G. On the one hand, it provides a way of exploiting the symmetry of X, as manifested by the G-action, to understand H^*X; on the other hand, appropriate choices of X are useful in studying representations of G.

Defined by A. Borel in his 1958–9 seminar on transformation groups, equivariant cohomology arose in the context of a problem of interest to topologists: Given some cohomological information about X, what can be said about the group actions X admits? Must there be fixed points? How many? By constructing an auxiliary space, Borel built a framework for answering these questions in special situations, for example, when G is a torus and X is a compact manifold satisfying a technical hypothesis (now known as *equivariant formality*).

It took several decades for ideas of equivariant cohomology to enter mainstream algebraic geometry. By 2000, though, localization had become a standard technique in Gromov–Witten theory and applications to enumerative geometry. Equivariant methods were also used in producing degeneracy locus formulas and in proving Littlewood–Richardson rules in Schubert calculus.

One reason for the lag may be the role of infinite-dimensional spaces. Indeed, Borel's construction produces a certain fiber bundle over the classifying space $\mathbb{B}G$, with fiber X. Classifying spaces are almost always infinite-dimensional, so they are certainly not algebraic varieties.

However, for the groups appearing most frequently in applications to algebraic geometry—linear algebraic groups, and especially torus groups—these spaces can be "approximated" by familiar finite-dimensional varieties. Such approximation spaces were introduced by Totaro in the late 1990s, building on ideas of Bogomolov, and they were incorporated into a theory of equivariant Chow groups by Edidin and Graham. The same ideas work equally well for cohomology, and in fact, some of the foundational notions are simpler for cohomology than for Chow groups.

Our aim in this text is to introduce the main ideas of equivariant cohomology to an audience with a general background in algebraic geometry. We therefore avoid using infinite-dimensional spaces in any essential way, relying instead on finite-dimensional approximations. A recurring theme is that studying the equivariant geometry of X is essentially the same as studying fiber bundles with fiber X. The fiber bundle point of view has a long tradition in algebraic geometry, and by emphasizing this, we hope that newcomers to equivariant cohomology will find that many of the constructions are already familiar.

In our choice of topics, we were guided by a desire to keep prerequisites minimal. Apart from a "Leray–Hirsch"-type lemma, and a few basic facts about Chern classes and cohomology classes of subvarieties, all that we need is standard material from first courses in algebraic topology and algebraic geometry. Projective spaces and Grassmannians are usually familiar to beginners, and they suffice to illustrate a broad range of equivariant phenomena. Toric varieties and homogeneous spaces are natural next steps, and here one already encounters the frontiers of current research.

On the other hand, this introductory text is not an all-inclusive reference, and we have left out many exciting topics, inevitably including ones which some researchers (even ourselves!) might consider essential. Readers will have to look elsewhere for the construction of equivariant cohomology via differential forms, for a detailed discussion of the moment map and the symplectic point of view, for applications to the cohomology of finite or discrete groups, and for equivariant K-theory and more exotic cohomology theories. Part of our aim is to prepare and encourage readers to explore the many excellent sources for learning about such things.

The book grew out of lectures, and we have tried to blend some of the organic character of a lecture series with the logical organization of a textbook. The first six chapters cover the basics, including a simple version of the localization theorem and an illustration of its application to the space of conics. This material is important for most users of equivariant cohomology. Refinements of the localization theorem, including the "GKM" description of equivariant cohomology, are given in Chapter 7. Here we employ some more technical arguments, and for the most part the results are not logically required elsewhere in the book.

The remainder of the text consists of examples and applications—to toric varieties, Grassmannians, flag varieties, and general homogeneous spaces.

Grassmannians and flag varieties are fascinating objects of study in their own right, and we give an account of their combinatorial structure and equivariant geometry in Chapters 9 and 10. These spaces also form part of the link between equivariant cohomology and degeneracy locus formulas: in a precise way, a formula for the cohomology class of a degeneracy locus is equivalent to one for the equivariant class of a certain Schubert variety. This connection motivated much of our perspective, and it is the subject of Chapter 11.

Projective spaces, Grassmannians, and flag varieties are examples of homogeneous spaces for the general linear group. Other classical groups —the symplectic and orthogonal groups—appear in a similar way, and their corresponding flag varieties are related to refined degeneracy locus problems. The problem of extending what is known for GL_n ("type A") to the other classical types has received much attention over the past few decades. For a complete telling of this story, putting all classical groups on equal footing, we must refer elsewhere. Chapters 13 and 14 provide a sample, describing the equivariant cohomology of symplectic flag varieties ("type C").

The type C degeneracy locus formulas require a new coefficient ring, and this raises a question: Where is the analogous coefficient ring in type A? The answer has become clear only in very recent work, involving a certain infinite-dimensional Grassmannian. (As usual, and in keeping with our general theme, it can also be understood via appropriate finite-dimensional approximations.) To provide a bridge between type A and type C, this is discussed rather briefly in Chapter 12.

Once one understands something about flag varieties for symplectic and orthogonal groups, it is natural to ask about general homogeneous spaces. These spaces play a key role in the story of equivariant cohomology, too: thanks to a theorem of Borel, if G is a reductive group with Borel subgroup B and maximal torus T, then the G-equivariant cohomology of any space on which G acts is related to its T-equivariant cohomology through the flag variety G/B. This is explained in Chapter 15, and further developed in Chapter 16.

There are several possible approaches to defining equivariant *homology*. One which is well suited to our theme of finite-dimensional approximation is presented in Chapter 17, based on ideas of Edidin, Graham, and Totaro. Equivariant Segre classes appear naturally in this context, as do the equivariant multiplicities introduced by Rossmann and Brion.

In Chapters 18 and 19, we conclude with a study of Schubert varieties in homogeneous spaces. Highlights include a formula for the restriction of a Schubert class to a fixed point (due to Andersen, Jantzen, Soergel, and Billey), a criterion for a Schubert variety to be nonsingular at a fixed point (following Kumar and Brion), and some formulas for multiplying equivariant Schubert classes, along with a theorem of Graham which asserts that such products always expand positively, in a suitable sense.

Each chapter ends with a "Notes" section, providing some limited historical and mathematical context, as well as references for material in the text. We have also included hints for many of the exercises, and complete solutions in a few cases.

Appendix A is a brief summary of basic results from algebraic topology which we need in the text. Much of this material is essential, and we advise the reader to review it before embarking on the main text. The other appendices may be perused as needed.

Early drafts of what became this book began with WF's Eilenberg Lectures at Columbia University in 2007, and DA's notes have been available online since then. In the meantime, both authors have given lectures augmenting and improving on these notes—in courses at the University of Michigan, the University of Washington, and the Ohio State University, and in lecture series at the Institute for Advanced Study in Princeton in 2007, at IMPANGA in Będlewo in 2010, and at IMPA in Rio de Janeiro in 2014. We heartily thank the many students, friends, and colleagues who attended these lectures and gave feedback on the notes. Special thanks go to P. Achinger, I. Cavey, J. de Jong, D. Genlik, O. Lorscheid, D. Speyer, and A. Zinger for their helpful comments, and most especially to M. Franz for his careful reading of an earlier draft. Finally, we thank M. Brion, D. Edidin, W. Graham, and B. Totaro for their influence on our understanding of the subject.

1

Preview

Before beginning in earnest, we offer a taste of the themes and topics this book will explore. Although we give some definitions and sketches of arguments here, the reader should rest assured that later chapters will provide more detail.

Throughout this book, G is a Lie group. Usually – though not always – it will be a complex linear algebraic group. Unless otherwise specified, cohomology is always taken with integer coefficients.

1.1   The Borel construction

Suppose a Lie group G acts on a space X (on the left). The standard definition of the G-equivariant cohomology of X, written $H_G^* X$, goes like this. Find a contractible space $\mathbb{E}G$ with G acting freely (on the right), and form the quotient

$$\mathbb{E}G \times^G X := (\mathbb{E}G \times X)/(e \cdot g, x) \sim (e, g \cdot x).$$

Then define

$$H_G^i X := H^i(\mathbb{E}G \times^G X).$$

The idea behind this definition is to have $H_G^i X = H^i(G \backslash X)$ when the action on X is free; replacing X by $\mathbb{E}G \times X$ leaves the homotopy type unchanged, but produces a free action, with quotient $\mathbb{E}G \times^G X$. This construction first appeared (unnamed) in Borel's 1958–9 seminar on transformation groups, so the space $\mathbb{E}G \times^G X$ is often called the *Borel construction*.

The space $\mathbb{B}G := \mathbb{E}G/G$ is a classifying space for G and it, along with the quotient map $\mathbb{E}G \to \mathbb{B}G$, is universal in an appropriate category, so this definition is independent of choices. We will not need this general topological machinery, though.

The case when X is a point is important. Here we are looking at

$$\Lambda_G := H_G^*(\mathrm{pt}) = H^*\mathbb{B}G.$$

Since $\mathbb{B}G$ usually has nontrivial cohomology, $H_G^*(\mathrm{pt}) \neq \mathbb{Z}$ in general! This is an essential feature of equivariant cohomology.

Example 1.1.1. For the multiplicative group $G = \mathbb{C}^*$, we can take $\mathbb{E}G = \mathbb{C}^\infty \smallsetminus \{0\}$. Certainly G acts freely, and it is a pleasant exercise to prove this space is contractible. The quotient is $\mathbb{B}G = \mathbb{CP}^\infty$. This lets us compute our first equivariant cohomology ring:

$$\Lambda_{\mathbb{C}^*} = H_{\mathbb{C}^*}^*(\mathrm{pt}) = H^*\mathbb{CP}^\infty = \mathbb{Z}[t],$$

where t is the Chern class of the tautological line bundle on $\mathbb{CP}^\infty$.

For the circle group $G = S^1$, regarded as the unit complex numbers, we can use $\mathbb{E}G = S^\infty$, regarded as the unit sphere in $\mathbb{C}^\infty$. This is contractible, since $\mathbb{C}^\infty \smallsetminus \{0\}$ retracts onto it, and we obtain the same quotient space $\mathbb{B}G = \mathbb{CP}^\infty$ as for $\mathbb{C}^*$. Alternatively, we could use the same space $\mathbb{E}G = \mathbb{C}^\infty \smallsetminus \{0\}$, since the subgroup $G = S^1 \subseteq \mathbb{C}^*$ acts freely here. Either way, we obtain

$$\Lambda_{S^1} = \Lambda_{\mathbb{C}^*} = \mathbb{Z}[t].$$

This is an instance of a general phenomenon: cohomology for a complex group is the same as for a maximal compact subgroup.

Example 1.1.2. Elaborating on the previous example, for the torus $T = (\mathbb{C}^*)^n$ we can take $\mathbb{E}T = (\mathbb{C}^\infty \smallsetminus \{0\})^n$ to get $\mathbb{B}T = (\mathbb{CP}^\infty)^n$. We find

$$\Lambda_T = \mathbb{Z}[t_1, \ldots, t_n],$$

where t_i comes from the tautological bundle on the ith factor of $(\mathbb{P}^\infty)^n$. As before, we get the same result for the compact torus $(S^1)^n \subseteq (\mathbb{C}^*)^n$.

Early applications of equivariant cohomology were topological, focusing on questions about how the cohomology of a space constrains the group actions it admits. Algebraic geometers were slower to realize its utility, perhaps because the spaces $\mathbb{E}G \times^G X$ are generally infinite-dimensional (as we've already seen). However, some of the core ideas of equivariant cohomology had been used in algebraic geometry for quite a while. The space $\mathbb{E}G \times^G X$ is a fiber bundle over the classifying space $\mathbb{B}G$, with fiber X, and the study of such bundles goes back at least to Ehresmann in the 1940's. In algebraic geometry, fiber bundle constructions are familiar and ubiquitous – we are used to going from a vector space to a vector bundle, projective space to projective bundle, or Grassmannian to Grassmann bundle. A key theme for us is that equivariant cohomology is intimately linked to the study of general fiber bundles.

In fact, we will work with an alternative (but equivalent) definition of H_G^* which stays within the realm of finite-dimensional spaces. This involves using "approximations" $\mathbb{E}_m$ to $\mathbb{E}G$, and each $\mathbb{E}_m \times^G X$ will be a finite-dimensional algebraic manifold whenever X is. (A technical assumption on G or X may be necessary, to guarantee algebraicity of the quotient, but it will be automatic in most applications.) For instance, we'll use $\mathbb{E}_m = \mathbb{C}^m \smallsetminus 0 \to \mathbb{B}_m = \mathbb{P}^{m-1}$ to approximate $\mathbb{B}\mathbb{C}^*$. In the next chapter, we'll prove lemmas that show this leads to a well-defined theory.

As we'll see, equivariant cohomology shares many familiar properties with ordinary (singular) cohomology: it is functorial (contravariant for equivariant maps), has Chern classes (for equivariant vector bundles), and fundamental classes (for invariant subvarieties of a nonsingular variety). Most of these properties are verified by doing the analogous construction for ordinary cohomology on $\mathbb{E}G \times^G X$ (or an approximation).

1.2 Fiber bundles

The Borel construction produces a certain fiber bundle from the action of G on X: the fiber is X, and the base is $\mathbb{B}G$ (or an approximation). It is helpful to think of the diagram

$$
\begin{array}{ccc}
X & \longhookrightarrow & \mathbb{E} \times^G X \\
\downarrow & & \downarrow \\
\mathrm{pt} & \longhookrightarrow & \mathbb{B}
\end{array}
$$

with the vertical arrow on the right coming from the projection on the first factor. Pullback along the horizontal arrows – that is, restriction to a fiber – defines a forgetful homomorphism $H_G^* X \to H^* X$, from equivariant to ordinary cohomology. Pullback along the vertical arrows gives homomorphisms

$$\mathbb{Z} = H^*(\mathrm{pt}) \to H^* X \quad \text{and} \quad \Lambda_G = H_G^*(\mathrm{pt}) \to H_G^* X.$$

The first of these is trivial, but the second endows $H_G^* X$ with the richer structure of a Λ_G-algebra, at least when this ring is commutative. In some cases, this structure is rich enough to determine X itself! (If one includes the data of the equivariant first Chern class, this happens when X is a compact toric manifold – see Chapter 8, Exercise 8.4.3.)

Example.   The standard action of $T = (\mathbb{C}^*)^n$ on $\mathbb{C}^n$ (by scaling coordinates) defines an action on $\mathbb{P}^{n-1} = \mathbb{P}(\mathbb{C}^n)$, giving the universal quotient line bundle $\mathcal{O}(1)$ an equivariant structure. Writing $\zeta = c_1^T(\mathcal{O}(1))$ for its equivariant Chern class in $H_T^2 \mathbb{P}^{n-1}$, we have

$$H_T^* \mathbb{P}^{n-1} = \Lambda_T[\zeta] / \prod_{i=1}^n (\zeta + t_i).$$

We will work this out in detail soon; it follows easily from the general formula for the cohomology of a projective bundle. Note that sending $t_i \mapsto 0$ for all i defines a surjection $H_T^* \mathbb{P}^{n-1} \to H^* \mathbb{P}^{n-1} = \mathbb{Z}[\bar\zeta]/(\bar\zeta^n)$, where $\bar\zeta = c_1(\mathcal{O}(1))$ is the ordinary Chern class.

1.3   The localization package

The possibility of carrying out global computations using only local information at fixed points provides one of the most powerful applications of equivariant cohomology. This works best when $G = T$ is a torus. Two notions underwrite this technique. The first is that *equivariant cohomology should determine ordinary cohomology*: in good situations,

   $H_T^* X \to H^* X$ is *surjective*, with kernel generated by the kernel of $\Lambda_T \to \mathbb{Z}$.

The second notion is that *equivariant cohomology should be determined by fixed points*: in good situations, for $\iota \colon X^T \hookrightarrow X$,

$H_T^* X \xrightarrow{\iota^*} H_T^* X^T$ is *injective*, and becomes an isomorphism after inverting enough elements of Λ_T.

We will also see theorems characterizing the image of ι^*.

Both of these are desired properties, but they can certainly fail for a given action of T on X. (For example, if X has no fixed points, it will be difficult for the second notion to hold; if $X = T$, acting on itself by translation, then both properties fail. A useful exercise is to look for other examples where one or both of these properties fail.) In plenty of common situations, though, both do hold true: for example, whenever X is a nonsingular projective variety with finitely many fixed points. Theorems about when these properties hold form the core of the *localization package*.

Another component of this package is an *integration formula* which computes the pushforward along a proper map of nonsingular varieties, $f \colon X \to Y$, via restriction to fixed points. In the especially useful case of $\rho \colon X \to \mathrm{pt}$, with X^T finite, this takes the form

$$\int_X \alpha := \rho_*(\alpha) = \sum_{p \in X^T} \frac{\alpha|_p}{c_{\mathrm{top}}^T(T_p X)},$$

where the right-hand side is a finite sum of elements of the fraction field of Λ_T – that is, rational functions in the variables t_i.

All of this package consists of essentially equivariant phenomena: for any space X with nontrivial positive-degree cohomology and finitely many fixed points, you could never have an injection $H^* X \to H^* X^T$, by degree! Similarly, the right-hand side of the integration formula is only defined equivariantly, since the denominators are positive-degree elements of $H_T^*(\mathrm{pt})$.

Localization fits into the fiber bundle picture via sections: in terms of the previous diagram, we have

$$\iota_p \left(\begin{array}{ccc} X & \hookrightarrow & \mathbb{E} \times^G X \\ \downarrow & & \downarrow \\ \mathrm{pt} & \longleftarrow & \mathbb{B} \end{array} \right)$$

with the inclusion $\iota_p \colon \{p\} \hookrightarrow X$ of a fixed point inducing a section of the fiber bundle $\mathbb{E} \times^G X \to \mathbb{B}$. Pulling back along this section gives the restriction homomorphism $\iota_p^* \colon H_G^* X \to H_G^*(p) = \Lambda_G$.

1.4 Schubert calculus and Schubert polynomials

Our two main thematic strands – fiber bundles and localization – braid together nicely in modern Schubert calculus. Here X is a projective homogeneous space, for example, $\mathbb{P}^{n-1}$, $Gr(d, \mathbb{C}^n)$, $Fl(\mathbb{C}^n)$, or more generally, G/P for a reductive group G and parabolic subgroup P. The cohomology ring has a basis of Schubert classes $[\Omega_w]$, where $\Omega_w \subseteq X$ is a Schubert variety, defined by certain incidence conditions. These subvarieties are invariant for the action of a torus, and in fact their equivariant classes $\sigma_w = [\Omega_w]^T$ form a Λ_T-basis for $H_T^* X$. (The set W indexing the Schubert basis is a quotient of the Weyl group of G. For $G = GL_n$, this is the symmetric group S_n.)

A central problem is to understand these classes σ_w. In particular, one would like expressions for them as polynomials in ring generators for $H_T^* X$, formulas for their restrictions to fixed points, and combinatorial rules for their multiplication. The last of these is a long-standing open problem: one can write

$$\sigma_u \cdot \sigma_v = \sum_w c_{uv}^w \, \sigma_w,$$

for some homogeneous polynomials c_{uv}^w in $\Lambda_T = \mathbb{Z}[t_1, \ldots, t_n]$. What are these polynomials?

The structure constants c_{uv}^w satisfy a positivity property: when written in appropriate variables, these polynomials have nonnegative coefficients. (When there is no torus, the structure constants are nonnegative integers, by an application of the Kleiman–Bertini transversality theorem.) The problem is to find a combinatorial formula for c_{uv}^w manifesting this positivity. Good answers are known for some spaces – Grassmannians, cominuscule varieties, 3-step flag varieties – but even the non-equivariant question remains open in most cases, despite much recent progress. A key theme in recent advances is that equivariant techniques aid in proving non-equivariant theorems.

One can say more about the other problems. There is an elegant formula for restricting σ_w to a fixed point p_u, expressed as a sum over certain reduced words in the Weyl group. And there are good formulas for representing σ_w as a polynomial in Chern classes, at least in classical types.

We will focus on "type A", and in particular the complete flag variety $Fl(\mathbb{C}^n)$. For each permutation $w \in S_n$, there are Schubert classes $[\Omega_w]$ in $H^* Fl(\mathbb{C}^n)$ and $[\Omega_w]^T$ in $H^*_T Fl(\mathbb{C}^n)$. In 1982, Lascoux and Schützenberger defined and initiated the study of *Schubert polynomials* $\mathfrak{S}_w(x)$ in $\mathbb{Z}[x_1, \ldots, x_n]$, which are homogeneous polynomials mapping to $[\Omega_w]$ under a ring presentation $\mathbb{Z}[x] \twoheadrightarrow H^* Fl(\mathbb{C}^n)$.

There are also *double Schubert polynomials*

$$\mathfrak{S}_w(x; y) \in \mathbb{Z}[x_1, \ldots, x_n, y_1, \ldots, y_n],$$

and it was later proved that these map to $[\Omega_w]^T$ in $H^*_T Fl(\mathbb{C}^n) = \mathbb{Z}[x, y]/I$. Since $H^*_T Fl(\mathbb{C}^n)$ is a quotient of a polynomial ring, there are necessarily many choices for polynomials representing $[\Omega_w]^T$, but it is generally agreed that $\mathfrak{S}_w(x; y)$ are the best ones. They have many wonderful combinatorial and geometric properties, and we will study them in detail later.

Briefly, here is a different way the polynomials $\mathfrak{S}_w(x; y)$ arise, which would have been familiar to mathematicians working over 100 years earlier. We will place rank conditions on $n \times n$ matrices, and compute the degree of the corresponding variety defined by the vanishing of certain minors. This sort of problem was studied by nineteenth-century geometers, especially Cayley, Salmon, Roberts, and Giambelli. For a permutation $w \in S_n$, consider the (transposed) permutation matrix $A^\dagger_w$ having 1's in the $w(i)$th column of the ith row (position $(i, w(i))$) and 0's elsewhere. For example, the permutation $w = 2\,3\,1$ has matrix

$$A^\dagger_{2\,3\,1} = \begin{pmatrix} 0 & 1 & 0 \\ 0 & 0 & 1 \\ 1 & 0 & 0 \end{pmatrix}.$$

Let $A[p, q]$ denote the upper-left $p \times q$ submatrix of any matrix A, and define

$$D_w = \left\{ A \in M_{n,n} \mid \mathrm{rk}(A[p, q]) \leq \mathrm{rk}(A^\dagger_w[p, q]) \text{ for all } 1 \leq p, q \leq n \right\}.$$

This is an irreducible subvariety of $M_{n,n} \cong \mathbb{A}^{n^2}$, of codimension $\ell(w) = \{i < j \mid w(i) > w(j)\}$. It is invariant for an action of $T = (\mathbb{C}^*)^n \times (\mathbb{C}^*)^n$ which scales rows and columns: viewing each factor of T as diagonal matrices, $(u, v) \cdot A = u\, A\, v^{-1}$. This means there is a class

$$[D_w]^T \in H^*_T M_{n,n} = \Lambda_T = \mathbb{Z}[x_1, \ldots, x_n, y_1, \ldots, y_n],$$

homogeneous of degree $\ell(w)$ in the variables. (Since $M_{n,n}$ is contractible, its equivariant cohomology is that of a point.)

Theorem. *This class equals the Lascoux–Schützenberger double Schubert polynomial:* $[D_w]^T = \mathfrak{S}_w(x; y)$.

This theorem is one piece of evidence of the naturality of Schubert polynomials, as well as the advantage of working equivariantly: there are no relations in the polynomial ring $H_T^* M_{n,n}$, so no choices.

Example. The locus $D_{2\,3\,1}$ is defined by two equations, $a_{11} = a_{21} = 0$. Each coordinate a_{ij} comes with T-weight $x_i - y_j$, so Bézout's theorem implies $\mathfrak{S}_{2\,3\,1}(x; y) = (x_1 - y_1)(x_2 - y_1)$.

Notes

In addition to the original construction, many of the core ideas of equivariant cohomology appear in Borel's seminar on transformation groups (Borel, 1960). This includes the fiber bundle and localization perspective, as well as the idea of approximating by finite dimensional spaces. (They used CW complexes, not algebraic varieties.) In modern language, the Borel construction can be regarded as a "homotopy quotient" of X by G, since it is the homotopy colimit of a diagram $G \times X \rightrightarrows X$. Alternatively, one can view $H_G^* X$ as the cohomology of the quotient stack $[G\backslash X]$. See (Behrend, 2004) for an introduction to the stack perspective.

See (Milnor and Stasheff, 1974, §14) for the computation of $H^* \mathbb{CP}^\infty$, used in Example 1.1.1. A proof of the contractibility of $\mathbb{C}^\infty \smallsetminus \{0\}$ can be found in (Hatcher, 2002, Ex. 1B.3).

Much of the current work on equivariant cohomology in algebraic geometry has roots in the story of modern Schubert calculus. Recent breakthroughs in the structure constant problem begin with Knutson and Tao's puzzle rule for Grassmannians (Knutson and Tao, 2003), which we will see in Chapter 9. Since then, formulas for two-step flag varieties have been found (Coşkun, 2009; Buch, 2015; Buch et al., 2016), as well as very recent formulas for two- and three-step flags (Knutson and Zinn-Justin, 2020). There are also some rules for classical Schubert calculus on certain spaces G/P for groups G other than GL_n. Pragacz (1991) showed that a formula of Stembridge computes the structure constants of the Lagrangian Grassmannian, and more generally, Thomas and Yong (2009) have found type-uniform formulas for all cominuscule flag varieties.

The restriction formula for equivariant Schubert classes at fixed points is due to Andersen et al. (1994) and Billey (1999); we will prove it in Chapter 18. The relationship between double Schubert polynomials and equivariant classes was established in the 1990s (Fulton, 1992; Graham, 1997; Fehér and Rimányi, 2003; Knutson and Miller, 2005).

2

Defining Equivariant Cohomology

We will introduce our definition of equivariant cohomology using finite-dimensional algebraic varieties, constructing a contravariant functor from spaces with G-action to rings, and compute several examples of Λ_G from this definition. First we need some basic facts about principal bundles, which predate equivariant cohomology and to some extent motivate its original construction.

2.1   Principal bundles

Before discussing the general setup, here is a special case which may be familiar. Suppose E is a complex vector bundle of rank n on a space Y, so it is trivialized by some open cover U_α. The transition functions (from $U_\alpha \cap U_\beta$ to GL_n) can be used to construct a principal GL_n-bundle. Explicitly, let

$$p \colon \mathrm{Fr}(E) \to Y$$

be the *frame bundle* of E, whose fiber over $y \in Y$ is the set of all ordered bases $(v_1, \ldots, v_n)$ of E_y.   (This is also known as the *Stiefel variety* of E.) There is a natural right action of GL_n on $\mathrm{Fr}(E)$, given by

$$(v_1, \ldots, v_n) \cdot g = (w_1, \ldots, w_n), \quad \text{where } w_j = \sum_{i=1}^{n} g_{ij} \, v_i,$$

and over an open set $U \subseteq Y$ where E is trivial, the isomorphism

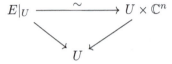

gives rise to

$$\mathrm{Fr}(E|_U) = p^{-1}(U) \xrightarrow{\ \sim\ } U \times GL_n,$$

so $\mathrm{Fr}(E)/GL_n = Y$. This bundle $p \colon \mathrm{Fr}(E) \to Y$, together with its GL_n action, is called the *associated principal bundle* to the vector bundle E. One can recover E from its associated principal bundle via isomorphisms

$$
\begin{array}{ccc}
\mathrm{Fr}(E) \times^{GL_n} \mathbb{C}^n & \xrightarrow{\ \sim\ } & E \\
\downarrow & & \downarrow \\
\mathrm{Fr}(E) \times^{GL_n} \mathrm{pt} & \xrightarrow{\ \sim\ } & Y,
\end{array}
\qquad (v_1, \ldots, v_n) \times (z_1, \ldots, z_n) \mapsto \sum_{i=1}^{n} z_i v_i.
$$

Here we are using the *balanced product* notation introduced in Chapter 1 when describing the Borel construction: in general, if G acts on the right on a space X, and on the left on a space Y, then

$$X \times^G Y$$

is the quotient of $X \times Y$ by the relation $(x \cdot g, y) \sim (x, g \cdot y)$.

The associated bundle can be used to construct other bundles. Multilinear constructions on the standard GL_n-representation $\mathbb{C}^n$ lead to analogous ones on E. For instance, one has

$$\mathrm{Fr}(E) \times^{GL_n} (\mathbb{C}^n)^\vee \cong E^\vee,$$
$$\mathrm{Fr}(E) \times^{GL_n} \textstyle\bigwedge^d \mathbb{C}^n \cong \textstyle\bigwedge^d E,$$
$$\mathrm{Fr}(E) \times^{GL_n} \mathrm{Sym}^d \mathbb{C}^n \cong \mathrm{Sym}^d E.$$

Using the (left) action of GL_n on projective space $\mathbb{P}^{n-1} = \mathbb{P}(\mathbb{C}^n)$, the Grassmannian $Gr(d, \mathbb{C}^n)$, or flag variety $Fl(\mathbb{C}^n)$, one obtains projective bundles, Grassmann bundles, and flag bundles:

$$\mathrm{Fr}(E) \times^{GL_n} \mathbb{P}(\mathbb{C}^n) \cong \mathbb{P}(E),$$
$$\mathrm{Fr}(E) \times^{GL_n} Gr(d, \mathbb{C}^n) \cong \mathbf{Gr}(d, E),$$
$$\mathrm{Fr}(E) \times^{GL_n} Fl(\mathbb{C}^n) \cong \mathbf{Fl}(E).$$

In fact, any space X with a left GL_n-action produces a bundle

$$\mathrm{Fr}(E) \times^{GL_n} X \to Y$$

which is locally trivial with fiber X. Special cases of this construction will give us $H_{GL_n}^k(X)$, at least if $\widetilde{H}^i(\mathrm{Fr}(E)) = 0$ for $i \le k$. Often it is simpler and more natural to study bundles in general, keeping in mind that this special case recovers equivariant cohomology.

Exercise 2.1.1. For $d \le n$, let $\mathrm{Fr}(d, E) \to Y$ be the bundle whose fiber over y is

$$\Big\{ (v_1, \ldots, v_d) \,\big|\, v_1, \ldots, v_d \text{ are linearly independent in the fiber } E_y \Big\}.$$

There is a right GL_d-action, as before. Show that $\mathrm{Fr}(d, E) \times^{GL_d} \mathbb{C}^d$ is naturally identified with the tautological rank d subbundle $\mathbb{S} \subseteq E_{\mathbf{Gr}}$ on the Grassmann bundle $\pi \colon \mathbf{Gr}(d, E) \to Y$, where $E_{\mathbf{Gr}} = \pi^* E$ is the pullback vector bundle.

Exercise 2.1.2. Note that $\mathrm{Fr}(d, E)$ is an open subspace of the Hom bundle $\mathrm{Hom}(\mathbb{C}_Y^d, E)$, where $\mathbb{C}_Y^d = Y \times \mathbb{C}^d$ is the trivial bundle. Use a similar open subset of $\mathrm{Hom}(E, \mathbb{C}_Y^{n-d})$ to construct the tautological rank $n - d$ quotient bundle $E_{\mathbf{Gr}} \twoheadrightarrow \mathbb{Q} = E_{\mathbf{Gr}}/\mathbb{S}$ on $\mathbf{Gr}(d, E)$.

Generally, for a Lie group G, a *(right) principal G-bundle* is

$$p \colon \mathbb{E} \to \mathbb{B},$$

where G acts freely on $\mathbb{E}$ (on the right) and the map p is isomorphic to the quotient map $\mathbb{E} \to \mathbb{E}/G$. We will always assume such bundles are *locally trivial*, so that $\mathbb{B}$ is covered by open sets U, with G-equivariant isomorphisms $p^{-1}U \cong U \times G$, where G acts on $U \times G$ by right multiplication on itself.

Exercise 2.1.3. With G acting by right multiplication on itself, trivially on a space $\mathbb{B}$, and on the left on a space X, show that there is a canonical isomorphism

$$(\mathbb{B} \times G) \times^G X \cong \mathbb{B} \times X.$$

Exercise 2.1.4. Suppose G acts on the right on $\mathbb{E}$ and on the left on X, and H acts on the right on X and on the left on Y, compatibly so that

$$g \cdot (x \cdot h) = (g \cdot x) \cdot h$$

for all $g \in G$, $x \in X$, and $h \in H$. Show that there is a canonical isomorphism

$$(\mathbb{E} \times^G X) \times^H Y \cong \mathbb{E} \times^G (X \times^H Y).$$

Remark. If one restricts to a category of paracompact and Hausdorff spaces, there is a *universal principal G-bundle* $\mathbb{E}G \to \mathbb{B}G$, with the property that any principal bundle $\mathbb{E} \to \mathbb{B}$ comes from the universal one by a pullback

$$
\begin{array}{ccc}
\mathbb{E} & \longrightarrow & \mathbb{E}G \\
\downarrow & & \downarrow \\
\mathbb{B} & \longrightarrow & \mathbb{B}G
\end{array}
$$

for some map $\mathbb{B} \to \mathbb{B}G$, uniquely defined up to homotopy. The base $\mathbb{B}G$ of such a bundle is called a *classifying space* for G. In fact, a principal bundle $\mathbb{E} \to \mathbb{B}$ is universal if and only if $\mathbb{E}$ is contractible.

The conditions paracompact and Hausdorff guarantee that partitions of unity exist, which is what is needed to construct the classifying map $\mathbb{B} \to \mathbb{B}G$. Any complex algebraic variety has these properties. On the other hand, for most groups G, there is no (finite-dimensional) algebraic variety $\mathbb{E}$ which is contractible and admits a free G-action, so the classifying space $\mathbb{B}G$ cannot be represented by any algebraic variety. See Appendix E for an algebraic approach to this universal property.

We will not need the universal construction in our approach to equivariant cohomology. Instead, we construct finite-dimensional algebraic varieties which "approximate" $\mathbb{E}G \to \mathbb{B}G$, and suffice to compute cohomology in any finite degree.

2.2   Definitions

The equivariant cohomology groups H_G^i will be contravariant functors for G-equivariant maps $f \colon X \to Y$, and $H_G^* X = \bigoplus_{i \geq 0} H_G^i X$ will be a ring. To define $H_G^i X$ in any range $i < N$ (with N a positive integer or infinity), it suffices to find a principal G-bundle $\mathbb{E} \to \mathbb{B}$ with $\widetilde{H}^i \mathbb{E} = 0$ for $i < N$. (That is, $\mathbb{E}$ is path-connected and $H^i \mathbb{E} = 0$ for $0 < i < N$.) Then we set

$$H_G^i X := H^i(\mathbb{E} \times^G X) \qquad \text{for} \quad i < N.$$

To use this definition, we must show it is independent of choices, and we must also find spaces $\mathbb{E}$ with N arbitrarily large.

For any G which embeds as a closed subgroup of GL_n, we have an answer to the second point.

Lemma 2.2.1. *Let G be any complex linear algebraic group, and $N > 0$ an integer. There are nonsingular finite-dimensional algebraic varieties $\mathbb{E}$ and $\mathbb{B}$, with $\widetilde{H}^i\mathbb{E} = 0$ for $i < N$, and G acting freely on $\mathbb{E}$ so that $\mathbb{E} \to \mathbb{B} = \mathbb{E}/G$ is a principal G-bundle which is locally trivial in the complex topology.*

In §2.4 we will give an explicit construction of $\mathbb{E}$ making the proof of the lemma clear. Let us grant this for now, and check that the definition does not depend on the choice of $\mathbb{E}$.

Proposition 2.2.2. *If $\mathbb{E} \to \mathbb{B}$ and $\mathbb{E}' \to \mathbb{B}'$ are principal G-bundles with $\widetilde{H}^i\mathbb{E} = \widetilde{H}^i\mathbb{E}' = 0$ for $i < N$, then there are canonical isomorphisms of cohomology groups*

$$H^i(\mathbb{E} \times^G X) \cong H^i(\mathbb{E}' \times^G X)$$

for all $i < N$, and these are compatible with cup products in this range.

Proof Consider the product space $\mathbb{E} \times \mathbb{E}'$, with the diagonal action of G, so $(e, e') \cdot g = (e \cdot g, e' \cdot g)$. The projections are equivariant and give a commuting diagram

$$
\begin{array}{ccccc}
\mathbb{E} \times X & \longleftarrow & \mathbb{E} \times \mathbb{E}' \times X & \longrightarrow & \mathbb{E}' \times X \\
\downarrow & & \downarrow & & \downarrow \\
\mathbb{E} \times^G X & \longleftarrow & (\mathbb{E} \times \mathbb{E}') \times^G X & \longrightarrow & \mathbb{E}' \times^G X.
\end{array}
$$

The horizontal maps to the left are locally trivial bundles with fiber $\mathbb{E}'$, and those to the right are locally trivial with fiber $\mathbb{E}$. A special case of the Leray–Hirsch theorem says that such bundle maps determine group isomorphisms

$$H^i(\mathbb{E} \times^G X) \xrightarrow{\sim} H^i((\mathbb{E} \times \mathbb{E}') \times^G X) \xleftarrow{\sim} H^i(\mathbb{E}' \times^G X)$$

for $i < N$ (see Appendix A, §A.4). Since these come from ring homomorphisms

$$H^*(\mathbb{E} \times^G X) \to H^*((\mathbb{E} \times \mathbb{E}') \times^G X) \leftarrow H^*(\mathbb{E}' \times^G X),$$

they respect cup products. $\qquad\square$

Exercise 2.2.3.   Verify that for a third principal bundle $\mathbb{E}'' \to \mathbb{B}''$ such that $\widetilde{H}^i \mathbb{E}'' = 0$ for $i < N$, the canonical isomorphisms are compatible: there is a commuting triangle

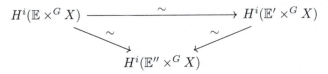

for $i < N$.

Exercise 2.2.4.   With $\mathbb{E}$ and $\mathbb{E}'$ as above, suppose there is a G-equivariant continuous map $\varphi \colon \mathbb{E}' \to \mathbb{E}$, so $\varphi(e' \cdot g) = \varphi(e') \cdot g$ for all $e' \in \mathbb{E}'$, $g \in G$. This defines a continuous map $\mathbb{E}' \times^G X \to \mathbb{E} \times^G X$, and a pullback homomorphism $H^i(\mathbb{E} \times^G X) \to H^i(\mathbb{E}' \times^G X)$. Show that this is the same as the canonical isomorphism given above when $i < N$.

Any G-equivariant continuous map $f \colon X \to Y$ determines a continuous map $\mathbb{E} \times^G X \to \mathbb{E} \times^G Y$, by $[e, x] \mapsto [e, f(x)]$, so we get homomorphisms

$$f^* \colon H_G^i Y \to H_G^i X.$$

In particular, from the projection $X \to \mathrm{pt}$, we obtain a ring homomorphism

$$\Lambda_G := H_G^*(\mathrm{pt}) \to H_G^* X,$$

making $H_G^* X$ a graded-commutative Λ_G-algebra. (If Λ_G^{odd} is nonzero, then one needs to use the convention that for $a \in \Lambda_G^p$ and $b \in H_G^q X$, one has $b \cdot a = (-1)^{pq} a \cdot b$.) Functoriality of cohomology means that the pullback f^* is a homomorphism of Λ_G-algebras. So we have constructed a contravariant functor

$$H_G^* \colon (G\text{-spaces}) \to (\Lambda_G\text{-algebras}).$$

In Chapter 3, we will construct more general pullbacks, allowing the group to vary as well.

Exercise 2.2.5.   Check that the isomorphisms verifying independence of the choice of $\mathbb{E}$ are functorial: given an equivariant map $X \to Y$, and spaces $\mathbb{E}$ and $\mathbb{E}'$ with $\tilde{H}^i\mathbb{E} = \tilde{H}^i\mathbb{E}' = 0$ for $i < N$, show that the diagram

$$\begin{array}{ccc} H^i(\mathbb{E} \times^G Y) & \xrightarrow{\sim} & H^i(\mathbb{E}' \times^G Y) \\ \downarrow & & \downarrow \\ H^i(\mathbb{E} \times^G X) & \xrightarrow{\sim} & H^i(\mathbb{E}' \times^G X) \end{array}$$

commutes.

As a simple and fundamental example, consider $G = \mathbb{C}^*$. This acts freely on $\mathbb{E}_m = \mathbb{C}^m \smallsetminus 0$, by $(z_1, \dots, z_m) \cdot s = (z_1 s, \dots, z_m s)$. The quotient is $\mathbb{B}_m = \mathbb{P}^{m-1}$. Since $\tilde{H}^i\mathbb{E}_m = \tilde{H}^i S^{2m-1} = 0$ for $i < 2m - 1$, any space X with a $\mathbb{C}^*$-action has

$$H^i_{\mathbb{C}^*}X = H^i((\mathbb{C}^m \smallsetminus 0) \times^{\mathbb{C}^*} X) \qquad \text{for } i < 2m - 1.$$

In particular, for the range $i < 2m - 1$, one has

$$H^i_{\mathbb{C}^*}(\mathrm{pt}) = H^i(\mathbb{P}^{m-1}) = \begin{cases} \mathbb{Z} & \text{if } i \text{ is even,} \\ 0 & \text{if } i \text{ is odd.} \end{cases}$$

Each $H^*(\mathbb{P}^{m-1})$ is a truncated polynomial ring isomorphic to $\mathbb{Z}[t]/(t^m)$, so $H^*_{\mathbb{C}^*}(\mathrm{pt})$ is a polynomial ring:

$$\Lambda_{\mathbb{C}^*} = \mathbb{Z}[t], \qquad \text{for } t \text{ a variable of degree 2.}$$

There are two possibilities for t, differing by a sign. In fact, there is a canonical choice of sign, as we will see in the next section.

For $G = (\mathbb{C}^*)^n$, one can take $\mathbb{E}_m = (\mathbb{C}^m \smallsetminus 0)^n$, so $\mathbb{B}_m = (\mathbb{P}^{m-1})^n$ and $\Lambda_G = \mathbb{Z}[t_1, \dots, t_n]$.

In these examples one already sees a key feature of our definition of equivariant cohomology: it takes place within the world of finite-dimensional varieties.

Proposition 2.2.6. *Let G be a complex linear algebraic group acting algebraically on a variety X. For any integer $N > 0$, there is a nonsingular algebraic variety $\mathbb{E}$ so that $H^i_G X = H^i(\mathbb{E} \times^G X)$ for $i < N$, where $\mathbb{E} \times^G X$ is a complex analytic space, nonsingular whenever X is.*

Proof Quite generally, suppose Z is a complex analytic space, and $Y \to Z$ is a continous map of topological spaces which is a locally trivial fiber bundle. If the fibers F are complex analytic spaces, and the transition functions are holomorphic maps $\varphi_{\alpha\beta}\colon U_\alpha \cap U_\beta \to G$ for some complex subgroup $G \subseteq \mathrm{Aut}(F)$ of holomorphic automorphisms, then Y inherits a canonical complex analytic structure by glueing. If both Z and F are complex manifolds, so is Y.

The proposition is the special case where $Y = \mathbb{E} \times^G X$ and $Z = \mathbb{B}$, where $\mathbb{E} \to \mathbb{B}$ is chosen as in Lemma 2.2.1. $\qquad\qquad\square$

2.3   Chern classes and fundamental classes

A *G-equivariant vector bundle* on X is a vector bundle $E \to X$ with G acting linearly on fibers, so that the projection is equivariant. (That is, G acts on E, and for all $g \in G$ and $x \in X$, and $e \in E_x$, the map $e \mapsto g \cdot e$ is a linear map of vector spaces $E_x \to E_{g \cdot x}$.) An equivariant vector bundle produces an ordinary vector bundle $\mathbb{E} \times^G E \to \mathbb{E} \times^G X$. Choosing $\mathbb{E}$ so that $\widetilde{H}^i \mathbb{E} = 0$ for $i \leq 2k$, we take the Chern classes of this bundle on $\mathbb{E} \times^G X$ to define the *equivariant Chern classes* of E:

$$c_k^G(E) := c_k(\mathbb{E} \times^G E) \qquad \text{in} \qquad H_G^{2k}X = H^{2k}(\mathbb{E} \times^G X).$$

Similarly, a G-invariant subvariety V of codimension d in a nonsingular variety X determines a subvariety $\mathbb{E} \times^G V \subseteq \mathbb{E} \times^G X$ of codimension d, and therefore an *equivariant fundamental class*

$$[V]^G = [\mathbb{E} \times^G V] \qquad \text{in} \qquad H_G^{2d}X = H^{2d}(\mathbb{E} \times^G X).$$

(Here we assume G is a complex linear algebraic group and $\mathbb{E}$ is a nonsingular algebraic variety. Then Proposition 2.2.6 says that $\mathbb{E} \times^G V$ is a complex analytic subvariety of the complex manifold $\mathbb{E} \times^G X$.)

Exercise 2.3.1.   Using arguments from before, show that these definitions are independent of choices. More precisely,

$$c_k(\mathbb{E} \times^G E) \mapsto c_k(\mathbb{E}' \times^G E) \text{ under } H^{2k}(\mathbb{E} \times^G X) \xrightarrow{\sim} H^{2k}(\mathbb{E}' \times^G X),$$

when $\widetilde{H}^i\mathbb{E} = \widetilde{H}^i\mathbb{E}' = 0$ for $i \leq 2k$; and

$$[\mathbb{E} \times^G V] \mapsto [\mathbb{E}' \times^G V] \text{ under } H^{2d}(\mathbb{E} \times^G X) \xrightarrow{\sim} H^{2d}(\mathbb{E}' \times^G X),$$

when $\widetilde{H}^i\mathbb{E} = \widetilde{H}^i\mathbb{E}' = 0$ for $i \leq 2d$.

Exercise 2.3.2. Show that multilinear constructions on vector bundles are preserved by the Borel construction. For instance, if E and F are G-equivariant vector bundles on X, verify that

$$\mathbb{E} \times^G (E \oplus F) \cong (\mathbb{E} \times^G E) \oplus (\mathbb{E} \times^G F)$$

as vector bundles on $\mathbb{E} \times^G X$, where $\mathbb{E} \to \mathbb{B}$ is a principal G-bundle. Do the same for tensor products $E \otimes F$, $\bigwedge^k E$, and $\mathrm{Sym}^k E$.

The basic properties of equivariant Chern classes and fundamental classes follow directly from the corresponding properties of ordinary classes on approximation spaces; details and references can be found in Appendix A, §A.3 and §A.5. For instance, one has the following:

- For equivariant line bundles L and M, equivariant Chern classes are additive: $c_1^G(L \otimes M) = c_1^G(L) + c_1^G(M)$.
- When $0 \to E' \to E \to E'' \to 0$ is an equivariant short exact sequence, there is a Whitney formula $c^G(E) = c^G(E') \cdot c^G(E'')$.
- If E has rank e on a nonsingular variety X, and s is an equivariant section, then $Z(s) \subseteq X$ is an invariant subvariety of codimension at most e. If $\mathrm{codim}(Z(s)) = e$, then $[Z(s)]^G = c_e^G(E)$ in $H_G^{2e} X$.
- If G is connected, and two invariant subvarieties V and W of a nonsingular variety X intersect properly, with $V \cdot W = \sum m_i Z_i$ as cycles, then $[V]^G \cdot [W]^G = \sum m_i [Z_i]^G$ in $H_G^* X$. In particular, if $V \cap W = \emptyset$, then $[V]^G \cdot [W]^G = 0$.

(In the last item, connectedness of G is needed to guarantee that each Z_i is also G-invariant.)

As usual, the basic case $X = \mathrm{pt}$ offers plenty to study. Here a G-equivariant vector bundle is just a representation of G, so each representation V of G has Chern classes $c_i^G(V) \in H_G^{2i}(\mathrm{pt}) = \Lambda_G^{2i}$.

Example 2.3.3. For each integer a, $\mathbb{C}^*$ has the one-dimensional representation $\mathbb{C}_a$, where $\mathbb{C}^*$ acts on $\mathbb{C}$ by $z \cdot v = z^a v$. So $\mathbb{C}_1$ is the *standard representation*. In Exercise 2.1.1, we saw that

$$
\begin{array}{ccc}
(\mathbb{C}^m \smallsetminus 0) \times^{\mathbb{C}^*} \mathbb{C}_1 & \xrightarrow{\sim} & \mathcal{O}(-1) \\
\downarrow & & \downarrow \\
(\mathbb{C}^m \smallsetminus 0) \times^{\mathbb{C}^*} \mathrm{pt} & \xrightarrow{\sim} & \mathbb{P}^{m-1},
\end{array}
$$

so $\mathbb{E} \times^{\mathbb{C}^*} \mathbb{C}_1$ gets identified with the tautological bundle $\mathcal{O}(-1)$ on $\mathbb{B}$. Taking $t = c_1^{\mathbb{C}^*}(\mathbb{C}_1) = c_1(\mathcal{O}(-1))$ as a generator for $\Lambda_{\mathbb{C}^*} = \mathbb{Z}[t]$, we see

> $\Lambda_{\mathbb{C}^*}$ is a polynomial ring generated by the Chern class of the standard representation.

More generally, since $\mathbb{C}_a \otimes \mathbb{C}_b = \mathbb{C}_{a+b}$, we have $c_1^{\mathbb{C}^*}(\mathbb{C}_a) = at$. One can also see this from an identification $\mathbb{E} \times^{\mathbb{C}^*} \mathbb{C}_a \cong \mathcal{O}(-a)$.

Example 2.3.4. Consider $T = (\mathbb{C}^*)^n$ acting on $\mathbb{C}^n = V$ by the standard action scaling coordinates. For $1 \le i \le n$, we have one-dimensional representations $\mathbb{C}_{t_i}$, where $z \cdot v = z_i v$. Then

$$c_i^T(V) = e_i(t_1, \ldots, t_n) \quad \text{in} \quad \Lambda_T = \mathbb{Z}[t_1, \ldots, t_n],$$

where $t_i = c_1^T(\mathbb{C}_{t_i})$ and e_i is the elementary symmetric polynomial. Using $\mathbb{E} = (\mathbb{C}^m \setminus 0)^n$ and $\mathbb{B}_m = (\mathbb{P}^{m-1})^n$, the class t_i is identified with the Chern class of the tautological bundle from the ith factor of $\mathbb{B}_m$.

Example 2.3.5. In the equivariant setting, it is harder to move G-invariant subvarieties so that they intersect properly. For example, consider $G = \mathbb{C}^*$ acting on $\mathbb{C}$ in the standard way. Then the only invariant subvarieties are $\{0\}$ and $\mathbb{C}$. In ordinary cohomology, one could move 0 to 1 to see $[0]^2 = [0] \cdot [1] = [\{0\} \cap \{1\}] = 0$, but this is not possible equivariantly. Indeed, $([0]^T)^2 = t^2 \neq 0$ in $H_{\mathbb{C}^*}^*(\mathbb{C})$.

2.4   The general linear group

Now we will consider $G = GL(V)$, for an n-dimensional vector space V. This has its standard representation on V itself, so there are Chern classes $c_i^G(V) \in H_G^{2i}(\mathrm{pt}) = \Lambda_G^{2i}$. Our main calculation is the following.

Proposition 2.4.1. *We have*

$$\Lambda_G = \mathbb{Z}[c_1, \ldots, c_n],$$

where $c_i = c_i^G(V)$.

In other words,

> $\Lambda_{GL(V)}$ is a polynomial ring generated by the Chern classes of the standard representation.

To prove this, we will use $\mathbb{E}_m = \mathrm{Emb}(V, \mathbb{C}^m)$, the space of linear embeddings $V \hookrightarrow \mathbb{C}^m$, for $m \geq n$. Choosing a basis, so $V \cong \mathbb{C}^n$, one identifies $\mathbb{E}_m$ with $M_{m,n}^\circ$, the space of full-rank $m \times n$ matrices. Let $\Omega_{n-1} = \mathrm{Hom}(V, \mathbb{C}^m) \smallsetminus \mathbb{E}_m$; choosing a basis identifies $\Omega_{n-1} \subseteq M_{m,n}$ with the locus of $m \times n$ matrices of rank at most $n-1$. A standard exercise in algebraic geometry computes its dimension.

Exercise 2.4.2. Consider the locus $\Omega_r \subseteq M_{m,n}$ of matrices of rank at most r. Show that this is irreducible of codimension $(m-r)(n-r)$.

Lemma 2.4.3. *We have* $\widetilde{H}^i \mathbb{E}_m = 0$ *for* $i \leq 2(m-n)$.

Proof From the long exact sequence in cohomology, we have $\widetilde{H}^i \mathbb{E}_m = H^{i+1}(\mathrm{Hom}(V, \mathbb{C}^m), \mathbb{E}_m) =: \overline{H}_{2mn-i-1} \Omega_{n-1}$. By the above exercise, Ω_{n-1} has (real) dimension $2mn - 2(m-n+1)$. When $i \leq 2(m-n)$, we have $2mn - i - 1 > 2mn - 2(m-n+1)$, so this Borel–Moore homology group vanishes. (See Appendix A, §A.3, for the relevant properties of Borel–Moore homology.) $\qquad\square$

An alternative way of proving the lemma is given in Appendix A, §A.7.

We also need a coarse description of the cohomology of the Grassmannian, which says it is generated by Chern classes, with no relations in small degree.

Lemma 2.4.4. *We have*

$$H^* Gr(n, \mathbb{C}^m) = \mathbb{Z}[c_1(\mathbb{S}), \ldots, c_n(\mathbb{S})]/(R_{m-n+1}, \ldots, R_m),$$

where R_k *is a relation of degree* k.

The lemma can be found in standard algebraic topology texts, and it also follows from computations we will do later (see §4.5).

Now we can prove Proposition 2.4.1. Observe that $\mathbb{B}_m = \mathbb{E}_m/G = Gr(n, \mathbb{C}^m)$, where G acts on $\mathrm{Emb}(V, \mathbb{C}^m)$ by $(\varphi \cdot g)(v) = \varphi(g \cdot v)$. By Exercise 2.1.1, $\mathbb{E}_m \to \mathbb{B}_m$ is the frame bundle $\mathrm{Fr}(\mathbb{S}) \to Gr(n, \mathbb{C}^m)$ associated to the tautological $\mathbb{S} \subseteq \mathbb{C}^m_{Gr}$, and so the vector bundle $\mathbb{E}_m \times^G V$ identifies with the tautological bundle $\mathbb{S}$ itself. (The map is $(\varphi, v) \mapsto (\varphi, \varphi(v))$.) Thus $c_i^G(V)$ is identified with $c_i(\mathbb{S})$, and the proposition follows from Lemma 2.4.4. $\qquad\square$

2.5 Some other groups

Any closed subgroup $G \subseteq GL(V)$ acts freely on $\mathbb{E}_m = \mathrm{Emb}(V, \mathbb{C}^m)$, so we can use these same approximation spaces for such G. (For computations, it is sometimes helpful to make other choices.) Let us see how far we can get using this explicit construction.

Exercise 2.5.1. Consider $G = SL(V) \subseteq GL(V)$ as the subgroup preserving the determinant $\bigwedge^n V \xrightarrow{\sim} \mathbb{C}$. Show that

$$\Lambda_{SL(V)} = \mathbb{Z}[c_1, \dots, c_n]/(c_1) = \mathbb{Z}[c_2, \dots, c_n],$$

where $c_i = c_i^G(V)$. (Note that $\bigwedge^n V$ is the trivial representation, so $c_1^G(V) = c_1^G(\bigwedge^n V) = 0$.)

For now, let us fix a basis, so $V = \mathbb{C}^n$. Our main example going forward will be $T = (\mathbb{C}^*)^n$, and we have already seen two possibilities for constructing its equivariant cohomology. Using $T = (GL_1)^n$, we get

$$\mathbb{E}_m = (\mathbb{C}^m \smallsetminus 0)^n = \big\{ A \in M_{m,n} \,|\, \text{no column is zero} \big\},$$

with $\mathbb{B}_m = (\mathbb{P}^{m-1})^n$.

On the other hand, considering $T \subseteq GL_n$ as diagonal matrices, we have

$$\mathbb{E}_m = M_{m,n}^\circ = \big\{ A \in M_{m,n} \,|\, \text{columns are linearly independent} \big\}.$$

Using this choice, we get

$$\mathbb{B}_m = M_{m,n}^\circ/T = \left\{ \begin{array}{c} V \subseteq \mathbb{C}^m \text{ of dimension } n, \\ \text{with a decomposition } V = L_1 \oplus \cdots \oplus L_n \end{array} \right\}$$

by sending a matrix to the tuple $(L_1, \dots, L_n)$, with L_i the span of the ith column. Call this space the "split Grassmannian" $Gr^{\mathrm{split}}(n, \mathbb{C}^m)$; it comes with tautological line bundles $\mathbb{L}_1, \dots, \mathbb{L}_n$, whose classes $t_i = c_1(\mathbb{L}_i)$ generate the cohomology ring.

There is a projection map $\pi\colon Gr^{\mathrm{split}}(n, \mathbb{C}^m) \to Gr(n, \mathbb{C}^m)$ sending $(L_1, \dots, L_n)$ to $V = L_1 \oplus \cdots \oplus L_n \subseteq \mathbb{C}^m$.

Exercise 2.5.2. Taking m sufficiently large, show that the corresponding pullback map on cohomology gives

$$\Lambda_{GL_n} = \mathbb{Z}[c_1, \dots, c_n] \to \mathbb{Z}[t_1, \dots, t_n] = \Lambda_T,$$

defined by $c_i \mapsto e_i(t_1, \ldots, t_n)$, so Λ_{GL_n} embeds in Λ_T as the ring of symmetric polynomials.

Remark. The inclusion $\Lambda_{GL_n} \hookrightarrow \Lambda_T$ is a manifestation of the *splitting principle*: given a vector bundle E on a space X, one can find a map $f \colon X' \to X$, such that f^*E splits into a direct sum of line bundles on X', and the pullback homomorphism $f^* \colon H^*X \to H^*X'$ is injective. For any $d \leq n = \operatorname{rk} E$, there is a "split Grassmann" bundle $\mathbf{Gr}^{\mathrm{split}}(d, E) \to X$, constructed as before by taking a quotient of the frame bundle, so

$$\operatorname{Fr}(E) \times^{GL_n} Gr^{\mathrm{split}}(d, \mathbb{C}^n) \cong \mathbf{Gr}^{\mathrm{split}}(d, E).$$

Taking $d = n = \operatorname{rk} E$ and $X' = Gr^{\mathrm{split}}(n, E)$, the pullback of E from X to X' splits, and the cohomology of X embeds into that of X'.

Using functorial pullbacks in equivariant cohomology, exactly the same construction establishes the analogous equivariant splitting principle: for a G-equivariant vector bundle $E \to X$, there is an equivariant map $f \colon X' \to X$, such that f^*E splits into equivariant line bundles, and such that $f^* \colon H_G^*X \to H_G^*X'$ is injective.

In between the torus and GL_n, there is the Borel group B of upper-triangular matrices. Using $\mathbb{E}_m = M_{m,n}^\circ$ again, we have

$$\mathbb{B}_m = M_{m,n}^\circ / B = \left\{ \begin{array}{c} V \subseteq \mathbb{C}^m \text{ of dimension } n, \\ \text{with a filtration } V_1 \subset V_2 \subset \cdots \subset V_n = V \end{array} \right\}$$

$$= Fl(1, 2, \ldots, n; \mathbb{C}^m),$$

the partial flag variety parametrizing chains $V_1 \subset \cdots \subset V_n \subseteq \mathbb{C}^m$, with $\dim V_i = i$. (The projection $\mathbb{E}_m \to \mathbb{B}_m$ sends a matrix to the flag where V_i is the span of the first i columns.) This comes with a tautological flag of bundles $\mathbb{S}_1 \subset \cdots \subset \mathbb{S}_n \subseteq \mathbb{C}_{Fl}^m$. The flag variety sits between $Gr^{\mathrm{split}}(n, \mathbb{C}^m)$ and $Gr(n, \mathbb{C}^m)$, with maps

$$Gr^{\mathrm{split}}(n, \mathbb{C}^m) \to Fl(1, \ldots, n; \mathbb{C}^m) \to Gr(n, \mathbb{C}^m),$$

sending $(L_1, \ldots, L_n)$ to the flag with $V_i = L_1 \oplus \cdots \oplus L_i$, and projecting a flag $V_\bullet$ to $V = V_n \subseteq \mathbb{C}^m$.

Exercise 2.5.3. Show that $Gr^{\mathrm{split}}(n, \mathbb{C}^m)$ is a locally trivial affine bundle, so the pullback map induces a ring isomorphism $\Lambda_B \xrightarrow{\sim} \Lambda_T$.

The isomomorphism in this exercise is part of a general phenomenon, as we will see in the next chapter, since the inclusion $T \hookrightarrow B$ is a deformation retract. On the other hand, one can also compute directly that $H^*Fl(1,\ldots,n;\mathbb{C}^m)$ is generated by the Chern classes $t_i = c_1(\mathbb{S}_i/\mathbb{S}_{i-1})$, with relations in degrees $2(m-n+1),\ldots,2m$, so that $\Lambda_B = \mathbb{Z}[t_1,\ldots,t_n]$.

Exercise 2.5.4. Let $\chi_i \colon B \to \mathbb{C}^*$ be the character which picks out the ith diagonal entry of a matrix in B, and let $\mathbb{C}_{\chi_i}$ be the corresponding representation. Show that $t_i = c_1^B(\mathbb{C}_{\chi_i})$.

For other groups, the rings Λ_G can be much more complicated. For instance, the answer for PGL_n is not completely known!

In the case of the symplectic group $G = Sp_{2n} \subseteq GL_{2n}$, with its standard representation $V = \mathbb{C}^{2n}$, there is a simple answer:

$$\Lambda_{Sp_{2n}} = \mathbb{Z}[c_2, c_4, \ldots, c_{2n}], \quad \text{where} \quad c_{2k} = c_{2k}^G(V),$$

so here again Λ_G is generated by the Chern classes of the standard representation.

Here is the easy half of this computation. Using $\mathbb{E}_m = M_{m,n}^\circ$, we find

$$\mathbb{B}_m = M_{m,2n}^\circ/Sp_{2n} = \left\{ (V,\omega) \;\middle|\; \begin{array}{l} V \subseteq \mathbb{C}^m \text{ has dimension } 2n, \text{ and} \\ \omega \text{ is a symplectic form on } V \end{array} \right\}.$$

Using the projection to $M_{m,2n}^\circ/GL_{2n} = Gr(2n,\mathbb{C}^m)$, one can pull back the tautological bundle $\mathbb{S}$. On $\mathbb{B}_m$, this pullback bundle acquires a tautological symplectic form, identifying it with its dual. So whenever i is odd, $2c_i \mapsto 0$ under the map $\Lambda_{GL_{2n}} \to \Lambda_{Sp_{2n}}$. This comes from the general fact that $c_i(E) + (-1)^i c_i(E^\vee) = 0$, for any bundle E. To complete the argument, one must show that $H^*\mathbb{B}_m$ has no torsion, and that $\Lambda_{GL_{2n}} \to \Lambda_{Sp_{2n}}$ is surjective. (See Example 15.5.2.)

Similar arguments show that $\Lambda_{GL_n} \to \Lambda_{SO_n}$ sends $2c_i$ to 0 for i odd, but in this case it is not true that $c_i \mapsto 0$ (there is 2-torsion on Λ_{SO_n}), and the map is not surjective in general.

Exercise 2.5.5. Show that $\Lambda_{\mathbb{Z}/2\mathbb{Z}} = \mathbb{Z}[t]/(2t)$, where t is a class in degree 2. For the additive group $\mathbb{Z}$, show that $\Lambda_{\mathbb{Z}} = \mathbb{Z}[t]/(t^2)$, where t has degree 1.

Remark. For a finite group G, there is another construction, which gives rise to an explicit cochain complex computing $\Lambda_G = H_G^*(\text{pt})$. Let $C^\bullet = C^\bullet(G, \mathbb{Z})$ be the complex with

$$C^i = \{\text{functions } \varphi \colon G^i \to \mathbb{Z}\}$$

and for $\varphi \in C^i$ define the differential $d\varphi \in C^{i+1}$ by

$$d\varphi(g_0, \ldots, g_i) = \varphi(g_1, \ldots, g_i)$$
$$+ \sum_{j=0}^{i-1} (-1)^{j+1} \varphi(g_1, \ldots, g_{j-1}, g_j g_{j+1}, \ldots, g_i)$$
$$+ (-1)^{i+1} \varphi(g_0, \ldots, g_{i-1}).$$

Then $H_G^k(\text{pt})$ is the cohomology $H^k(C^\bullet)$ of this complex. One way to prove this goes through the *Milnor construction* for the universal principal bundle $\mathbb{E}G \to \mathbb{B}G$.

In the context of group theory, $H_G^*(\text{pt}) = H^*\mathbb{B}G = H^*(G, \mathbb{Z})$ is known as the *group cohomology* of G with coefficients in the trivial G-module $\mathbb{Z}$.

2.6 Projective space

Let G be any group acting linearly on an n-dimensional vector space V. Then G also acts on the projective space $\mathbb{P}(V)$, as well as the tautological subbundle $\mathcal{O}(-1)$ and its dual $\mathcal{O}(1)$. Let $\zeta = c_1^G(\mathcal{O}(1))$ be the Chern class in $H_G^2 \mathbb{P}(V)$.

Proposition 2.6.1. *We have*

$$H_G^* \mathbb{P}(V) = \Lambda_G[\zeta]/(\zeta^n + c_1 \zeta^{n-1} + \cdots + c_n),$$

where $c_i = c_i^G(V)$ are the Chern classes of the given representation.

Proof This is a special case of the general formula computing the cohomology of a projective bundle in terms of that of the base. In our circumstance, the relevant identification is

$$\mathbb{E} \times^G \mathbb{P}(V) = \mathbb{P}(\mathbb{E} \times^G V)$$
$$\downarrow \qquad\qquad\qquad \downarrow$$
$$\mathbb{B} =\!=\!=\!=\!= \mathbb{B},$$

compatibly with identifications of $\mathcal{O}(1)$. Thus ζ is the hyperplane class for the projective bundle, and $c_i^G(V) = c_i(\mathbb{E} \times^G V)$ are the Chern classes of this vector bundle on $\mathbb{B}$.   □

Example 2.6.2.   For $G = GL(V)$, we have

$$H_G^* \mathbb{P}(V) = \mathbb{Z}[c_1, \ldots, c_n][\zeta]/(\zeta^n + c_1 \zeta^{n-1} + \cdots + c_n).$$

For $T = (\mathbb{C}^*)^n$ acting on V via the standard action, we have

$$H_T^* \mathbb{P}(V) = \mathbb{Z}[t_1, \ldots, t_n][\zeta]/\prod_{i=1}^n (\zeta + t_i).$$

(This comes from the computation $c_i^T(V) = e_i(t_1, \ldots, t_n)$.)

Notes

Our definition of equivariant cohomology, using approximations by algebraic varieties, is modelled on the analogous construction for Chow groups. This technique was pioneered by Totaro (1999) and further developed by Edidin and Graham (1998), who defined equivariant Chow groups.

Algebraic versions of Lemma 2.2.1 appear in Totaro's construction of the Chow ring of a classifying space; see (Totaro, 1999, Remark 1.4) or (Totaro, 2014, §2). In algebraic geometry, the method of proving Proposition 2.2.2 (establishing independence of choice of approximation) was used by mathematicians studying invariant theory; see especially Bogomolov's definition of the Brauer group (Bogomolov, 1987, §3). In topology, this argument goes back to Borel's foundational papers; see (Borel, 1953, §18).

An alternative argument for Proposition 2.2.6 showing that the quotient $\mathbb{E} \times^G X = (\mathbb{E} \times X)/G$ is a complex analytic space can be given using a general statement about analytic structures on quotients, proved by Cartan (1957) and generalized by Holmann (1960).

Even when X is a nonsingular variety, the space $\mathbb{E} \times^G X$ may not exist as a scheme (although it is always an algebraic space). Some general criteria guaranteeing that it does exist are given by Edidin and Graham (1998, Proposition 23). Sufficient conditions include: X is quasi-projective, with a linearized G-action; or G is a *special* group such as GL_n, SL_n, a torus, or products of such groups.

We will work out the cohomology rings of Grassmannians and flag varieties in Chapter 4. Alternative arguments for the computation in Lemma 2.4.4 can be found in many algebraic topology texts – for example, the book by Dold (1980, Proposition 12.17).

Using coefficients in a field, there are classical computations of $H^* \mathbb{B}G$ by Borel (1953). Some computations of integral cohomology for orthogonal groups were carried out in (Brown, 1982; Feshbach, 1983).

In many other cases, the integral cohomology (or Chow) rings of $\mathbb{B}G$ are either unknown, or were computed rather recently. The Chow ring for SO_{2n} was computed by Field in her 2000 Ph. D. thesis (Field, 2012). Computations for PGL_p, with p prime, were done in both cohomology and Chow rings by Vistoli (2007), whose paper also serves as a good survey for other work on the subject. More recent progress can be found in (Gu, 2021).

The "Milnor construction" for $\mathbb{B}G$ was given in (Milnor, 1956); see also (Husemoller, 1975, §4.11).

Hints for Exercises

Exercise 2.1.1. For the fiber over $y \in Y$, the map to the tautological bundle of $Gr(d, E_y)$ is simply $(v_1, \ldots, v_d) \times (z_1, \ldots, z_d) \mapsto (\text{span}\{v_1, \ldots, v_d\}, \sum z_i v_i)$.

Exercise 2.2.3. Use the triple product $\mathbb{E} \times \mathbb{E}' \times \mathbb{E}''$.

Exercise 2.2.4. The equivariant map φ induces a section of the projection $(\mathbb{E} \times \mathbb{E}') \times^G X \to \mathbb{E}' \times^G X$.

Exercise 2.3.1. For Chern classes, this just uses the pullback of the vector bundle E to $\mathbb{E} \times \mathbb{E}'$. For classes of subvarieties, one needs the smooth pullback property; see Appendix A, Proposition A.3.2.

Exercise 2.4.2. Use a Grassmannian correspondence to parametrize the kernel of such a matrix. See (Harris, 1992, Proposition 12.2).

Exercise 2.5.1. Use the same $\mathbb{E}_m$, and identify $\mathbb{E}_m/SL(V) \to \mathbb{E}_m/GL(V)$ with the variety $\text{Iso}(\bigwedge^n \mathbb{S}, \mathbb{C})$ over $Gr(n, \mathbb{C}^m)$, parametrizing subspaces $V \subseteq \mathbb{C}^m$ equipped with an isomorphism $\bigwedge^n V \to \mathbb{C}$. Explicitly, this is the variety cut by Plücker equations in $\bigwedge^n \mathbb{C}^m \smallsetminus 0$. It is also the complement of the 0-section in the line bundle $\text{Hom}(\bigwedge^n \mathbb{S}, \mathbb{C}) \cong \bigwedge^n \mathbb{S}^\vee$. Then one can use the following general fact, which is an easy application of the Gysin sequence: for a vector bundle E of rank r on X, if the homomorphism

$$H^{i-2r+1}X \xrightarrow{c_r(E)\cdot} H^{i+1}X$$

is injective, then $H^i(E \smallsetminus 0) = (H^i X)/(c_r(E) \cdot H^{i-2r}X)$. See (Milnor and Stasheff, 1974, Theorem 12.2).

Exercise 2.5.2. Use $\pi^* \mathbb{S} = \mathbb{L}_1 \oplus \cdots \oplus \mathbb{L}_n$.

Exercise 2.5.5. For $\mathbb{Z}/2\mathbb{Z}$, use $\mathbb{E}_m = S^m$, so $\mathbb{B}_m = \mathbb{RP}^m$. For $\mathbb{Z}$, use $\mathbb{E} = \mathbb{R}$, which is already contractible, with $\mathbb{B} = S^1$.

3

Basic Properties

In this chapter, we will develop the basic properties of equivariant cohomology. Some are the expected analogues from ordinary cohomology, such as functoriality, homotopy-invariance, and Mayer–Vietoris sequences. Some are more intrinsically equivariant. We'll begin with a detailed study of tori, where functoriality can be seen explicitly.

3.1   Tori

A *torus* T is an algebraic group which is isomorphic to $(\mathbb{C}^*)^n$, so we already know $\Lambda_T = \mathbb{Z}[t_1, \ldots, t_n]$ – for example, by using $\mathbb{E}_m = (\mathbb{C}^m \smallsetminus 0)^n$ and $\mathbb{B}_m = (\mathbb{P}^{m-1})^n$. We want an intrinsic description, though, without choosing the isomorphism $T \cong (\mathbb{C}^*)^n$.

Let M be the group of *characters* of T,

$$M = \mathrm{Hom}_{\text{alg. gp.}}(T, \mathbb{C}^*).$$

The group operation is written additively: $(\chi_1 + \chi_2)(z) = \chi_1(z)\chi_2(z)$. Each $\chi \in M$ determines an equivariant line bundle on a point, denoted $\mathbb{C}_\chi$, by the action $z \cdot v = \chi(z)v$ for $z \in T$ and $v \in \mathbb{C}$. This correspondence is compatible with the group structure on M: we have $\mathbb{C}_{\chi_1 + \chi_2} = \mathbb{C}_{\chi_1} \otimes \mathbb{C}_{\chi_2}$, so we get a group homomorphism

$$M \to \Lambda_T^2 = H_T^2(\mathrm{pt}), \qquad \chi \mapsto c_1^T(\mathbb{C}_\chi).$$

In fact, this is an isomorphism, as one can easily check by choosing a basis, so that $T \cong (\mathbb{C}^*)^n$ and $M \cong \mathbb{Z}^n$. We therefore have a canonical isomorphism of rings

$$\mathrm{Sym}^* M \xrightarrow{\sim} \Lambda_T.$$

Example 3.1.1. Consider $T = \{(z_1, \ldots, z_n) \in (\mathbb{C}^*)^n \mid \prod z_i = 1\}$, so there are exact sequences

$$1 \to T \to (\mathbb{C}^*)^n \to \mathbb{C}^* \to 1 \quad \text{and} \quad 0 \to \mathbb{Z} \to \mathbb{Z}^n \to M \to 0,$$

realizing $M \cong \mathbb{Z}^n / \mathbb{Z} \cdot (1, \ldots, 1)$. (This T can be viewed as the diagonal torus in SL_n.)

Example 3.1.2. Consider $T = (\mathbb{C}^*)^n / \mathbb{C}^*$, with exact sequences

$$1 \to \mathbb{C}^* \to (\mathbb{C}^*)^n \to T \to 1 \quad \text{and} \quad 0 \to M \to \mathbb{Z}^n \to \mathbb{Z} \to 0,$$

where $\mathbb{C}^* \hookrightarrow (\mathbb{C}^*)^n$ is the diagonal $(z, \ldots, z)$, and

$$M = \left\{ (a_1, \ldots, a_n) \in \mathbb{Z}^n \mid \sum a_i = 0 \right\}.$$

(This T is the image of the diagonal torus in $PGL_n = GL_n / \mathbb{C}^*$.)

In both of these examples, $M \cong \mathbb{Z}^{n-1}$, but not canonically!

In fact, we will see that the isomorphism $\mathrm{Sym}^* M = \Lambda_T$ is *natural* for homomorphisms of tori $\varphi \colon T \to T'$. Such a homomorphism corresponds to a homomorphism $M' \to M$ of character groups. (Given φ, one gets $M' \to M$ by pulling back homomorphisms. Conversely, there is a canonical identification $T = \mathrm{Spec}\,\mathbb{C}[M]$, so a group algebra homomorphism $\mathbb{C}[M'] \to \mathbb{C}[M]$ determines a homomorphism of tori $\varphi \colon T \to T'$.) Thus a homomorphism of tori gives a ring homomorphism

$$\mathrm{Sym}^* M' \to \mathrm{Sym}^* M.$$

On the other hand, we will construct ring homomorphisms $\Lambda_{G'} \to \Lambda_G$ associated to any group homomorphism $\varphi \colon G \to G'$. For tori, these two maps are compatible:

$$\begin{CD}
\mathrm{Sym}^* M' @>\sim>> \Lambda_{T'} \\
@VVV @VVV \\
\mathrm{Sym}^* M @>\sim>> \Lambda_T.
\end{CD} \qquad (*)$$

3.2   Functoriality

We have already seen that for a fixed group G acting on spaces X and X', with an equivariant map $f\colon X \to X'$, there is a pullback homomorphism $f^*\colon H_G^* X' \to H_G^* X$. In fact, equivariant cohomology is functorial in the group, as well. Here is the general setup. Let $\varphi\colon G \to G'$ be a continuous homomorphism of groups. Suppose G acts on X and G' acts on X'. A continuous map $f\colon X \to X'$ is *equivariant with respect to φ* if

$$f(g \cdot x) = \varphi(g) \cdot f(x)$$

for all $g \in G$ and $x \in X$. Given such a map, we have a pullback homomorphism (of rings)

$$f^*\colon H_{G'}^* X' \to H_G^* X.$$

(This depends on φ as well, but we suppress that in the notation.) In particular, a group homomorphism $G \to G'$ determines a ring map $\Lambda_{G'} \to \Lambda_G$.

To construct these pullbacks, take principal bundles $\mathbb{E} \to \mathbb{B}$ for G and $\mathbb{E}' \to \mathbb{B}'$ for G'. Then G acts on $\mathbb{E} \times \mathbb{E}'$ by $(e, e') \cdot g = (e \cdot g, e' \cdot \varphi(g))$. This action is free, and the projection to $\mathbb{E}'$ is equivariant (with respect to φ). Thus there are maps

$$(\mathbb{E} \times \mathbb{E}') \times^G X \to (\mathbb{E} \times \mathbb{E}') \times^G X' \to \mathbb{E}' \times^{G'} X',$$

where G acts on X' via φ in the middle space. If $\widetilde{H}^i \mathbb{E} = \widetilde{H}^i \mathbb{E}' = 0$ for $i < N$, then $H_G^i X = H^i((\mathbb{E} \times \mathbb{E}') \times^G X)$ for $i < N$, and the pullback via $(\mathbb{E} \times \mathbb{E}') \times^G X \to \mathbb{E}' \times^{G'} X'$ produces the desired homomorphism $H_{G'}^i X' \to H_G^i X$. Since it comes from a pullback, it is compatible with cup products in the same range ($i < N$), so these homomorphisms of groups define a homomorphism of graded rings.

Exercise 3.2.1.   Check that the result of this construction is independent of choices of $\mathbb{E}$ and $\mathbb{E}'$.

If there is an equivariant map $\mathbb{E} \to \mathbb{E}'$ (with respect to φ), check that pullback via $\mathbb{E} \times^G X \to \mathbb{E}' \times^{G'} X'$ agrees with the one we constructed.

Exercise 3.2.2.   For $G = G' = \mathbb{C}^*$, consider $\varphi(z) = z^a$, some $a \in \mathbb{Z}$. Using $\mathbb{E} = \mathbb{C}^m \smallsetminus 0$, find an equivariant map $\mathbb{E} \to \mathbb{E}$ (with respect to φ).

Show that the corresponding endomorphism of $\Lambda_{\mathbb{C}^*} = \mathbb{Z}[t]$ is given by $t \mapsto at$.

If $E' \to X'$ is a G'-equivariant vector bundle, it acquires a G-equivariant structure via φ, and the pullback $E = f^*E'$ is a G-equivariant vector bundle on X. Under the corresponding homomorphism $H_{G'}^{2k} X' \to H_G^{2k} X$, we have

$$c_k^{G'}(E') \mapsto c_k^G(E).$$

That is, equivariant Chern classes commute with pullback.

Since the isomorphism $\operatorname{Sym}^* M \to \Lambda_T$ was constructed by sending $\chi \in M$ to $c_1^T(L(\chi))$, functoriality of equivariant Chern classes establishes the naturality claimed at the end of the last section, in diagram $(*)$.

Exercise 3.2.3. Suppose $\varphi \colon (\mathbb{C}^*)^n \to (\mathbb{C}^*)^r$ is given by

$$(z_1, \ldots, z_n) \mapsto \left(\prod_{i=1}^n z_i^{a_{i,1}}, \ldots, \prod_{i=1}^n z_i^{a_{i,r}} \right)$$

for some $n \times r$ integer matrix $A = (a_{i,j})$. (So A gives a linear map $\mathbb{Z}^r \to \mathbb{Z}^n$.) Show that the homomorphism

$$\mathbb{Z}[s_1, \ldots, s_r] = \Lambda_{(\mathbb{C}^*)^r} \to \Lambda_{(\mathbb{C}^*)^n} = \mathbb{Z}[t_1, \ldots, t_n]$$

sends s_j to $\sum_{i=1}^n a_{ij} t_i$. By choosing bases so that $T = (\mathbb{C}^*)^n$ and $T' = (\mathbb{C}^*)^r$, this gives an alternative argument for the naturality of the isomorphism $\Lambda_T = \operatorname{Sym}^* M$.

Exercise 3.2.4. If $\mathbb{E} \to \mathbb{B}$ is a principal G-bundle, and $\mathbb{E}' \to \mathbb{B}'$ is a principal bundle for G', show that $\mathbb{E} \times \mathbb{E}' \to \mathbb{B} \times \mathbb{B}'$ is a principal bundle for $G \times G'$. If Λ_G or $\Lambda_{G'}$ is free over $\mathbb{Z}$, deduce that the natural map $\Lambda_G \otimes \Lambda_{G'} \to \Lambda_{G \times G'}$ is an isomorphism of rings.

3.3 Invariance

When the maps $f \colon X \to X'$ and $\varphi \colon G \to G'$ induce isomorphisms on ordinary cohomology, the same is true for equivariant cohomology. To prove this, we can break the problem into simple steps.

Proposition 3.3.1. *Let $f\colon X \to X'$ be G-equivariant, and suppose $f^*\colon H^i X' \to H^i X$ is an isomorphism for $i < N$. Then $f^*\colon H_G^i X' \to H_G^i X$ is an isomorphism for $i < N$.*

Proof Take a principal G-bundle $\mathbb{E} \to \mathbb{B}$ with $\widetilde{H}^i \mathbb{E} = 0$ for $i < N$. Then the map $\mathbb{E} \times^G X \to \mathbb{E} \times^G X'$ induces isomorphisms

$$H^i(\mathbb{E} \times^G X') \to H^i(\mathbb{E} \times^G X)$$

for $i < N$, by a general fact about fiber bundles. Indeed, we have

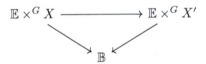

with fiber maps $X \to X'$, so the claim is an instance of Lemma A.4.3 of Appendix A. $\qquad\square$

Corollary 3.3.2. *If $\varphi\colon G \to G'$ is a continuous homomorphism such that $H^i G' \to H^i G$ is an isomorphism for $i < N$, then for any principal G-bundle $\mathbb{E} \to \mathbb{B}$, the map $H^i(\mathbb{E} \times^G G') \to H^i \mathbb{E}$ is an isomorphism for $i < N$.*

In particular, if $\widetilde{H}^i \mathbb{E} = 0$ for $i < N$ then also $\widetilde{H}^i(\mathbb{E} \times^G G') = 0$ for $i < N$.

(Apply the Proposition with $X = G$ and $X' = G'$.)

Putting these together, we get the claimed invariance for equivariant cohomology.

Theorem 3.3.3. *Suppose $f\colon X \to X'$ is equivariant with respect to $\varphi\colon G \to G'$, and $H^i G' \to H^i G$ and $H^i X' \to H^i X$ are isomorphisms for $i < N$. Then $H_{G'}^i X' \to H_G^i X$ is an isomorphism for $i < N$.*

Proof Take a principal G-bundle $\mathbb{E} \to \mathbb{B}$ so that $\widetilde{H}^i \mathbb{E} = 0$ for $i < N$. By the Corollary, $\mathbb{E}' = \mathbb{E} \times^G G'$ works as an approximation space for G', that is, $\widetilde{H}^i \mathbb{E}' = 0$ for $i < N$. Now we have

$$H_{G'}^i X' = H^i((\mathbb{E} \times^G G') \times^{G'} X') = H^i(\mathbb{E} \times^G X') = H_G^i X',$$

and we can apply the Proposition. $\qquad\square$

We will often encounter situations where the maps are homotopy equivalences, so N can be taken to be ∞.

Corollary 3.3.4. *If $\varphi \colon G \to G'$ and $f \colon X \to X'$ are (weak) homotopy equivalences, then $H^*_{G'} X' \cong H^*_G X$ as algebras over $\Lambda_{G'} \cong \Lambda_G$.*

(In fact, one can use the Whitehead theorem and the Hurewicz isomorphism to show $\mathbb{E}' \times^{G'} X'$ and $\mathbb{E} \times^G X$ are weakly homotopy equivalent, but we do not need this.)

Example 3.3.5. The inclusion of the diagonal torus $T = (\mathbb{C}^*)^n$ in upper-triangular matrices $B \subseteq GL_n$ is a deformation retract, since $B/T \cong \mathbb{A}^N$ (where $N = \binom{n}{2}$). This gives a general reason for the identification $\Lambda_T = \Lambda_B$, which we have already observed.

Example 3.3.6. A *unipotent group* is a closed algebraic subgroup U of the unitriangular matrices in GL_n (meaning the diagonal entries are 1). As an algebraic variety, any such group is isomorphic to affine space, so it is contractible, and $\Lambda_U = \mathbb{Z}$.

More generally, any linear algebraic group G has a maximal normal unipotent subgroup, the *unipotent radical* $R_u(G)$, with reductive quotient G^{red}, so there is an exact sequence

$$1 \to R_u(G) \to G \xrightarrow{\varphi} G^{\mathrm{red}} \to 1.$$

Such a sequence makes φ a locally trivial fiber bundle with fibers isomorphic to $R_u(G)$, so φ is a homotopy equivalence and we see $\Lambda_G = \Lambda_{G^{\mathrm{red}}}$. Comparing with the previous example, $U = R_u(B)$ is the group of unitriangular matrices, with $T \cong B/U$.

Example 3.3.7. Let $P \subseteq GL_n$ be the subgroup stabilizing the subspace spanned by the first d basis vectors $e_1, \dots, e_d$, so

$$P = \left[\begin{array}{c|c} * & * \\ \hline 0 & * \end{array} \right]$$

in block form. The unipotent radical and reductive quotient are

$$R_u(P) = \left[\begin{array}{c|c} I & * \\ \hline 0 & I \end{array} \right] \quad \text{and} \quad P^{\mathrm{red}} = \left[\begin{array}{c|c} * & 0 \\ \hline 0 & * \end{array} \right] \cong GL_d \times GL_{n-d},$$

so $\Lambda_P = \Lambda_{GL_d \times GL_{n-d}} \cong \mathbb{Z}[c_1, \dots, c_d, c'_1, \dots, c'_{n-d}]$.

3.4 Free and trivial actions

Proposition 3.4.1. *Suppose G acts freely on X, so that $X \to G\backslash X$ is a locally trivial principal bundle. Then $H_G^* X = H^*(G\backslash X)$ (as rings).*

Proof Take $\mathbb{E}$ with $\widetilde{H}^i\mathbb{E} = 0$ for $i < N$, and consider the diagram

$$
\begin{array}{ccc}
\mathbb{E} \times X & \longrightarrow & X \\
\downarrow & & \downarrow \\
\mathbb{E} \times^G X & \longrightarrow & G\backslash X,
\end{array}
$$

where the horizontal arrows are locally trivial with fiber $\mathbb{E}$. Applying the Leray–Hirsch theorem, $H^i(G\backslash X) \xrightarrow{\sim} H^i(\mathbb{E} \times^G X) = H_G^i X$ for $i < N$. (See Corollary A.4.2 of Appendix A.) $\qquad\square$

Proposition 3.4.2. *Suppose G acts trivially on X, and $\mathbb{E} \to \mathbb{B}$ is a principal G-bundle with $\widetilde{H}^i\mathbb{E} = 0$ for $i < N$. Then*

$$
H_G^i X = H^i(\mathbb{B} \times X)
$$

for $i < N$.

When $N = \infty$ and the Künneth theorem applies – for instance, when Λ_G is free over $\mathbb{Z}$, or when $\Lambda_G^{\mathrm{odd}} = H^{\mathrm{odd}} X = 0$, we have

$$
H_G^* X = \Lambda_G \otimes H^* X
$$

as Λ_G-algebras.

Proof There is a canonical homeomorphism $\mathbb{E} \times^G X \to \mathbb{B} \times X$, by $[e, x] \mapsto (b, x)$, where $e \mapsto b$ under $\mathbb{E} \to \mathbb{B}$. $\qquad\square$

Example 3.4.3. If $H \subseteq G$ is a closed subgroup, acting on X (on the left), then G acts on $G \times^H X$ and there is a canonical isomorphism

$$
H_G^i(G \times^H X) = H_H^i X,
$$

natural with respect to H-equivariant maps $X \to Y$. In particular, taking $X = \mathrm{pt}$, we have

$$
H_G^*(G/H) = H_H^*(\mathrm{pt}) = \Lambda_H.
$$

To see this, use the isomorphism $\mathbb{E} \times^G G \times^H X = \mathbb{E} \times^H X$ from Exercise 2.1.4 of Chapter 2.

Example 3.4.4. In the situation of the previous example, if G acts on X extending the action of H, then there is a G-equivariant isomorphism

$$G \times^H X \xrightarrow{\sim} G/H \times X,$$

with G acting on $G \times^H X$ via left multiplication, and diagonally on $G/H \times X$. (The map is given by $[g, x] \mapsto (gH, g \cdot x)$.) So

$$H_G^*(G/H \times X) = H_H^* X$$

in this case.

Exercise 3.4.5. Projective space $\mathbb{P}^{n-1}$ can be realized as the homogeneous space G/P, where $G = GL_n$ and P is the subgroup of block upper-triangular matrices stabilizing the coordinate line spanned by the first standard basis vector e_1; that is, invertible matrices of the form

$$\begin{bmatrix} * & * & \cdots & * \\ \hline 0 & * & \cdots & * \\ \vdots & \vdots & & \vdots \\ 0 & * & \cdots & * \end{bmatrix}.$$

In Example 2.6.2, we saw

$$H_G^* \mathbb{P}^{n-1} = \Lambda_G[\zeta]/(\zeta^n + c_1 \zeta^{n-1} + \cdots + c_n).$$

On the other hand,

$$\begin{aligned} H_G^*(G/P) &= \Lambda_P \\ &= \Lambda_{\mathbb{C}^* \times GL_{n-1}} \\ &= \mathbb{Z}[t, c_1', \ldots, c_{n-1}']. \end{aligned}$$

Find an explicit ring isomorphism between these two rings.

3.5 Exact sequences

If U and V are G-invariant open subsets of X, there is a natural long exact *Mayer–Vietoris sequence*

$$\cdots \to H_G^k(U \cup V) \to H_G^k U \oplus H_G^k V \to H_G^k(U \cap V) \to H_G^{k+1}(U \cup V) \to \cdots,$$

which can be useful for computations.

There are also exact sequences for triples of G-invariant subspaces $Z \subseteq Y \subseteq X$. One defines $H_G^k(X, Y)$ in the usual way, by choosing $\mathbb{E}$ so that $\widetilde{H}^i\mathbb{E} = 0$ for all $i < N$, and setting

$$H_G^k(X, Y) := H^k(\mathbb{E} \times^G X, \mathbb{E} \times^G Y)$$

for $k < N$. Then

$$\cdots \to H_G^k(X, Y) \to H_G^k(X, Z) \to H_G^k(Y, Z) \to H_G^{k+1}(X, Y) \to \cdots$$

is exact.

Exercise 3.5.1. Let $\mathbb{C}^*$ act on $\mathbb{P}^1$ by $z \cdot [a, b] = [a, zb]$, fixing the points $0 = [1, 0]$ and $\infty = [0, 1]$. Let X be the nodal curve obtained by identifying $0 \sim \infty$, with the induced $\mathbb{C}^*$-action. Show that

$$H_{\mathbb{C}^*}^* X \cong \Lambda_{\mathbb{C}^*}[\alpha]/(\alpha^2, t\alpha)$$

for a class α in $H_{\mathbb{C}^*}^1 X$, where $\Lambda_{\mathbb{C}^*} = \mathbb{Z}[t]$.

3.6 Gysin homomorphisms

Consider a linear algebraic (or complex Lie) group G acting on nonsingular varieties X and Y. For a G-equivariant proper morphism $f \colon X \to Y$, there is an *equivariant Gysin homomorphism*

$$f_* \colon H_G^i X \to H_G^{i+2d} Y,$$

where $d = \dim Y - \dim X$. This is constructed, as usual, from the ordinary Gysin homomorphism for $\mathbb{E} \times^G X \to \mathbb{E} \times^G Y$, where $\mathbb{E}$ is a finite-dimensional nonsingular variety. (See §A.6 of Appendix A for the construction and properties. Our conditions on G, X, Y, and f mean that this is a proper map of complex manifolds, with $d = \dim \mathbb{E} \times^G Y - \dim \mathbb{E} \times^G X$.)

Exercise 3.6.1. Check that this definition is compatible for different choices of $\mathbb{E}$. That is, if $\mathbb{E} \to \mathbb{B}$ and $\mathbb{E}' \to \mathbb{B}'$ are principal G-bundles with $\widetilde{H}^i\mathbb{E} = \widetilde{H}^i\mathbb{E}'$ for $i < N$, then the maps $\mathbb{E} \times^G X \to \mathbb{E} \times^G Y$ and $\mathbb{E}' \times^G X \to \mathbb{E}' \times^G Y$ determine the same Gysin homomorphism $H_G^i X \to H_G^i Y$ for any $i < N$.

Basic examples which we will often use are the inclusion of a G-invariant nonsingular subvariety, $\iota\colon X \hookrightarrow Y$, and the projection of a complete nonsingular variety to a point, $\rho\colon X \to \mathrm{pt}$.

Here are some key properties of equivariant Gysin homomorphisms. They all follow directly from the analogous properties of ordinary Gysin maps, applied to approximation spaces. Here all varieties are assumed to be nonsingular unless otherwise indicated.

(1) (Functoriality) For proper G-equivariant maps $X \xrightarrow{f} Y \xrightarrow{g} Z$, we have $(g \circ f)_* = g_* f_*$.

(2) (Projection formula) For $b \in H_G^* Y$ and $a \in H_G^* X$, we have $f_*(f^* b \cdot a) = b \cdot f_* a$.

As a particular case, considering $b \in \Lambda_G$ mapping to $H_G^* Y$, the projection formula says that f_* is a homomorphism of Λ_G-modules. (In contrast to pullbacks, Gysin pushforwards are generally not homomorphisms of rings.)

(3) (Naturality) For a fiber square of equivariant maps

$$
\begin{array}{ccc}
X' & \xrightarrow{\;g'\;} & X \\
{\scriptstyle f'}\big\downarrow & & \big\downarrow{\scriptstyle f} \\
Y' & \xrightarrow{\;g\;} & Y,
\end{array}
$$

with f (and f') proper, and $\dim Y - \dim X = \dim Y' - \dim X'$, we have $g^* f_* = f'_* (g')^*$.

As with the corresponding property for ordinary Gysin maps, our convention is that the dimension condition is automatically satisfied whenever $X' = \emptyset$ in the fiber square. So $g^* f_* = 0$ whenever the images of g and f are disjoint.

(4) (Self-intersection) If $\iota\colon X \hookrightarrow Y$ is an equivariant closed embedding with normal bundle N of rank d, then we have $\iota^* \iota_*(a) = c_d^G(N) \cdot a$ for any $a \in H_G^* X$.

The equivariant embedding gives an inclusion of equivariant vector bundles $T_X \hookrightarrow \iota^* T_Y$, so it induces a canonical G-equivariant structure on $N = \iota^* T_Y / T_X$.

(5) (Finite cover) If $f\colon X \to Y$ is a proper equivariant map, $V \subseteq X$ is a G-invariant irreducible subvariety (possibly singular), and $W = f(V) \subseteq Y$, then

$$f_*[V]^G = \begin{cases} \deg(V/W)[W]^G & \text{if } \dim W = \dim V; \\ 0 & \text{otherwise.} \end{cases}$$

Equivariant Gysin maps are compatible with change-of-groups homomorphisms. Suppose $f\colon X \to Y$ is a proper G-equivariant morphism of nonsingular varieties, so it has a Gysin map. A group homomorphism $\varphi\colon G' \to G$ determines actions of G' on X and Y, for which f is also equivariant, and the corresponding diagram

$$\begin{array}{ccc} H_G^* X & \longrightarrow & H_{G'}^* X \\ \downarrow{\scriptstyle f_*} & & \downarrow{\scriptstyle f_*} \\ H_G^* Y & \longrightarrow & H_{G'}^* Y \end{array}$$

commutes. In particular, taking G' to be the trivial group, ordinary Gysin homomorphisms are compatible with equivariant ones. (This is proved by applying the non-equivariant version of the naturality property to a diagram

$$\begin{array}{ccc} \mathbb{E}' \times^{G'} X & \longrightarrow & \mathbb{E} \times^G X \\ \downarrow & & \downarrow \\ \mathbb{E}' \times^{G'} Y & \longrightarrow & \mathbb{E} \times^G Y \end{array}$$

of approximation spaces.)

Example 3.6.2. Let a torus T act on $\mathbb{C}^2$ by characters χ_1 and χ_2, inducing an action on $\mathbb{P}^1$. (So $z\cdot[a,b] = [\chi_1(z)a, \chi_2(z)b]$.) Then $0 = [1,0]$ and $\infty = [0,1]$ are fixed points. Setting $\chi = \chi_2 - \chi_1$, the tangent space $T_0\mathbb{P}^1$ has weight χ, since

$$z \cdot [1,b] = [\chi_1(z), \chi_2(z)b] = [1, \chi(z)b].$$

Similarly, $T_\infty\mathbb{P}^1$ has weight $-\chi$. Writing ι_0 and ι_∞ for the inclusions of 0 and ∞, and $\zeta = c_1^T(\mathcal{O}(1))$, we have

$$\iota_0^*\zeta = -\chi_1 = \chi - \chi_2 \quad \text{and} \quad \iota_\infty^*\zeta = -\chi_2 = -\chi - \chi_1.$$

Since $(\iota_0)_*(1) = [0]^T$ and $(\iota_\infty)_*(1) = [\infty]^T$, the self-intersection formula says

$$\iota_0^*[0]^T = \chi \quad \text{and} \quad \iota_\infty^*[\infty]^T = -\chi.$$

Naturality says

$$\iota_0^*[\infty]^T = \iota_\infty^*[0]^T = 0$$

(since $\{0\} \cap \{\infty\} = \emptyset$). Putting these together, we see

$$[0]^T = \zeta + \chi_2 \quad \text{and} \quad [\infty]^T = \zeta + \chi_1$$

in $H_T^2 \mathbb{P}^1$.

Remark. We have pullbacks with respect to $\varphi \colon G \to G'$ and $f \colon X \to X'$, so $f^* \colon H_{G'}^* X' \to H_G^* X$, as well as Gysin pushforwards for $G = G'$. Under what conditions is there a functorial Gysin pushforward associated to a map f which is equivariant with respect to φ?

For example, when $X = X' = \mathrm{pt}$ and $\varphi \colon B \hookrightarrow GL_n$ is the inclusion, one has

$$\Lambda_B = \mathbb{Z}[t_1, \ldots, t_n] \to \Lambda_{GL_n} = \mathbb{Z}[c_1, \ldots, c_n]$$

given by $f \mapsto \partial_{w_\circ}(f)$, where $\partial_{w_\circ}$ is the divided difference operator for the longest element $w_\circ \in S_n$. (In Chapter 10, we will see these operators in a different context.)

3.7 Poincaré duality

Suppose $f \colon X \to Y$ is a fiber bundle, with f proper and X, Y, and the fiber F being oriented manifolds. If $\{a_i\}$ are classes in $H^* X$ restricting to a basis of $H^* F$ – so they form a basis for $H^* X$ over $H^* Y$, by the Leray–Hirsch theorem – then there is a unique *Poincaré dual basis* $\{b_i\}$ such that

$$f_*(a_i \cdot b_j) = \delta_{ij}$$

in $H^* Y$.

Exercise 3.7.1. Prove this, using ordinary Poincaré duality and the graded Nakayama lemma. Note that demanding $f_*(a_i \cdot b_j) = 0$ for $i \neq j$

is a stronger requirement than in the case where Y is a point, because here H^*Y may be nonzero in positive degree.

For such dual bases $\{a_i\}$ and $\{b_i\}$, any class $c \in H^*X$ is expressed uniquely as

$$c = \sum f_*(c \cdot b_i)\, a_i,$$

with coefficients $f_*(c \cdot b_i) \in H^*Y$. Furthermore, there is a relative Künneth decomposition of the diagonal in this context.

Proposition 3.7.2. *Assume X, Y, and F are complex manifolds. Then for $\delta\colon X \to X \times_Y X$, we have*

$$\delta_*(1) = \sum_i b_i \times a_i$$

in $H^*(X \times_Y X)$.
  Further assuming $H^{\mathrm{odd}} F = 0$, we have

$$\delta_*(a_k) = \sum_{i,j} c_{ij}^k\, a_i \times a_j \quad \Leftrightarrow \quad b_i \cdot b_j = \sum_k c_{ij}^k\, b_k$$

for certain coefficients $c_{ij}^k \in H^*Y$.

That is, the same coefficients express both the Künneth decomposition of $\delta_*(a_k)$ in $H^*(X \times_Y X)$ and the expansion of the cup product $b_i \cdot b_j$ in H^*X. The proof is a straightforward calculation, using the fact that $\{a_i \times b_j\}$ and $\{b_i \times a_j\}$ are Poincaré dual bases for $H^*(X \times_Y X)$ over H^*Y, as are $\{a_i \times a_j\}$ and $\{b_i \times b_j\}$.

Consider the case where $X = \mathbb{E} \times^G F$ and $Y = \mathbb{B}$, for G acting on F and $\mathbb{E} \to \mathbb{B}$ a principal G-bundle with $\tilde{H}^k \mathbb{E} = 0$ for $k \leq 2\dim_\mathbb{R} F$. In this setting, the proposition applies to Poincaré dual bases $\{a_i\}$ and $\{b_i\}$ for $H_G^* F$ over Λ_G.

More generally, suppose $f\colon X \to Y$ is a G-equivariant fiber bundle with fiber F, and elements $\{a_i\}$ in $H_G^* X$ restricting to a basis for $H^* F$. Then the a_i form a basis for $H_G^* X$ over $H_G^* Y$, there is a unique Poincaré dual basis $\{b_i\}$, and the proposition holds for these bases. (Apply it to the fiber bundle $\mathbb{E} \times^G X \to \mathbb{E} \times^G Y$, also with fiber F.)

Exercise 3.7.3. If $H^{\mathrm{odd}} F \neq 0$, the coefficients for the Künneth decomposition and those in the cup product are equal up to sign. Show that

$$\delta_*(a_k) = \sum_{i,j} c_{ij}^k \, a_i \times a_j \quad \Leftrightarrow \quad b_i \cdot b_j = \sum_k d_{ij}^k \, b_k$$

where $d_{ij}^k = (-1)^{\deg(b_i)\deg(a_j)} c_{ij}^k$.

Hints for Exercises

Exercise 3.2.1. As before, use a section of $\mathbb{E} \times \mathbb{E}' \to \mathbb{E}$.

Exercise 3.2.2. The induced map on $\mathbb{B} = \mathbb{P}^{m-1}$ will pull back the line bundle $\mathcal{O}(1)$ to $\mathcal{O}(a)$. If $a < 0$, such a map cannot be holomorphic! In this case,

$$(x_1, \ldots, x_m) \mapsto \|x\|^a \left(\frac{1}{\bar{x}_1^a}, \ldots, \frac{1}{\bar{x}_m^a} \right)$$

works, where $\bar{x}_i$ is complex conjugation and $\|x\|$ is the hermitian norm.

Exercise 3.4.5. Use $c_i \mapsto c_i' + t c_{i-1}'$ for $i < n$, $c_n \mapsto t c_{n-1}'$, and $\zeta \mapsto -t$. In fact, this comes from identifying $t = c_1^G(\mathcal{O}(-1))$, $c_i' = c_i^G(\mathbb{C}^n/\mathcal{O}(-1))$, and using the Whitney sum relations.

Exercise 3.5.1. Replace $\mathbb{C}^*$ by S^1 without changing cohomology. Then use the Mayer–Vietoris sequence on a contractible open neighborhood U of the node, with V being the complement of the node.

Exercise 3.6.1. Use the naturality property of Gysin maps: with respect to a fiber square

$$
\begin{array}{ccc}
(\mathbb{E} \times \mathbb{E}') \times^G X & \xrightarrow{\ f'\ } & (\mathbb{E} \times \mathbb{E}') \times^G Y \\
\downarrow{\scriptstyle \pi'} & & \downarrow{\scriptstyle \pi} \\
\mathbb{E} \times^G X & \xrightarrow{\ f\ } & \mathbb{E} \times^G Y,
\end{array}
$$

one has $\pi^* f_* = f'_* (\pi')^*$.

4

Grassmannians and flag varieties

Our goal in this chapter is to compute and study the equivariant cohomology rings H_G^*X, where X is a variety of partial flags in a vector space V, and G is a group acting linearly on V. We will also see some first examples of equivariant Poincaré duality.

4.1   Schur polynomials

For each partition $\lambda = (\lambda_1 \geq \cdots \geq \lambda_d \geq 0)$, there is a *Schur polynomial* s_λ in variables $x_1, x_2, \ldots, x_n$. There are many ways to define these polynomials; we will review three equivalent definitions.

Bialternants. Schur functions were first studied by Cauchy, who considered ratios of two determinants:

$$s_\lambda(x_1, \ldots, x_n) = \frac{\left| x_i^{\lambda_j + n - j} \right|_{1 \leq i,j \leq n}}{\left| x_i^{n-j} \right|_{1 \leq i,j \leq n}}.$$

Here we assume $n = d$, which can always be achieved by appending 0's to either λ or x. Since both the numerator and denominator are alternating functions of the x variables, the ratio is symmetric. Furthermore, both numerator and denominator vanish when $x_i = x_j$ for any pair $i \neq j$; since the denominator is the Vandermonde determinant $\prod_{i<j}(x_i - x_j)$, it must divide the numerator, and the ratio is a polynomial.

Young tableaux. We identify a partition λ with its Young diagram – a left-justified collection of boxes with λ_i boxes in the ith row. A *semistandard Young tableau* (SSYT) of shape λ is a filling $\mathcal{T}$ of the boxes of λ with positive integers, so that entries weakly increase along rows, and strictly increase down columns. To such a tableau, one assigns a monomial weight $x^{\mathcal{T}} = \prod_{(i,j) \in \lambda} x_{\mathcal{T}(i,j)}$. For instance, with $\lambda = (5, 4, 3, 1)$,

the SSYT
$$\begin{array}{|c|c|c|c|c|}\hline 1 & 1 & 2 & 2 & 3 \\\hline 2 & 3 & 3 & 3 \\\cline{1-4} 4 & 4 & 5 \\\cline{1-3} 5 \\\cline{1-1}\end{array}$$
has weight $\quad x^{\mathcal{T}} = x_1^2\, x_2^3\, x_3^4\, x_4^2\, x_5^2.$

A combinatorial definition of Schur functions says

$$s_\lambda(x_1, \ldots, x_n) = \sum_{\mathcal{T} \in SSYT(\lambda)} x^{\mathcal{T}},$$

the sum over all SSYT $\mathcal{T}$ of shape λ with entries in $\{1, \ldots, n\}$. From this definition it is obvious that a Schur polynomial has nonnegative coefficients when written in terms of monomials in x (although it is less obvious that it is symmetric).

Jacobi–Trudi determinant. Let $e_k = e_k(x_1, \ldots, x_n)$ be the elementary symmetric polynomial, and $h_k = h_k(x_1, \ldots, x_n)$ the complete homogeneous symmetric polynomial. (That is, e_i is the sum of all squarefree monomials of degree k, and h_k is the sum of all monomials of degree k.) The Jacobi–Trudi formula expresses a Schur polynomial as a determinant:

$$s_\lambda = \left| h_{\lambda_i + j - i} \right|_{1 \le i, j \le d}. \tag{4.1}$$

There is a dual formulation (also known as the Nägelsbach–Kostka formula):

$$s_\lambda = \left| e_{\lambda_i' + j - i} \right|_{1 \le i, j \le d'}, \tag{4.2}$$

where $\lambda' = (\lambda_1' \ge \cdots \ge \lambda_{d'}' \ge 0)$ is the conjugate partition, that is, the partition whose diagram is the transpose of that of λ. For example, if $\lambda = (5, 4, 3, 1)$ then $\lambda' = (4, 3, 3, 2, 1)$. Note that $(\lambda')' = \lambda$, and λ_1' is the number of nonzero parts of λ.

Both determinants are unchanged if one appends 0's to the partition λ, since $e_0 = h_0 = 1$, and $e_k = h_k = 0$ for $k < 0$.

Basic results in algebraic combinatorics demonstrate the equivalence of these three definitions.

When manipulating Chern class formulas, the Jacobi–Trudi determinantal expression will often be the most useful version of the Schur polynomial. We will introduce notation based on this.

Definition 4.1.1. Given a series $c = 1 + c_1 + c_2 + \cdots$ and a partition λ, the *Schur determinant* $\Delta_\lambda(c)$ is defined as

$$\Delta_\lambda(c) = |c_{\lambda_i + j - i}|_{1 \le i, j \le d},$$

where λ has at most d parts.

Example 4.1.2. Let $c = c(E)$ be the total Chern class of a vector bundle. Then $\Delta_{(k)}(c) = c_k(E)$ is the kth Chern class, and $\Delta_{(1^k)}(c) = s_k(E^\vee)$ is the kth Segre class of $E^\vee$.

More generally, taking $c = c(E - F)$ to be the total Chern class of a virtual bundle, we have

$$\Delta_\lambda(c(E - F)) = \Delta_{\lambda'}(c(F^\vee - E^\vee)). \tag{4.3}$$

To see this, write $h_k = c_k(E - F)$ and $e_k = c_k(F^\vee - E^\vee) = (-1)^k c_k(F - E)$. These satisfy the same basic relation as the complete homogeneous and elementary symmetric functions, namely

$$\left(\sum_{k \ge 0} h_k \, u^k \right) \left(\sum_{k \ge 0} (-1)^k \, e_k \, u^k \right) = 1.$$

So the identity (4.3) follows from the Jacobi–Trudi formula and its dual (4.1)–(4.2).

Generally, when E is a vector bundle with Chern roots $x_1, \ldots, x_n$,

$$\Delta_\lambda(c(E)) = s_{\lambda'}(x_1, \ldots, x_n).$$

(The switch between λ and λ' can be avoided by using Segre classes in the Schur determinant, since $\Delta_\lambda(s(E^\vee)) = s_\lambda(x_1, \ldots, x_n)$.)

A fundamental fact about symmetric functions is that the Schur polynomials s_λ form an additive basis for the ring of symmetric polynomials in $x_1, \ldots, x_n$, as λ ranges over partitions with at most n parts. (By taking an appropriate graded limit as $n \to \infty$, one can suppress the dependence on the number of variables.) A recurring theme in algebraic geometry, combinatorics, and representation theory is that the Schur basis is often the most natural one to use when expanding a given symmetric polynomial.

Example 4.1.3. Griffiths asked which polynomials in Chern classes are *positive* for all ample vector bundles. That is, for which polynomials $P = P(c_1, \ldots, c_n)$, homogeneous of (weighted) degree n, does one have $\int_X P > 0$ whenever one evaluates $c_r = c_r(E)$ for an ample vector bundle E on an n-dimensional variety X?

Bloch showed that the Chern classes $c_1, \ldots, c_n$ are positive in this sense. Griffiths gave other examples, e.g.,

$$c_1^2 - c_2, \quad \begin{vmatrix} c_1 & c_2 & c_3 \\ 1 & c_1 & c_2 \\ 0 & 1 & c_1 \end{vmatrix}, \quad \text{etc.}$$

The complete answer was given by Fulton and Lazarsfeld: Write $P = \sum a_\lambda \Delta_\lambda(c)$, where $\Delta_\lambda(c)$ is as above, and a_λ is an integer. Then P is positive if and only if $a_\lambda \geq 0$ for all λ (and at least one $a_\lambda > 0$). For example, $c_1^2 - 2c_2 = \Delta_{(1,1)} - \Delta_{(2)}$ is not positive.

Example 4.1.4. For each partition $\lambda = (\lambda_1 \geq \cdots \geq \lambda_n \geq 0)$, there is an irreducible polynomial representation V_λ of $GL(V) \cong GL_n$. These are called *Schur modules*, and they interpolate between $\operatorname{Sym}^k \mathbb{C}^n$ – the case $\lambda = (k, 0, \ldots, 0)$, a single row with k boxes – and $\bigwedge^k \mathbb{C}^n$, where $\lambda = (1, \ldots, 1, 0, \ldots, 0)$, a single column with k boxes.

Considering V_λ as a GL_n-equivariant vector bundle on a point, there are classes

$$c_r^{GL_n}(V_\lambda) \in \Lambda_{GL_n} = \mathbb{Z}[c_1, \ldots, c_n]$$

for $1 \leq r \leq \dim V_\lambda$. Restricting to the diagonal torus $T = (\mathbb{C}^*)^n \subseteq GL_n$ gives an inclusion $\Lambda_{GL_n} \hookrightarrow \Lambda_T$, sending $c_k \mapsto e_k(t_1, \ldots, t_n)$, and one obtains a symmetric function $c_r^T(V_\lambda)$. Thus, one has an expression

$$c_r^T(V_\lambda) = \sum_\mu a_{r,\lambda}^\mu \, s_\mu, \tag{4.4}$$

the sum over partitions μ of size r having at most n parts. By the Jacobi–Trudi formula, we have $s_\mu = s_\mu(t_1, \ldots, t_n) = \left| c_{\mu_i' + j - i} \right|_{1 \leq i, j \leq m}$. (Here $\mu_1 \leq m$, so μ' has at most m parts.)

Semistandard tableaux arise naturally here, as well: they index the T-weights of the representation V_λ, so the total Chern class can be expressed as

$$c^{GL_n}(V_\lambda) = \prod_{T \in SSYT(\lambda)} \left(1 + \sum_{i \in T} t_i\right).$$

So the expansion (4.4) can be computed explicitly in any given case.

A remarkable fact, observed by Pragacz, is that the integers $a_{r,\lambda}^\mu$ are nonnegative. The proof of this combinatorial statement involves deep theorems in algebraic geometry! Lascoux gave positive formulas for $c_r^{GL_n}(\bigwedge^2 V)$ and $c_r^{GL_n}(\mathrm{Sym}^2 V)$. Beyond these cases, however, few explicit general formulas for these polynomials are known. See the Notes for further discussion of this question.

4.2   Flag bundles

Let V be a vector bundle of rank n on a variety Y, and fix an integer $0 \le d \le n$. We have the *Grassmann bundle*

$$\mathbf{Gr}(d, V) \to Y$$

representing the functor which assigns to a morphism $f \colon Z \to Y$ the set of rank-d subbundles $S \subset V_Z := f^*V$. (By *subbundle*, we mean that the inclusion is locally split, so V_Z/S is also a vector bundle; this is a stronger requirement than asking that S be a locally free subsheaf.) We saw a construction of $\mathbf{Gr}(d, V)$ using the frame bundle of V in Chapter 2. Here we construct it as a scheme using local data.

For a split vector bundle $V = A \oplus B$, with A of rank d and B of rank $e = n - d$, there is an open subset $U \subseteq \mathbf{Gr}(d, V)$, defined by the additional condition that the composition of linear maps

$$S \hookrightarrow V \twoheadrightarrow A$$

be an isomorphism. There is a natural identification $U \xrightarrow{\sim} \mathrm{Hom}(A, B)$, making U a vector bundle over Y, given by the map

$$(S \subset V) \mapsto (A \cong S \hookrightarrow V \twoheadrightarrow B).$$

The inverse morphism sends a local section $\varphi \in \mathrm{Hom}(A, B)$ to its graph $\Gamma_\varphi \subseteq A \oplus B$.

Since any vector bundle V splits locally on Y, these open sets U give an affine covering of $\mathbf{Gr} = \mathbf{Gr}(d, V)$, constructing it as a scheme. It also shows that the projection $\rho\colon \mathbf{Gr} \to Y$ is smooth of relative dimension $d(n - d) = \operatorname{rk}\operatorname{Hom}(A, B)$.

The functorial description equips $\mathbf{Gr}(d, V)$ with a tautological subbundle $\mathbb{S} \subset V_{\mathbf{Gr}} = \rho^*V$, as well as a quotient bundle $\mathbb{Q} = V_{\mathbf{Gr}}/\mathbb{S}$, so there is a universal sequence

$$0 \to \mathbb{S} \to V_{\mathbf{Gr}} \to \mathbb{Q} \to 0$$

on $\mathbf{Gr}(d, V)$. The relative tangent bundle is

$$T_{\mathbf{Gr}/Y} = \operatorname{Hom}(\mathbb{S}, \mathbb{Q}),$$

as one sees from the local description of the open sets U.

There are several dualities among Grassmannians. For $e = n - d$, there is a canonical isomorphism

$$\mathbf{Gr}(d, V) = \mathbf{Gr}(e, V^\vee),$$

by sending a subspace $S \subset V$ to the subspace $(V/S)^\vee \subseteq V^\vee$. Often it is also useful to consider the Grassmannian of quotients,

$$\mathbf{Gr}(d, V) = \mathbf{Gr}(V, e) = \{V \twoheadrightarrow Q\}/\cong,$$

where two rank e quotients of V are identified if there is an isomorphism $Q \cong Q'$ commuting with the projection from V.

Example 4.2.1. For $d = 0$ or $d = n$, one has $\mathbf{Gr}(d, V) = Y$.

For $d = 1$, one has the projective bundle $\mathbb{P}(V) = \mathbf{Gr}(1, V)$, and the dual projective bundle $\mathbb{P}^\vee(V) = \mathbf{Gr}(V, 1)$. The first has tautological subbundle $\mathbb{S} = \mathcal{O}(-1)$, and the second has tautological quotient bundle $\mathbb{Q} = \mathcal{O}(1)$.

In general, sending $S \subseteq V$ to $\bigwedge^d S \subseteq \bigwedge^d V$ defines the *Plücker embedding*

$$\mathbf{Gr}(d, V) \lhook\joinrel\longrightarrow \mathbb{P}(\textstyle\bigwedge^d V)$$
$$\rho \searrow \qquad \swarrow$$
$$Y,$$

a closed embedding locally defined by the Plücker relations, so ρ is projective.

More generally, given a sequence $\mathbf{d} = (0 \le d_1 < \cdots < d_r \le n)$, we have a *partial flag bundle*

$$\mathbf{Fl}(\mathbf{d}, V) \xrightarrow{\rho} Y,$$

parametrizing flags $S_1 \subset \cdots \subset S_r \subset V_Z$ for any $f\colon Z \to Y$, where the rank of S_i is d_i.

A partial flag bundle can be constructed inductively as a sequence of Grassmann bundles, in $r!$ ways. For instance, if one starts by choosing S_1, so $\mathbf{Gr}(d_1, V) \to Y$ has a tautological bundle $\mathbb{S}_1 \subset V_{\mathbf{Gr}}$ of rank d_1, one obtains a factorization

$$
\begin{array}{ccc}
\mathbf{Fl}(d_2 - d_1, \ldots, d_r - d_1, V_{\mathbf{Gr}}/\mathbb{S}_1) & \longrightarrow & \mathbf{Gr}(d_1, V) \\
\| & & \downarrow \\
\mathbf{Fl}(d_1, \ldots, d_r, V) & \longrightarrow & Y.
\end{array}
$$

Alternatively, for any $1 \le s \le r$, one can start by choosing the subspace S_s, and obtain a factorization

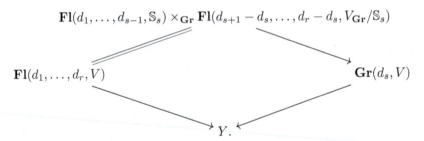

A partial flag bundle is equipped with tautological subbundles and quotient bundles,

$$\mathbb{S}_1 \subset \cdots \subset \mathbb{S}_r \subset V_{\mathbf{Fl}} \twoheadrightarrow \mathbb{Q}_1 \twoheadrightarrow \cdots \twoheadrightarrow \mathbb{Q}_r,$$

just as for Grassmannians. Also as before, these are identified with quotient flag bundles

$$\mathbf{Fl}(d_1, \ldots, d_r, V) = \mathbf{Fl}(V, e_1, \ldots, e_r),$$

where $n \ge e_1 > \cdots > e_r \ge 0$ and $e_i = n - d_i$. These are related in the evident way, by sending $S_i \subset V$ to $V \twoheadrightarrow Q_i = V/S_i$.

The *complete flag bundle* is the case $\mathbf{Fl}(V) = \mathbf{Fl}(1, \ldots, n-1, V)$. One choice of the inductive construction realizes this as a tower of projective bundles,

$$Y \leftarrow \mathbb{P}(V) \leftarrow \mathbb{P}(V/\mathbb{S}_1) \leftarrow \cdots \leftarrow \mathbb{P}(V/\mathbb{S}_{n-2}) = \mathbf{Fl}(V).$$

4.3 Projective space

Consider G acting linearly on an n-dimensional vector space V. In §2.6, we have already seen a computation

$$H_G^*\mathbb{P}(V) = \Lambda_G[\zeta]/(\zeta^n + c_1\zeta^{n-1} + \cdots + c_n),$$

where $\zeta = c_1^G(\mathcal{O}(1))$ and $c_i = c_i^G(V)$. This was a special case of the general formula for cohomology of a projective bundle $\mathbb{P}(V) \to Y$, using the Leray–Hirsch theorem and the fact that $1, \zeta, \ldots, \zeta^{n-1}$ form a basis for $H^*\mathbb{P}(V)$ over H^*Y.

The relations come from the Whitney formula for the exact sequence

$$0 \to \mathcal{O}(-1) \to V \to \mathbb{Q} \to 0,$$

so

$$c^G(\mathbb{Q}) = \frac{c^G(V)}{c^G(\mathcal{O}(-1))} = \frac{1 + c_1 + \cdots + c_n}{1 - \zeta},$$

and $c_n^G(\mathbb{Q}) = 0$.

Before moving on to other flag varieties (and bundles), it is worth investigating a second basis and presentation suggested by symmetry. The classes $1, c_1^G(\mathbb{Q}), \ldots, c_{n-1}^G(\mathbb{Q})$ also form a basis for $H^*\mathbb{P}(V)$ over H^*Y, since $c_k^G(\mathbb{Q})$ restricts to ζ^k on fibers. What are the relations?

Proposition 4.3.1. *We have*

$$H_G^*\mathbb{P}(V) = \Lambda_G[\xi_1, \ldots, \xi_{n-1}]/(s_2, \ldots, s_n),$$

where ξ_i maps to $c_i^G(\mathbb{Q})$ and s_k is the degree k term in the expansion of

$$\frac{1 + c_1 + c_2 + \cdots + c_n}{1 + \xi_1 + \xi_2 + \cdots + \xi_{n-1}}.$$

Proof Let R be the algebra on the right-hand side. There is a well-defined homomorphism $R \to H_G^*\mathbb{P}(V)$, since s_k maps to $c_k^G(V - \mathbb{Q}) =$

$c_k^G(\mathcal{O}(-1))$, which vanishes for $k > 1$. Since the classes $c_k^G(\mathbb{Q})$ form a basis for the free Λ_G-module $H_G^*\mathbb{P}(V)$, the sequence

$$0 \to I \to R \to H_G^*\mathbb{P}(V) \to 0$$

splits, and it suffices to show that the classes $1, \xi_1, \ldots, \xi_{n-1}$ span R as a Λ_G-module. By the graded Nakayama lemma, it is enough to do this modulo Λ_G^+, i.e., after setting $c_k = c_k^G(V) = 0$ for $k > 0$. This case is a basic fact about symmetric functions, stated in the exercise below. $\qquad\square$

Exercise 4.3.2. Let Λ be an associative ring, with commuting indeterminates $e_1, \ldots, e_d$, for some $0 < d < n$. For $k > 0$, let h_k be defined inductively by the relation $h_k - h_{k-1}e_1 + \cdots + (-1)^k e_k = 0$, using the conventions $e_0 = h_0 = 1$ and $e_k = 0$ for $k > d$. Show that h_k lies in the two-sided ideal $(h_{n-d+1}, \ldots, h_n) \subseteq \Lambda[e_1, \ldots, e_d]$ for all $k > n - d$. Taking $d = n - 1$, conclude that the elements $1, e_1, \ldots, e_{n-1}$ span $\Lambda[e_1, \ldots, e_{n-1}]/(h_2, \ldots, h_n)$ as a Λ-module.

The same argument leads to presentations for equivariant projective bundles.

Exercise 4.3.3. Suppose $V \to Y$ is a G-equivariant vector bundle of rank n, so G acts equivariantly on the projective bundle $\mathbb{P}(V) \to Y$, with universal line bundle $\mathcal{O}(1)$ and quotient bundle $\mathbb{Q} = V/\mathcal{O}(-1)$. Show that

$$H_G^*\mathbb{P}(V) = (H_G^*Y)[\zeta]/(\zeta^n + c_1\zeta^{n-1} + \cdots + c_n)$$
$$= (H_G^*Y)[\xi_1, \ldots, \xi_{n-1}]/(s_2, \ldots, s_n),$$

where ζ maps to $c_1^G(\mathcal{O}(1))$, ξ_i maps to $c_i^G(\mathbb{Q})$, $c_k = c_k^G(V)$, and s_k is defined as in Proposition 4.3.1.

4.4   Complete flags

A partial flag variety $Fl(\mathbf{d}, V) = Fl(d_1, \ldots, d_r, V)$ has quotient bundles $A_i = \mathbb{S}_i/\mathbb{S}_{i-1}$ of rank $a_i = d_i - d_{i-1}$. (By convention, $d_0 = 0$ and $d_{r+1} = n$, so $\mathbb{S}_0 = 0$ and $\mathbb{S}_{r+1} = V$.) Equivalently, these bundles are $A_i = \ker(\mathbb{Q}_{i-1} \to \mathbb{Q}_i) = \ker(V/\mathbb{S}_{i-1} \to V/\mathbb{S}_i)$, so $a_i = e_{i-1} - e_i$. From the Whitney formula, we have

$$c^G(V) = \prod_{i=1}^{r+1} c^G(A_i). \tag{4.5}$$

We will see that $H_G^* Fl(\mathbf{d}, V)$ is generated by the Chern classes of A_i, with relations coming from (4.5).

The first step is to compute the cohomology of the complete flag variety $Fl(V)$. Here the A_i are line bundles, with $x_i = c_1^G(A_i)$.

Proposition 4.4.1. *We have*

$$H_G^* Fl(V) = \Lambda_G[x_1, \ldots, x_n]/(e_i(x) - c_i)_{i=1,\ldots,n},$$

where $c_i = c_i^G(V)$. A basis over Λ_G is given by

$$\{x_1^{m_1} \cdots x_n^{m_n} \mid 0 \leq m_i \leq n - i\},$$

so $H_G^* Fl(V)$ has rank $n!$ as a Λ_G-module.

The presentation is symmetric in the variables $x_1, \ldots, x_n$, so any permutation of them gives another basis.

Proof Let R be the algebra on the right-hand side. The basis for $H_G^* Fl(V)$ comes directly from writing $Fl(V)$ as a tower of projective bundles. The relations hold by the Whitney formula (4.5), so we have a homomorphism $R \to H_G^* Fl(V)$. As for projective space, it suffices to show the monomials $x_1^{m_1} \cdots x_n^{m_n}$ span R as a Λ_G-module; by graded Nakayama, this reduces to showing the same monomials span the $\mathbb{Z}$-module $A = \mathbb{Z}[x_1, \ldots, x_n]/(e_1(x), \ldots, e_n(x))$.

To see this, observe that in $A[u]$ we have $\prod_{i=1}^n (1 - x_i u) = 1$, so for any $1 \leq \ell \leq n$, the polynomial

$$\sum_{k \geq 0} h_k(x_1, \ldots, x_\ell) u^k = \prod_{i=1}^\ell \frac{1}{1 - x_i u} = \prod_{i=\ell+1}^n (1 - x_i u)$$

has degree $n - \ell$. It follows that $h_{n-\ell+1}(x_1, \ldots, x_\ell) = 0$, which gives a relation expressing $x_\ell^{n-\ell+1}$ in terms of the given monomials by induction on ℓ. $\qquad\square$

As in §4.3, the same argument leads to presentations for equivariant flag bundles. For a G-equivariant vector bundle $V \to Y$, we have G

acting equivariantly on the associated flag bundle $\mathbf{Fl}(V) \to Y$, with equivariant tautological line bundles $A_i = \mathbb{S}_i/\mathbb{S}_{i-1}$. Then

$$H_G^*\mathbf{Fl}(V) = (H_G^*Y)[x_1, \ldots, x_n]/(e_i(x) - c_i)_{i=1,\ldots,n},$$

where x_i maps to $c_1^G(A_i)$ and $c_i = c_i^G(V)$.

4.5 Grassmannians and partial flag varieties

Consider a G-equivariant vector bundle V of rank n on Y. The Grassmann bundle can be factored as

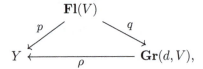

where both p and q are towers of projective bundles. We will use this to compute $H_G^*\mathbf{Gr}(d, V)$. Since a partial flag bundle is a tower of Grassmann bundles, this will also lead to a computation of $H_G^*\mathbf{Fl}(\mathbf{d}, V)$.

Let $X = \mathbf{Gr}(d, V)$, with tautological sequence

$$0 \to \mathbb{S} \to V_X \to \mathbb{Q} \to 0,$$

so $A_1 = \mathbb{S}$ has rank d, and $A_2 = \mathbb{Q}$ has rank $e = n - d$.

Proposition 4.5.1. *We have*

$$\begin{aligned}
H_G^*X &= (H_G^*Y)[c_1, \ldots, c_d, \widetilde{c}_1, \ldots, \widetilde{c}_e]/(c \cdot \widetilde{c} = c^G(V)) \\
&= (H_G^*Y)[c_1, \ldots, c_d]/(s_{e+1}, \ldots, s_n) \\
&= (H_G^*Y)[\widetilde{c}_1, \ldots, \widetilde{c}_e]/(\widetilde{s}_{d+1}, \ldots, \widetilde{s}_n),
\end{aligned}$$

under the evaluations $c \mapsto c^G(\mathbb{S})$ and $\widetilde{c} \mapsto c^G(\mathbb{Q})$. The elements s_m are defined by

$$s = c^G(V)/c = 1 + (c_1^G(V) - c_1) + (c_2^G(V) - c_1^G(V)c_1 + c_1^2 - c_2) + \cdots,$$

and the elements $\widetilde{s}_m$ are defined similarly, by $\widetilde{s} = c^G(V)/\widetilde{c}$.

A basis over H_G^*Y is given by monomials

$$\prod_{j=1}^{d} c_j^G(\mathbb{S}^\vee)^{m_j}, \quad \text{with } \sum m_j \le e.$$

(There are $\binom{n}{d}$ such monomials.) Another basis is given by Schur determinants

$$\Delta_\lambda(c^G(\mathbb{S}^\vee)) \quad \text{for } \lambda \subseteq \boxed{}\begin{smallmatrix}d\\ \\e\end{smallmatrix},$$

i.e., $d \geq \lambda_1 \geq \cdots \geq \lambda_e \geq 0$.

Similarly, there are bases of monomials

$$\prod_{j=1}^{e} c_j^G(\mathbb{Q})^{m_j}, \quad \text{with } \sum m_j \leq d,$$

and of Schur determinants

$$\Delta_\lambda(c^G(\mathbb{Q})) \quad \text{for } \lambda \subseteq \boxed{}\begin{smallmatrix}e\\ \\d\end{smallmatrix},$$

i.e., $e \geq \lambda_1 \geq \cdots \geq \lambda_d \geq 0$.

As before, we will only discuss the case where Y is a point, so $X = Gr(d, V)$ and $H_G^* Y = \Lambda_G$. The general case is no different.

In the proof, we show that monomials in $c_j^G(\mathbb{S})$, and determinants $\Delta_\lambda(c^G(\mathbb{S}))$, form bases. (Replacing $\mathbb{S}$ by $\mathbb{S}^\vee$ only changes the sign, but it ensures these classes are positive.) The argument for $\mathbb{Q}$ is similar.

Proof The cases $d = 1$ and $e = 1$ have been done, since these are projective spaces. In general, one knows the relations hold, because $c_m^G(V - \mathbb{Q}) = c_m^G(\mathbb{S})$ vanishes for $m > d$, and $c_m^G(V - \mathbb{S}) = c_m^G(\mathbb{Q})$ vanishes for $m > e$. So, as before, we have a homomorphism from each algebra on the right-hand side to $H_G^* X$.

Now we use a trick that goes back to Grothendieck. Consider the two-step flag variety

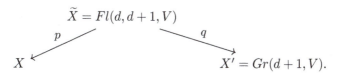

By descending induction on d, with the base case $e = 1$, we may assume we know the presentation for X'. Let $\mathbb{S} = \mathbb{S}_d$ be the tautological bundle on X, and $\mathbb{S}_{d+1}$ the tautological on X', so on $\widetilde{X}$ we have $\mathbb{S}_d \subset \mathbb{S}_{d+1}$. (We suppress notation for pullbacks.) Let $a_i = c_i^G(\mathbb{S}_d)$, $b_i = c_i^G(\mathbb{S}_{d+1})$, and $x = c_1^G(\mathbb{S}_{d+1}/\mathbb{S}_d)$. In $H_G^* \widetilde{X}$, we have

$$b_i = a_i + x a_{i-1} \quad \text{for } 1 \le i \le d+1, \tag{4.6}$$

coming from the relation $c^G(\mathbb{S}_{d+1}) = c^G(\mathbb{S}_d)(1+x)$.

Since $\widetilde{X} = \mathbb{P}^\vee(\mathbb{S}_{d+1}) = \mathbf{Gr}(d, \mathbb{S}_{d+1}) \to X'$ is an equivariant projective bundle, with $x = c_1^G(\mathcal{O}(1))$, we see that $H_G^*\widetilde{X}$ is free over H_G^*X', with basis $\{1, x, \ldots, x^d\}$. By the inductive assumption, H_G^*X' is spanned over Λ_G by monomials in b_i's, so we conclude that $H_G^*\widetilde{X}$ is spanned over Λ_G by monomials in a_i's and x, using (4.6).

On the other hand, $\widetilde{X} = \mathbb{P}(V/\mathbb{S}_d) \to X$, with $x = c_1^G(\mathcal{O}(-1))$, so $H_G^*\widetilde{X}$ is free over H_G^*X with basis $1, x, \ldots, x^{e-1}$, and relation

$$x^e - c_1^G(V/\mathbb{S}_d)x^{e-1} + \cdots + (-1)^e c_e^G(V/\mathbb{S}_d) = 0.$$

Each $c_i^G(V/\mathbb{S}_d)$ lies in the subalgebra $A \subseteq H_G^*X$ generated over Λ_G by $a_1, \ldots, a_d$, so each element of $H_G^*\widetilde{X}$ can be expressed uniquely as $\sum_{i=0}^{e-1} \alpha_i x^i$, for some $\alpha_i \in A$. It follows that $H_G^*X = A$, so the map

$$A' = \Lambda_G[c_1, \ldots, c_d]/(s_{e+1}, \ldots, s_n) \to H_G^*X, \quad c_i \mapsto a_i,$$

is surjective, and this surjection splits. (The Λ_G-module H_G^*X is projective, since it embeds as a direct summand in the free Λ_G-module $H_G^*\widetilde{X}$.)

As an aside, the above arguments show

$$\mathrm{rk}_{\Lambda_G} H_G^*\widetilde{X} = (d+1) \cdot \mathrm{rk}_{\Lambda_G} H_G^*X' = (d+1) \cdot \binom{n}{d+1}$$

and

$$\mathrm{rk}_{\Lambda_G} H_G^*X = \frac{1}{e} \cdot \mathrm{rk}_{\Lambda_G} H_G^*\widetilde{X} = \frac{d+1}{e} \cdot \binom{n}{d+1} = \binom{n}{d}.$$

To conclude the proof, it suffices to show that the algebra A' is spanned over Λ_G by the $\binom{n}{d}$ elements $\Delta_\lambda(c)$, for $\lambda \subseteq \boxed{}\begin{smallmatrix} d \\ \\ \end{smallmatrix}{}^e$. Since these elements belong to A, it will follow that the split surjection $A' \to A$ is an isomorphism.

Using graded Nakayama, it is enough to show that these elements span in the case where G is trivial, so $\Lambda_G = \mathbb{Z}$, and we will assume this for the rest of the argument. As shown in Exercise 4.5.2, the set $\{\Delta_\lambda(c) \mid \lambda_1 \le d\}$ is a basis for the polynomial ring $\mathbb{Z}[c_1, \ldots, c_d]$, so we just need to see that $\Delta_\lambda(c)$ maps to zero in A' whenever λ has more than e nonzero parts.

For this, we use the duality formula (Eq. (4.3) from §4.1) to write $\Delta_\lambda(c) = \Delta_{\lambda'}(c')$, where c' is defined inductively by relations

$$c'_k - c'_{k-1}c_1 + \cdots + (-1)^k c_k = 0$$

for $k \geq 1$. Note that $c'_k = (-1)^k s_k$, and by Exercise 4.3.2, c'_k lies in the ideal $(s_{e+1}, \ldots, s_n)$ for all $k > e$. We can write

$$\Delta_{\lambda'}(c') = \begin{vmatrix} c'_{\lambda'_1} & c'_{\lambda'_1+1} & \cdots \\ \vdots & c'_{\lambda'_2} & \\ & & \ddots \end{vmatrix},$$

so if $\lambda'_1 > e$, each entry in the top row maps to zero in A', and hence $\Delta_{\lambda'}(c')$ also maps to zero in A'.

Finally, using Exercise 4.5.2 again, the monomials c^m can be ordered so that $c_{\lambda_1} \cdots c_{\lambda_e}$ is the leading term of $\Delta_\lambda(c)$, and it follows that the monomials $c_1^{m_1} \cdots c_d^{m_d}$ with $\sum m_j \leq e$ map to another basis of A'. $\qquad\square$

Exercise 4.5.2. Find a monomial order on $c^m = c_1^{m_1} \cdots c_d^{m_d}$ so that the diagonal term $c_{\lambda_1} \cdots c_{\lambda_e}$ is the smallest term in $\Delta_\lambda(c)$, for any partition λ and any d, e. Use this to conclude that the determinants $\Delta_\lambda(c)$ (with no restrictions on λ) form a basis for the polynomial ring in infinitely many variables $\Lambda[c_1, c_2, \ldots]$, where Λ is an associative ring, and that the kernel of the projection $\Lambda[c_1, c_2, \ldots] \to \Lambda[c_1, \ldots, c_d]$ (sending c_i to 0 if $i > d$) has basis $\{\Delta_\lambda(c) \mid \lambda_1 > d\}$. Conclude that $\{\Delta_\lambda(c) \mid \lambda_1 \leq d\}$ is a basis for $\Lambda[c_1, \ldots, c_d]$.

Example 4.5.3. Consider $G = GL(V)$ acting on $X = Gr(d, V)$, so $X \cong G/P$, where

$$P = \left[\begin{array}{c|c} * & * \\ \hline 0 & * \end{array} \right]$$

is block upper-triangular, with diagonal blocks of size d and e; in other words, it is the stabilizer of the subspace $E \subseteq V$ spanned by the first d standard basis vectors. So we have two kinds of presentations: on one hand, Proposition 4.5.1 says that

$$H_G^* X = H_G^* Gr(d, V)$$
$$= \Lambda_G[c_1^G(\mathbb{S}), \ldots, c_d^G(\mathbb{S}), c_1^G(\mathbb{Q}), \ldots, c_e^G(\mathbb{Q})]/I,$$

where I is the ideal generated by the relations $c^G(\mathbb{S}) \cdot c^G(\mathbb{Q}) = c^G(V)$; on the other hand, we have

$$H_G^* X = H_G^*(G/P) = \Lambda_P = \mathbb{Z}[a_1, \ldots, a_d, b_1, \ldots, b_e],$$

using Examples 3.3.7 and 3.4.3 from Chapter 3.

The isomorphism between these rings can be written quite simply, by $a_i \mapsto c_i^G(\mathbb{S})$ and $b_i \mapsto c_i^G(\mathbb{Q})$. We also have $H_G^* X \cong \Lambda_P$ as a Λ_G-algebra via an inclusion

$$\Lambda_G = \mathbb{Z}[c_1, \ldots, c_n] \hookrightarrow \Lambda_P[a_1, \ldots, a_d, b_1, \ldots, b_e].$$

Associated to $\rho \colon X \to \mathrm{pt}$, there is the Gysin pushforward homomorphism $\rho_* \colon H_G^* X \to \Lambda_G$. Under the above isomorphisms, ρ_* becomes a $\mathbb{Z}[c]$-module homomorphism $\mathbb{Z}[a, b] \to \mathbb{Z}[c]$ which drops degree by de. (What is an explicit formula?)

Finally, we can deduce presentations and bases for partial flag varieties. To take advantage of inductive structure – a partial flag variety is an iterated Grassmann bundle – it helps to state the theorem for equivariant flag bundles. (As we have already seen several times, no additional work is needed.)

Corollary 4.5.4. *Let $V \to Y$ be a G-equivariant vector bundle, with associated partial flag bundle $X = \mathbf{Fl}(\mathbf{d}, V) \to Y$, and tautological bundles $\mathbb{S}_1 \subset \cdots \subset \mathbb{S}_r \subset V$. Setting $A_i = \mathbb{S}_i/\mathbb{S}_{i-1}$, we have*

$$H_G^* X = (H_G^* Y)[c_j^G(A_i)] / \left(\prod_{i=1}^{r+1} c^G(A_i) = c^G(V) \right),$$

where the generators $c_j^G(A_i)$ run over $1 \le i \le r+1$ and $1 \le j \le d_i - d_{i-1}$. There is a basis of products of Schur determinants,

$$\left\{ \prod_{i=1}^r \Delta_{\lambda^{(i)}}(c^G(\mathbb{S}_i^\vee)) \right\}, \qquad \text{for } \lambda^{(i)} \subseteq \boxed{} \begin{array}{c} d_i \\ d_{i+1} - d_i. \end{array}$$

Proof Factor $X \to Y$ as

$$X = \mathbf{Fl}(d_1, \ldots, d_{r-1}, \mathbb{S}_r) \to \mathbf{Gr}(d, V) \to Y,$$

and use induction on r together with our calculation for Grassmannians. $\qquad \square$

Exercise 4.5.5. Prove the dual version of the theorem: the products

$$\prod_{i=1}^{r} \Delta_{\mu^{(i)}}(c^{G}(\mathbb{Q}_i)), \quad \text{for } \mu^{(i)} \subseteq \boxed{}\!\!\begin{smallmatrix}e_i\\ \\ \end{smallmatrix} \ e_i - e_{i+1},$$

form a basis for $H_G^* \mathbf{Fl}(\mathbf{d}, V)$ as a module over $H_G^* Y$.

Remark. The Λ_G-module $H_G^* Fl(\mathbf{d}, V)$ has rank

$$\mathrm{rk}_{\Lambda_G} H_G^* Fl(\mathbf{d}, V) = \prod \binom{d_{i+1}}{d_i} = \frac{n!}{a_1! \cdots a_{r+1}!} = \binom{n}{a_1, \ldots, a_{r+1}},$$

where $a_i = \mathrm{rk}\, A_i = d_i - d_{i-1}$.

4.6   Poincaré dual bases

The bases we have computed for Grassmann bundles are Poincaré dual, in the sense of §3.7. Given a partition λ in the $d \times e$ rectangle, its *complement* in the $e \times d$ rectangle is the conjugate to $(e - \lambda_d, \ldots, e - \lambda_1)$. For example, if $\lambda = (5, 3, 2, 2, 2)$ in the 5×6 rectangle, then $\mu = (5, 4, 4, 3, 0, 0)$ is its complement.

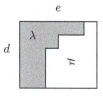

Theorem 4.6.1. *Let V be a G-equivariant vector bundle of rank n on Y, with Grassmann bundle $\rho\colon X = \mathbf{Gr}(d, V) \to Y$. The bases $\{\Delta_\lambda(c^G(\mathbb{Q}))\}$ and $\{\Delta_\mu(c^G(\mathbb{S}^\vee))\}$ are Poincaré dual, where*

$$\lambda \subseteq \boxed{}\!\begin{smallmatrix}e\\ \\d\end{smallmatrix} \quad \text{and} \quad \mu \subseteq \boxed{}\!\begin{smallmatrix}d\\ \\e\end{smallmatrix}$$

are complementary partitions.

Proof By Proposition 3.7.2, the assertion is equivalent to the formula

$$[\delta(X)]^{G} = \sum_{\lambda, \mu \text{ complementary}} \Delta_\lambda(c^G(\mathbb{Q})) \times \Delta_\mu(c^G(\mathbb{S}^\vee)), \qquad (4.7)$$

where $\delta(X) \subseteq X \times_Y X$ is the diagonal, so $[\delta(X)]^{G} = \delta_*(1)$.

On $X \times_Y X$, we have the universal bundles $\mathbb{Q}^{(1)}$ and $\mathbb{S}^{(2)}$ pulled back from the first and second factors, respectively. Writing V also for the pullback of this vector bundle to $X \times_Y X$, we have $\mathbb{S}^{(2)} \subseteq V$ and $V \twoheadrightarrow \mathbb{Q}^{(1)}$, and the diagonal is the locus in $X \times_Y X$ where the composition $\mathbb{S}^{(2)} \to \mathbb{Q}^{(1)}$ vanishes. The class of $\delta(X)$ is therefore the top Chern class of the corresponding Hom bundle:

$$[\delta(X)]^G = c_{de}^G((\mathbb{S}^{(2)})^\vee \otimes \mathbb{Q}^{(1)}).$$

There are several ways to show this is equal to the right-hand side. One way is to appeal to the splitting principle and argue using Chern roots $y_1, \ldots, y_d$ of $(\mathbb{S}^{(2)})^\vee$ and $x_1, \ldots, x_e$ of $\mathbb{Q}^{(1)}$. Formula (4.7) is then equivalent to the symmetric function identity

$$\prod_{i=1}^{e} \prod_{j=1}^{d} (x_i + y_j) = \sum_{\lambda, \mu \text{ complementary}} s_{\lambda'}(x_1, \ldots, x_e) \cdot s_{\mu'}(y_1, \ldots, y_d), \quad (4.8)$$

which follows from a well-known "Cauchy identity" for Schur functions. $\qquad \square$

Example 4.6.2. When $d = 1$, the theorem says that the bases

$$\{1, c_1^G(\mathbb{Q}), \ldots, c_{n-1}^G(\mathbb{Q})\} \quad \text{and} \quad \{\zeta^{n-1}, \ldots, \zeta, 1\}$$

are Poincaré dual, where as usual $\zeta = c_1^G(\mathcal{O}(1))$. We can see this directly, using the relation $\zeta^n + c_1^G(V)\zeta^{n-1} + \cdots + c_n^G(V) = 0$. We can write

$$\zeta^i c_j(\mathbb{Q}) = \zeta^i c_j^G(V - \mathbb{S}) = \zeta^i c_j^G(V) + \zeta^{i+1} c_{j-1}^G(V) + \cdots + \zeta^{i+j}.$$

By the projection formula, we have

$$\rho_*(\zeta^k c_m^G(V)) = \rho_*(\zeta^k) c_m^G(V),$$

so this is 0 if $k < n - 1$. On the other hand, for $k \geq 0$, we can use

$$\begin{aligned}
\zeta^{n-1+k} &= c_1^G(V^\vee - \mathbb{Q}^\vee)^{n-1+k} \\
&= c_{n-1+k}^G(\mathbb{Q} - V) \\
&= c_{n-1}^G(\mathbb{Q}) c_k^G(-V) + c_{n-2}^G(\mathbb{Q}) c_{k+1}^G(-V) + \cdots,
\end{aligned}$$

so $\rho_*(\zeta^{n-1+k} c_m^G(V)) = c_k^G(-V) \cdot c_m^G(V)$. Putting these calculations together, we see

$$\rho_*(\zeta^i c_j^G(\mathbb{Q})) = c_{i+j-n+1}^G(V) + c_1^G(-V)\,c_{i+j-n}^G(V) + \cdots + c_{i+j-n+1}^G(-V)$$
$$= c_{i+j-n+1}^G(V - V)$$
$$= \begin{cases} 1 & \text{if } i + j = n - 1, \\ 0 & \text{otherwise.} \end{cases}$$

4.7 Bases and duality from subvarieties

When looking at $H_G^* X$, we have not yet seen "geometric" classes coming from invariant subvarieties. Indeed, for $G = GL(V)$ acting on $X = Fl(\mathbf{d}, V)$, there are no invariant subvarieties, except X itself! For other group actions, we will often have invariant subvarieties, so we can compare their classes with bases we have already seen.

The first tool is an equivariant cell-decomposition lemma.

Proposition 4.7.1. *Suppose G acts on a nonsingular variety X, and there are G-invariant closed algebraic subsets*

$$\emptyset \subset X_0 \subset X_1 \subset \cdots \subset X_m = X,$$

such that each $X_p \smallsetminus X_{p-1} = \coprod_j U_{p,j}$ is a disjoint union of finitely many irreducible nonsingular subvarieties with $H_G^* U_{p,j} \cong \Lambda_G$. Let $V_{p,j} = \overline{U}_{p,j}$. The classes $[V_{p,j}]^G$ form a basis for $H_G^* X$ over Λ_G.

Proof This follows by applying the fibered cell decomposition lemma, Proposition A.3.4 of Appendix A, to the bundle $\mathbb{E} \times^G X \to \mathbb{B}$. $\square$

A common case where the proposition holds is when all $U_{p,j}$ are affine spaces $\mathbb{C}^{n(p,j)}$.

Example 4.7.2. Consider the Borel subgroup $B \subset GL(V)$ of upper-triangular matrices acting on $X = \mathbb{P}(V)$. This subgroup fixes a flag $E_\bullet$, $0 \subset E_0 \subset \cdots \subset E_{n-1} = V$, with $\dim E_i = i + 1$. (In the standard basis, $E_i = \text{span}\{e_1, \ldots, e_{i+1}\}$.) The B-invariant subvarieties are $X_i = \mathbb{P}(E_i)$, with $\dim X_i = i$. We have $X_i \smallsetminus X_{i-1} \cong \mathbb{C}^i$, so the classes

$$1 = [X_{n-1}]^B, [X_{n-2}]^B, \ldots, [X_0]^B$$

form a basis for $H_B^* \mathbb{P}(V)$ over Λ_B.

When two bases come from sufficiently transverse invariant subvarieties, they are Poincaré dual.

Proposition 4.7.3.   *Suppose $\{X_i\}$ and $\{Y_j\}$ are closed G-invariant subvarieties of X giving two Λ_G-bases for $H^*_G X$. If $X_i \cap Y_i = \{pt\}$ transversally, and $X_i \cap Y_j = \emptyset$ whenever $i \neq j$ and $\operatorname{codim} Y_j \geq \dim X_i$, then $[X_i]^G$ and $[Y_i]^G$ define Poincaré dual bases.*

Proof   The assumptions imply $[X_i]^G \cdot [Y_i]^G = [pt]^G$, so we have $\rho_*([X_i]^G \cdot [Y_i]^G) = 1$, and they also imply $[X_i]^G \cdot [Y_j]^G = 0$ whenever $i \neq j$ and $\operatorname{codim} Y_j \geq \dim X_i$, so that $\rho_*([X_i]^G \cdot [Y_j]^G) = 0$ in this case. In the remaining case, when $\operatorname{codim} Y_j < \dim X_i$, we have $\rho_*([X_i]^G \cdot [Y_j]^G) = 0$ automatically, since it lies in degree

$$2 \operatorname{codim} X_i + 2 \operatorname{codim} Y_j - 2 \dim X < 0,$$

and Λ_G is zero in negative degrees.   □

To see geometric Poincaré dual bases in $\mathbb{P}(V)$, we need to reduce the group further, to a torus.

Example 4.7.4.   Let T act on an n-dimensional vector space V by characters $\chi_1, \ldots, \chi_n$, so $V = L_1 \oplus \cdots \oplus L_n$. Consider flags

$$E_\bullet : L_1 \subset L_1 \oplus L_2 \subset \cdots \subset V,$$

writing $E_i = L_1 \oplus \cdots \oplus L_{i+1}$ as before, as well as

$$\widetilde{E}^\bullet : L_n \subset L_n \oplus L_{n-1} \subset \cdots \subset V,$$

with $0 = \widetilde{E}^n \subset \widetilde{E}^{n-1} \subset \cdots \subset \widetilde{E}^0 = V$, so $\widetilde{E}^i$ has dimension $n - i$.

Let $X_i = P(E_i)$ as before, so $\dim X_i = i$ and the classes $[X_i]^T$ form a basis. Then $Y_i = \mathbb{P}(\widetilde{E}^i)$ has $\operatorname{codim} Y_i = i$, and the classes $[Y_i]^T$ are the Poincaré dual basis. Indeed, one sees

$$X_i = \big\{ [\underbrace{*, \ldots, *}_{i+1}, 0, \ldots, 0] \big\} \quad \text{and} \quad Y_j = \big\{ [\underbrace{0, \ldots, 0}_{j}, *, \ldots, *] \big\},$$

so

$$X_i \cap Y_j = \begin{cases} \{[0, \ldots, 0, 1, 0, \ldots, 0]\} = \{p_{i+1}\} & \text{if } i = j; \\ \{[0, \ldots, 0, \underbrace{*, \ldots, *}_{i+1-j}, 0, \ldots, 0]\} & \text{if } j < i; \\ \{[0, \ldots, 0]\} = \emptyset & \text{if } j > i. \end{cases}$$

It is easy to compute the classes $x_k = [X_k]^T$ and $y_k = [Y_k]^T$ in $H_T^* \mathbb{P}(V) = \Lambda_T[\zeta]/\prod(\zeta + \chi_i)$. Since $Y_k = \{a_1 = \cdots = a_k = 0\}$ is the zeroes of a map

$$\mathcal{O}(-1) \to V/\widetilde{E}^k = L_1 \oplus \cdots \oplus L_k,$$

we have

$$y_k = c_k^T(\mathcal{O}(1) \otimes V/\widetilde{E}^k) = \prod_{i=1}^{k}(\zeta + \chi_i),$$

by basic properties of Chern classes (§2.3, and Appendix A, §A.5). Similarly, X_k is the zeroes of

$$\mathcal{O}(-1) \to V/E_k = L_{k+2} \oplus \cdots \oplus L_n,$$

so

$$x_k = c_k^T(\mathcal{O}(1) \otimes V/E_k) = \prod_{i=k+2}^{n}(\zeta + \chi_i).$$

Furthermore, for each $k = 1, \ldots, n$, the point

$$p_k = [0, \ldots, 0, 1, 0, \ldots] = \{a_i = 0 \text{ for } i \neq k\}$$

is defined by $a_i = 0$ for $i \neq k$, so

$$[p_k]^T = \prod_{i \neq k}(\zeta + \chi_i).$$

Putting this together, we can see the Poincaré dual bases algebraically:

$$\rho_*(x_i \cdot y_j) = 0 \text{ if } j < i \text{ by degree;}$$

$$\rho_*(x_i \cdot y_j) = 0 \text{ if } j > i \text{ by the relation } \prod_{i=1}^{n}(\zeta + \chi_i) = 0;$$

$$\rho_*(x_k \cdot y_k) = 1 \text{ since } x_k \cdot y_k = \prod_{i \neq k+1}(\zeta + \chi_i) = [p_{k+1}]^T.$$

The classes x_i and y_i are simple examples of *Schubert bases* in equivariant cohomology – they are defined by incidence conditions on geometric figures. (In this case, the condition is that a line be contained in a subspace.) A major goal of *equivariant Schubert calculus* is to compute the multiplication of elements in such a basis.

Exercise 4.7.5. With y_k as above, show that

$$y_i \cdot y_j = y_{i+j} + \sum_{j \leq k < i+j} c_{ij}^k y_k,$$

where

$$c_{ij}^k = \sum_{1 \leq p_1 < \cdots < p_r \leq i} \prod_{s=1}^{r} (\chi_{p_s} - \chi_{p_s+j+1-s})$$

and $r = i + j - k$.

With a moment's thought, one can see from the formula that the structure constants c_{ij}^k have an interesting *positivity* feature: each c_{ij}^k is a nonnegative sum of monomials in $\chi_a - \chi_b$, with $a < b$. That is,

$$c_{ij}^k \in \mathbb{Z}_{\geq 0}[\chi_1 - \chi_2, \chi_2 - \chi_3, \ldots, \chi_{n-1} - \chi_n].$$

For example, one computes

$$y_1^2 = (\zeta + \chi_1)(\zeta + \chi_1) = (\zeta + \chi_1)((\zeta + \chi_2) + (\zeta + \chi_1))$$
$$= y_2 + (\chi_1 - \chi_2)\, y_1.$$

A similar computation shows $y_1 \cdot y_p = y_{p+1} + (\chi_1 - \chi_{p+1})\, y_p$, which is a simple instance of an *equivariant Pieri rule*.

An interesting challenge is to find a formula for c_{ij}^k which makes both the positivity and the symmetry $c_{ij}^k = c_{ji}^k$ evident.

This positivity feature is common to Schubert bases in all homogeneous spaces G/P by a theorem of Graham (see §19.3), although no such explicit rule for the structure constants is known in this generality.

Notes

Pragacz's theorem on the nonnegativity of the coefficients $a_{r,\lambda}^{\mu}$ of (4.4) can be proved as follows. Given an ample vector bundle E, the Schur power $S^\lambda E$ is also ample, so by the Bloch–Gieseker theorem (Bloch and Gieseker, 1971), the Chern class $c_r(S^\lambda E)$ is a Griffiths–positive polynomial in E; now it follows from (Fulton and Lazarsfeld, 1983, Theorem 1) that it is Schur-positive. Applying this to the situation where E is the ample bundle $L_{-t_1} \oplus \cdots \oplus L_{-t_n}$ on the approximation space $\mathbb{B}_m T = (\mathbb{P}^{m-1})^n$ yields the positivity.

This argument applies more generally to the coefficient $a_{\nu,\lambda}^{\mu}$ in the expansion of any Schur functor of V_λ:

$$s_\nu(c^T(V_\lambda)) = \sum_\mu a_{\nu,\lambda}^\nu s_\mu.$$

See (Pragacz, 1996, Corollary 7.2), and also (Lazarsfeld, 2004, Example 8.3.13). In the cases $\lambda = (k, 0, \ldots, 0)$ or $\lambda = (1, \ldots, 1, 0, \ldots, 0)$, Lascoux's formulas for the coefficients $a_{r,\lambda}^\mu$ are in (Lascoux, 1978); see also (Macdonald, 1995, §I.3, Ex. 10 and §I.4, Ex. 5). More recently, a version of this problem has found applications in other parts of combinatorics (Billera et al., 2018).

For a more detailed, functorial version of the Grassmann bundles described in §4.2, see (Grothendieck and Dieudonné, 1971, §9.7).

The Schur polynomial identity (4.8), used in our proof of the Poincaré duality theorem for Grassmann bundles, can be found in (Macdonald, 1995, §I.4, Ex. 5).

The rule for the structure constants c_{ij}^k in $H_T^* \mathbb{P}^{n-1}$ is the simplest closed formula we know. Other formulas, including ones for the structure constants for weighted projective spaces, can be found in (Bahri et al., 2009; Tymoczko, 2008c).

Hints for Exercises

Exercise 4.3.2. Use the relation $(\sum_{k\geq0}(-1)^k h_k u^k)(1 + e_1 u + \cdots + e_d u^d) = 1$.

Exercise 4.5.2. Use graded lexicographic order, i.e., for two monomials of the same degree, define $c^m < c^n$ if the smallest i so that $m_i \neq n_i$ has $m_i < n_i$.

Exercise 4.7.5. Show that $c_{ij}^k = c_{i-1,j}^{k-1} + (\chi_i - \chi_{k+1})c_{i-1,j}^k$ for $i \leq j \leq k \leq i+j$, and use induction on i.

5

Localization I

The possibility of restricting attention to fixed points is a key feature of equivariant cohomology. The technique works best when the group is a torus T, and we will see some examples indicating why. There are three basic pieces of the *localization package*:

(1) the *main localization theorem*, which says when the restriction homomorphism $\iota^*\colon H_T^*X \to H_T^*X^T$ is injective, or an isomorphism after inverting elements of Λ_T;

(2) the *integration formula*, which computes a Gysin pushforward homomorphism $f_*\colon H_T^*X \to H_T^*Y$ in terms of a corresponding map on fixed loci; and

(3) the *image theorem*, describing the image of ι^* as a subring of $H_T^*X^T$ defined by divisibility conditions.

We will return to the third component in Chapter 7, and focus on the first two pieces here.

In this chapter, the group is usually a torus T. When the group is understood, we often write $\Lambda = \Lambda_T$.

5.1 The main localization theorem (first approach)

The main theorem says the restriction homomorphism

$$\iota^*\colon H_T^*X \to H_T^*X^T$$

becomes an isomorphism after inverting classes in $\Lambda_T = \mathrm{Sym}^* M$ coming from characters $\chi \in M$. This is true for any algebraic variety X, as we will see later. A very simple proof can be given for nonsingular varieties, though, so we consider that case first. The main idea is to prove this statement about restriction to the fixed locus by considering the Gysin pushforward from the fixed locus.

Example 5.1.1. Let T act on $\mathbb{P}(V) = \mathbb{P}^{n-1}$ by characters $\chi_1, \ldots, \chi_n$. We have computed

$$H_T^* \mathbb{P}^{n-1} = \Lambda[\zeta]/\prod(\zeta + \chi_i),$$

where $\zeta = c_1^T(\mathcal{O}(1))$. If the characters $\chi_1, \ldots, \chi_n$ are distinct, the fixed points are the coordinate lines $p_i = [0, \ldots, 0, 1, 0, \ldots, 0]$, for $i = 1, \ldots, n$. The tangent spaces are

$$T_{p_i} \mathbb{P}^{n-1} = \mathrm{Hom}(L_i, V/L_i) \cong \bigoplus_{j \neq i} L_i^\vee \otimes L_j,$$

where L_i is the coordinate line, isomorphic to $\mathbb{C}_{\chi_i}$ as a T-representation. In coordinates, one sees this by computing

$$z \cdot [a_1, \ldots, a_{i-1}, 1, a_{i+1}, \ldots, a_n] = [\chi_1(z) \, a_1, \ldots, \chi_i(z), \ldots, \chi_n(z) \, a_n]$$

$$= [\frac{\chi_1(z)}{\chi_i(z)} a_1, \ldots, 1, \ldots, \frac{\chi_n(z)}{\chi_i(z)} a_n].$$

So $c_{n-1}^T(T_{p_i} \mathbb{P}^{n-1}) = \prod_{j \neq i}(\chi_j - \chi_i)$.

The self-intersection formula then says $(\iota_{p_i})^*(\iota_{p_i})_*$ is multiplication by $\prod_{j \neq i}(\chi_j - \chi_i)$. One can also see this directly. The Gysin pushforward $(\iota_{p_i})_* \colon H_T^*(p_i) \to H_T^* \mathbb{P}^{n-1}$ sends 1 to $[p_i]^T = \prod_{j \neq i}(\zeta + \chi_j)$, and the restriction of the tautological bundle is $\mathcal{O}(-1)|_{p_i} = L_i$, so ζ restricts to $c_1^T(L_i^\vee) = -\chi_i$.

Exercise 5.1.2. Using the basis $\{1, \zeta, \ldots, \zeta^{n-1}\}$ for $H_T^* \mathbb{P}^{n-1}$ and the standard basis for $\Lambda^{\oplus n}$, compute the matrix of the restriction homomorphism

$$\iota^* \colon H_T^* \mathbb{P}^{n-1} \to H_T^*(\mathbb{P}^{n-1})^T \cong \Lambda^{\oplus n}.$$

Compute its determinant, and conclude that the map is injective.

Exercise 5.1.3. If the characters $\chi_1, \ldots, \chi_n$ are not distinct, the fixed locus $(\mathbb{P}^{n-1})^T$ has positive-dimensional components. Identify the fixed locus, and show that the restriction homomorphism is still injective.

The slice theorem provides a useful tool for linearizing group actions near fixed points or orbits: For any reductive (or compact) group G acting on X, there is an invariant neighborhood of p in X which is equivariantly isomorphic to an invariant neighborhood of 0 in T_pX. More generally, we have the following:

Theorem 5.1.4 (Slice theorem). *Let X be a nonsingular complex algebraic variety.*

(1) Suppose K is a compact Lie group acting on X, with an orbit $O = K \cdot x \subseteq X$. Then there is a K-invariant open neighborhood $U \subseteq X$ of O which is equivariantly isomorphic to an open neighborhood of the zero section in the normal bundle $N_{O/X}$.

(2) Suppose X is affine, and G is a reductive group acting on X, with a closed orbit $O = G \cdot x$. Then there is a G-equivariant étale neighborhood $U \to X$ of O which is equivariantly isomorphic to an étale neighborhood of the zero section of the normal bundle $N_{O/X}$.

The first statement, for compact groups, is easily proved: by averaging any hermitian metric over K, one can find a K-invariant hermitian metric on X. A tubular neighborhood of the orbit $K \cdot x$ with respect to this metric provides the desired K-invariant open neighborhood. References with more details can be found in the Notes.

Often we will assume that T acts with finitely many fixed points. This has a characterization in terms of tangent spaces. A fixed point $p \in X^T$ is *isolated* if it is a connected component of X^T.

Lemma 5.1.5. *Let G be a connected reductive linear algebraic group (or compact connected Lie group) acting on a nonsingular algebraic variety X, with a fixed point $p \in X^G$. The point p is isolated if and only if the trivial representation does not occur in T_pX.*

Proof By the slice theorem, we can reduce to the case where $X = V$ is a representation of G, and $p = 0$ is the origin. In this case, the lemma is immediate, since for any representation V of a connected group, the origin $0 \in V$ is an isolated fixed point if and only if V contains no copy of the trivial representation. □

The reductive (or compact) hypothesis is necessary.

Example 5.1.6. Let the additive group $G = \mathbb{C}$ act on $\mathbb{C}^2$ by the matrix $\begin{bmatrix} 1 & a \\ 0 & 1 \end{bmatrix}$, inducing an action on $\mathbb{P}^1$. The point $p = [1, 0]$ is the unique fixed point, but the representation on $T_p\mathbb{P}^1$ is trivial.

When $G = T$ is a torus and $\dim X = n$, the lemma says that $p \in X^T$ is isolated if and only if $c_n^T(T_pX) \neq 0$. This formulation is particular to tori, and is not true for other reductive groups.

Example 5.1.7. Consider $G = SL_n = X$ acting on itself by conjugation. The fixed points are the center of G, so there are finitely many; in particular, the identity element $e \in G$ is isolated. The action of G on $T_eG = \mathfrak{sl}_n$ is the adjoint representation. Restricting this to the diagonal torus $T \subset G$ one sees an $(n-1)$-dimensional space of weight zero, namely $\mathfrak{t} \subseteq \mathfrak{sl}_n$, so $c_{top}^T(T_eG) = 0$. Since this is the image of $c_{top}^G(T_eG)$ under the injective map $\Lambda_G \to \Lambda_T$, it follows that $c_{top}^G(T_eG) = 0$, as well.

We can now state our first localization theorem.

Theorem 5.1.8 (Localization Theorem, finite fixed locus). *Consider a d-dimensional nonsingular variety X with finitely many fixed points. Let*

$$c = \prod_{p \in X^T} c_d^T(T_pX) \in \Lambda,$$

and let $S \subseteq \Lambda$ be a multiplicative set containing c (which is nonzero, since all fixed points are isolated). Assume there are $m \leq \#X^T$ classes in H_T^*X restricting to a basis of H^*X.

Then $m = \#X^T$, the homomorphisms

$$S^{-1}H_T^*X \xrightarrow{S^{-1}\iota^*} S^{-1}H_T^*X^T \quad and \quad S^{-1}H_T^*X^T \xrightarrow{S^{-1}\iota_*} S^{-1}H_T^*X$$

are isomorphisms, and $\iota^* : H_T^*X \to H_T^*X^T$ is injective.

Most of the hypotheses can be omitted, and we will see a stronger form of the localization theorem in Chapter 7. However, this simple version suffices for all the examples we will study, and it has the advantage of being very easy to prove. The main idea is to use the Gysin pushforward, as we saw for projective space.

Proof Let us temporarily write $n = \#X^T$, so we have

$$\Lambda^{\oplus n} = H_T^*X^T \xrightarrow{\iota_*} H_T^*X \xrightarrow{\iota^*} H_T^*X^T = \Lambda^{\oplus n}.$$

By basic properties of Gysin maps, the composition $\iota^* \circ \iota_*\colon \Lambda^{\oplus n} \to \Lambda^{\oplus n}$ is diagonal, and on the summand corresponding to $p \in X^T$ it is multiplication by $c_d^T(T_p X)$. So $\det(\iota^* \iota_*) = c$, and the cokernel of ι^* is annihilated by c. In particular, $S^{-1} H_T^* X \to S^{-1} H_T^* X^T$ is surjective.

The assumption that m elements restrict to a basis of $H^* X$ means that $H_T^* X$ is a free Λ-module of rank m (by Leray–Hirsch or graded Nakayama). Since Λ is a domain, we can pass to its fraction field to conclude that $m = n$. It follows that $S^{-1} H_T^* X \to S^{-1} H_T^* X^T$ is an isomorphism, and that ι^* is injective. $\qquad\qquad\square$

Example 5.1.9. When T acts on $V \cong \mathbb{C}^n$ by distinct characters $\chi_1, \ldots, \chi_n$, the localization theorem for $X = \mathbb{P}(V) = \mathbb{P}^{n-1}$ is simply the Chinese Remainder Theorem. Indeed, with

$$A = S^{-1} H_T^* \mathbb{P}^{n-1} = (S^{-1}\Lambda)[\zeta] / (\textstyle\prod(\zeta + \chi_i)),$$

the localization theorem says that the ring homomorphism

$$A \to A/(\zeta + \chi_1) \times \cdots \times A/(\zeta + \chi_n)$$

is an isomorphism. Algebraically, this is true because the ideals $(\zeta + \chi_i)$ are pairwise comaximal.

Example 5.1.10. Again suppose T acts on $V \cong \mathbb{C}^n$ by distinct characters $\chi_1, \ldots, \chi_n$. Then $X = Gr(d, V)$ has finitely many fixed points, corresponding to coordinate subspaces:

$$X^T = \{ p_I \mid I = \{i_1 < \cdots < i_d\} \subseteq \{1, \ldots, n\} \},$$

where $p_I = [E_I]$ is the subspace $E_I = \langle e_{i_1}, \ldots, e_{i_d} \rangle = \langle e_i \mid i \in I \rangle$.

Indeed, each tangent space

$$T_{p_I} X = \mathrm{Hom}(E_I, V/E_I) \cong \bigoplus_{\substack{i \in I \\ j \notin I}} L_i^\vee \otimes L_j$$

has weights $\chi_j - \chi_i$, for $i \in I$ and $j \notin I$, which are all nonzero. We see

$$c_{top}^T(T_{p_I} X) = \prod_{\substack{i \in I \\ j \notin I}} (\chi_j - \chi_i).$$

There are $\binom{n}{d}$ fixed points, and we know bases of $H_T^* X$ with $\binom{n}{d}$ elements, restricting to bases of $H^* X$. So $H_T^* X \hookrightarrow H_T^* X^T \cong \Lambda^{\oplus \binom{n}{d}}$.

An explicit coordinate description of this action is as follows. Given a subset I, let $J = \{1, \ldots, n\} \smallsetminus I$ be the complement, so $V/E_I \cong E_J$ and there is a decomposition $V = E_I \oplus E_J$. Corresponding to this decomposition, there is an open neighborhood $U \cong \mathrm{Hom}(E_I, E_J)$ of p_I (cf. §4.2). For instance, let us take $n = 6 \; d = 3$, and $I = \{2, 4, 5\}$, and the standard action of $T = (\mathbb{C}^*)^6$ on $V = \mathbb{C}^6$. The induced action on U can be represented in matrix form as

$$
z \cdot
\begin{bmatrix}
* & * & * \\
1 & 0 & 0 \\
* & * & * \\
0 & 1 & 0 \\
0 & 0 & 1 \\
* & * & *
\end{bmatrix}
=
\begin{bmatrix}
z_1* & z_1* & z_1* \\
z_2 & 0 & 0 \\
z_3* & z_3* & z_3* \\
0 & z_4 & 0 \\
0 & 0 & z_5 \\
z_6* & z_6* & z_6*
\end{bmatrix}
=
\begin{bmatrix}
\frac{z_1}{z_2}* & \frac{z_1}{z_4}* & \frac{z_1}{z_5}* \\
1 & 0 & 0 \\
\frac{z_3}{z_2}* & \frac{z_3}{z_4}* & \frac{z_3}{z_5}* \\
0 & 1 & 0 \\
0 & 0 & 1 \\
\frac{z_6}{z_2}* & \frac{z_6}{z_4}* & \frac{z_6}{z_5}*
\end{bmatrix}.
$$

This description makes the tangent weights visible.

With only a little more care, we can relax the hypothesis that the fixed locus be finite. We still assume that X is nonsingular in this case. A basic fact is that X^T is always nonsingular. In fact, this is true more generally of fixed loci for actions by *diagonalizable groups*, i.e., $G \cong (\mathbb{C}^*)^r \times A$ for some finite abelian group A:

Lemma 5.1.11. *When a diagonalizable group G acts on a nonsingular variety X, the fixed locus X^G is nonsingular.*

(In topology, this can be deduced easily from the slice theorem, and it holds more generally for the action of any compact group G. A stronger version of this lemma in algebraic geometry was proved by Iversen.)

We will need another lemma about the characters of the torus acting on the normal bundle to a fixed component.

Lemma 5.1.12. *Let X be a nonsingular variety, and let $Z \subseteq X^T$ be a connected component of the fixed locus, of codimension d in X. Write $N = N_{Z/X}$ for its normal bundle, an equivariant vector bundle of rank d on Z. Then there are nonzero characters $\chi_1, \ldots, \chi_d$ so that for any point $p \in Z$, the fiber $N_p = T_pX/T_pZ$ has T acting by these weights. The action of T on T_pZ is trivial.*

Proof Use the slice theorem to find a neighborhood $U \subseteq X$ of p which is equivariantly isomorphic to a neighborhood of 0 in T_pX. Then $Z \cap U$

maps to an open subset of the 0-weight space of $T_p X$ (where T acts trivially), since this is $T_p Z \subseteq T_p X$. It follows that the characters on $N_p = T_p X / T_p Z$ are all nonzero. Since Z is connected, these characters are the same for any other point $q \in Z$. $\qquad\qquad\qquad\qquad\square$

For any connected component $Z \subseteq X^T$ of codimension d, the self-intersection formula says that the composition

$$H_T^* Z \xrightarrow{\iota_*} H_T^* X \xrightarrow{\iota^*} H_T^* Z$$

is multiplication by the top Chern class $c_d^T(N_{Z/X})$. In $H_T^* Z = \Lambda \otimes_{\mathbb{Z}} H^* Z$, this class can be written as

$$c_d^T(N_{Z/X}) = \chi_1 \cdots \chi_d + \sum_{i=1}^{d} a_{d-i} c_i,$$

for some classes $a_j \in \Lambda^{2j}$ and $c_i \in H^{2i} Z$, where $\chi_1, \ldots, \chi_d$ are the characters of T on the normal bundle, as in the previous lemma. Since $H^* Z$ is a finite-dimensional ring, the elements c_i are nilpotent, so $c_d^T(N_{Z/X})$ becomes invertible in $S^{-1} H_T^* Z$, for any multiplicative set S containing $\chi_1 \cdots \chi_d$.

With these observations, the proof of the following goes just as in the case where X^T is finite.

Theorem 5.1.13 (Localization Theorem, nonsingular varieties). *Let X be a nonsingular variety, and $S \subseteq \Lambda$ a multiplicative set containing all nonzero characters appearing in $T_p X$, for all $p \in X^T$. Assume $H^* X^T$ is free (as a $\mathbb{Z}$-module), and there are m elements of $H_T^* X$ restricting to a basis of $H^* X$, with $m \leq \operatorname{rk} H^* X^T$.*

Then $m = \operatorname{rk} H^* X^T$, the homomorphisms

$$S^{-1} H_T^* X \xrightarrow{S^{-1} \iota^*} S^{-1} H_T^* X^T \quad \text{and} \quad S^{-1} H_T^* X^T \xrightarrow{S^{-1} \iota_*} S^{-1} H_T^* X$$

are isomorphisms, and $\iota^* \colon H_T^* X \to H_T^* X^T$ is injective.

Exercise 5.1.14. Prove Theorem 5.1.13, using the Gysin homomorphism as before.

Exercise 5.1.15. Consider T acting on $X = \mathbb{P}^2$ by characters $0, \chi, \chi$. What is X^T? Work out the weights on each tangent space.

Exercise 5.1.16. Suppose the T-action on V decomposes as $V = \bigoplus_{i=1}^{m} V_i$, where V_i is the χ_i-isotypic component, and $\chi_1, \ldots, \chi_m$ are distinct. Say $\dim V_i = n_i$. Show that $X = Gr(d, V)$ has fixed locus

$$X^T \cong \coprod_{\substack{d_1 + \cdots + d_m = d \\ 0 \le d_i \le n_i}} Gr(d_1, V_1) \times \cdots \times Gr(d_m, V_m).$$

Note that $\operatorname{rk} H^* X = \binom{n}{d} = \sum \prod_{i=1}^{m} \binom{n_i}{d_i} = \operatorname{rk} H^* X^T$. The normal bundle to a component $Z_{\mathbf{d}} = Gr(d_1, V_1) \times \cdots \times Gr(d_m, V_m)$ is

$$N_{\mathbf{d}} = \bigoplus_{j \ne i} \operatorname{Hom}(\mathbb{S}_i, \mathbb{Q}_j).$$

What are the characters of T acting on the restriction of $N_{\mathbf{d}}$ to a fixed point?

5.2 Integration formula

From now on, we will assume that $S \subseteq \Lambda$ is a multiplicative set such that the maps

$$S^{-1} H_T^* X^T \xrightarrow{S^{-1} \iota_*} S^{-1} H_T^* X \xrightarrow{S^{-1} \iota^*} S^{-1} H_T^* X^T$$

are isomorphisms. (We have proved this in the case where X is nonsingular, with $H_T^* X$ free over Λ of rank equal to that of $H^* X^T$. In fact, $S^{-1} \iota^*$ is an isomorphism for any X, for a suitable S, as we will see in Chapter 7.)

Consider a proper T-equivariant map of nonsingular varieties $f \colon X \to Y$. For each connected component $P \subseteq X^T$, $f(P)$ is contained in a unique connected component $Q \subseteq Y^T$; let $f_P \colon P \to Q$ be the restriction of f. For any class $u \in H_T^* X$, we will write $u|_P \in H_T^* P$ for the restriction of this class to P, and similarly for the restriction classes in $H_T^* Y$ to Q.

Being components of the fixed locus for actions on nonsingular varieties, both P and Q are nonsingular, and the map f_P is proper, so both vertical maps in the diagram

$$\begin{array}{ccc} P & \hookrightarrow & X \\ f_P \downarrow & & \downarrow f \\ Q & \hookrightarrow & Y \end{array}$$

have associated Gysin homomorphisms. Our goal is to compute f_* in terms of $(f_P)_*$. More precisely, we compute the restrictions $f_*(u)|_Q$, for any $u \in H_T^* X$.

Theorem 5.2.1 (Integration formula). *For any $u \in H_T^* X$ and any connected component $Q \subseteq Y^T$, we have*

$$f_*(u)|_Q = c_{top}^T(N_{Q/Y}) \cdot \sum_{P: f(P) \subseteq Q} (f_P)_* \left(\frac{u|_P}{c_{top}^T(N_{P/X})} \right),$$

where the sum is over connected components $P \subset X^T$ mapping to Q.

In general, the formula takes place in the image of $\Lambda \otimes H^* Q = H_T^* Q$ in $S^{-1} H_T^* Q = S^{-1} \Lambda \otimes H^* Q$. When $H^* Q$ is free over $H^*(\text{pt})$ – for example, if Q is a point, or if one uses $\mathbb{Q}$ coefficients for cohomology – the homomorphism $\Lambda \otimes H^* Q \to S^{-1} \Lambda \otimes H^* Q$ is injective, and the formula holds in $H_T^* Q = \Lambda \otimes H^* Q$. This will be the case in all our applications.

Proof Since the Gysin map $S^{-1} \iota_* : S^{-1} H_T^* X^T \to S^{-1} H_T^* X$ is an isomorphism, it suffices to prove the formula for $u = (\iota_P)_*(z)$, for some component $P \subseteq X^T$ and $z \in H_T^* P$. By functoriality and the self-intersection formula, the left-hand side is

$$(\iota_Q^* \circ f_* \circ (\iota_P)_*)(z) = (\iota_Q^*(\iota_Q)_*(f_P)_*)(z)$$

$$= \begin{cases} c_{top}^T(N_{Q/Y}) \cdot (f_P)_*(z) & \text{if } f(P) \subseteq Q; \\ 0 & \text{otherwise.} \end{cases}$$

On the right-hand side, using the same properties of Gysin maps, we have

$$u|_P = (\iota_P^* \circ (\iota_P)_*)(z) = c_{top}^T(N_{P/X}) \cdot z,$$

and $u|_{P'} = 0$ for $P' \neq P$. So the sum on this side reduces to the single term

$$c_{top}^T(N_{Q/Y}) \cdot (f_P)_* \left(\frac{c_{top}^T(N_{P/X}) \cdot z}{c_{top}^T(N_{P/X})} \right) = c_{top}^T(N_{Q/Y}) \cdot (f_P)_*(z),$$

agreeing with the left-hand side. $\qquad \square$

Example 5.2.2. When Y is a point, we get an integration formula for $\rho: X \to \text{pt}$:

$$\rho_*(u) = \sum_{P \subseteq X^T} (\rho_P)_* \left(\frac{u|_P}{c_{top}^T(N_{P/X})} \right),$$

where $(\rho_P)_* \colon H_T^* P \to \Lambda$ is integration over the connected component $P \subset X^T$.

Example 5.2.3. Suppose X and Y have finitely many fixed points, and $f \colon X \to Y$ is a smooth morphism with relative tangent bundle $T_{X/Y}$. For each $q \in Y^T$ we have

$$f_*(u)|_q = \sum_{p \in f^{-1}(q)^T} \frac{u|_p}{c_{top}^T(T_{X/Y}|_p)},$$

since each f_p is an isomorphism.

When $P = \{p\}$ is a point, the Chern class appearing in the corresponding summand is $c_d^T(T_pX) = \chi_1(p)\cdots\chi_d(p)$, where $d = \dim X$ and the $\chi_i(p)$ are the characters of T acting on the tangent space T_pX. Combining the two previous examples gives a particularly useful and simple case:

Corollary 5.2.4. *Let X be a d-dimensional nonsingular compact algebraic variety with finitely many fixed points. Then*

$$\rho_*(u) = \sum_{p \in X^T} \frac{u|_p}{c_d^T(T_pX)}$$

for any class $u \in H_T^* X$.

Example 5.2.5. For T acting on $\mathbb{P}^{n-1}$ via distinct characters $\chi_1, \ldots, \chi_n$, with $\zeta = c_1^T(\mathcal{O}(1))$, we know

$$\rho_*(\zeta^k) = \begin{cases} 0 & \text{if } k < n-1, \\ 1 & \text{if } k = n-1, \end{cases}$$

by degree considerations in the first case, and by the classical fact that $n-1$ hyperplanes intersect in a point in the second case. On the other hand, the integration formula computes this as

$$\rho_*(\zeta^k) = \sum_{i=1}^n \frac{(-\chi_i)^k}{\prod_{j \neq i}(\chi_j - \chi_i)}.$$

Comparing the two yields a nontrivial algebraic identity!

Example 5.2.6. Consider $T = \mathbb{C}^*$ acting on $\mathbb{P}^2$ by the characters $0, t, 2t$, so $z \cdot [a, b, c] = [a, zb, z^2c]$. The fixed points are the usual coordinate points p_1, p_2, p_3. For $u \in H_T^* \mathbb{P}^2$, let $u_i = u|_{p_i}$. The integration formula says

$$\rho_*(u) = \frac{u_1}{2t^2} + \frac{u_2}{-t^2} + \frac{u_3}{2t^2} = \frac{u_1 - 2u_2 + u_3}{2t^2}.$$

This must be a class in $\Lambda = \mathbb{Z}[t]$, so the integration formula implies a *divisibility condition* relating the restrictions to the three fixed points: $2t^2$ must divide the polynomial $u_1 - 2u_2 + u_3$.

When computing via localization, it is often convenient to represent the fixed points of X as the vertices of a graph, with edges connecting vertices when the corresponding fixed points are connected by a T-invariant curve. This graph is called the *moment graph* of X, and we will see several examples in the next few chapters. (Symplectic geometry explains the way these graphs are drawn; see the Notes in Chapter 7.) The image of a class under the restriction $H_T^* X \hookrightarrow H_T^* X^T$ is given by labeling the vertices of the moment graph with characters.

Example 5.2.7. We will compute the number of lines meeting four general lines in $\mathbb{P}^3$. Let $X = Gr(2, \mathbb{C}^4)$ be the space of lines on $\mathbb{P}^3$, with an action of T induced by characters $\chi_1, \ldots, \chi_4$.

Fix the line ℓ_0 corresponding to the subspace $E_{12} = \mathrm{span}\{e_1, e_2\} \subseteq \mathbb{C}^4$, and consider the locus $\Omega \subseteq X$ of lines ℓ meeting ℓ_0, i.e.,

$$\Omega = \Big\{ E \subseteq \mathbb{C}^4 \,\Big|\, \dim(E \cap E_{12}) \geq 1 \Big\}.$$

This is defined by the condition that $\mathbb{S} \to \mathbb{C}^4/E_{12}$ has rank at most 1, where $\mathbb{S}$ is the tautological bundle on X. In other words, the determinant homomorphism

$$\bigwedge^2 \mathbb{S} \to \bigwedge^2 (\mathbb{C}^4/E_{12})$$

is zero. So $\Omega = Z(s)$ is the zeroes of a section of the line bundle

$$\mathrm{Hom}(\bigwedge^2 \mathbb{S}, \bigwedge^2 (\mathbb{C}^4/E_{12})) = \bigwedge^2 \mathbb{S}^\vee \otimes \mathbb{C}_{\chi_3 + \chi_4},$$

and $[\Omega]^T$ is equal to its equivariant first Chern class. We will compute its restriction to the fixed points p_I.

We have $c_1^T(\bigwedge^2 \mathbb{S}^\vee \otimes \mathbb{C}_{\chi_3 + \chi_4})|_{p_{ij}} = -\chi_i - \chi_j + \chi_3 + \chi_4$. The class $[\Omega]^T$ is shown as a labeled moment graph in Figure 5.1.

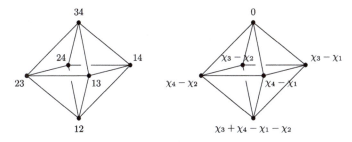

Figure 5.1 The fixed points in $X = Gr(2, \mathbb{C}^4)$, and the class $[\Omega]^T$ restricted to $H_T^* X^T$.

To address the four-lines problem, first note that the assumption that the given lines be general means that the intersection

$$\Omega_{\ell_1} \cap \Omega_{\ell_2} \cap \Omega_{\ell_3} \cap \Omega_{\ell_4} = \{\ell \,|\, \ell \text{ meets } \ell_1, \ell_2, \ell_3, \text{ and } \ell_4\}$$

is transverse and zero-dimensional, and we wish to compute the number of points – that is,

$$\int_X [\Omega_{\ell_1}] \cdot [\Omega_{\ell_2}] \cdot [\Omega_{\ell_3}] \cdot [\Omega_{\ell_4}],$$

where $\int_X$ is the (non-equivariant) pushforward $H^* X \to H^*(\mathrm{pt}) = \mathbb{Z}$.

Any line ℓ' in $\mathbb{P}^3$ can be translated to ℓ_0 by an element $g \in GL_4$. So

$$\Omega_{\ell'} = \{\ell \,|\, \ell \cap \ell' \neq \emptyset\} = g^{-1}\Omega,$$

and since GL_4 is a connected group, we have $[\Omega_{\ell'}] = [\Omega]$ in $H^* X$. So it is equivalent to compute $\int_X [\Omega]^4$.

By basic properties of Gysin homomorphisms (see §3.6), $\int_X [\Omega]^4$ is equal to the image of $\rho_*(([\Omega]^T)^4)$ under $H_T^*(\mathrm{pt}) \to H^*(\mathrm{pt})$. The class is in degree 0, and $H_T^0(\mathrm{pt}) = H^0(\mathrm{pt}) = \mathbb{Z}$. So this non-equivariant push-forward is the same as the equivariant one, and we can compute it using the integration formula:

$$\rho_*(([\Omega]^T)^4) = \frac{(\chi_3 + \chi_4 - \chi_1 - \chi_2)^4}{(\chi_3 - \chi_1)(\chi_3 - \chi_2)(\chi_4 - \chi_1)(\chi_4 - \chi_2)}$$
$$+ \frac{(\chi_4 - \chi_1)^4}{(\chi_2 - \chi_1)(\chi_2 - \chi_3)(\chi_4 - \chi_1)(\chi_4 - \chi_3)}$$
$$+ \text{(four more terms, one of which is zero).}$$

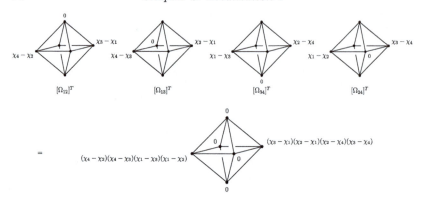

Figure 5.2 The product $[\Omega_{12}]^T \cdot [\Omega_{13}]^T \cdot [\Omega_{34}]^T \cdot [\Omega_{24}]^T$ in $H_T^* Gr(2, \mathbb{C}^4)$, represented by its localizations at fixed points.

This expression can be evaluated quickly by computer algebra, but to carry out the calculation by hand, it is useful to employ another simplification.

Let us write $\Omega_{ij} = \{ E \mid \dim(E \cap E_{ij}) \geq 1 \}$, so $\Omega = \Omega_{12}$. By the same reasoning as before, we can compute with any four choices of ij; in particular, we may choose them so that many terms in the integration formula are zero. For example, the product $[\Omega_{12}]^T \cdot [\Omega_{13}]^T \cdot [\Omega_{34}]^T \cdot [\Omega_{24}]^T$ has nonzero localizations at only two fixed points, p_{14} and p_{23}. (See Figure 5.2.) Using the integration formula for this product, one sees

$$\int_X [\Omega]^4 = \rho_* \left([\Omega_{12}]^T \cdot [\Omega_{13}]^T \cdot [\Omega_{34}]^T \cdot [\Omega_{24}]^T \right)$$

$$= \frac{(\chi_3 - \chi_1)(\chi_2 - \chi_1)(\chi_2 - \chi_4)(\chi_3 - \chi_4)}{(\chi_2 - \chi_1)(\chi_3 - \chi_1)(\chi_2 - \chi_4)(\chi_3 - \chi_4)}$$

$$+ \frac{(\chi_4 - \chi_2)(\chi_4 - \chi_3)(\chi_1 - \chi_3)(\chi_1 - \chi_2)}{(\chi_1 - \chi_2)(\chi_4 - \chi_2)(\chi_1 - \chi_3)(\chi_4 - \chi_3)}$$

$$= 1 + 1 = 2,$$

so there are two lines through the four given lines.

Exercise 5.2.8. How many lines in $\mathbb{P}^4$ meet six general planes?

5.3 Equivariant formality

There are general criteria which imply the hypotheses of the localization theorems – in particular, freeness of H_T^*X as a Λ-module. As noted earlier, we will be able to verify these hypotheses directly for our main examples and applications, so the results of this section are not logically necessary. However, it is sometimes useful to know when to expect the localization package to work, and the terminology appears frequently in the literature.

In what follows, we have a Lie group G acting on X, and we take cohomology with coefficients in a ring R (usually $\mathbb{Z}$ or a field). For an integer $m > 0$, consider the following condition:

$(*_m)$ For $0 \leq i \leq m$, H^iX is finitely generated and free over R, and there are elements $x_{ij} \in H_G^iX$ that restrict to a basis for H^iX.

The space X is called *(cohomologically) equivariantly formal* with respect to the action of G and the coefficient ring R if $(*_m)$ holds for all $m > 0$. The main reason for introducing this condition is the following direct consequence of the Leray–Hirsch theorem (Appendix A, §A.4):

Proposition 5.3.1.

(1) Assume $(*_m)$ holds for some $m > 0$. Every element of H_G^mX has a unique expression as $\sum_{i,j} c_{ij} x_{ij}$, for some $c_{ij} \in H^{m-i}\mathbb{B}G$.

(2) If X is equivariantly formal, then H_G^*X is a free Λ_G-module with basis $\{x_{ij}\}$, and the forgetful homomorphism

$$H_G^*X \otimes_{\Lambda_G} R \to H^*X$$

is an isomorphism. In fact, for any G' acting on X through a homomorphism $G' \to G$, the corresponding homomorphism

$$H_G^*X \otimes_{\Lambda_G} \Lambda_{G'} \to H_{G'}^*X$$

is an isomorphism.

We are most interested in the case where $G = T$ is a torus. For nonsingular projective varieties with finitely many fixed points, a general theorem provides a cell decomposition.

Theorem 5.3.2 (Białynicki–Birula).   *Suppose a torus T acts on a non-singular projective variety X with finitely many fixed points. Then there is a filtration by T-invariant closed subsets $X = X_n \supseteq X_{n-1} \supseteq \cdots \supseteq X_0 \supseteq \emptyset$, with $X_i \smallsetminus X_{i-1} = \coprod U_{ij}$ and $U_{ij} \cong \mathbb{A}^i$. Moreover, the total number of cells U_{ij} is equal to $\#X^T$.*

This implies such varieties are always equivariantly formal, since the classes of the invariant subvarieties $\overline{U}_{ij}$ form bases for $H_T^* X$ and $H^* X$, over Λ and R, respectively.

Corollary 5.3.3.   *Let a torus T act on a nonsingular projective variety X, with finitely many fixed points. Then X is equivariantly formal with integral coefficients. In particular,*

(1) $H_T^* X \to H^* X$ is surjective, with kernel generated by the kernel of $\Lambda_T \to \mathbb{Z}$; and

(2) $H_T^* X \to H_T^* X^T$ is injective, and becomes an isomorphism after inverting finitely many characters in Λ_T.

Furthermore, the rank of $H^* X$ (as a $\mathbb{Z}$-module) and the rank of $H_T^* X$ (as a Λ-module) are both equal to $\#X^T$.

Proof   With cells U_{ij} as in the Białynicki–Birula decomposition, the equivariant class $[\overline{U}_{ij}]^T$ restricts to the nonequivariant class $[\overline{U}_{ij}]$, so X is equivariantly formal. Injectivity of the restriction homomorphism comes from the diagram

$$
\begin{array}{ccc}
H_T^* X & \xrightarrow{\ \iota^* \ } & H_T^* X^T \\
\big\downarrow & & \big\downarrow \\
S^{-1} H_T^* X & \xrightarrow{\ \sim \ } & S^{-1} H_T^* X^T,
\end{array}
$$

for a suitable multiplicative set $S \subseteq \Lambda$, where the vertical arrows are injective since $H_T^* X$ and $H_T^* X^T$ are free over Λ, and the bottom arrow is an isomorphism by the basic localization theorem (Theorem 5.1.8).

The claim about ranks follows from (2), or from the bijection between cells and fixed points. $\square$

Thus nonsingular projective varieties with finitely many fixed points provide a large class of examples where one sees the "two notions" about equivariant cohomology described in Chapter 1.

Applying the general localization theorem to be proved in Chapter 7, similar reasoning shows that if a T-variety X is equivariantly formal with respect to a coefficient ring R, and H^*X^T is also free over R, then the restriction homomorphism $\iota^* \colon H_T^*X \to H_T^*X^T$ is injective.

Notes

Luna's étale slice theorem is explained in (Mumford et al., 1994, p. 198). The topological slice theorem is apparently due to Koszul (1953), and can be found in Audin's book (2004, Chapter I). We learned Example 5.1.7 from Johan de Jong.

Iversen's theorem on the nonsingularity of the fixed locus (Lemma 5.1.11) applies more generally for actions of linearly reductive groups, i.e., those for which all finite-dimensional representations are completely reducible; in positive characteristic this amounts to considering diagonalizable groups (Iverson, 1972). Iversen also includes a formula for Euler characteristics which gives $\operatorname{rk} H^*X = \#X^T$ in the case when X has finitely many fixed points and no odd-dimensional cohomology. Again, the novelty is mainly the algebraic proof and the application to positive characteristic; as Iversen points out, in topology it can be deduced from the Lefschetz trace formula.

The idea of proving localization theorems using Gysin pushforwards can be traced to Quillen (1971) and Quart (1979), who used similar techniques in cobordism and K-theory, respectively.

The integration formula, especially in the finite fixed point case of Corollary 5.2.4, is known by many names in the literature. Names commonly attached to it include Atiyah–Bott (after their paper (Atiyah and Bott, 1984)), Berline–Vergne (1982), Duistermaat–Heckman (1982), and "stationary phase formula" (especially in the physics literature).

Example 5.2.5 is one case of a family of identities due to Sylvester, and rediscovered by many other mathematicians. A short review of the history, along with an elementary proof, can be found in (Bhatnaga, 1999). Many such identities can be obtained by equivariant localization on other spaces.

The usage of the term "equivariantly formal" in the sense of §5.3 appears to originate in the seminal article of Goresky, Kottwitz, and MacPherson (1998). In this paper (and in much of the literature stemming from it), an equivariantly formal space is defined to be one for which the Serre spectral sequence for the fibration $\mathbb{E}G \times^G X \to \mathbb{B}G$,

$$E_2^{p,q} = H^p(\mathbb{B}G; H^q X) \Rightarrow H_G^{p+q}X,$$

degenerates at the E_2 term. This condition was considered earlier by Borel (1960, §XII.3–6).

Using coefficients in $\mathbb{Q}$, nine sufficient conditions for equivariant formality are given in (Goresky et al., 1998, Theorem 14.1), including the following.

– $H^*(X;\mathbb{Q})$ vanishes in odd degrees, and G is a connected Lie group.

- X is a nonsingular complete variety, and G is a connected linear algebraic group (see (Weber, 2003)).
- X is a possibly singular projective algebraic variety, $G = T$ is a torus, and for all $q \geq 0$, $H^q(X; \mathbb{Q})$ is pure of weight q (in the sense of mixed Hodge theory).

The use of field coefficients is essential in all of these conditions. Variations of these conditions are also given for *intersection cohomology*; applied in this context, the last one includes all toric varieties (Weber, 2003).

A different notion of equivariant formality is used in rational homotopy theory, where it involves an isomorphism between $H^*(X; \mathbb{Q})$ and a certain differential graded algebra. In order to disambiguate the terminology, Allday, Franz, and Puppe propose to add the modifier "cohomological" to the equivariant formality we consider. (They also point out that the abbreviation CEF also stands for "cohomology extension of the fiber," which nicely captures the geometry.)

Białynicki–Birula (1973) proved a stronger version of Theorem 5.3.2, where the fixed locus X^T may have positive-dimensional components; see also (Brion, 1997b, §3.1).

Hints for Exercises

Exercise 5.1.2. Here is another way to prove injectivity. The composition

$$\Lambda^{\oplus n} = H_T^*(\mathbb{P}^{n-1})^T \xrightarrow{\iota_*} H_T^*\mathbb{P}^{n-1} \xrightarrow{\iota^*} H_T^*(\mathbb{P}^{n-1})^T = \Lambda^{\oplus n}$$

is diagonal. What is its determinant? (This shows the maps $S^{-1}H_T^*(\mathbb{P}^{n-1})^T \to S^{-1}H_T^*\mathbb{P}^{n-1} \to S^{-1}H_T^*(\mathbb{P}^{n-1})^T$ are isomorphisms, for an appropriate multiplicative set S.)

Exercise 5.1.16. The tangent bundle TX restricts to $Z_\mathbf{d}$ as $\mathrm{Hom}(\mathbb{S}, \mathbb{Q}) = \bigoplus_{i,j} \mathrm{Hom}(\mathbb{S}_i, \mathbb{Q}_j)$, and $TZ_\mathbf{d}$ accounts for the diagonal summands; this explains the computation of $N_\mathbf{d}$. The characters are $\chi_j - \chi_i$ for $i \neq j$, appearing with multiplicity $d_i(n_j - d_j)$.

Exercise 5.2.8. The locus $\Omega \subseteq Gr(2, \mathbb{C}^5)$ of lines meeting the plane $\mathbb{P}(E_{123})$ is given by the vanishing of $\bigwedge^2 \mathbb{S} \to \bigwedge^2(\mathbb{C}^5/E_{123})$. So its class is $[\Omega]^T = c_1^T(\bigwedge^2 \mathbb{S}^\vee \otimes \mathbb{C}_{\chi_4 + \chi_5})$.

6

Conics

The problem of determining the number of conics tangent to five given conics is a famous example in intersection theory. In this chapter, we work out the answer as a sample computation using localization in equivariant cohomology.

6.1   Steiner's problem

In 1848, Steiner asked for the number of conics which are tangent to five fixed general conics. One approach is to compactify the space of conics as the projective space of coefficients of an equation $aX^2 + bY^2 + cZ^2 + dXY + eYZ + fXZ = 0$; that is, using $\mathbb{P}(V) \cong \mathbb{P}^5$, where $V = \mathrm{Sym}^2 \mathbb{C}^3$. For a given conic C, the set Z_C of all C' tangent to C forms a sextic hypersurface in $\mathbb{P}(V)$. Steiner observed this by examining the equations. It can also be seen geometrically by taking C to be an ellipse, and drawing the pencil of circles tangent to a fixed line at a fixed general point in the interior of C: one sees six circles tangent to C in this pencil.

Naively, one might use $[Z_C] = 6H$ to compute the desired number as

$$\int_{\mathbb{P}(V)} [Z_{C_1}] \cdot [Z_{C_2}] \cdot [Z_{C_3}] \cdot [Z_{C_4}] \cdot [Z_{C_5}] = \int_{\mathbb{P}^5} (6H)^5 = 7776,$$

by Bézout's theorem. This gives the wrong answer, though: each of the hypersurfaces Z_{C_i} contains the Veronese surface S of double lines, so the intersection $Z_{C_1} \cap Z_{C_2} \cap Z_{C_3} \cap Z_{C_4} \cap Z_{C_5}$ is not transverse.

The correct answer, 3264, was first computed by de Jonquières in 1859, and Chasles in 1864. From a modern point of view, the basic idea is to modify the moduli space by blowing up the Veronese surface $S \subseteq \mathbb{P}(V)$, working instead with the space of *complete conics* $\mathscr{C} = \mathrm{Bl}_S\mathbb{P}(V)$.

Exercise 6.1.1. Consider the simpler problem of conics tangent to five general lines. Show that the set $Z_\ell \subseteq \mathbb{P}(V)$ of conics tangent to a given line is a quadric hypersurface. On the other hand, every such Z_ℓ contains the Veronese surface S; show that the number of conics tangent to five lines is 1.

6.2   Cohomology of a blowup

We first consider the general setting of blowing up a nonsingular variety along a nonsingular subvariety.

Let G be a linear algebraic group acting on a nonsingular variety X, with $S \subseteq X$ a nonsingular G-invariant subvariety. The blowup of X along S, written $\widetilde{X} = \mathrm{Bl}_S X$, is equipped with a natural G-action. Let d be the codimension of S in X, so the normal bundle $N = N_{S/X}$ has rank d, and the exceptional divisor $E = \mathbb{P}(N) \to S$ is a $\mathbb{P}^{d-1}$-bundle. These fit into a diagram

$$\begin{array}{ccc} E & \overset{j}{\longhookrightarrow} & \widetilde{X} \\ \downarrow{\scriptstyle q} & & \downarrow{\scriptstyle p} \\ S & \overset{i}{\longhookrightarrow} & X, \end{array}$$

and $j^*\mathcal{O}(E) = \mathcal{O}(-1) \subseteq p^*N$ is the normal bundle to E in $\widetilde{X}$. All these maps and bundles are naturally G-equivariant.

Proposition 6.2.1. *There is an isomorphism*

$$H_G^k X \oplus \bigoplus_{\ell=1}^{d-1} H_G^{k-2\ell} S \overset{\sim}{\to} H_G^k \widetilde{X},$$

$$(a,\, b_1,\dots,b_{d-1}) \mapsto p^*a + j_*\left(\sum_{\ell=1}^{d-1} \zeta^{\ell-1} q^* b_\ell\right),$$

where $\zeta = c_1^G(\mathcal{O}(1))$.

The product is determined by three formulas:

(1) $p^*(a) \cdot p^*(a') = p^*(a \cdot a')$;

(2) $j_*(b) \cdot j_*(b') = -j_*(b \cdot b' \cdot \zeta)$; and

(3) $p^*(a) \cdot j_*(b) = j_*(q^* i^*(a) \cdot b)$.

The inverse to the isomorphism of the proposition maps a class c to $(a, b_1, \ldots, b_{d-1})$, with $a = p_* c$ and $b_\ell = q_* \big(c_{d-1-\ell}^G(Q) \cdot j^*(c - p^* p_* c) \big)$. Here Q is the quotient bundle on $E = \mathbb{P}(N)$, that is, there is an exact sequence $0 \to \mathcal{O}(-1) \to q^* N \to Q \to 0$ on E. There is also a split exact sequence

$$0 \to H_G^k \widetilde{X} \to H_G^k X \oplus H_G^k E \to H_G^k S \to 0,$$

where the first map is given by $c \mapsto (p_* c, j^* c)$, and the second by $(a, b) \mapsto i^* a - q_*(c_{d-1}^G(Q) \cdot b)$.

The proofs of all these facts are exactly the same as in the non-equivariant setting.

Another useful formula, whose proof is the same as in the non-equivariant case, is the following equivariant analogue of a theorem of Keel:

Proposition 6.2.2 (Keel). *Suppose $i^* \colon H_G^* X \to H_G^* S$ is surjective, with kernel I. Choose lifts $a_k \mapsto c_k^G(N)$, for $k = 1, \ldots, d-1$. Then*

$$H_G^* \widetilde{X} = (H_G^* X)[E]/J,$$

where the ideal J is generated by $[E] \cdot I$ and $[E]^d - a_1 [E]^{d-1} + \cdots + (-1)^d [S]$.

Exercise 6.2.3. Consider vector bundles $F \subseteq V$ on a variety Y. There is a rational map

$$\mathbb{P}(V) \dashrightarrow \mathbb{P}(V/F)$$

of projective bundles, whose indeterminacy is resolved by

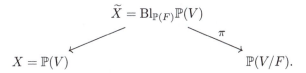

Show that π identifies $\widetilde{X} \cong \mathbb{P}(F') \to \mathbb{P}(V/F)$, where $F'/F \cong \mathcal{O}(-1) \subseteq V/F$ is the tautological bundle on $\mathbb{P}(V/F)$. Compute $H^* \widetilde{X}$ using the projective bundle formula.

Exercise 6.2.4. Let $G = GL_3$ act on $S = \mathbb{P}^2$ via its standard action on $\mathbb{C}^3$, and on $\mathbb{P}^5$ via the representation $\mathrm{Sym}^2 \mathbb{C}^3$. Let $i \colon S \hookrightarrow \mathbb{P}^5$ be the Veronese embedding, so it is G-equivariant. Then $i^* \colon H^*\mathbb{P}^5 \to H^*S$ is surjective if one uses $\mathbb{Q}$ coefficients (but not $\mathbb{Z}$ coefficients). Compute the Keel presentation of $H_G^* \mathrm{Bl}_S\mathbb{P}^5$ (with $\mathbb{Q}$ coefficients).

For our purposes, local information on tangent spaces will often suffice. It follows from the definition of the blowup that there are canonical isomorphisms $T_x\widetilde{X} \cong T_xX$ for $x \in \widetilde{X} \smallsetminus E = X \smallsetminus S$. Similarly, for $x \in E \subseteq \widetilde{X}$, there are exact sequences

$$0 \to T_xE \to T_x\widetilde{X} \to L \to 0$$

and

$$0 \to \mathrm{Hom}(L, N/L) \to T_xE \to T_sS \to 0,$$

where $s = q(x) \in S \subseteq X$, $N = N_{S/X}$, and $L = \mathcal{O}(-1)|_x \subseteq (q^*N)_x = N_s$. (So $L \subseteq N_s$ is the line corresponding to the point $x \in E = \mathbb{P}(N)$.)

Now suppose $G = T$ is a torus acting on X, with $S \subseteq X$ a T-invariant subvariety. The above isomorphisms and exact sequences are T-equivariant, for a fixed point x. That is, there are isomorphisms of T-modules

$$T_x\widetilde{X} = T_xX$$

for $x \in \widetilde{X}^T \smallsetminus E^T = X^T \smallsetminus S^T$, and

$$\begin{aligned} T_x\widetilde{X} &= T_xE \oplus L \\ &= \mathrm{Hom}(L, N_s/L) \oplus T_sS \oplus L, \end{aligned}$$

for $x \in E^T$ mapping to $s \in S^T$. These observations lead to a criterion for $\widetilde{X}$ to have finitely many fixed points.

Proposition 6.2.5. *Suppose X^T is finite. Then $\widetilde{X}^T$ is finite if and only if for all $s \in S^T$, the characters on $N_s = N_{S/X}|_s$ are distinct.*

Proof As we saw in Lemma 5.1.5, a fixed point x is isolated if and only if all characters at x are nonzero. There is nothing to check for $x \in \widetilde{X} \smallsetminus E = X \smallsetminus S$, since X^T is finite. At a point $x \in E^T \subseteq \widetilde{X}^T$ corresponding to a T-invariant line $L \subseteq N_s$ of nonzero weight χ, the characters on $T_x\widetilde{X}$ are

$$\{\text{characters of } T_sS\} \cup \{\chi\} \cup \{\chi_i - \chi \mid \chi_i \text{ is a character of } N_s/L\}.$$

All the characters in the first two sets are nonzero; those of the third are nonzero for any choice of L exactly when all characters of N_s are distinct. □

Example 6.2.6. Consider $X = \mathbb{A}^2$ with T acting by nonzero characters χ_1, χ_2. Let $S = \{0\}$ be the origin, and $\widetilde{X} = \mathrm{Bl}_0 \mathbb{A}^2$. Take $x = [1,0] \in E \cong \mathbb{P}^1$, corresponding to the horizontal tangent line at 0, that is, $L = \langle e_1 \rangle \subseteq \mathbb{C}^2 = T_0 X$. The weight on $T_x E \subseteq T_x \widetilde{X}$ comes from $T_x E = \mathrm{Hom}(L, \mathbb{C}^2/L)$, so it is $\chi_2 - \chi_1$. The weights on $T_x \widetilde{X}$ are therefore $\{\chi_1, \chi_2 - \chi_1\}$.

Exercise 6.2.7. Verify the above calculation of weights by working directly in coordinates on $\widetilde{X}$.

6.3 Complete conics

Given a nonsingular plane curve $C \subset \mathbb{P}^2$, the *dual curve* $C^\vee \subseteq (\mathbb{P}^2)^\vee = \{\ell \subseteq \mathbb{P}^2\}$ is
$$C^\vee = \{\ell \subseteq \mathbb{P}^2 \,|\, \ell \text{ is tangent to } C\}.$$
If C is singular and irreducible, $C^\vee$ is defined to be the closure of the locus of tangents to C at nonsingular points. When C is a nonsingular conic in $\mathbb{P}^2$, the dual curve $C^\vee$ is a nonsingular conic in $(\mathbb{P}^2)^\vee$.

Recall that $\mathbb{P}(V) \cong \mathbb{P}^5$ parametrizes degree two equations, where $V = \mathrm{Sym}^2 \mathbb{C}^3$. Taking the dual curve defines a rational map
$$\mathbb{P}(V) \dashrightarrow \mathbb{P}(V^\vee)$$
$$[C] \mapsto [C^\vee].$$

In fact, this map is regular on $\mathbb{P}(V) \smallsetminus S$, the complement of the Veronese surface of double lines: a reduced degenerate conic, the union of two distinct lines $C = \ell_1 \cup \ell_2$, maps to the double line through points $[\ell_1]$ and $[\ell_2]$ in $(\mathbb{P}^2)^\vee$.

The blowup $\mathscr{C} = \widetilde{X} = \mathrm{Bl}_S \mathbb{P}(V)$ resolves the indeterminacy of this rational map:

$$\mathscr{C} = \overline{\{([C],[C^\vee]) \,|\, C, C^\vee \text{ are nonsingular conics}\}} \subset \mathbb{P}(V) \times \mathbb{P}(V^\vee)$$

$$\mathbb{P}(V) \qquad \xdashrightarrow{\hspace{4cm}} \qquad \mathbb{P}(V^\vee).$$

The space $\mathscr{C}$ is a moduli space of *complete conics*. It parametrizes four types of geometric figures in $\mathbb{P}^2$:

(1) nonsingular conics C;
(2) unions of two lines $\ell_1 \cup \ell_2$;
(3) lines with two marked points $(\ell \ni p_1, p_2)$; and
(4) lines with one marked point $(\ell \ni p)$.

Duality preserves types (1) and (4), and exchanges types (2) and (3).

For any point $p \in \mathbb{P}^2$, there is a divisor $\Sigma_p \subseteq \mathbb{P}(V)$ of conics containing p; this is a hyperplane. Similarly, for any line $\ell \subseteq \mathbb{P}^2$, there is a divisor $\Theta_\ell \subseteq \mathbb{P}(V)$ of conics tangent to ℓ; this is a (singular) quadric. The two are exchanged under duality: the rational map $\mathbb{P}(V) \dashrightarrow \mathbb{P}(V^\vee)$ sends Σ_p to $\Theta_{[p]}^\vee$ and Θ_ℓ to $\Sigma_{[\ell]}^\vee$, where $[p] \subseteq (\mathbb{P}^2)^\vee$ and $[\ell] \in (\mathbb{P}^2)$ are the line and point corresponding to p and ℓ, respectively.

Let $\sigma_p \subseteq \mathscr{C}$ and $\tau_\ell \subseteq \mathscr{C}$ be the proper transforms of Σ_p and Θ_ℓ, respectively. We will abuse notation by using the same symbols for their (equivariant) cohomology classes. Non-equivariantly, these classes are independent of p and ℓ, so we may omit the subscripts. One has $\pi_* \sigma = \Sigma = H$ and $\pi_* \tau = \Theta = 2H$, where H is the hyperplane class in $H^2 \mathbb{P}(V)$; similarly, $\varphi_* \sigma = \Theta^\vee = 2H^\vee$ and $\varphi_* \tau = \Sigma^\vee = H^\vee$ in $H^2 \mathbb{P}(V^\vee)$.

The class of the proper transform $\tilde{Z}_C$ of Z_C – the sextic hypersurface of conics tangent to C – is also independent of C, and in $H^* \mathscr{C}$ there is a relation $\tilde{Z} = 2\sigma + 2\tau$. One can prove that for five general conics $C_1, \ldots, C_5$, the corresponding subvarieties $\tilde{Z}_{C_i} \subseteq \mathscr{C}$ do intersect transversally, so the solution to Steiner's problem is given by the integral

$$\int_{\mathscr{C}} (2\sigma + 2\tau)^5 = 32 \left(\int \sigma^5 + 5 \int \sigma^4 \tau + 10 \int \sigma^3 \tau^2 \right.$$

$$\left. + 10 \int \sigma^2 \tau^3 + 5 \int \sigma \tau^4 + \int \tau^5 \right). \qquad (*)$$

There are many ways to compute the numbers $\int \sigma^i \tau^{5-i}$. We will do this equivariantly, as an example of the localization techniques from the previous chapter.

Let $T = (\mathbb{C}^*)^3$ act in the standard way on $\mathbb{P}^2$, with characters t_1, t_2, t_3. The induced action on $V = \operatorname{Sym}^2 \mathbb{C}^3$ has characters

$$2t_1, 2t_2, 2t_3, t_1 + t_2, t_2 + t_3, t_1 + t_3,$$

corresponding to the basis

$$X^2, Y^2, Z^2, XY, YZ, XZ.$$

It is often useful to record this in a weight diagram:

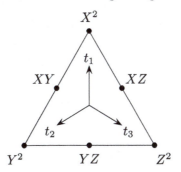

The tangent spaces to $\mathbb{P}(V)$ at X^2 and XY have weights

$$T_{X^2}\mathbb{P}(V) : \{t_2 + t_3 - 2t_1,\, t_2 - t_1,\, 2(t_2 - t_1),\, t_3 - t_1,\, 2(t_3 - t_1)\} \quad \text{and}$$

$$T_{XY}\mathbb{P}(V) : \{2t_3 - t_1 - t_2,\, t_1 - t_2,\, t_2 - t_1,\, t_3 - t_1,\, t_3 - t_2\}.$$

The others can be obtained from these by symmetry.

Products of tangent weights can be represented by arrows on the weight diagram: each arrow represents a weight (considered as a vector in the character lattice), and when several arrows are drawn together, they are multiplied. An example is shown in Figure 6.1.

The Veronese surface $S \subseteq \mathbb{P}(V)$ has fixed points X^2, Y^2, Z^2. The tangent space $T_{X^2}S$ has weights $\{t_2 - t_1, t_3 - t_1\}$, from the standard

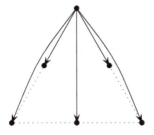

Figure 6.1 $c_5^T(T_{X^2}\mathbb{P}(V)) = 4(t_2 - t_1)^2(t_3 - t_1)^2(t_2 + t_3 - 2t_1)$, represented via arrows.

action of T on $S = \mathbb{P}^2$. So the normal weights to S at X^2 are

$$N_{X^2} : \big\{ t_2 + t_3 - 2t_1,\, 2(t_2 - t_1),\, 2(t_3 - t_1) \big\}.$$

This is all we need in order to compute weights on $\mathscr{C} = \mathrm{Bl}_S \mathbb{P}(V)$.

First, since the normal weights are distinct at each fixed point of S, the fixed locus $\mathscr{C}^T$ is finite, by Proposition 6.2.5. In fact, there are 6 fixed points in $\mathbb{P}(V)$, 3 of which lie in S. Each fiber of $\pi \colon \mathscr{C} \to \mathbb{P}(V)$ over S is a $\mathbb{P}^2$, so there are 3 fixed points in $\mathscr{C}$ mapping to each fixed point in S. Thus we have

$$\#\mathscr{C}^T = \#(\mathbb{P}(V) \smallsetminus S)^T + 3 \cdot \#S^T$$
$$= 3 + 3 \cdot 3 = 12.$$

On the other hand, from the blowup exact sequence

$$0 \to H^*\mathscr{C} \to H^*\mathbb{P}(V) \oplus H^*E \to H^*S \to 0,$$

we see $H^*\mathscr{C}$ is free, of rank $6 + 9 - 3 = 12$. (The exceptional divisor E is a $\mathbb{P}^2$-bundle over $S = \mathbb{P}^2$, so its cohomology has rank $3 \cdot 3 = 9$.) Using the basic localization theorem (Theorem 5.1.8), we know that

$$H_T^*\mathscr{C} \hookrightarrow H_T^*\mathscr{C}^T.$$

While not logically necessary for computing integrals, this allows us to determine classes in $H_T^*\mathscr{C}$ from their localizations.

What are the fixed points of $\mathscr{C}$? They map to fixed points in $\mathbb{P}(V)$, so to points of the form XY or X^2.

(1) Over XY, the projection π is an isomorphism, so the corresponding fixed point is just a pair of lines.

(2) Over the double line X^2, we need to add the data of one or two points, and these points must be T-fixed.

Torus-fixed points and lines in $\mathbb{P}^2$ can be represented as a triangle, and T-fixed complete conics can be drawn on such a triangle. Here is $\mathbb{P}^2$, together with the complete conics $XY = 0$ and $(X = 0, [0,0,1])$.

Figure 6.2 displays a diagram showing all 12 fixed points of $\mathscr{C}$.

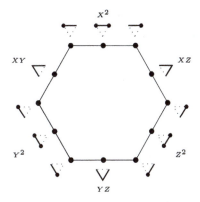

Figure 6.2 The 12 fixed points in the space of complete conics.

The tangent weights to $\mathscr{C}$ at ∇ are the same as those on $T_{XY}\mathbb{P}(V)$:

$$T_{\nabla}\mathscr{C} : \{2t_3 - t_1 - t_2,\ t_1 - t_2,\ t_2 - t_1,\ t_3 - t_1,\ t_3 - t_2\}.$$

Next we consider points in the exceptional divisor, in the fiber over $X^2 \in S$. As noted earlier, the normal weights to S at the point X^2 are $\{t_2 + t_3 - 2t_1,\ 2(t_2 - t_1),\ 2(t_3 - t_1)\}$, each corresponding to a T-invariant line $L \subseteq N$ and therefore a fixed point in E^T. Taking L with character $2(t_2 - t_1)$ corresponds to the point $\cdots$, and we see weights

$$T_{\cdot}E : \Big\{ \underbrace{t_2 - t_1,\ t_3 - t_1}_{\text{from } T_{X^2}S} \Big\}$$

$$\cup \Big\{ \underbrace{(t_2 + t_3 - 2t_1) - 2(t_2 - t_1),\ 2(t_3 - t_1) - 2(t_2 - t_1)}_{\text{from } L^{\vee} \otimes N/L} \Big\}.$$

Including L (the normal space to E), we see weights

$$T_{\cdot}\mathscr{C} : \Big\{ \underbrace{t_2 - t_1,\ t_3 - t_1,\ t_3 - t_2,\ 2(t_3 - t_2)}_{\text{from } E},\ \underbrace{2(t_2 - t_1)}_{\text{from } L} \Big\}.$$

Starting with $L \subseteq N$ of character $t_2 + t_3 - 2t_1$, which corresponds to the point $\cdots$, a similar calculation gives

$$T_{\cdot}\mathscr{C} : \{ t_2 - t_1,\ t_3 - t_1,\ t_2 - t_3,\ t_3 - t_2,\ t_2 + t_3 - 2t_1 \}.$$

Diagrams for these three examples are in Figure 6.3. Weights at the other fixed points can be obtained by symmetry.

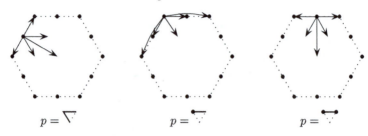

Figure 6.3 Top Chern classes of $T_p\mathscr{C}$ at three points.

We can compute the classes σ_p and τ_ℓ by localizing. For the point $p = [1,0,0] = \cdot\,$, the hyperplane $\Sigma_p \subseteq \mathbb{P}(V)$ is the zeroes of a map $\mathcal{O}(-1) \to V/F$, where $F = \langle Y^2, Z^2, XY, XZ, YZ \rangle$. That is, it is the zeroes of a section of $\mathcal{O}(1) \otimes \mathbb{C}_{2t_1}$, so its class in $H_T^2\mathbb{P}(V)$ is

$$[\Sigma_p]^T = c_1^T(\mathcal{O}(1) \otimes \mathbb{C}_{2t_1}).$$

We know the restriction of $\mathcal{O}(1)$ to each fixed point, from Example 5.1.1. Since $\sigma_{\cdot} = \pi^*\Sigma_{\cdot}$, we have

$$\sigma_{\cdot}|_{\leftarrow} = \sigma_{\cdot}|_{\rightarrow} = \sigma_{\cdot}|_{\leftrightarrow} = \Sigma_{\cdot}|_{\leftarrow} = 0,$$

$$\sigma_{\cdot}|_{\triangledown} = \Sigma_{\cdot}|_{\triangledown} = t_1 - t_2,$$

etc.

Similarly, $\tau_{\diagdown} = \varphi^*\Sigma_{\diagdown}^*$ is the proper transform of a hyperplane in the dual projective space $\mathbb{P}(V^*)$, and we have

$$\tau_{\diagdown}|_{\rightarrow} = \tau_{\diagdown}|_{7} = \tau_{\diagdown}|_{\diagup} = 0,$$

$$\tau_{\diagdown}|_{\leftrightarrow} = t_3 - t_2,$$

etc.

The restrictions of the exceptional divisor E are even easier to compute, since one simply records the character of $L \subseteq N$ at each fixed point. Figure 6.4 shows the complete data.

Exercise 6.3.1. For any fixed points $p \neq q$ in $\mathbb{P}^2$, show $\sigma_p + \sigma_q - E = \tau_{\overline{pq}}$ in $H_T^2\mathscr{C}$.

Now we can compute the numbers $\int_{\mathscr{C}} \sigma^i \tau^{5-i}$ by localization. As we saw with the Grassmannian, convenient choices of equivariant lifts will simplify the calculation.

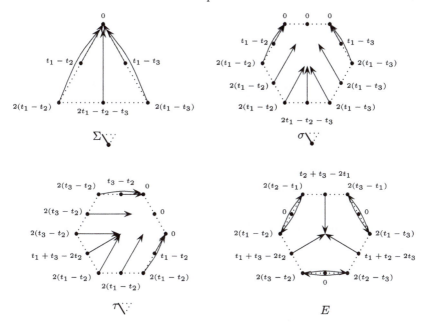

Figure 6.4 Localizations of divisor classes.

We know $\int \sigma^5 = \int \tau^5 = 1$, since these come from hyperplanes in $\mathbb{P}(V)$ and $\mathbb{P}(V^\vee)$. More generally, $\int \sigma^i \tau^{5-i} = \int \sigma^{5-i} \tau^i$ by duality, so there are only two more integrals to compute.

To compute $\int_{\mathscr{C}} \sigma^4 \tau$, we use localization on the product

$$\alpha = (\sigma_{\because})^2 \, \sigma_{\bullet} \, \sigma_{\because} \cdot \tau_{\because}.$$

Diagrammatically, this is

$$\left(\begin{smallmatrix} \end{smallmatrix} \right)^2 \left(\begin{smallmatrix} \end{smallmatrix} \right) \left(\begin{smallmatrix} \end{smallmatrix} \right) \left(\begin{smallmatrix} \end{smallmatrix} \right)$$

$$= \left(\begin{smallmatrix} \end{smallmatrix} \right),$$

so the integration formula gives

$$\int_{\mathscr{C}} \sigma^4 \tau = \frac{\alpha|_{\nabla}}{c_5^T(T_{\nabla}\mathscr{C})} + \frac{\alpha|_{\vee}}{c_5^T(T_{\vee}\mathscr{C})}$$

$$= \frac{(t_1 - t_2)^2(2t_3 - t_1 - t_2)(t_2 - t_1)(2(t_3 - t_2))}{(2t_3 - t_1 - t_2)(t_1 - t_2)(t_2 - t_1)(t_3 - t_2)(t_3 - t_1)}$$

$$+ \frac{(2t_1 - t_2 - t_3)^2(t_3 - t_2)(t_2 - t_3)(2(t_1 - t_2))}{(2t_1 - t_2 - t_3)(t_1 - t_2)(t_1 - t_3)(t_3 - t_2)(t_2 - t_3)}$$

$$= \frac{2(t_1 - t_2)}{t_3 - t_1} + \frac{2(2t_1 - t_2 - t_3)}{t_1 - t_3}$$

$$= 2.$$

Similarly, to compute $\int_{\mathscr{C}} \sigma^3 \tau^2$, one can apply the integration formula to $\beta = \sigma \cdot \sigma_\bullet \cdot \sigma \cdot \tau \cdot \tau$ and obtain

$$\int_{\mathscr{C}} \sigma^3 \tau^2 = \frac{\beta|_{\vee}}{c_5^T(T_{\vee}\mathscr{C})}$$

$$= \frac{(2t_1 - t_2 - t_3)(2(t_1 - t_2))(2(t_1 - t_3))(t_3 - t_2)(t_2 - t_3)}{(2t_1 - t_2 - t_3)(t_1 - t_2)(t_1 - t_3)(t_3 - t_2)(t_2 - t_3)}$$

$$= 4.$$

With these numbers in Equation (∗), we obtain

$$\int_{\mathscr{C}} (2\sigma + 2\tau)^5 = 32(1 + 5 \cdot 2 + 10 \cdot 4 + 10 \cdot 4 + 5 \cdot 2 + 1)$$

$$= 3264$$

conics tangent to five (general) conics.

Notes

A sketch of the six circles tangent to the ellipse, along with a solution to Steiner's problem using modern intersection theory, can be found in (Fulton and MacPherson, 1978). The claim about 7776 conics appears near the end of Steiner's paper (1848). A more detailed history of this problem was assembled by Kleiman (1980).

The first serious applications of the equivariant integration formula as a tool to solve classical problems in enumerative geometry appeared in the 1990s, especially in work of Ellingsrud and Strømme (1996) and Kontsevich (1995). For example, Ellingsrud and Strømme use localization on Hilbert schemes to show that there are $317, 206, 375$ twisted cubic curves on a quintic threefold in $\mathbb{P}^4$.

Proposition 6.2.1 is proved in (Fulton, 1998, §6.7). Proposition 6.2.2 is in (Keel, 1992). A generalization of Keel's theorem to Chow motives was given by Li (2009).

The space of complete conics is a prototypical example of a *spherical variety*, and a description of its equivariant cohomology ring from this point of view can be found in Brion's expository articles (1989, 1998).

We learned the notation for representing (localized) equivariant classes as arrows on moment graphs from Allen Knutson.

Hints for Exercises

Exercise 6.1.1. Use projective duality: this is equivalent to the number of conics passing through five general points.

Exercise 6.2.4. Let $z = c_1^G(\mathcal{O}_{\mathbb{P}^5}(1))$ and $\zeta = c_1^G(\mathcal{O}_{\mathbb{P}^2}(1))$, and write $\Lambda_G = \mathbb{Z}[c_1, c_2, c_3]$. We have presentations

$$H_G^*\mathbb{P}^5 = \Lambda_G[z]/\big((z^3 + 2c_1 z^2 + (c_1^2 + c_2)z + c_1 c_2 - c_3)(z^3 + 2c_1 z^2 + 4c_2 z + 8c_3)\big)$$

and

$$H_G^* S = \Lambda_G[\zeta]/(\zeta^3 + c_1\zeta^2 + c_2\zeta + c_3),$$

and the homomorphism $H_G^*\mathbb{P}^5 \to H_G^* S$ is given by $z \mapsto 2\zeta$.

One computes

$$c_1^G(N_{S/\mathbb{P}^5}) = 9\zeta + 3c_1 \quad \text{and} \quad c_2^G(N_{S/\mathbb{P}^5}) = 30\zeta^2 + 8c_1\zeta + 4c_2$$

so, now using $\mathbb{Q}$ coefficients, these classes are lifted to $H_G^*\mathbb{P}^5$ by

$$a_1 = \frac{9}{2}z + 3c_1 \quad \text{and} \quad a_2 = \frac{15}{2}z^2 + 4c_1 z + 4c_2$$

respectively. Since we have

$$[S] = 4(z^3 + 2c_1 z^2 + (c_1^2 + c_2)z + c_1 c_2 - c_3),$$

setting $e = [E]$, the Keel presentation is $H_G^*\mathrm{Bl}_S\mathbb{P}^5 = \Lambda_G[z, e]/(f, g, h)$, where the ideal is generated by elements

$$f = (z^3 + 2c_1 z^2 + (c_1^2 + c_2)z + c_1 c_2 - c_3)(z^3 + 2c_1 z^2 + 4c_2 z + 8c_3),$$
$$g = (z^3 + 2c_1 z^2 + 4c_2 z + 8c_3) \cdot e, \text{ and}$$
$$h = e^3 - (\frac{9}{2}z + 3c_1)e^2 + (\frac{15}{2}z^2 + 4c_1 z + 4c_2)e$$
$$\quad - 4(z^3 + 2c_1 z^2 + (c_1^2 + c_2)z + c_1 c_2 - c_3).$$

(One way to verify these computations is by using the calculations of local tangent and normal weights from §6.3.)

Exercise 6.2.7. Use coordinates x, y on X, and x, y' on $\widetilde{X}$, where $y = xy'$. Since the characters of x, y are χ_1, χ_2, one sees the character of y' is $\chi_2 - \chi_1$.

Exercise 6.3.1. Non-equivariantly, $\sigma_p = \sigma_q = \sigma$ in $H^*\mathscr{C}$, so this says $2\sigma - E = \tau$. Since $2\sigma - E$ is the class of the proper transform of the quadric hypersurface parametrizing conics tangent to a given line, one has

$$\int (2\sigma - E)^5 = \int \tau^5 = 1,$$

resolving the "conics tangent to 5 lines" problem.

7

Localization II

In this chapter, we refine the main localization theorem from Chapter 5 to see that it applies to all algebraic varieties. We then turn to the third piece of the localization package, which characterizes the image of the restriction homomorphism: we will see a version of a theorem due to Chang and Skjelbred, refined to allow integer coefficients. Along the way, we give a criterion for a nonsingular variety to have finitely many T-invariant curves.

Most of the results of this chapter are not needed elsewhere in the book. However, it is often useful – at least psychologically – to know a particular case of the image theorem, where there are finitely many fixed points and finitely many invariant curves. This is often known as the "GKM" description of equivariant cohomology, after Goresky, Kottwitz, and MacPherson.

7.1   The general localization theorem

We first set up some notation, which will be useful later in the chapter as well. Given a subgroup $L \subseteq M$ of the character lattice $M = \mathrm{Hom}(T, \mathbb{C}^*)$, let $T(L) \subseteq T$ be the subgroup of T corresponding to the quotient M/L; that is,

$$T(L) = \bigcap_{\chi \in L} \ker(\chi).$$

Let $S(L) \subseteq \Lambda = \mathrm{Sym}^* M$ be the multiplicative set generated by $M \smallsetminus L$. By Lemma 5.1.11, the fixed locus $X^{T(L)}$ is nonsingular whenever X is, because $T(L)$ is a diagonalizable group.

Theorem 7.1.1 (Localization). *Let X be an algebraic variety with the action of a torus T. The restriction homomorphism*

$$S(L)^{-1} H_T^* X \xrightarrow{\; S(L)^{-1} \iota^* \;} S(L)^{-1} H_T^* X^{T(L)}$$

is an isomorphism, where $\iota \colon X^{T(L)} \hookrightarrow X$ is the inclusion.

In fact, $S(L)$ may be replaced by a smaller multiplicative set, generated by a finite set of characters depending on X.

Taking $L = 0$, the theorem says that whenever S contains all nonzero characters, $S^{-1} H_T^* X \to S^{-1} H_T^* X^T$ is an isomorphism.

In proving the localization theorem, we will use some fundamental (but nontrivial) facts about torus actions.

Theorem 7.1.2. *Let a torus T act on an algebraic variety X.*

(1) (Sumihiro) If X is normal, then it is covered by (finitely many) T-invariant Zariski open affine sets $U \subseteq X$.

(2) (Alper–Hall–Rydh) For any X, there is a T-equivariant affine étale cover $U \to X$.

References can be found in the Notes at the end of the chapter.

We also use a very easy fact about equivariant cohomology:

Lemma 7.1.3. *Suppose $Y' \to Y$ is a T-equivariant map. If $c \in \Lambda$ annihilates $H_T^* Y$, then it also annihilates $H_T^* Y'$.*

(This is immediate from the functoriality of cohomology, because $H_T^* Y \to H_T^* Y'$ is a homomorphism of Λ-algebras.)

Proof of Theorem 7.1.1   First we treat the case where $X^{T(L)} = \emptyset$, in three steps.

Case 1. Suppose V is an affine space with a linear T-action, and $X = V \smallsetminus V^{T(L)}$. We can write

$$V = \bigoplus_{\chi \in M} V_\chi = V^+ \oplus V^-,$$

where V^+ is the sum of V_χ with $\chi \in L$, and V^- is the sum of V_χ with $\chi \notin L$. Then $V^+ = V^{T(L)}$ is the fixed locus for $T(L)$. Regarding $X = V \smallsetminus V^{T(L)}$ as the complement of the zero section of a vector bundle over $V^{T(L)}$, it follows from the Gysin sequence that $H_T^* X = \Lambda/(c)$, where $c = \prod_{\chi \notin L} \chi^{\dim V_\chi}$. Since $c \in S(L)$, we have $S(L)^{-1} H_T^* X = 0$. (Note that it suffices to invert the finitely many characters $\chi \notin L$ such that $V_\chi \neq 0$.)

Case 2. Next suppose X is any affine variety with $X^{T(L)} = \emptyset$. It is a basic fact about linear algebraic group actions that one can find an equivariant embedding $X \hookrightarrow V$ in an affine space V with linear T-action. By the assumption $X^{T(L)} = \emptyset$, we have $X \hookrightarrow V \smallsetminus V^{T(L)}$. Taking c as in the previous case, Lemma 7.1.3 says that c annihilates $H_T^* X$, so $S(L)^{-1} H_T^* X = 0$.

Case 3. Now consider any variety X with $X^{T(L)} = \emptyset$. Using Theorem 7.1.2, we can find a finite cover of X by invariant affines $U_i \to X$. From the previous case, we have elements c_i annihilating $H_T^* U_i$, for each i. It follows from the Mayer–Vietoris sequence that $c = \prod c_i$ annihilates $H_T^* X$.

For the general case where $X^{T(L)}$ may be nonempty, proving the theorem is equivalent to showing that $S(L)^{-1} H_T^*(X, X^{T(L)}) = 0$. We will do this by working on approximation spaces.

Fix k, and choose a principal T-bundle $\mathbb{E} \to \mathbb{B}$ so that
$$H_T^{\leq k}(X, X^{T(L)}) = H^{\leq k}(\mathbb{E} \times^T X, \, \mathbb{E} \times^T X^{T(L)}).$$

Let $\mathcal{U}$ be an open neighborhood of $\mathbb{E} \times^T X^{T(L)}$ in $\mathbb{E} \times^T X$. By the case where the fixed locus $X^{T(L)}$ is empty, we have an element $c \in H^* \mathbb{B}$ which annihilates $H^{\leq k}((\mathbb{E} \times^T X) \smallsetminus (\mathbb{E} \times^T X^{T(L)}))$. It follows that c also annihilates $H^{\leq k}(\mathcal{U} \smallsetminus (\mathbb{E} \times^T X^{T(L)}))$, and therefore it annihilates $H^{\leq k}((\mathbb{E} \times^T X) \smallsetminus (\mathbb{E} \times^T X^{T(L)}), \, \mathcal{U} \smallsetminus (\mathbb{E} \times^T X^{T(L)}))$, using the long exact sequence. By tautness and excision, we have

$$\begin{aligned} H_T^{\leq k}(X, X^{T(L)}) &= H^{\leq k}(\mathbb{E} \times^T X, \, \mathbb{E} \times^T X^{T(L)}) \\ &= \varinjlim_{\mathcal{U}} H^{\leq k}(\mathbb{E} \times^T X, \, \mathcal{U}) \\ &= \varinjlim_{\mathcal{U}} H^{\leq k}((\mathbb{E} \times^T X) \smallsetminus (\mathbb{E} \times^T X^{T(L)}), \\ &\quad\quad \mathcal{U} \smallsetminus (\mathbb{E} \times^T X^{T(L)})), \end{aligned}$$

so this group is also annihilated by c.

Finally, observe the element c is a product of finitely many characters in $M \smallsetminus L$, and $S(L)$ may be replaced by any multiplicative set
containing c.   □

The localization theorem has many consequences. Here is one.

Corollary 7.1.4.   *Let $S \subseteq \Lambda$ be the multiplicative set generated by
$M \smallsetminus 0$. Suppose $Y \subseteq X$ is a T-invariant subvariety, and assume the
restriction $H^* X^T \to H^* Y^T$ is surjective. If $\{\alpha\}$ is a set of generators for
$H_T^* X$ as a Λ-algebra, then their restrictions to $H_T^* Y$ generate $S^{-1} H_T^* Y$
as an $S^{-1}\Lambda$-algebra.*

A special case of the corollary gives a strong statement about varieties
with finitely many attractive fixed points. An isolated fixed point $p \in X^T$
is *attractive* if all the weights in the (Zariski) tangent space $T_p X$ lie in an
open half-space of $M \otimes \mathbb{R}$; that is, there is some dual vector $\lambda \in (M \otimes \mathbb{R})^\vee$
such that $\langle \lambda, \chi \rangle > 0$ for all weights χ on $T_p X$.

Corollary 7.1.5.   *If X is an irreducible projective variety with finitely
many fixed points, all of which are attractive, then the $S^{-1}\Lambda$-algebra
$S^{-1} H_T^* X$ is generated by the equivariant Chern class of an ample line
bundle.*

This is a direct consequence of the previous corollary, using the following lemma.

Lemma 7.1.6.   *Suppose X is an irreducible projective variety, and let
$X \hookrightarrow \mathbb{P}^N$ be an equivariant embedding. Suppose $p \in X^T$ is an attractive
fixed point mapping to a connected component $Z \subseteq (\mathbb{P}^N)^T$. Then p is
the only point of X which maps to Z.*

Proof Note that Z is a linear subspace with trivial T-action. We need to
show $Z \cap X^T$ consists of at most one point. If Z is a point, this is obvious,
so we may assume Z is positive-dimensional. One can find a hyperplane
in Z avoiding any finite collection of points, and any hyperplane in Z
may be written as $Z \cap H$, for some T-invariant hyperplane $H \subseteq \mathbb{P}^N$. By
choosing H so that it avoids $Z \cap X^T$, we may replace X by the affine
variety $U = X \smallsetminus H$.

We have reduced to proving that if an irreducible affine T-variety U
contains an attractive fixed point p, then $U^T = \{p\}$. To see this, let

$A = \mathcal{O}(U)$ be the coordinate ring, and consider the action of $\mathbb{C}^*$ on $T_p U$ via $\lambda \colon \mathbb{C}^* \to T$, where $\langle \lambda, \chi \rangle > 0$ for all characters χ of T acting on $T_p U$. Lifting a basis of eigenvectors from $(T_p U)^\vee = \mathfrak{m}_p / \mathfrak{m}_p^2$ to generators of $\mathfrak{m}_p$, we obtain a grading

$$A = \bigoplus_{d \geq 0} A_d.$$

It follows that U^T is defined by the ideal of positive-degree elements, and its coordinate ring is isomorphic to A_0. Since $A_0 \subseteq A$ is a subring, it is a domain; since p is an isolated fixed point, it then follows that $A_0 = \mathbb{C}$, so $U^T = \{p\}$ as claimed. $\qquad\square$

In the situation of Corollary 7.1.5, if X is also nonsingular, the Białynicki–Birula decomposition (Theorem 5.3.2) implies that $H_T^* X$ is a free Λ-module, so that it embeds in $S^{-1} H_T^* X$. This leads to a remarkable and useful property of such varieties:

If X is an irreducible nonsingular projective variety with finitely many attractive fixed points, then the ring structure of $H_T^* X$ is determined by multiplication by divisors.

This principle applies to standard torus actions on Grassmannians and flag varieties, and more generally, homogeneous spaces G/P – although for most such spaces, $H_T^* X$ is not generated by divisor classes! Later we will see an alternative, algorithmic proof of this principle for $X = G/P$.

Without the assumption on attractive fixed points, however, the lemma and corollary may fail.

Exercise 7.1.7. Let T act on $\mathbb{P}^3$ by characters $\chi, -\chi, 0, 0$, for some nonzero character χ. Using homogeneous coordinates x, y, z, w, the quadric hypersurface $X = \{xy = z^2 - w^2\}$ is T-invariant. Check that X^T is finite, determine the tangent weights on each fixed point, and show that two fixed points of X lie in the same fixed component of $\mathbb{P}^3$. Show that the homomorphism $H_T^* \mathbb{P}^3 \to H_T^* X$ does not become surjective after tensoring with the fraction field of Λ.

7.2 Invariant curves

Much information can be gleaned from the T-invariant curves in a variety X. In the next section we will see how they determine the image of the restriction homomorphism $\iota^*\colon H_T^*X \to H_T^*X^T$. First, we will need some notation and basic facts about such curves.

Suppose T acts on $\mathbb{P}^1$ by distinct characters χ_1 and χ_2, so the fixed points are $0 = [1,0]$ and $\infty = [0,1]$. Writing $\chi = \chi_2 - \chi_1$, we have seen $T_0\mathbb{P}^1 = \mathbb{C}_\chi$ and $T_\infty\mathbb{P}^1 = \mathbb{C}_{-\chi}$.

More generally, if T acts on a nonsingular curve C with two fixed points, $C^T = \{p, q\}$, then there is an equivariant isomorphism $C \cong \mathbb{P}^1$ sending p to 0 and q to ∞. Indeed, for any $x \in C \smallsetminus C^T$, the action map

$$T \to T \cdot x = T/T_x,$$

realizes $C \smallsetminus C^T$ as a one-dimensional quotient of T. One sees that C is rational; a nonsingular rational curve containing $T/T_x \cong \mathbb{C}^*$ is either $\mathbb{C}^*$, $\mathbb{A}^1$, or $\mathbb{P}^1$, and only the latter has two fixed points. Choosing the isomorphism $T/T_x \cong \mathbb{C}^*$ so that $p = \lim_{z \to 0} z \cdot x$ defines an equivariant isomorphism $C \to \mathbb{P}^1$, with $p \mapsto 0$ and $q \mapsto \infty$. This also identifies the action map $T \to T \cdot x \cong \mathbb{C}^*$ as a character χ.

There is one other choice of isomorphism, swapping p with q, z with z^{-1}, and χ with $-\chi$. Up to sign, then, the character χ depends only on the T-action on C. We will call $\pm\chi$ the *character* of T acting on C.

Similarly, the *character* of T acting on any (possibly singular) curve C is defined to be $\pm\chi$ if a choice of isomorphism $T \cdot x \cong \mathbb{C}^*$ identifies $T \to T \cdot x$ with $\chi\colon T \to \mathbb{C}^*$, for some $x \in C \smallsetminus C^T$; and it is defined to be 0 if the action is trivial.

For T acting on a variety X, a *T-curve* $C \subseteq X$ is the closure of a one-dimensional T-orbit in X, so $C = \overline{T \cdot x}$. By choosing $T \cdot x \cong \mathbb{C}^*$, each T-curve has an associated nonzero character $\pm\chi$.

Example 7.2.1. Even when the fixed locus X^T is finite, there are often infinitely many T-curves.

Suppose T acts on $\mathbb{P}^2$ by characters $0, \chi, 2\chi$, for some nonzero character χ. The fixed points are the standard coordinate points $p_1 = [1, 0, 0]$, $p_2 = [0, 1, 0]$, and $p_3 = [0, 0, 1]$. The T-curves are the coordinate lines

$$\{X_1 = 0\} \text{ with character } \pm\chi,$$
$$\{X_2 = 0\} \text{ with character } \pm 2\chi, \text{ and}$$
$$\{X_3 = 0\} \text{ with character } \pm\chi,$$

together with the conics

$$\{X_2^2 - \lambda X_1 X_3 = 0\} \text{ with character } \pm \chi,$$

for $\lambda \neq 0$.

Invariant curves may be singular. For example, if T acts on $\mathbb{P}^2$ via characters $0, a\chi, b\chi$, with $0 < a < b$, then the curves defined by $X_2^b - \lambda X_1^{b-a} X_3^a$ are invariant; these have cuspidal singularities if, for example, $a = 1$ and $b \geq 3$.

However, there are limitations on the singularities that can occur.

Example 7.2.2. Let $C = \mathbb{P}^1/(0 \sim \infty)$ be a nodal curve with a nontrivial T-action induced from an action on $\mathbb{P}^1$. Such a curve cannot occur as a T-curve in any nonsingular, or even normal, algebraic variety X. This is because Sumihiro's theorem (Theorem 7.1.2) provides a T-invariant affine cover of such an X, but there is no T-invariant affine neighborhood of the singular point of C.

The local structure of T-curves on nonsingular varieties can be classified. We will use some terminology for characters and curves. Any nonzero character $\chi \in M$ is uniquely $c \cdot \eta$, where c is a positive integer and η is a *primitive* character – that is, the only expression $\eta = c' \cdot \eta'$, with c' a positive integer, is $c' = 1$ and $\eta' = \eta$. Writing a character χ this way, we call c its *coefficient* and η its *direction*. Two characters χ_1, χ_2 are *parallel* if their directions are the same or opposite, that is, if $\eta_1 = \pm \eta_2$.

Proposition 7.2.3. *Let T act on an n-dimensional nonsingular algebraic variety X, and let $p \in X^T$ be an isolated fixed point, so the tangent weights $\chi_1, \ldots, \chi_n$ on $T_p X$ are all nonzero.*

(1) If no two characters at p are parallel, then there are finitely many T-curves in X through p. In fact, there are n such curves, all nonsingular at p, with characters $\chi_1, \ldots, \chi_n$.

(2) If two characters have the same direction, then there are infinitely many T-curves through p.

(3) If two characters have opposite directions, then there are infinitely many T-curves through any T-invariant neighborhood of p.

Proof Using the slice theorem as in Chapter 5, we find a T-invariant neighborhood of p, equivariantly isomorphic to a neighborhood of $0 \in T_p X$, thereby reducing to the case where $p = 0$ in $X = \mathbb{C}^n$, with T acting via characters $\chi_1, \ldots, \chi_n$.

If, say, χ_1 and χ_2 are parallel, Example 7.2.1 shows that there are infinitely many T-curves in the corresponding plane; if χ_1 and χ_2 also have the same direction, all these curves go through 0. (More precisely, suppose $\chi_1 = a\chi$ and $\chi_2 = b\chi$, with a and b relatively prime, and let (x_1, x_2) be coordinates on the plane $\mathbb{C}^2 \subseteq \mathbb{C}^n$ where T acts by χ_1 and χ_2. If $a, b > 0$ then for all $\lambda \neq 0$, the curves $\{x_1^b = \lambda x_2^a\}$ are invariant and pass through 0. If $a > 0 > b$, then for all $\lambda \neq 0$, the curves $\{x_1^{-b} x_2^a = \lambda\}$ are invariant, and by taking λ small, they pass arbitrarily close to 0.) This proves (2) and (3).

For (1), if no two characters are parallel, any point $x \in \mathbb{C}^n$ with at least two nonzero coordinates has a T-orbit of dimension at least two. So in this case, the only T-curves are the n coordinate axes. $\square$

Exercise 7.2.4. Let X be a (possibly singular) variety of dimension n, with an isolated fixed point $p \in X^T$. Show that the number of T-curves through p is at least n. If X has finitely many T-curves, but there are more than n through p, then p is a singular point.

7.3 Image of the restriction map

Our final aim is to characterize the image of the homomorphism $\iota^* \colon H_T^* X \to H_T^* X^T$. Throughout this section, we use cohomology with coefficients in a UFD R. (Typically one takes R to be $\mathbb{Z}$, $\mathbb{Q}$, or $\mathbb{F}_p$.) Using the terminology of the previous section, two characters correspond to relatively prime elements of Λ if and only if they are non-parallel, and their coefficients are relatively prime in R.

Before turning to the general theorems, we consider three illustrative examples. The strongest results will characterize the image of ι^* using the characters of T-curves, so let us first consider the case where X itself is a curve.

Example 7.3.1. Let T act on $\mathbb{P}^1$ by distinct characters χ_1, χ_2, so the fixed points are $0 = [1, 0]$ and $\infty = [0, 1]$. Writing $\chi = \chi_2 - \chi_1$ for the character of this action, we have $T_0 \mathbb{P}^1 = \mathbb{C}_\chi$ and $T_\infty \mathbb{P}^1 = \mathbb{C}_{-\chi}$. The image of $H_T^* \mathbb{P}^1$ in $H_T^*(\mathbb{P}^1)^T = \Lambda \oplus \Lambda$ consists of pairs (u_0, u_∞) such that $u_\infty - u_0$ is divisible by χ.

To see this, recall that $H_T^* \mathbb{P}^1 = \Lambda[\zeta]/(\zeta + \chi_1)(\zeta + \chi_2)$ maps to $H_T^*(\mathbb{P}^1)^T = \Lambda \oplus \Lambda$ by $\zeta \mapsto (-\chi_1, -\chi_2)$. The image satisfies the divisibility condition, because the image of ζ does. On the other hand, using the basis $\{1, [\infty]^T\}$ for $H_T^* \mathbb{P}^1$, we see the divisibility condition is also sufficient to characterize the image: if (u_0, u_∞) satisfies it, we can write

$$(u_0, u_\infty) = u_0 \cdot (1,1) + u'_\infty \cdot (0, \chi)$$
$$= \iota^*(u_0 \cdot 1 - u'_\infty \cdot [\infty]^T),$$

where $u_\infty - u_0 = \chi \cdot u'_\infty$.

Next, we consider the general case of a projective space with finitely many T-curves.

Example 7.3.2. Let T act on $\mathbb{P}^{n-1}$ by distinct characters $\chi_1, \ldots, \chi_n$, and for each i, assume the $n-1$ characters $\chi_j - \chi_i$ (for $j \neq i$) are pairwise relatively prime in Λ. By Proposition 7.2.3, this means there are finitely many T-curves, namely the coordinate lines C_{ij} connecting points p_i and p_j.

Claim. An element $(u_1, \ldots, u_n) \in H_T^*(\mathbb{P}^{n-1})^T = \Lambda^{\oplus n}$ belongs to the image of $\iota^* \colon H_T^* \mathbb{P}^{n-1} \to H_T^*(\mathbb{P}^{n-1})^T$ if and only if $u_i - u_j$ is divisible by $\chi_i - \chi_j$ for all $i \neq j$.

First, observe that divisibility is necessary. The restriction map factors as

$$H_T^* \mathbb{P}^{n-1} \to H_T^* C_{ij} \xrightarrow{\iota_{ij}^*} \Lambda \oplus \Lambda,$$

and any (u_i, u_j) in the image of ι_{ij}^* must have $u_i - u_j$ divisible by $\chi_i - \chi_j$, by the $\mathbb{P}^1$ case discussed in Example 7.3.1.

To see that it is also sufficient, we can proceed inductively. For any $u_1 \in \Lambda$, the element $(u_1, \ldots, u_1) = u_1 \cdot (1, \ldots, 1)$ is certainly in the image of ι^*. To see $(u_1, u_2, \ldots, u_n)$ is in the image, it suffices to show that $(0, u_2 - u_1, \ldots, u_n - u_1)$ is in the image – that is, we may assume the first entry is zero. By the divisibility condition, we can write such an element as $(0, (\chi_1 - \chi_2)v_2, (\chi_1 - \chi_3)v_3, \ldots, (\chi_1 - \chi_n)v_n)$.

The element $(\zeta + \chi_1)v_2 \in H_T^* \mathbb{P}^{n-1}$ restricts to

$$(0, (\chi_1 - \chi_2)v_2, (\chi_1 - \chi_3)v_2, \ldots),$$

and by subtracting this, we reduce to the case where the first two entries are zero. So it suffices to prove that

$$(0, 0, (\chi_1-\chi_3)(\chi_2-\chi_3)w_3, (\chi_1-\chi_4)(\chi_2-\chi_4)w_4, \ldots, (\chi_1-\chi_n)(\chi_2-\chi_n)w_n)$$

lies in the image. In using the divisibility condition to extract the factors $(\chi_1 - \chi_i)(\chi_2 - \chi_i)$, we have used that R (and hence Λ) is a UFD.

Continuing in this way, we reduce to proving that

$$\left(0, \ldots, 0, \prod_{i=1}^{n-1} (\chi_i - \chi_n)z_n \right)$$

lies in the image. By the self-intersection formula, this is the restriction of $z_n \cdot \prod_{i=1}^{n-1}(\zeta + \chi_i) = z_n \cdot [p_n]^T \in H_T^* \mathbb{P}^{n-1}$ (§4.7). □

Without the condition that characters be relatively prime – so with infinitely many T-curves – the divisibility criterion is more complicated.

Example 7.3.3.   Let T act on $\mathbb{P}^2$ by characters $0, \chi, 2\chi$, where χ is a primitive (nonzero) character, and assume 2 is not a zerodivisor in R. The image of

$$H_T^* \mathbb{P}^2 = \Lambda[\zeta]/\zeta(\zeta + \chi)(\zeta + 2\chi) \to \Lambda^{\oplus 3} = H_T^*(\mathbb{P}^2)^T$$
$$\zeta \mapsto (0, -\chi, -2\chi)$$

is the subring of triples (u_1, u_2, u_3) such that

(1) $u_2 - u_1$ and $u_3 - u_2$ are divisible by χ, $u_3 - u_1$ is divisible by 2χ; and

(2) $u_1 - 2u_2 + u_3$ is divisible by $2\chi^2$.

Necessity of the first condition is just as before, by factoring through $H_T^* C_{ij}$. Necessity of the second condition follows from the integration formula, as we saw in Example 5.2.6. Sufficiency of the conditions is an exercise—one imitates the argument from the previous example to construct a class.

To state and prove the general theorems about the image of the restriction map, we introduce one more piece of notation. We say f is an *irreducible factor* if it is either a prime in $\mathbb{Z}$ or a primitive character in M. The images of such f under the canonical homomorphisms $\mathbb{Z} \to \Lambda$ and $M \to \Lambda$ are the elements which occur as factors of characters in Λ. When referring to f as an element of Λ, we mean its image under the canonical homomorphism.

Given an irreducible factor f, we will write $L_f \subseteq M$ for the subgroup of characters divisible by f, and $T(f) = T(L_f) \subseteq T$ for the corresponding subgroup. If $f = \chi$ is a primitive character in M, then $T(f) \subseteq T$ is a subtorus of codimension one; if $f = p$ is a prime in $\mathbb{Z}$, then $T(f) \subseteq T$ is the finite subgroup of all elements of order p (a "p-torus"). Note that $X^T \subseteq X^{T(f)} \subseteq X$.

The main theorem on the image of the restriction homomorphism gives a characterization in terms of the fixed loci $X^{T(f)}$.

Theorem 7.3.4 (Chang–Skjelbred). *Let T act on a variety X, and assume that H^*X^T is a free R-module and H_T^*X is a free Λ-module. Then an element $\alpha \in H_T^*X^T$ lies in the image of $\iota^* \colon H_T^*X \to H_T^*X^T$ if and only if it lies in the image of $H_T^*X^{T(f)} \to H_T^*X^T$ for all irreducible factors f.*

Letting $Z \subseteq X^T$ range over connected components, so

$$H_T^*X^T = \bigoplus_{Z \subseteq X^T} \Lambda \otimes_R H^*Z,$$

one can write $\alpha = (\alpha_Z)$ as a tuple of elements of $\Lambda \otimes_R H^*Z$. This is especially useful when the fixed points are isolated, where (α_Z) is just a tuple of polynomials. (See Corollary 7.4.3.)

Proof Since $X^{T(f)} \supseteq X^T$, the "only if" direction is clear. For the other direction, assume $\alpha \in H_T^*X^T$ lies in the images of all $H_T^*X^{T(f)}$. Since $S^{-1}H_T^*X = S^{-1}H_T^*X^T$ for $S \subseteq \Lambda$ generated by all nonzero characters, we can find an element $g \in S$ such that $g\alpha \in H_T^*X$. Take g to be minimal, so that for any proper divisor g' of g, $g'\alpha \notin H_T^*X$. Let $e_1, \ldots, e_r$ be a basis for H_T^*X over Λ, and write $g\alpha = a_1 e_1 + \cdots + a_r e_r$.

Suppose for contradiction that α is not in the image of H_T^*X. Then g is not a unit in Λ. Let f be an irreducible factor of g. By minimality of g, some a_i is relatively prime to f; say this is a_1.

By the general localization theorem, $S(f)^{-1}H_T^*X = S(f)^{-1}H_T^*X^{T(f)}$, where $S(f)$ is the multiplicative set generated by characters not divisible by f. So there is an element $\psi_f \in S(f)$ so that $\psi_f\alpha \in H_T^*X$. Write $\psi_f\alpha = b_1e_1 + \cdots + b_re_r$. From the definition of $S(f)$, we know f does not divide ψ_f. On the other hand, the coefficient of e_1 in $(g\psi_f)\alpha$ is $\psi_f a_1 = gb_1$. Since f divides the right-hand side, we reach a contradiction. $\qquad\square$

A version of the theorem was first proved by Chang and Skjelbred, using $\mathbb{Q}$ coefficients. In fact, one can relax the requirement that H_T^*X be a free Λ-module, as shown in the following exercises. More comments on this story are in the Notes.

Exercise 7.3.5. Show that one can strengthen the statement of Theorem 7.3.4, by requiring only that the quotient of H_T^*X modulo Λ-torsion is free as a Λ-module. (In the case where $T = \mathbb{C}^*$ and R is a field, this hypothesis holds whenever H_T^*X is finitely generated, since Λ is a PID.)

Exercise 7.3.6. Let X be the nodal curve obtained by identifying the points $0 = [1,0]$ and $\infty = [0,1]$ on $\mathbb{P}^1$, with a $\mathbb{C}^*$ action induced by the action $z \cdot [a,b] = [a, zb]$ on $\mathbb{P}^1$. (This was considered in Exercise 3.5.1.) Compute the restriction map $H_T^*X \to H_T^*X^T = H_T^*(p)$, where $p \in X^T$ is the node, and verify the conditions of Theorem 7.3.4, as refined by Exercise 7.3.5. On the other hand, show that the forgetful map $H_T^*X \to H^*X$ is not surjective.

Exercise 7.3.7. Let $X \subseteq \mathbb{P}^2$ be the union of the three coordinate lines. (If X_1, X_2, X_3 are homogeneous coordinates on $\mathbb{P}^2$, X is defined by the equation $X_1X_2X_3 = 0$.) Let $T = \mathbb{C}^*$, and let $K = S^1 \subset T$ be the (compact) circle subgroup, so $\Lambda_T = \Lambda_K = \mathbb{Z}[t]$. Consider the action of T given by $z \cdot [x_1, x_2, x_3] = [z^2x_1, zx_2, x_3]$, so X is T- and K-invariant.

Using the Mayer–Vietoris sequence, compute $H_T^*X = H_K^*X$. In particular, show that H_T^iX is zero for $i > 1$ odd, but $H_T^1X = \mathbb{Z}$. Therefore $t \cdot H_T^1X = 0$, so H_T^*X is not a free Λ-module.

On the other hand, H_T^*X is free modulo Λ-torsion. Check that the image of $H_T^*X \to H_T^*X^T = \Lambda^{\oplus 3}$ is characterized by the condition of Theorem 7.3.4 (using the refinement from Exercise 7.3.5).

7.4 The image theorem for nonsingular varieties

Theorem 7.3.4 describes the image of $H_T^* X$ as an intersection of $H_T^* X^{T(f)}$ over all irreducible factors f, but in fact it suffices to consider only finitely many fixed loci. When X is nonsingular, further refinements are possible.

Let $\mathscr{S} \subseteq M$ be the (finite) set of all characters occurring as weights on $T_p X$, for fixed points $p \in X^T$. Given an irreducible factor f, take $L_f \subseteq M$ to be the subgroup of characters divisible by f (as before). Let $\mathscr{S}_f^+ = \mathscr{S} \cap L_f$ be the set of characters which occur as tangent weights and are divisible by f, and let $\mathscr{S}_f^- = \mathscr{S} \setminus \mathscr{S}_f^+$ be those which are not divisible by f. Finally, let

$$T_f = \bigcap_{\chi \in \mathscr{S}_f^+} \ker(\chi) \quad \text{and} \quad X^f = X^{T_f},$$

and let $S_f \subseteq \Lambda$ be the multiplicative set generated by $\mathscr{S}_f^-$.

Since $T_f \subseteq T$ is a diagonalizable group and X is nonsingular, X^f is also nonsingular. In general, $T(f) \subseteq T_f$, $X^{T(f)} \subseteq X^f$, and $S(f) \supseteq S_f$. If f does not divide any characters occurring at fixed points, then $X^f = X^T$.

Exercise 7.4.1. Suppose X is nonsingular. Show that in Theorem 7.3.4, one can replace $X^{T(f)}$ and $S(f)$ by X^f and S_f, respectively. Furthermore, in the "if" statement of the theorem, it suffices to consider only those f which occur as irreducible factors of tangent weights at fixed points.

When X has finitely many T-curves, the sets X^f have a more concrete description.

Lemma 7.4.2. *Suppose X is nonsingular, X^T is finite, and at each fixed point, the tangent weights are relatively prime, so there are finitely many T-curves in X. Then X^f is the union of all T-curves C whose character is divisible by f, together with all isolated fixed points $p \in X^T$ where no weight of $T_p X$ is divisible by f.*

Proof We may assume $R = \mathbb{Z}$ and $\dim X > 1$. Certainly, X^f contains this union: T_f fixes all T-curves whose character χ is divisible by f, because $T_f \subseteq \ker(\chi)$ by definition.

For the other direction, we must show that T-curves whose characters are not divisible by f are not fixed by T_f. That is, when $\chi \in \mathscr{S}_f^-$, the restriction of χ to T_f is nontrivial. Note that $\mathscr{S}_f^-$ is nonempty – since the characters at a given fixed point are pairwise relatively prime, those which are divisible by f form a proper subset $\mathscr{S}_f^+ \subsetneq \mathscr{S}$.

Let K_f be the sublattice generated by $\mathscr{S}_f^+$, so $K_f \subseteq L_f \subseteq M$, and $M_f = M/K_f$ is the character group of T_f. Any character $\chi \in \mathscr{S}_f^-$ lies in $M \smallsetminus L_f \subseteq M \smallsetminus K_f$. Its restriction to T_f is therefore nonzero, because this is given by the surjection $M \twoheadrightarrow M_f = M/K_f$. $\square$

A particularly vivid instance of the image theorem was popularized by Goresky, Kottwitz, and MacPherson.

Corollary 7.4.3 (GKM).   *Let X be a nonsingular variety with X^T finite, and assume $H_T^* X$ is free over Λ. Suppose that for each $p \in X^T$, the weights on $T_p X$ are relatively prime. Then a tuple*

$$(u_p)_{p \in X^T} \in H_T^* X^T$$

lies in the image of $\iota^*\colon H_T^* X \to H_T^* X^T$ if and only if for each T-curve $C_{pq} \cong \mathbb{P}^1$ connecting distinct points $p, q \in X^T$, the difference $u_p - u_q$ is divisible by the character $\pm\chi_{pq}$ of C_{pq}.

Proof   The divisibility condition is always necessary: as before, we can factor ι^* as

$$
\begin{array}{ccc}
H_T^* X & \xrightarrow{\ \iota^*\ } & H_T^* X^T = \bigoplus_{p \in X^T} \Lambda \\
\downarrow & & \\
H_T^* C_{pq} & \longrightarrow & H_T^* C_{pq}^T = \Lambda \oplus \Lambda,
\end{array}
$$

and apply the $\mathbb{P}^1$ case.

For sufficiency, we will show that the divisibility condition means (u_p) lifts to $H_T^* X^f$, for each irreducible factor f, and apply Theorem 7.3.4. Using the lemma, X^f is the union of T-curves whose character is divisible by f, together with isolated fixed points. Such T-curves are nonsingular and disjoint, so each component must be (1) $C_{pq} \cong \mathbb{P}^1$, (2) $C_p \cong \mathbb{A}^1$, or (3) $C_\emptyset \cong \mathbb{C}^*$. Now consider the restriction homomorphism $H_T^* X^f \to H_T^* X^T$, on components of each type. Divisibility guarantees that (u_p) lifts to each summand $H_T^* C_{pq}$ of type (1), by the $\mathbb{P}^1$ case. Summands of type (2) map isomorphically, $H_T^* C_p = H_T^* \mathbb{A}^1 \xrightarrow{\sim} H_T^*(p) = \Lambda$, so lifting is trivial for such summands. Finally, summands of type (3) map to zero, since $H_T^* C_\emptyset$ is a torsion module.

Putting this together, we see that (u_p) lifts to $H_T^* X^f$, so the corollary follows. □

The relatively prime condition on torus weights implies that X has finitely many T-curves. In fact, when the coefficients are $R = \mathbb{Q}$ (or any field), having relatively prime weights at all fixed points is equivalent to having finitely many T-curves; such varieties are often called *GKM varieties*.

The hypothesis of the "GKM" corollary also implies that ι^* is injective, so it gives an appealing characterization of $H_T^* X$ as tuples of polynomials satisfying divisibility conditions. As we have seen in the last two chapters, this information is often organized into a *moment graph*: vertices correspond to fixed points $p \in X^T$, and edges correspond to T-curves C_{pq} for $p \neq q$. Each edge is labeled by the character $\pm \chi_{pq}$ of C_{pq}. When X is quasi-projective, a basic fact from symplectic geometry says that the moment graph can be embedded in Euclidean space so that edges with parallel characters embed as parallel lines. We have seen examples of this in the last two chapters.

Example 7.4.4. Consider $X = Gr(2, \mathbb{C}^4)$, with the standard action of $T = (\mathbb{C}^*)^4$. The fixed points are $p_{ij} = \langle e_i, e_j \rangle$, that is, the coordinate subspaces. We have already computed the weights at fixed points, and this is enough to determine the characters of the T-curves. (Half of the 12 edge labels are shown in Figure 7.1; the remaining labels are determined by parallel edges.)

Inside X, we have

$$\Omega = \left\{ E \subseteq \mathbb{C}^4 \,\middle|\, \dim(E \cap E_{12}) \geq 1 \right\} = \overline{\Omega^\circ},$$

where Ω° is a neighborhood of p_{24} in Ω. Using columns of a matrix to represent points of X,

$$\Omega^\circ = \begin{bmatrix} * & * \\ 1 & 0 \\ 0 & * \\ 0 & 1 \end{bmatrix}.$$

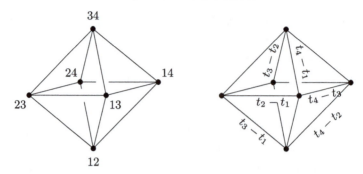

Figure 7.1 The moment graph for $T = (\mathbb{C}^*)^4$ acting on $Gr(2, \mathbb{C}^4)$, with edges labeled by characters of T-curves.

We know that $p_{34} \notin \Omega$, and can see that $p_{24} \in \Omega$ is a nonsingular point with normal character $t_3 - t_2$. These observations show that

$$[\Omega]^T|_{p_{34}} = 0 \qquad \text{and} \qquad [\Omega]^T|_{p_{24}} = t_3 - t_2.$$

The restrictions to all other fixed points are determined by the divisibility condition! (We saw another way to compute this in Example 5.2.7.)

Exercise 7.4.5. Use divisibility to work out all restrictions $[\Omega]^T|_{p_{ij}}$.

Using the integration formula, divisibility criteria extend to varieties with possibly infinite families of T-curves.

Corollary 7.4.6. *Let X be a nonsingular variety, with T acting so that H_T^*X is a free Λ-module. For all irreducible factors f, assume X^f is compact and $H_T^*X^f$ is free over Λ. Letting $Z \subseteq X^T$ range over connected components, an element $(\alpha_Z) \in H_T^*X^T$ lies in the image of H_T^*X if and only if for all irreducible factors f and all $\beta \in H_T^*X^f$, we have*

$$\sum_{Z \subseteq X^T} \frac{\alpha_Z\, \beta|_Z}{c_{top}^T(N_{Z/X^f})} \in \Lambda.$$

In fact, it suffices to let f vary over irreducible factors of tangent weights at fixed points in X^T, and to choose β from a Λ-module basis of $H_T^*X^f$.

Proof If (α_Z) is the image of $\alpha \in H_T^*X$, and α_f is the restriction of α to $H_T^*X^f$, then by the integration formula, the sum equals $\rho_*^f(\alpha_f \cdot \beta)$, where $\rho^f: X^f \to \mathrm{pt}$ is the projection. Thus the condition is necessary.

To see that the condition is sufficient, suppose it holds for some f. We will show that (α_Z) lies in the image of $H_T^* X^f \to H_T^* X^T$, and conclude by Theorem 7.3.4.

Let $x_1, \ldots, x_r$ be a Λ-module basis for $H_T^* X^f$, with Poincaré dual basis $y_1, \ldots, y_r$. Choose S so that $S^{-1} H_T^* X^f = S^{-1} H_T^* X^T$, and write $\alpha = (\alpha_Z) = a_1 x_1 + \cdots + a_r x_r$, for some $a_i \in S^{-1}\Lambda$. So α lies in $H_T^* X^f$ exactly when all a_i lie in Λ. By Poincaré duality, we know $a_i = \rho_*^f(\alpha \cdot y_i)$, where ρ_*^f is extended linearly to a homomorphism $S^{-1} H_T^* X^f \to S^{-1}\Lambda$. By the observation above, this is

$$a_i = \rho_*^f(\alpha \cdot y_i) = \sum_{Z \subseteq X^T} \frac{\alpha_Z \cdot y_i|_Z}{c_{top}^T(N_{Z/X^f})},$$

which lies in Λ by hypothesis. $\qquad\square$

When X^T is finite and the tangent weights at each fixed point are relatively prime, so each X^f is a union of $\mathbb{P}^1$'s and isolated fixed points, we recover the case of the GKM theorem (Corollary 7.4.3) where all T-curves are complete. Here we can take $\beta = 1$. For a component $C_{pq} \cong \mathbb{P}^1$ of X^f where T acts with character χ, the condition of Corollary 7.4.6 says

$$\frac{\alpha_p}{\chi} + \frac{\alpha_q}{-\chi} = \frac{\alpha_p - \alpha_q}{\chi} \in \Lambda.$$

That is, χ divides $\alpha_p - \alpha_q$.

The description of $H_T^* \mathbb{P}^2$ given in Example 7.3.3 may be obtained using Corollary 7.4.6. Similar examples are given in the following exercises.

Exercise 7.4.7. Let T act on $\mathbb{P}^1$ by characters $(0, \chi)$ (for nonzero χ), and diagonally on $X = \mathbb{P}^1 \times \mathbb{P}^1$. Check that X^T consists of the four fixed points $p_1 = (0, 0)$, $p_2 = (0, \infty)$, $p_3 = (\infty, 0)$, and $p_4 = (\infty, \infty)$, but there are infinitely many T-curves. Use Corollary 7.4.6 to compute the image of $H_T^* X \to H_T^* X^T$.

Exercise 7.4.8. Let T act on $X = \mathbb{P}^{n-1}$ by characters $0, \chi, 2\chi, \ldots, (n-1)\chi$, where χ is some nonzero character of T. Generalize Example 7.3.3 to compute the image of $H_T^* X \to H_T^* X^T$. (Assume the coefficient ring is $\mathbb{Z}$.)

Notes

Topological versions of the main localization theorem can be found in textbooks on transformation groups (Hsiang, 1975, §III.2), (tom Dieck, 1987, §III.3), (Allday and Puppe, 1993, §3.1). A simple proof in this context was given by Brion and Vergne (1997b).

Sumihiro's theorem (1974, Corollary 2), (1975, Corollary 3.11) was refined by Brion (2015) to include non-normal quasi-projective varieties, and vastly generalized by Alper, Hall, and Rydh (2020, Theorem 2.4) to the setting of algebraic spaces (and beyond).

Special cases of Corollaries 7.1.4 and 7.1.5 appear in (Ciocan-Fontanine et al., 2008, Lemma 4.1.3) and (Buch et al., 2018, Remark 5.11).

The image theorem can be phrased quite generally, but less explicitly, in terms of natural exact sequences. Chang and Skjelbred (1974, 2.3, 2.4) proved one of the first theorems of this type:

Theorem. *Suppose X is equivariantly formal with respect to a T-action and $\mathbb{Q}$ coefficients. Let X_i be the union of all T-orbits of complex dimension at most i, so $X_0 = X^T$. Then the sequence*

$$0 \to H_T^*(X;\mathbb{Q}) \to H_T^*(X_0;\mathbb{Q}) \to H_T^{*+1}(X_1, X_0;\mathbb{Q}) \qquad (*)$$

is exact.

The last map in $(*)$ comes from the long exact sequence of the pair (X_1, X_0), so the theorem says the image of $H_T^* X$ under the restriction map is equal to that of $H_T^* X_1$; that is, equivariant cohomology of X is determined by equivariant cohomology of one-dimensional orbits (its "1-skeleton"). There is no hypothesis on $\dim(X_i)$, so in this form the theorem applies to the situation of infinitely many fixed points or T-curves. Our proof of Theorem 7.3.4 is similar to the one given in (Chang and Skjelbred, 1974).

Also using $\mathbb{Q}$ coefficients, there is a refinement of the Chang–Skjelbred theorem when X is a compact manifold (or rational Poincaré duality space), due to Allday, Franz, and Puppe: the sequence $(*)$ is exact if and only if $H_T^* X$ is a reflexive Λ-module, or equivalently, if the Poincaré pairing $H_T^* X \otimes_\Lambda H_T^* X \to \Lambda$ is perfect (Allday et al., 2014, Corollary 1.3).

For a compact nonsingular variety X with the action of an n-dimensional torus T, and a T-equivariant ample line bundle L, there is a *moment map* $\mu_L \colon X \to \mathfrak{t}_\mathbb{R}^\vee \cong \mathbb{R}^n$. The convexity theorem (Atiyah, 1982; Guillemin and Sternberg, 1982) says that the image of the moment map is convex, and its fibers are connected. The fact that the moment graph embeds so that edges are parallel to their labels is a consequence of the convexity theorem, but it also follows from two easier properties of the moment map: (1) if $Y \subseteq X$ is a T-invariant subvariety, then $\mu_{L|_Y} \colon Y \to \mathfrak{t}_\mathbb{R}^\vee$ is the restriction of μ_L to Y; and (2) the image of the moment map of a T-curve is a line segment parallel to its character.

These theorems belong to a more general story about Hamiltonian actions on symplectic manifolds; see (Audin, 2004, §IV.4).

The formulation of the image theorem (Theorem 7.3.4) in terms of explicit divisibility conditions – at least in the case of finitely many T-curves, and using coefficients in $\mathbb{Q}$, $\mathbb{R}$, or $\mathbb{C}$ – is due to Goresky, Kottwitz, and MacPherson (1998, Theorems 1.2.2 and 7.2). The idea of encoding equivariant cohomology in a moment graph also appears in this paper. (As the authors note, the Euclidean embedding of the moment graph is not necessarily a graph embedding, since edges may cross or overlap.)

The idea of exploiting exactness of the sequence (∗) to describe equivariant cohomology in terms of fixed points can be traced to work of Atiyah (1974, Lecture 7) in K-theory and Bredon (1974, Main Lemma) in cohomology. For an n-dimensional torus, Bredon proves exactness of the longer sequence

$$0 \to H_T^*(X) \to H_T^*(X_0) \to H_T^{*+1}(X_1, X_0) \to \cdots \to H_T^{*+n}(X_n, X_{n-1}) \to 0,$$
$$(**)$$

again using $\mathbb{Q}$ coefficients.

Franz and Puppe (2007, 2011) have carried out a detailed study of the role of equivariant formality in localization theorems, emphasizing the "Atiyah–Bredon sequence" (∗∗). They give extensions of the Chang–Skjelbred theorem to allow integer coefficients, as well as weaker conditions which imply partial exactness of the sequence (∗∗).

A version of Corollary 7.4.6 was proved by Evain (2007), developing ideas of Brion. Some variations are discussed by Braden, Chen, and Sottile (2008, §5.C).

We thank Matthias Franz and Volker Puppe for explaining the history of these ideas.

Hints for Exercises

Exercise 7.1.7. In the patch $\mathbb{A}^3 = \{w \neq 0\} \subseteq \mathbb{P}^3$, X is given by $\{xy = z^2 - 1\}$. The torus fixes the z-axis of $\mathbb{A}^3$. The fixed points of X are $(0, 0, \pm 1)$ in this patch.

Example 7.3.3. Imitate the argument from Example 7.3.2, noting that an element of the form $(0, 0, w_3)$ which satisfies the second divisibility condition must have w divisible by $2\chi^2$, so $(0, 0, w_3)$ is a multiple of the point class $[p_3]^T$.

Exercise 7.3.7. $H_T^* X \cong \Lambda[x, y, z, \alpha]/I$, where α has degree 1 and the other variables have degree 2, and I is generated by

$$\alpha^2,\ x + y + z - t,\ xy,\ xz,\ yz,\ x^2 - tx,\ y^2 - ty,\ z^2 - tz,\ x\alpha,\ y\alpha,\ z\alpha.$$

The groups are $H_T^0 X = \mathbb{Z}$, $H_T^1 X = \mathbb{Z}$, $H_T^{2i} X = \mathbb{Z}^3$, and $H_T^{2i+1} X = 0$ (for $i > 0$). They can be computed via the Mayer–Vietoris sequence for two K-invariant open sets. This forces most of the relations in the ring; to see $x + y + z = t$, use restriction to fixed points.

Exercise 7.4.7. The image consists of tuples (u_1, u_2, u_3, u_4) such that

$$u_1 - u_2, \; u_1 - u_3, \; u_2 - u_4, \; \text{ and } \; u_3 - u_4$$

are divisible by χ, and

$$u_1 - u_3 + u_2 - u_4$$

is divisible by χ^2.

Exercise 7.4.8. For each k from 2 to n, there is a condition that

$$(k-1)! \cdot \chi^{k-1} \quad \text{divides} \quad \sum_{i=1}^{k} (-1)^{i-1} \binom{k-1}{i-1} u_i.$$

8

Toric Varieties

Toric varieties provide a rich source of examples of equivariant geometry. In this chapter, we will describe the equivariant cohomology of a complete nonsingular toric variety. There are many ways to present this ring, and we will see several of them. With finitely many fixed points and invariant curves, one can apply the "GKM" package developed in the last chapter to obtain one such description.

In this chapter, all toric varieties are complete and nonsingular.

8.1 Equivariant geometry of toric varieties

We begin by quickly reviewing some basic notions. Some prior experience with toric geometry will be helpful in this chapter; for readers new to the subject, references are given in the Notes.

Let T be an n-dimensional torus with character group M, and let $N = \mathrm{Hom}_{\mathbb{Z}}(M, \mathbb{Z})$ be the dual lattice, with pairing denoted $\langle \, , \, \rangle$. A complete nonsingular toric variety $X = X(\Sigma)$ corresponds to a *complete nonsingular fan* Σ. This means Σ is a collection of convex polyhedral cones σ in the vector space $N_{\mathbb{R}} = N \otimes_{\mathbb{Z}} \mathbb{R}$ such that any two cones meet along a face of each; each cone must be generated by part of a basis for N (so Σ is nonsingular), and the union of the cones is all of $N_{\mathbb{R}}$ (so Σ is complete).

For any convex cone $\sigma \subseteq N_{\mathbb{R}}$, the *dual cone* in $M_{\mathbb{R}}$ is

$$\sigma^{\vee} = \{u \in M_{\mathbb{R}} \mid \langle u, v \rangle \geq 0 \text{ for all } v \in \sigma\}.$$

Intersecting with the lattice, one obtains a semigroup $\sigma^\vee \cap M$, with corresponding semigroup algebra $\mathbb{C}[\sigma^\vee \cap M]$. For any $u \in M$, we will write $e^u \in \mathbb{C}[M]$ for the corresponding element of the semigroup algebra.

The toric variety X is covered by T-invariant open affine sets $U_\sigma = \operatorname{Spec} \mathbb{C}[\sigma^\vee \cap M]$. In fact, $U_\sigma \cong \mathbb{C}^k \times (\mathbb{C}^*)^{n-k}$, where $k = \dim \sigma$. The affines U_σ for n-dimensional (maximal) cones σ suffice to cover X. At the other extreme, $U_{\{0\}} = \operatorname{Spec} \mathbb{C}[M] = T$. In general, intersection of cones corresponds to intersection of open sets: $U_\sigma \cap U_\tau = U_{\sigma \cap \tau}$.

Each cone τ of the fan Σ also determines a closed T-invariant subvariety $V(\tau) \subseteq X$. This is again nonsingular, and its codimension in X is the dimension of τ. On open affines, it is given by

$$V(\tau) \cap U_\sigma = \operatorname{Spec} \mathbb{C}[\tau^\perp \cap \sigma^\vee \cap M],$$

with the containment in U_σ given by

$$\mathbb{C}[\sigma^\vee \cap M] \twoheadrightarrow \mathbb{C}[\tau^\perp \cap \sigma^\vee \cap M],$$

$$e^u \mapsto \begin{cases} e^u & \text{if } u \in \tau^\perp; \\ 0 & \text{otherwise.} \end{cases}$$

(This is a homomorphism, because $\tau^\perp \cap \sigma^\vee$ is a face of $\sigma^\vee$.) Thus $V(\tau)$ is a nonsingular toric variety, for the torus with character group $\tau^\perp \cap M$ (a quotient of T); it corresponds to a fan in N/N_τ, where N_τ is the sublattice generated by τ.

The $V(\tau)$ are all the T-invariant subvarieties of X. In particular, the T-fixed points are $p_\sigma = V(\sigma)$ for maximal cones σ (so $\dim \sigma = n$). The invariant curves are $V(\tau)$ for cones τ of dimension $n-1$. Each such τ lies in exactly two maximal cones, with $\tau = \sigma \cap \sigma'$, and the corresponding invariant curve $V(\tau)$ is isomorphic to $\mathbb{P}^1$, with $V(\tau)^T = \{p_\sigma, p_{\sigma'}\}$.

The tangent weights to X at a fixed point p_σ are the primitive lattice vectors generating $\sigma^\vee$.

A toric variety X is projective if and only if there is a lattice polytope $P \subseteq M_\mathbb{R}$ such that Σ is the *(inward) normal fan* to P. A *polytope* in $M_\mathbb{R}$ is the convex hull of finitely many points. A *lattice polytope* is one whose vertices are in M. More precisely, for each face F of P, the corresponding cone in Σ is

$$\sigma_F = \{v \mid \langle u', v \rangle \geq \langle u, v \rangle \text{ for all } u' \in P, u \in F\}.$$

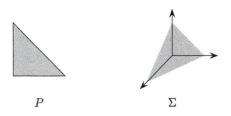

Figure 8.1 Polytope and fan for $X = \mathbb{P}^2$.

This correspondence reverses dimensions and inclusions: $\dim \sigma_F = \operatorname{codim} F$, and $\sigma_F \subseteq \sigma_{F'}$ if and only if $F \supseteq F'$. The normal fan to P is nonsingular if and only if P is a *simple* lattice polytope: at each vertex of P, the primitive lattice vectors along incident edges form a basis for M. (This is stronger than the condition often used in convex geometry, which only requires that these vectors be a basis for the vector space $M_{\mathbb{R}}$.)

Example. Let $u_1, \ldots, u_n$ be a basis for $M \cong \mathbb{Z}^n$. For the standard n-dimensional simplex, with vertices at $0, u_1, \ldots, u_n$, the corresponding toric variety is $\mathbb{P}^n$, with $T \cong (\mathbb{C}^*)^n$ acting by $(z_1, \ldots, z_n) \cdot [a_0, a_1, \ldots, a_n] = [a_0, z_1 a_1, \ldots, z_n a_n]$. (See Figure 8.1.)

The n-dimensional cube, with vertices at $\pm u_1, \ldots, \pm u_n$, corresponds to $(\mathbb{P}^1)^n$.

8.2 Cohomology rings

Suppose $X = X(\Sigma)$ is projective, with P a polytope whose normal fan is Σ. Choosing a general vector $v \in N_{\mathbb{R}}$, one obtains an ordering of the vertices $u_1, \ldots, u_s$ by $\langle u_1, v \rangle < \cdots < \langle u_s, v \rangle$. Via the correspondence between faces of P and cones of Σ, we get an ordering of maximal cones, $\sigma_1, \ldots, \sigma_s$. For $1 \leq i \leq s$, let

$$\tau_i = \bigcap_{\substack{j > i \\ \dim(\sigma_j \cap \sigma_i) = n-1}} \sigma_i \cap \sigma_j,$$

so $\tau_1 = \{0\}$, $\tau_s = \sigma_s$, and generally $\tau_p \subseteq \tau_q$ implies $p \leq q$. Such an ordering of cones is called a *shelling* of the fan.

A shelling gives a cellular decomposition of X, with the closures of cells being $V(\tau_1), \ldots, V(\tau_s)$. It follows that the classes

$$[V(\tau_1)], \ldots, [V(\tau_s)]$$

form a basis for H^*X (over $\mathbb{Z}$), and the corresponding equivariant classes $[V(\tau_1)]^T, \ldots, [V(\tau_s)]^T$ form a basis for H_T^*X over $\Lambda = \Lambda_T$.

If X is not projective, by subdividing cones one can always find a refinement Σ' of Σ, giving a surjective birational T-equivariant morphism $\pi\colon X' \to X$, with $X' = X(\Sigma')$ projective and nonsingular. (This is a toric version of Chow's lemma.) Under π, a subvariety $V(\tau')$ maps onto $V(\tau)$, where τ is the smallest cone of Σ containing τ'. The map $V(\tau') \to V(\tau)$ is birational if $\dim \tau' = \dim \tau$.

The composition $\pi_* \circ \pi^*$ is the identity on H^*X and on H_T^*X (e.g., by the projection formula), and it follows that π^* is injective and π_* is surjective. This also shows:

Proposition. *For any complete nonsingular toric variety X, the cohomology ring H^*X is generated by classes $[V(\tau)]$ as a $\mathbb{Z}$-module, and H_T^*X is generated by $[V(\tau)]^T$ as a Λ-module. There is an isomorphism $H_T^*X \otimes_\Lambda \mathbb{Z} \xrightarrow{\sim} H^*X$.*

In the next section, we will see that H^*X and H_T^*X are always free of rank s, the number of maximal cones. Unlike the projective case, however, we don't know if the cohomology always has a basis of classes of invariant varieties.

Let $D_1, \ldots, D_d$ be the T-invariant divisors, with $D_i = V(\rho_i)$ for rays $\rho_1, \ldots, \rho_d$ of Σ. Let $v_i \in N$ be the minimal generator of the ray ρ_i. For $u \in M$, the element $e^u \in \mathbb{C}[M]$ determines a rational function on X. The corresponding divisor is

$$\operatorname{div}(e^u) = \sum \langle u, v_i \rangle D_i.$$

Equivariantly, e^u is a rational section of the (topologically trivial) line bundle L_u with character u, so we have a relation

$$u = c_1^T(L_u) = [\operatorname{div}(e^u)]^T = \sum \langle u, v_i \rangle [D_i]^T$$

in H_T^2X.

If two cones σ and τ span a cone γ, then $V(\sigma) \cap V(\tau) = V(\gamma)$. If $\dim \gamma = \dim \sigma + \dim \tau$, the intersection is transverse, so

$$[V(\sigma)]^T \cdot [V(\tau)]^T = [V(\gamma)]^T$$

in $H_T^* X$. If σ and τ are not contained in a common cone of Σ, then $V(\sigma) \cap V(\tau) = \emptyset$, and in this case $[V(\sigma)]^T \cdot [V(\tau)]^T = 0$. In particular, given distinct rays $\rho_{i_1}, \ldots, \rho_{i_r}$, we have

$$[D_{i_1}]^T \cdots [D_{i_r}]^T = [V(\tau)]^T$$

if the rays span a cone τ of Σ, and the product is zero otherwise. (The same is true for non-equivariant products.)

8.3 The Stanley–Reisner ring

Let $x_1, \ldots, x_d$ be variables, one for each ray of Σ. In the polynomial ring $\mathbb{Z}[x] = \mathbb{Z}[x_1, \ldots, x_d]$, we have two ideals:

- I is generated by all monomials $x_{i_1} \cdots x_{i_r}$ such that the corresponding rays $\rho_{i_1}, \ldots, \rho_{i_r}$ do not span a cone.

- J is generated by all elements $\sum \langle u, v_i \rangle x_i$, ranging over all $u \in M$.

To generate $I \subseteq \mathbb{Z}[x]$, it suffices to consider minimal sets of rays not spanning a cone (so any proper subset does span a cone). To generate $J \subseteq \mathbb{Z}[x]$, it suffices to let u run over a basis for M. The ring $\mathbb{Z}[x]/I$ is called the *Stanley–Reisner ring* of Σ.

We have a homomorphism

$$\mathbb{Z}[x]/(I + J) \to H^* X, \qquad (*)$$

given by $x_i \mapsto [D_i]$. Indeed, we have seen that I and J map to zero, so the homomorphism is well defined. It is surjective, because $[V(\tau)] = [D_{i_1}] \cdots [D_{i_r}]$, where $\rho_{i_1}, \ldots, \rho_{i_r}$ are the rays spanning τ. In fact, $(*)$ is an isomorphism, and we will deduce this from the corresponding equivariant statement.

Turning to equivariant cohomology, in $\Lambda[x] = \Lambda[x_1, \ldots, x_d]$ again we have two ideals:

- I' has the same generators as I, all monomials $x_{i_1} \cdots x_{i_r}$ such that the corresponding rays $\rho_{i_1}, \ldots, \rho_{i_r}$ do not span a cone.

- J' is generated by elements $u - \sum \langle u, v_i \rangle x_i$, ranging over all u in M (or a basis for M).

We have a homomorphism

$$\Lambda[x]/(I' + J') \to H_T^*X, \qquad\qquad (*_T)$$

by $x_i \mapsto [D_i]^T$. Again, we have seen that I' and J' map to zero, so the homomorphism is well defined; it is surjective for similar reasons. We will prove that it is an isomorphism.

Theorem 8.3.1.   *The homomorphisms* $(*)$ *and* $(*_T)$ *are isomorphisms, presenting the cohomology rings as*

$$H^*X \cong \mathbb{Z}[x]/(I + J) \quad and \quad H_T^*X \cong \Lambda[x]/(I' + J').$$

This identifies the equivariant cohomology of $X = X(\Sigma)$ with the Stanley–Reisner ring of Σ, thanks to the following simple lemma.

Lemma 8.3.2.   *The canonical homomorphism*

$$\mathbb{Z}[x]/I \to \Lambda[x]/(I' + J')$$

is an isomorphism.

Proof   Let $u_1, \ldots, u_n$ be a basis for M. Since $\Lambda = \mathbb{Z}[u_1, \ldots, u_n]$, the elements

$$u_1 - \sum_i \langle u_1, v_i \rangle x_i, \ \ldots, \ u_n - \sum_i \langle u_n, v_i \rangle x_i$$

form a regular sequence in $\Lambda[x]$ and generate J', with quotient $\Lambda[x]/J' \cong \mathbb{Z}[x]$. Since $I' = I \cdot \Lambda[x]$ by definition, we have the asserted isomorphism.   □

There are several ways to prove the theorem. We will construct a complex which is useful for studying general toric varieties, including singular or non-compact ones, and use this complex to give an algebraic argument that $\mathbb{Z}[x]/I$ is a free Λ-module. (Another proof, based on the GKM picture, is indicated in Exercise 8.4.1 below.)

For each cone τ, let $v_{i_1}, \ldots, v_{i_k}$ be its minimal generators, and set

$$\mathbb{Z}[\tau] := \mathbb{Z}[x_{i_1}, \ldots, x_{i_k}] = \mathbb{Z}[x]/(x_j \mid v_j \notin \tau).$$

We consider this both as a $\mathbb{Z}$-module, and as a $(\mathbb{Z}[x]/I)$-module. For each face γ of τ, there is a canonical surjection $\mathbb{Z}[\tau] \twoheadrightarrow \mathbb{Z}[\gamma]$.

Next, we set

$$C_k = \bigoplus_{\dim \tau = k} \mathbb{Z}[\tau],$$

and define a homomorphism $d_k \colon C_k \to C_{k-1}$ which maps $\mathbb{Z}[\tau]$ to the sum of $\mathbb{Z}[\gamma]$ over facets $\gamma \subseteq \tau$. More precisely, let $v_{i_1}, \ldots, v_{i_k}$ be the generators of τ, ordered so that $i_1 < \cdots < i_k$, and let γ be generated by $v_{i_1}, \ldots, \widehat{v_{i_p}}, \ldots, v_{i_k}$. Then d_k is $(-1)^p$ times the canonical surjection $\mathbb{Z}[\tau] \to \mathbb{Z}[\gamma]$. This defines a complex resolving the Stanley–Reisner ring.

Lemma 8.3.3. *We have an exact sequence of $\mathbb{Z}[x]/I$-modules*

$$C_\bullet \colon \quad 0 \to \mathbb{Z}[x]/I \to C_n \xrightarrow{d_n} C_{n-1} \xrightarrow{d_{n-1}} \cdots \xrightarrow{d_1} C_0 \to 0.$$

Proof Consider $\mathbb{Z}[x]$ as a multigraded ring, graded by $\mathbb{N}^d$, so it decomposes into a direct sum with one graded piece for each monomial $x_1^{m_1} \cdots x_d^{m_d}$. The map d_k is a homomorphism of graded modules over $\mathbb{Z}[x]$, so we may analyze the complex $C_\bullet$ by graded pieces.

Fix a monomial $x_1^{m_1} \cdots x_d^{m_d}$. Every term of $C_\bullet$ vanishes unless the set of v_i with $m_i > 0$ spans a cone γ in Σ. In this case, each C_k contributes one copy of $\mathbb{Z}$ for each τ that contains γ. The resulting complex is the one computing the reduced homology of a simplicial sphere in N/N_γ, so it is exact. $\qquad\square$

The composition $\Lambda \to \Lambda[x] \to \mathbb{Z}[x]/I$ takes $u \in M$ to $\sum \langle u, v_i \rangle x_i$. With this Λ-module structure, it follows that the complex $C_\bullet$ is an exact sequence of Λ-modules.

Proposition 8.3.4. *The Stanley–Reisner ring $\mathbb{Z}[x]/I \cong \Lambda[x]/(I' + J')$ is free over Λ, of rank $s = \#(\text{maximal cones})$.*

Proof Given a k-dimensional cone τ with primitive generators $v_{i_1}, \ldots, v_{i_k}$, choose any vectors $z_{k+1}, \ldots, z_n$ to complete a basis of N. Let $u_1, \ldots, u_n$ be the dual basis of M. Then $\mathbb{Z}[\tau] \cong \Lambda/(u_{k+1}, \ldots, u_n)$ as Λ-modules, so the projective dimension of C_k is $\mathrm{pd}_\Lambda C_k = n - k$. By induction, it follows that

$$\mathrm{pd}_\Lambda(\ker(C_k \to C_{k-1})) \le n - k,$$

and in particular, $\mathrm{pd}_\Lambda \mathbb{Z}[x]/I = 0$. So $\mathbb{Z}[x]/I$ is a projective Λ-module, and by the (easy) graded version of the Quillen–Suslin theorem, $\mathbb{Z}[x]/I$ is free.

Finally, consider the beginning of the complex $C_\bullet$:

$$0 \to \mathbb{Z}[x]/I \to C_n \to C_{n-1}.$$

We know C_n is free over Λ on s generator, since $\mathbb{Z}[\sigma] \cong \Lambda$ for n-dimensional cones, and C_{n-1} is a torsion Λ-module. It follows that $\mathbb{Z}[x]/I$ is a free Λ-module of rank s.   □

Now we complete the proof of the theorem.

Proof of Theorem 8.3.1   Consider the diagram

$$
\begin{array}{ccccccc}
0 & \longrightarrow & \mathbb{Z}[x]/I & \longrightarrow & C_n & \xrightarrow{\ d_n\ } & C_{n-1} \\
 & & \downarrow & & \downarrow{\scriptstyle\varphi} & & \\
 & & H_T^* X & \xrightarrow{\ \iota^*\ } & H_T^* X^T, & &
\end{array}
$$

where φ maps $\mathbb{Z}[\sigma]$ isomorphically to $H_T^*(p_\sigma)$, as follows. For each such maximal cone σ, suppose $v_{i_1}, \ldots, v_{i_n}$ are the primitive vectors spanning σ, and let $u_1(\sigma), \ldots, u_n(\sigma)$ be the dual basis for M. Using $\mathbb{Z}[\sigma] = \mathbb{Z}[x_{i_1}, \ldots, x_{i_n}]$, the map φ is given by the isomorphisms

$$\mathbb{Z}[x_{i_1}, \ldots, x_{i_n}] \to \Lambda$$

sending $x_{i_j} \mapsto u_j(\sigma)$.

The left vertical arrow is the composition

$$\mathbb{Z}[x]/I \to \Lambda[x]/(I' + J') \to H_T^* X,$$

taking x_i to $[D_i]^T$ for each i. We identify this with the homomorphism $(*_T)$.

The diagram commutes, because the restriction $H_T^* X \to H_T^*(p_\sigma)$ factors through $H_T^* U_\sigma$, and $U_\sigma \cong \mathbb{C}^n$ with T acting by weights $u_1(\sigma), \ldots, u_n(\sigma)$. For any $1 \le i \le d$, if $v_i \in \sigma$, say $i = i_j$, then $[D_i]^T$ restricts to $u_j(\sigma)$. (Indeed, $D_i \cap U_\sigma$ is defined by the equation $\{e^{u_j(\sigma)} = 0\}$, so its equivariant class restricts to $c_1^T(\mathbb{C}_{u_j(\sigma)}) = u_j(\sigma)$.) On the other hand, if $v_i \notin \sigma$, then $D_i \cap U_\sigma = \emptyset$, so $[D_i]^T \mapsto 0$ in $H_T^*(p_\sigma)$.

Since $\mathbb{Z}[x]/I \to C_n$ is injective, and φ is an isomorphism, it follows that the left vertical arrow is injective. Identifying this map with $(*_T)$, we have already seen that it is surjective. We conclude that $\mathbb{Z}[x]/I \cong \Lambda[x]/(I' + J') \cong H_T^* X$.

Using $\mathbb{Z} \cong \Lambda/M\Lambda$, we have

$$(\Lambda[x]/(I' + J')) \otimes_\Lambda \mathbb{Z} = \mathbb{Z}[x]/(I + J),$$

and similarly $H_T^* X \otimes_\Lambda \mathbb{Z} \xrightarrow{\sim} H^* X$, so the non-equivariant presentation $(*)$ follows. $\qquad\square$

The complex $C_\bullet$ leads quickly to combinatorial results in convex geometry. For $0 \leq i \leq n$, let a_i be the number of cones of dimension i in Σ. (So $a_0 = 1$ and $a_n = s$.) Considering the Stanley–Reisner ring as a usual (singly) graded ring, its Hilbert series is defined as

$$H(\mathbb{Z}[x]/I, t) = \sum_{m=0}^\infty \mathrm{rk}_\mathbb{Z}(\mathbb{Z}[x]/I)_m \, t^m.$$

Using additivity of Hilbert series with respect to exact sequences, together with the fact that $\mathbb{Z}[\tau]$ is a polynomial ring on k variables whenever $\dim \tau = k$, we obtain the following formula.

Corollary 8.3.5. *We have*

$$H(\mathbb{Z}[x]/I, t) = \sum_{i=0}^n \frac{(-1)^{n-i} a_i}{(1-t)^i} = \frac{P(t)}{(1-t)^n}.$$

where $p(t) = a_n - (1-t)a_{n-1} + (1-t)^2 a_{n-2} - \cdots + (-1)^n (1-t)^n a_0.$

The ideal $J \subset \mathbb{Z}[x]$ is generated by elements of degree 1, which form a regular sequence in $\mathbb{Z}[x]/I$. It follows that $P(t)$ is the Hilbert series of $\mathbb{Z}[x]/(I + J)$.

Corollary 8.3.6. *We have*

$$p(t) = \sum_{i=0}^n (-1)^i (1-t)^i a_{n-i}$$

$$= \sum_{m=0}^\infty \mathrm{rk}_\mathbb{Z}(\mathbb{Z}[x]/(I+J))_m \, t^m$$

$$= \sum_{m=0}^{2n} \mathrm{rk}_\mathbb{Z}(H^{2m} X) \, t^m.$$

Since $H^{\text{odd}} X = 0$, one obtains the Poincaré polynomial of X by sub-stituting $t = q^2$ in p. In particular, setting $q = -1$, so $t = 1$, the Euler characteristic of X is a_n, the number of n-dimensional cones, as we have already seen.

Poincaré duality for the complete nonsingular variety X implies

$$t^n p(t^{-1}) = p(t),$$

and this encodes nontrivial relations among the numbers a_i. The classical case is when X is projective, and Σ is the normal fan to a simple lattice polytope P. We have

$$a_i = f_{n-i} := \#(i\text{-dimensional faces of } P).$$

The h-numbers of P are defined by writing $p(t) = \sum h_k t^k$, so

$$h_k = \sum_{i=k}^{n} (-1)^{i-k} \binom{i}{k} f_{n-i-1},$$

where $f_{-1} = 1$ by convention. The identities $h_k = h_{n-k}$ express the *Dehn–Sommerville relations* for P.

8.4   Other presentations

Using an alternative description of $\mathbb{Z}[x]/I$ and the localization theorems of Chapter 7, one can give another proof of Theorem 8.3.1. For any cone $\tau \subset N_{\mathbb{R}}$, one has the sublattice $N_\tau \subseteq N$ spanned by τ, with corre-sponding quotient lattice $M \twoheadrightarrow M_\tau$. For $\gamma \subseteq \tau$, there is a corresponding projection $M_\tau \twoheadrightarrow M_\gamma$. We will write $f \mapsto f|_\gamma$ for the corresponding map $\text{Sym}^* M_\tau \to \text{Sym}^* M_\gamma$. For any rational polyhedral fan Σ in N, the ring of *piecewise polynomial functions* with respect to Σ is

$$PP^*(\Sigma) = \big\{ (f_\tau)_{\tau \in \Sigma} \mid f_\tau \in \text{Sym}^*_{\mathbb{Z}} M_\tau, \text{ and } f_\tau|_\gamma = f_\gamma \text{ for all } \gamma \subseteq \tau \big\}.$$

When Σ is a complete fan, $PP^*(\Sigma)$ is the ring of continuous functions on $N_{\mathbb{R}}$ which are given by polynomials in $\Lambda = \text{Sym}^* M$ on each maximal cone σ. The following exercise identifies this ring with $H^*_T X(\Sigma)$.

Exercise 8.4.1.   If Σ is a nonsingular complete fan, show that there are canonical ring isomorphisms

$$\mathbb{Z}[x]/I \cong PP^*(\Sigma)$$
$$\cong \big\{ (f_\sigma)_{\dim \sigma = n} \mid f_\sigma|_\tau = f_{\sigma'}|_\tau \text{ if } \tau \text{ is a facet of } \sigma \text{ and } \sigma' \big\}.$$

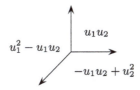

Figure 8.2 A piecewise polynomial function.

By computing the characters on the T-invariant curves $V(\tau)$, identify this last ring with the subring of

$$H_T^* X^T = \bigoplus_{\dim \sigma = n} \Lambda$$

defined by the GKM conditions. Using Corollary 7.4.3, conclude that $\mathbb{Z}[x]/I \cong PP^*(\Sigma) \cong H_T^* X$.

For example, the piecewise polynomial function shown in Figure 8.2 represents the equivariant top Chern class $c_2^T(T\mathbb{P}^2) \in H_T^* \mathbb{P}^2$.

Exercise 8.4.2. Show that the equivariant Chern class of the anti-canonical divisor, $c_1^T(\bigwedge^n TX)$, corresponds to the piecewise linear function φ_Σ which takes the value 1 on each ray of Σ.

In fact, a toric variety is nearly determined by its equivariant cohomology ring. One only needs the additional information of the Λ-algebra structure, along with the Chern class $c_1^T(\bigwedge^n TX)$.

Exercise 8.4.3. Let Σ be a complete nonsingular fan. Show how to recover Σ uniquely from the Λ-algebra $PP^*(\Sigma)$, along with the piecewise linear function φ_Σ.

There is a dual description of $H_T^* X$. For each $v \in N$, we will write $y^v \in \mathbb{Z}[N]$ for the corresponding element of the group ring; these form a $\mathbb{Z}$-basis for $\mathbb{Z}[N]$, with multiplication given by $y^v \cdot y^{v'} = y^{v+v'}$. (The element y^v may be regarded as a function on M; note that the above descriptions of $H_T^* X$ involved functions on N.) For any rational polyhedral fan Σ, the *deformed group ring* $\mathbb{Z}[N]^\Sigma$ has the same basis y^v, but multiplication is deformed so that

$$y^v \cdot y^{v'} = \begin{cases} y^{v+v'} & \text{if } v, v' \in \sigma \text{ for some cone } \sigma \in \Sigma; \\ 0 & \text{otherwise.} \end{cases}$$

Exercise 8.4.4. If Σ is a nonsingular complete fan, show that there is an isomorphism

$$\mathbb{Z}[x]/I \cong \mathbb{Z}[N]^{\Sigma},$$

and conclude that $H_T^* X \cong \mathbb{Z}[N]^{\Sigma}$.

Notes

We use (Fulton, 1993) as a general reference for basic facts about toric varieties; see also the book by Cox, Little, and Schenck (2011), which contains a short discussion of equivariant cohomology.

The results of this chapter generally hold for *simplicial fans*, where each k-dimensional cone is spanned by k vectors, if one uses $\mathbb{Q}$ coefficients in place of $\mathbb{Z}$ coefficients. The corresponding toric varieties $X(\Sigma)$ are orbifolds.

The Stanley–Reisner ring was studied by Stanley, Reisner, Hochster, and others in the 1970s. These authors usually consider simplicial complexes in general, rather than only nonsingular fans. Stanley applied Hilbert series to the combinatorics of polytopes and simplicial complexes, obtaining much finer results than the ones presented here. His book provides a fuller account of this story (Stanley, 1996).

The presentation (∗) for $H^* X$ was proved by Jurkiewicz (1980) in the projective case, and by Danilov (1978) in general. The isomorphism (∗$_T$) between the Stanley–Reisner ring and $H_T^* X$ is due to Bifet, De Concini, and Procesi (1990), and holds for any nonsingular (but possibly not complete) toric variety.

The complex $C_\bullet$ is related to the Bredon sequence for equivariant cohomology (see the notes to Chapter 7). The arguments we give are similar to those of Danilov (1978, §3), who attributes them to Kushnirenko. For general toric varieties, a version of this complex was studied by Schenck (2012).

For a nonsingular toric variety X, Brion (1996, 1997a) realized $H_T^* X$ as an algebra of piecewise polynomial functions; see also (Brion and Vergne, 1997a). For a general fan Σ, Payne (2006) showed that the ring $PP^*(\Sigma)$ is isomorphic to the (operational) equivariant Chow cohomology $A_T^* X(\Sigma)$.

Using rational coefficients, the deformed group ring $\mathbb{Q}[N]^{\Sigma}$ was introduced by Borisov, Chen, and Smith (2004), who computed orbifold cohomology of toric Deligne–Mumford stacks. A different perspective on the isomorphism $H_T^* X(\Sigma) \cong \mathbb{Z}[N]^{\Sigma}$, using the arc space of a toric variety, is explored in (Anderson and Stapledon, 2013).

Hints for Exercises

Exercise 8.4.1. Suppose $u = 0$ defines the common facet $\tau = \sigma \cap \sigma'$. Then $V(\tau)$ has character u, and the relation $f_\sigma|_\tau = f_{\sigma'}|_\tau$ is the same as requiring that u divide the difference $f_\sigma - f_{\sigma'}$.

Exercise 8.4.2. Consider a maximal cone σ, spanned by a basis $v_1, \ldots,$ v_n of N. At the corresponding fixed point p_σ, the tangent weights are given by the dual basis $u_1, \ldots, u_n$ of M, and $c_1^T(\bigwedge^n T_{p_\sigma} X) = u_1 + \cdots + u_n$.

Exercise 8.4.3. Localization provides a canonical homomorphism of Λ-algebras $f \colon PP^*(\Sigma) \hookrightarrow \bigoplus \Lambda$. (Intrinsically, the target is the normalization of $PP^*(\Sigma)$ inside its total quotient ring.) Formally index the direct summands of the target ring by "cones" σ, and consider the composition $f_\sigma \colon PP^*(\Sigma) \to \Lambda$ of f with the projection onto the σth factor. Show that the annihilator of the Λ-module $\Lambda/\mathrm{im}(f_\sigma)$ is a product of linear forms $u_1(\sigma) \cdots u_n(\sigma)$. The vanishing of this annihilator defines a hyperplane arrangement in $N_{\mathbb{R}}$, and the convex cone $\sigma \subset N_{\mathbb{R}}$ is recovered as the unique chamber where the given function $f_\sigma(\varphi_\Sigma)$ is nonnegative. Repeating this for each summand, one obtains all the maximal cones of Σ. The fact that equivariant cohomology determines a toric variety was observed by Masuda (2008), who gave a slightly different argument; see also the clarification in (Higashitani et al., 2022, Remark 2.5).

9

Schubert Calculus on Grassmannians

In Chapter 4, we computed $H_G^* Gr(d, V)$ in terms of Schur polynomials, using the tautological bundles on $Gr(d, V)$. Here we will study the geometry of this space in more detail. Our main focus is on Schubert varieties, especially ways of describing and multiplying their classes in equivariant cohomology.

9.1 Schubert cells and Schubert varieties

As in Chapter 4, we fix $d + e = n$, and consider the Grassmannian $Gr(d, V) = Gr(V, e)$. Now we also fix a flag

$$E_\bullet : E_1 \subset E_2 \subset \cdots \subset E_n = V,$$

with $\dim E_q = q$. Often we will write $E^q = E_{n-q}$, so subscripts indicate dimension, and superscripts indicate codimension in V.

Given a partition $\lambda = (e \geq \lambda_1 \geq \cdots \geq \lambda_d \geq 0)$, the *Schubert cell* $\Omega_\lambda^\circ = \Omega_\lambda^\circ(E_\bullet)$ is the set of subspaces $F \subset V$ satisfying the conditions

$$\dim(F \cap E_q) = k \ \text{ for } \ q \in [e + k - \lambda_k, e + k - \lambda_{k+1}], \ k = 0, \ldots, d.$$

Equivalently, given a subset $I = \{i_1 < \cdots < i_d\} \subset \{1, \ldots, n\}$, this is the same as defining

$$\Omega_I^\circ = \Big\{ F \ \big| \ \dim(F \cap E^{q-1}) = d - k \ \text{for} \ q \in (i_k, i_{k+1}], \ k = 0, \ldots, d \Big\}.$$

(By convention, we set $\lambda_0 = e$, $\lambda_{d+1} = 0$, $i_0 = 0$, $i_{d+1} = n + 1$.) The equivalence is by $i_k = k + \lambda_{d+1-k}$. The bijection between partitions λ

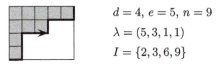

$$d = 4,\ e = 5,\ n = 9$$
$$\lambda = (5, 3, 1, 1)$$
$$I = \{2, 3, 6, 9\}$$

Figure 9.1 Partitions and k-subsets of $\{1, \ldots, n\}$.

inside the $d \times e$ rectangle and d-element subsets $I \subseteq \{1, \ldots, n\}$ can be seen graphically by recording the vertical steps when walking SW to NE along the border of λ, as shown in Figure 9.1.

Let us choose a standard basis $e_1, \ldots, e_n$ so that the fixed subspace E^q is the span of $e_{q+1}, \ldots, e_n$. The Borel subgroup $B^- \subseteq GL(V)$ preserving the flag $E_\bullet$ gets identified with lower-triangular matrices. For each partition λ (or subset I), there is a point $p_\lambda = p_I \in Gr(d, V)$, corresponding to the subspace $E_I \subseteq V$ spanned by standard basis vectors $\{e_i \mid i \in I\}$. The cell Ω_λ° can then be described as the B^--orbit of this point, so

$$\Omega_\lambda^\circ = B^- \cdot p_\lambda.$$

A simple exercise in Gaussian elimination shows that points in Ω_λ° are uniquely represented as column spans of matrices in "column echelon form" as

$$\Omega_\lambda^\circ = \Omega_I^\circ = \begin{bmatrix} 0 & 0 & 0 & 0 \\ 1 & 0 & 0 & 0 \\ 0 & 1 & 0 & 0 \\ * & * & 0 & 0 \\ * & * & 0 & 0 \\ 0 & 0 & 1 & 0 \\ * & * & * & 0 \\ * & * & * & 0 \\ 0 & 0 & 0 & 1 \end{bmatrix},$$

with the pivots appearing in rows I. This shows that Ω_λ° is an affine space of codimension $|\lambda|$ in $Gr(d, V)$, that is, $\Omega_\lambda^\circ \cong \mathbb{C}^{de - |\lambda|}$. It also shows that the Schubert cells decompose $Gr(d, V)$, that is,

$$Gr(d, V) = \coprod_\lambda \Omega_\lambda^\circ, \quad \text{union over } \lambda \subseteq \begin{array}{c} e \\ \boxed{} \end{array} d.$$

The *Schubert varieties* are

$$\Omega_\lambda = \Omega_\lambda(E_\bullet) = \overline{\Omega_\lambda^\circ} \subseteq Gr(d, V).$$

They can be described by replacing equalities by inequalities in the dimension conditions:

$$\Omega_\lambda = \Big\{ F \,\big|\, \dim(F \cap E_{e+k-\lambda_k}) \geq k \text{ for } k = 1, \ldots, d \Big\}.$$

(This is not difficult to see, but it is not obvious either!) It follows that each Schubert variety decomposes into Schubert cells:

$$\Omega_\lambda = \coprod_{\mu \supseteq \lambda} \Omega_\mu^\circ,$$

the union over partitions μ in the $d \times e$ rectangle which contain λ. As with Schubert cells, we will often write $\Omega_I = \Omega_\lambda$, when I is the d-element subset corresponding to the partition λ.

Not all the inequalities are needed to define a Schubert variety:

Exercise 9.1.1.   Show that the inequalities in the above definition of Ω_λ are equivalent to

$$\Omega_\lambda = \Big\{ F \,\big|\, \dim(F \cap E_{e+k-\lambda_k}) \geq k \text{ for } k \text{ such that } \lambda_k > \lambda_{k+1} \Big\}.$$

That is, the conditions coming from corners of the Young diagram suffice to define Ω_λ.

For example, if $\lambda = (p, \ldots, p, 0, \ldots, 0)$, with p occurring q times (so the Young diagram is a $q \times p$ rectangle), then Ω_λ is defined by the single condition $\dim(F \cap E_{e+q-p}) \geq q$.

9.2   Schubert classes and the Kempf–Laksov formula

The Schubert varieties $\Omega_\lambda(E_\bullet)$ are evidently B^--invariant subvarieties, and the Schubert cell decomposition implies their classes form a basis for cohomology.

Proposition 9.2.1.   *The classes $[\Omega_\lambda]^{B^-}$ form a basis for $H_{B^-}^* Gr(d, V)$ over $\Lambda = \Lambda_{B^-}$.*

Proof Let $X_i \subseteq Gr(d, V)$ be the union of all Ω_λ with $|\lambda| = de - i$, i.e., all Schubert varieties of dimension i. Then $X_i \setminus X_{i-1}$ is the disjoint union of Schubert cells Ω_λ° of dimension i, so the statement follows from the equivariant cell decomposition lemma (Proposition 4.7.1).   □

In Chapter 4, we saw a presentation

$$H_{B^-}^* Gr(d, V) = \Lambda_{B^-}[c_1, \dots, c_e]/(s_{d+1}, \dots, s_n),$$

with $c_k = c_k^{B^-}(\mathbb{Q})$ and $s_k = c_k^{B^-}(V - \mathbb{Q})$. Now that we have a basis of Schubert classes, a question naturally arises: How does one express $[\Omega_\lambda]^{B^-}$ in terms of the presentation, i.e., as polynomials in the Chern classes $c_k^{B^-}(\mathbb{Q})$?

To give such a formula for $[\Omega_\lambda]^{B^-}$, we introduce some notation. Given a partition $\lambda = (\lambda_1 \geq \cdots \geq \lambda_d \geq 0)$ and elements $c(1), c(2), \dots, c(d)$, where $c(i)$ is a graded series $1 + c_1(i) + c_2(i) + \cdots$, the *multi-Schur determinant* is

$$\Delta_\lambda(c) = \Delta_\lambda(c(1), \dots, c(d)) := \det\left(c_{\lambda_i + j - i}(i)\right)_{1 \leq i, j \leq d}$$

$$= \begin{vmatrix} c_{\lambda_1}(1) & c_{\lambda_1 + 1}(1) & \cdots & \\ c_{\lambda_2 - 1}(2) & c_{\lambda_2}(2) & \cdots & \\ \vdots & & \ddots & \\ & & & c_{\lambda_d}(d) \end{vmatrix}.$$

(To remember this formula, write $c_{\lambda_i}(i)$ down the diagonal, and make the subscripts increase by 1 across rows.) One may truncate zeroes in λ without changing the determinant $\Delta_\lambda(c)$. The Schur determinant considered in Chapter 4 is the case where $c = c(1) = \cdots = c(d)$.

Our first formula for equivariant Schubert classes was proved by Kempf and Laksov in the context of degeneracy loci. Special cases were found much earlier by Giambelli.

Theorem 9.2.2 (Kempf–Laksov). *For λ in the $d \times e$ rectangle, we have*

$$[\Omega_\lambda]^{B^-} = \Delta_\lambda(c(1), \dots, c(d)),$$

where $c(i) = c^{B^-}(\mathbb{Q} - E_{e+i-\lambda_i})$.

The entries $c(i)$ of the Schur determinant may be replaced by $c'(i) = \cdots = c'(k) = c^{B^-}(\mathbb{Q} - E_{e+k-\lambda_k})$ if $\lambda_i = \cdots = \lambda_k > \lambda_{k+1}$.

This follows from an easy property of multi-Schur determinants. Suppose $c(i-1) = c(i) \cdot (1+a)$, for some element a of degree 1. If $\lambda_{i-1} = \lambda_i$, then

$$\Delta_\lambda(\ldots, c(i-1), c(i), \ldots) = \Delta_\lambda(\ldots, c(i), c(i), \ldots).$$

(The left-hand side comes by adding a times the ith row to the $(i-1)$st row of the matrix on the right-hand side, and this operation leaves the determinant unchanged.)

One can prove the Kempf–Laksov formula by finding a desingularization of the locus Ω_λ and computing pushforwards; this was Kempf and Laksov's approach. In §9.4, we will give a different proof in T-equivariant cohomology (which is the same as B^--equivariant cohomology) via combinatorics of symmetric functions.

Example 9.2.3.   The formula says

$$[\Omega_\square(E_\bullet)]^{B^-} = c_1^{B^-}(\mathcal{Q} - E_e).$$

This is easy to see directly:

$$\begin{aligned}
\Omega_\square(E_\bullet) &= \Big\{ F \mid \dim(E_e \cap F) \geq 1 \Big\} \\
&= \Big\{ F \mid \mathrm{rk}(E_e \to \mathcal{Q}) < e \Big\} \\
&= \Big\{ F \mid \textstyle\bigwedge^e E_e \to \bigwedge^e \mathcal{Q} \text{ is } 0 \Big\},
\end{aligned}$$

so its class is the first Chern class of the line bundle $\mathrm{Hom}(\bigwedge^e E_e, \bigwedge^e \mathcal{Q})$, which is equal to $c_1^{B^-}(\mathcal{Q} - E_e)$. (We saw this for $Gr(2, \mathbb{C}^4)$ in Example 5.2.7.)

Example 9.2.4.   Other instances of the Kempf–Laksov formula are

$$[\Omega_\boxplus(E_\bullet)]^{B^-} = \begin{vmatrix} c_2^{B^-}(\mathcal{Q} - E_{e-1}) & c_3^{B^-}(\mathcal{Q} - E_{e-1}) \\ 1 & c_1^{B^-}(\mathcal{Q} - E_{e+1}) \end{vmatrix}$$

$$= c_2(1)\, c_1(2) - c_3(1)$$

and

$$[\Omega_\square(E_\bullet)]^{B^-} = \begin{vmatrix} c_1^{B^-}(\mathcal{Q} - E_e) & c_2^{B^-}(\mathcal{Q} - E_e) \\ 1 & c_1^{B^-}(\mathcal{Q} - E_{e+1}) \end{vmatrix}$$

$$= c_1(1)\, c_1(2) - c_2(1),$$

or alternatively,

$$[\Omega_{\boxbar}(E_\bullet)]^{B^-} = \begin{vmatrix} c_1^{B^-}(\mathbb{Q} - E_{e+1}) & c_2^{B^-}(\mathbb{Q} - E_{e+1}) \\ 1 & c_1^{B^-}(\mathbb{Q} - E_{e+1}) \end{vmatrix}$$

$$= c_1(2)\,c_1(2) - c_2(2).$$

9.3 Tangent spaces and normal spaces

In the rest of this chapter, we study the equivariant geometry of $Gr(d, \mathbb{C}^n)$ with respect to an action of a torus T by characters $\chi_1, \ldots, \chi_n$. Let $e_1, \ldots, e_n$ be a weight basis, so $z \cdot e_i = \chi(z)\,e_i$ for all $z \in T$. As we have seen before (Chapter 5, Example 5.1.10), $Gr(d, \mathbb{C}^n)$ has an open cover by T-invariant open sets, one for each $I = \{i_1 < \cdots < i_d\} \subset \{1, \ldots, n\}$. In matrix form, these open sets are represented by $n \times d$ matrices so that the submatrix on rows I is the identity matrix, and the remaining entries are free. For example, if $d = 4$, $n = 9$, and $I = \{2, 3, 6, 9\}$, the corresponding open set is shown at right. This illustrates a natural identification with the de-dimensional affine space $E_I^\vee \otimes E_{\{1, \ldots, n\} \setminus I}$, where T acts by the characters $\chi_j - \chi_i$ for $i \in I$ and $j \notin I$. There is an equivariant isomorphism between this open affine and the tangent space $T_{p_I} Gr(d, \mathbb{C}^n)$, identifying p_I with the origin $0 \in T_{p_I} Gr(d, \mathbb{C}^n)$. If

$$\begin{bmatrix} * & * & * & * \\ 1 & 0 & 0 & 0 \\ 0 & 1 & 0 & 0 \\ * & * & * & * \\ * & * & * & * \\ 0 & 0 & 1 & 0 \\ * & * & * & * \\ * & * & * & * \\ 0 & 0 & 0 & 1 \end{bmatrix}$$

all characters $\chi_1, \ldots, \chi_n$ are distinct, then all characters on each tangent space $T_{p_I} Gr(d, \mathbb{C}^n)$ are nonzero, and the fixed locus consists of the finitely many points p_I.

Comparing this with our description of the Schubert cell Ω_I°, we see that the weights of T on $T_{p_I}\Omega_I$ are $\{\chi_j - \chi_i \mid i \in I, j \notin I, i < j\}$. It follows that the weights of T on the normal space N_I to Ω_I at p_I are $\{\chi_j - \chi_i \mid i \in I, i \notin I, i > j\}$. From the self-intersection formula, this means

$$[\Omega_I]^T|_{p_I} = c_{top}^T(N_I) = \prod_{\substack{i \in I, j \notin I \\ i > j}} (\chi_j - \chi_i). \tag{$*$}$$

From the cell decomposition, we know

$$\Omega_\mu \subseteq \Omega_\lambda \quad \text{iff} \quad \mu \supseteq \lambda \quad \text{(as diagrams)}$$
$$\text{iff} \quad J \geq I \quad \text{(i.e., } j_k \geq i_k \text{ for all } k\text{)},$$

where I is the subset corresponding to the partition λ, and J is the one corresponding to μ. This means that $p_\mu = p_J$ lies in $\Omega_\lambda = \Omega_I$ if and only if $\mu \supseteq \lambda$ if and only if $J \geq I$, so

$$[\Omega_I]^T|_{p_J} = 0 \quad \text{unless} \quad J \geq I. \qquad (**)$$

Now let us assume the characters χ_i are all distinct, so the fixed points are isolated. It turns out that the two conditions $(*)$ and $(**)$ uniquely determine the class $[\Omega_I]^T$.

Lemma 9.3.1. *If a homogeneous element $\alpha \in H_T^* Gr(d, \mathbb{C}^n)$ satisfies $(*)$ and $(**)$, then $\alpha = [\Omega_I]^T$.*

Proof We have seen that $[\Omega_I]^T$ satisfies the conditions, so it is enough to show that if two classes α and α' satisfy $(*)$ and $(**)$, then they must be the same. Equivalently, we will show that $\beta = \alpha - \alpha'$ is zero.

We know $\beta|_{p_J} = 0$ unless $J \geq I$ by $(**)$. Let $K \geq I$ be a minimal element such that $\beta|_{p_K} \neq 0$. It must be that $K > I$, since $\alpha|_{p_I} = \alpha'|_{p_I}$ by $(*)$. The "GKM" divisibility conditions (Corollary 7.4.3) force $\beta|_{p_K}$ to be a multiple of

$$\prod_{\substack{i \in K, j \notin K \\ i > j}} (\chi_j - \chi_i),$$

which is equal to $[\Omega_K]^T|_{p_K}$. So $\beta|_{p_K}$ has degree at least $|\nu|$, where ν is the partition corresponding to K. Since $K > I$, we have $|\nu| > |\lambda|$, contradicting the homogeneity of the classes α and α'. $\qquad\square$

The conditions $(*)$ and $(**)$ can be regarded as an interpolation problem; the lemma says this problem has a unique solution. We will next tie this to symmetric functions, making it an explicit problem of polynomial interpolation.

9.4   Double Schur polynomials

We consider functions of two sets of variables, $x = (x_1, x_2, \ldots, x_d)$ and $y = (y_1, y_2, \ldots)$. (One can extend the y variables to be doubly infinite, allowing non-positive indices, but in practice only finitely many appear.) We define the "double monomial"

$$(x_i|y)^p = (x_i - y_1)(x_i - y_2)\cdots(x_i - y_p).$$

There are several equivalent definitions of *double Schur functions* $s_\lambda(x|y)$, generalizing corresponding definitions of the single Schur polynomials, which are recovered by setting $y = 0$.

Bialternants. Generalizing Cauchy's functions, we set

$$s_\lambda(x|y) = \frac{\left|(x_i|y)^{\lambda_j+d-j}\right|_{1\le i,j\le d}}{\left|(x_i|y)^{d-j}\right|_{1\le i,j\le d}},$$

where both determinants are $d \times d$. The numerator is an alternating function of x, and a pleasant exercise shows that the denominator is the Vandermonde

$$\left|(x_i|y)^{d-j}\right| = \left|x_i^{d-j}\right| = \prod_{i<j}(x_i - x_j),$$

so the ratio is a polynomial.

Tableaux. There is a formula in terms of semistandard Young tableaux:

$$s_\lambda(x|y) = \sum_{\mathcal{T}\in SSYT(\lambda)} \prod_{(i,j)\in\lambda} (x_{\mathcal{T}(i,j)} - y_{\mathcal{T}(i,j)+j-i}),$$

the sum over SSYT tableaux $\mathcal{T}$ of shape λ with entries in $\{1,\dots,d\}$.

For example, if $d = 3$, there are 8 semistandard Young tableaux of shape $\lambda = $ ⬚,

so the double Schur function is

$$\begin{aligned}
s_\lambda(x|y) = {}&(x_1 - y_1)(x_1 - y_2)(x_2 - y_1) + (x_1 - y_1)(x_1 - y_2)(x_3 - y_2) \\
&+ (x_1 - y_1)(x_2 - y_3)(x_2 - y_1) + (x_1 - y_1)(x_2 - y_3)(x_3 - y_2) \\
&+ (x_1 - y_1)(x_3 - y_4)(x_2 - y_1) + (x_1 - y_1)(x_3 - y_4)(x_3 - y_2) \\
&+ (x_2 - y_2)(x_2 - y_3)(x_3 - y_2) + (x_2 - y_2)(x_3 - y_4)(x_3 - y_2).
\end{aligned}$$

Jacobi–Trudi. The Jacobi–Trudi determinantal formula generalizes to

$$s_\lambda(x|y) = \left|h_{\lambda_i+j-i}(x|\tau^{1-j}y)\right|_{1\le i,j\le d},$$

where

$$h_p(x|y) = s_{(p)}(x|y) = \sum_{1 \le i_1 \le \cdots \le i_p \le d} (x_{i_1} - y_{i_1})(x_{i_2} - y_{i_2+1}) \cdots (x_{i_p} - y_{i_p+p-1}),$$

and τ is the shift operator defined by $(\tau^j y)_i = y_{i+j}$.

The main fact we need is a *vanishing theorem* for double Schur functions. Let $y_k^\lambda = y_{\lambda_{d+1-k}+k}$; or in terms of the corresponding subset I, $y_k^I = y_{i_k}$.

Lemma 9.4.1. *We have*

$$s_\lambda(y^\lambda|y) = \prod_{\substack{i \in I, j \notin I \\ i > j}} (y_j - y_i),$$

where $I \subseteq \{1, \ldots, n\}$ is the subset corresponding to λ, and

$$s_\lambda(y^\mu|y) = 0 \quad \text{if } \mu \not\supseteq \lambda.$$

Exercise 9.4.2. Prove Lemma 9.4.1.

After appropriately identifying the variables, the lemma says that double Schur functions satisfy conditions $(*)$ and $(**)$ from §9.3 – that is, the same interpolation problem solved uniquely by $[\Omega_\lambda]^T$! To make this precise, let $x_1, \ldots, x_d$ be (equivariant) Chern roots of the dual tautological bundle $\mathbb{S}^\vee$ on $Gr(d, \mathbb{C}^n)$, so

$$c_k^T(\mathbb{S}^\vee) = e_k(x_1, \ldots, x_d).$$

Then, specializing the y variables by $y_i = -\chi_i$,

$$c_k^T(\mathbb{S}^\vee)|_{p_\lambda} = c_k^T(E_I^\vee) = e_k(y_1^\lambda, \ldots, y_d^\lambda).$$

In other words, there is a commuting diagram

$$
\begin{array}{ccc}
c_k^T(\mathbb{S}^\vee) & H_T^* Gr(d, \mathbb{C}^n) & \\
\uparrow & \uparrow & \searrow^{\iota_{p_\lambda}^*} \\
c_k & \Lambda[c_1, \ldots, c_d] \longrightarrow & \Lambda. \\
\updownarrow & \updownarrow & \nearrow_{x_k \mapsto y_k^\lambda} \\
e_k(x) & \Lambda[x_1, \ldots, x_d] &
\end{array}
$$

The polynomials $s_\lambda(x|y)$ are symmetric in the x variables, so they lie in $\Lambda[c_1, \ldots, c_d]$. They satisfy $(*)$ and $(**)$ by Lemma 9.4.1, so it follows from Lemma 9.3.1 that $s_\lambda(x|y) \mapsto [\Omega_\lambda]^T$.

Invoking the Jacobi–Trudi formula, we obtain:

Corollary 9.4.3. *Evaluating $x_1, \ldots, x_d$ as equivariant Chern roots of $\mathbb{S}^\vee$, and $y_i = -\chi_i$, we have*

$$[\Omega_\lambda]^T = s_\lambda(x|y)$$
$$= \left| h_{\lambda_i + j - i}(x | \tau^{1-j} y) \right|_{1 \le i, j \le d}.$$

This proves the Kempf–Laksov formula (Theorem 9.2.2), once one knows the entries of the matrices are identical.

Exercise 9.4.4. With the specializations as in Corollary 9.4.3, show that $c^T_{\lambda_i + j - i}(\mathbb{Q} - E_{e+i-\lambda_i}) = h_{\lambda_i + j - i}(x | \tau^{1-j} y)$.

9.5 Poincaré duality

We have seen one basis for $H^*_T Gr(d, \mathbb{C}^n)$, the Schubert classes $\sigma_\lambda = [\Omega_\lambda]^T$. Our next goal is to describe the Poincaré dual basis. Let $\widetilde{E}_\bullet$ be the *opposite flag* to $E_\bullet$, so if E_k is spanned by $e_{n+1-k}, \ldots, e_n$, then $\widetilde{E}_k$ is spanned by $e_1, \ldots, e_k$. The flag $\widetilde{E}_\bullet$ is fixed by the Borel group B, which in this basis is the set of upper-triangular matrices in GL_n.

The *opposite Schubert cells* and *varieties* are defined as before, but with respect to the flag $\widetilde{E}_\bullet$:

$$\widetilde{\Omega}^\circ_\lambda := \Omega^\circ_\lambda(\widetilde{E}_\bullet) \qquad \text{and} \qquad \widetilde{\Omega}_\lambda := \Omega_\lambda(\widetilde{E}_\bullet).$$

These are B-invariant, so also T-invariant. To identify the T-fixed points contained in $\widetilde{\Omega}_\lambda$, it will help to introduce some more notation. Let $\lambda^\vee$ be the complement to λ in the $d \times e$ rectangle, also called the *dual partition*. In formulas, this is $\lambda^\vee_k = e - \lambda_{d+1-k}$. Let $I^\vee \subseteq \{1, \ldots, n\}$ be the corresponding d-element subset, so $I^\vee = \{i^\vee_1 < \cdots < i^\vee_d\}$, with $i^\vee_k = n + 1 - i_{d+1-k}$. This can be seen by reading the border of $\lambda \subseteq$ ⬚ d in the opposite direction, from NE to SW, as illustrated below.

$$\lambda = (5,3,1,1) \qquad \lambda^\vee = (4,4,2,0)$$

$$I = \{2,3,6,9\} \qquad I^\vee = \{1,4,7,8\}$$

Exercise 9.5.1. Verify that $p_I = p_\lambda$ is the unique T-fixed point in $\widetilde{\Omega}^\circ_{I^\vee} = \widetilde{\Omega}^\circ_{\lambda^\vee}$, so $\widetilde{\Omega}^\circ_{\lambda^\vee} = B \cdot p_\lambda$.

For example, with $d = 4$, $n = 9$, and $I = \{2,3,6,9\}$, we have

$$
\Omega^\circ_I =
\begin{bmatrix}
0 & 0 & 0 & 0 \\
1 & 0 & 0 & 0 \\
0 & 1 & 0 & 0 \\
* & * & 0 & 0 \\
* & * & 0 & 0 \\
0 & 0 & 1 & 0 \\
* & * & * & 0 \\
* & * & * & 0 \\
0 & 0 & 0 & 1
\end{bmatrix}
\quad \text{and} \quad
\widetilde{\Omega}^\circ_{I^\vee} =
\begin{bmatrix}
* & * & * & * \\
1 & 0 & 0 & 0 \\
0 & 1 & 0 & 0 \\
0 & 0 & * & * \\
0 & 0 & * & * \\
0 & 0 & 1 & 0 \\
0 & 0 & 0 & * \\
0 & 0 & 0 & * \\
0 & 0 & 0 & 1
\end{bmatrix},
$$

both inside the T-invariant affine neighborhood of p_I. The pivot 1's in $\widetilde{\Omega}^\circ_{I^\vee}$ are in the rows indicated by $I^\vee$, but read from bottom to top. So

$$\operatorname{codim} \Omega_\lambda = \dim \widetilde{\Omega}_{\lambda^\vee} = |\lambda|,$$

while

$$\dim \Omega_\lambda = \operatorname{codim} \widetilde{\Omega}_{\lambda^\vee} = |\lambda^\vee|.$$

The correspondence $\lambda \leftrightarrow \lambda^\vee$ reverses inclusions. Since the opposite Schubert variety decomposes as $\widetilde{\Omega}_\lambda = \coprod_{\mu \supseteq \lambda} \widetilde{\Omega}^\circ_\mu$ and $p_{\mu^\vee} \in \widetilde{\Omega}^\circ_\mu$, we see

$$p_\mu \in \widetilde{\Omega}_\lambda \quad \text{iff} \quad \mu^\vee \supseteq \lambda \quad \text{iff} \quad \mu \subseteq \lambda^\vee.$$

Proposition 9.5.2. *Let $\sigma_\lambda = [\Omega_\lambda]^T$ and $\widetilde{\sigma}_\lambda = [\widetilde{\Omega}_\lambda]^T$. Then $\{\widetilde{\sigma}_{\lambda^\vee}\}$ is the Poincaré dual basis to $\{\sigma_\lambda\}$.*

Proof We must show

$$\rho_*(\sigma_\lambda \cdot \widetilde{\sigma}_\mu) = \begin{cases} 1 & \text{if } \mu = \lambda^\vee; \\ 0 & \text{otherwise.} \end{cases}$$

When $\mu = \lambda^\vee$, the varieties Ω_λ and $\widetilde{\Omega}_{\lambda^\vee}$ meet transversally in the single point p_λ. In general, the above analysis of fixed points shows that

$$(\Omega_\lambda \cap \widetilde{\Omega}_\mu)^T = \{ p_\nu \,|\, \mu^\vee \supseteq \nu \supseteq \lambda \}.$$

The fact that $\Omega_\lambda \cap \widetilde{\Omega}_{\lambda^\vee}$ is transverse is apparent from a computation of tangent spaces – say, by using matrix descriptions of Ω_I° and $\widetilde{\Omega}_{I^\vee}^\circ$. One sees as before that $T_{p_\lambda} \widetilde{\Omega}_{\lambda^\vee}$ has weights $\{ \chi_j - \chi_i \,|\, i \in I, j \notin I, i > j \}$, so that

$$T_{p_\lambda} Gr(d, \mathbb{C}^n) = T_{p_\lambda} \Omega_\lambda \oplus T_{p_\lambda} \widetilde{\Omega}_{\lambda^\vee}.$$

This shows $\rho_*(\sigma_\lambda \cdot \widetilde{\sigma}_{\lambda^\vee}) = 1$.

If $\mu \neq \lambda^\vee$, there are two possibilities to consider. First, suppose $\mu \not\subseteq \lambda^\vee$, so $\mu^\vee \not\supseteq \lambda$. Then $(\Omega_\lambda \cap \widetilde{\Omega}_\mu)^T = \emptyset$, so the intersection is empty. (Any nonempty projective variety has a T-fixed point, by Borel's fixed point theorem.) So $\rho_*(\sigma_\lambda \cdot \widetilde{\sigma}_\mu) = 0$ in this case.

On the other hand, suppose $\mu \subsetneq \lambda^\vee$, so $\mu^\vee \supsetneq \lambda$. Then $|\mu^\vee| - |\lambda| > 0$. But this means $de - |\mu| - |\lambda| > 0$, that is, $|\mu| + |\lambda| - de < 0$. Since $\rho_*(\sigma_\lambda \cdot \widetilde{\sigma}_\mu) \in \Lambda_T^{2(|\mu|+|\lambda|-de)} = 0$, we are done. $\qquad\square$

We obtain a formula for the class of an opposite Schubert variety by replacing $E_\bullet$ by $\widetilde{E}_\bullet$ in the Kempf–Laksov formula.

Theorem 9.5.3. *In $H_T^* Gr(d, \mathbb{C}^n)$, we have*

$$[\widetilde{\Omega}_\lambda]^T = \Delta_\lambda(\widetilde{c}(1), \dots, \widetilde{c}(s))$$
$$= s_\lambda(x|\widetilde{y}),$$

where $\widetilde{c}(k) = c^T(\mathbb{Q} - \widetilde{E}_{e+k-\lambda_k})$, $x_1, \dots, x_d$ are equivariant Chern roots of $\mathbb{S}^\vee$, and $\widetilde{y}_k = y_{n+1-k} = -\chi_{n+1-k}$.

9.6 Multiplication

A major goal of equivariant Schubert calculus is to describe the coefficients $c_{\lambda\mu}^\nu$ appearing in the expansion

$$\sigma_\lambda \cdot \sigma_\mu = \sum_\nu c_{\lambda\mu}^\nu \, \sigma_\nu,$$

in $H_T^* Gr(d, \mathbb{C}^n)$. The same problem can be posed for other flag varieties, but Grassmannians are one of the very few cases where a complete and satisfying answer is known.

We will prove some basic facts about these coefficients, assuming throughout that the characters $\chi_1, \ldots, \chi_n$ are distinct. (The general case can be obtained from this one, by specializing the χ_i's.) Evidently, $c_{\lambda\mu}^\nu$ is a homogeneous polynomial of degree $|\lambda| + |\mu| - |\nu|$.

Lemma 9.6.1. *The coefficients $c_{\lambda\mu}^\nu$ satisfy the following properties:*

(i) $c_{\lambda\mu}^\nu = 0$ *unless* $\lambda \subseteq \nu$ *and* $\mu \subseteq \nu$.

(ii) $c_{\lambda\mu}^\mu = \sigma_\lambda|_\mu = [\Omega_\lambda]^T|_{p_\mu}$.

(iii) $c_{\lambda\lambda}^\lambda = \displaystyle\prod_{\substack{i \in I, j \notin I \\ i > j}} (\chi_j - \chi_i)$.

Proof For (i), the cell decomposition lemma shows that restrictions of classes σ_α for $\alpha \not\supseteq \lambda$ form a basis for $H^*(X \setminus \Omega_\lambda)$, because these are classes of the Schubert varieties Ω_α not contained in Ω_λ. The class σ_λ maps to 0 under $H_T^* X \to H_T^*(X \setminus \Omega_\lambda)$ – as one can see by using the long exact sequence of Borel–Moore homology – so $\sigma_\lambda \cdot \sigma_\mu$ also maps to 0, for any μ. It follows that $\sigma_\lambda \cdot \sigma_\mu = \sum_\nu c_{\lambda\mu}^\nu \sigma_\nu$ involves only those ν such that $\nu \supseteq \lambda$. By symmetry, one concludes that $\nu \supseteq \mu$, as well.

For (ii), we restrict the equation defining $c_{\lambda\mu}^\nu$ to p_μ, obtaining

$$\sigma_\lambda|_\mu \cdot \sigma_\mu|_\mu = \sum_\nu c_{\lambda\mu}^\nu \, \sigma_\nu|_\mu.$$

We know $\sigma_\nu|_\mu = 0$ unless $p_\mu \in \Omega_\nu$, that is, $\mu \supseteq \nu$; but by (i), we also know $c_{\lambda\mu}^\nu = 0$ unless $\mu \subseteq \nu$. The only term surviving on the right-hand side is $\nu = \mu$, so we find $\sigma_\lambda|_\mu \cdot \sigma_\mu|_\mu = c_{\lambda\mu}^\mu \sigma_\mu|_\mu$. We found the formula for $\sigma_\mu|_\mu$ in §9.3, and this is not a zerodivisor since all χ_i are distinct. Canceling these factors gives the claimed formula for $c_{\lambda\mu}^\mu$.

Formula (iii) follows from (ii), using the formula for $\sigma_\lambda|_\lambda$. $\qquad\square$

The *Chevalley–Pieri formula* gives a rule for multiplication by the divisor class $\sigma_\square$. The classical (non-equivariant) version says that in $H^* X$,

$$\sigma_\square \cdot \sigma_\lambda = \sum_{\lambda+} \sigma_{\lambda+},$$

the sum over all partitions λ^+ obtained from λ by adding one box. For example, in $H^*Gr(3, \mathbb{C}^7)$ we have

$$\sigma_\square \cdot \sigma_{(3,2)} = \sigma_{(4,2)} + \sigma_{(3,3)} + \sigma_{(3,2,1)}.$$

We will prove a very general version of this formula in Chapter 19. The reader may enjoy proving the corresponding formula for multiplying a Schur polynomial by $h_1 = x_1 + x_2 + \cdots + x_d$. (As usual, references are in the Notes.)

To state the equivariant version precisely, we need another formula:

$$\sigma_\square|_\lambda = \sum_{j \notin I} \chi_j - \sum_{i=d+1}^{n} \chi_i$$

$$= \sum_{j=1}^{d} \chi_j - \sum_{i \in I} \chi_i.$$

This follows from $\sigma_\square = c_1^T(\mathbb{Q} - E_e)$, using $c_1^T(\mathbb{Q})|_{p_\lambda} = \sum_{j \notin I} \chi_j$. The second line makes it clear that the formula is independent of n. Note that $\sigma_\square|_\lambda \neq \sigma_\square|_\mu$ if $\lambda \neq \mu$.

Theorem 9.6.2 (Equivariant Chevalley–Pieri). *In $H_T^* X$, we have*

$$\sigma_\square \cdot \sigma_\lambda = \sum_{\lambda^+} \sigma_{\lambda^+} + \left(\sum_{j=1}^{d} \chi_j - \sum_{i \in I} \chi_i \right) \sigma_\lambda,$$

the sum over λ^+ obtained from λ by adding one box.

Proof The sum is from the nonequivariant case; the equivariant coefficients must agree by degree. The other term has coefficient $c_{\square\lambda}^\lambda = \sigma_\square|_\lambda$ by Lemma 9.6.1(iii). No other terms appear, since $c_{\square\lambda}^\nu$ is nonzero only for $|\nu| \leq |\lambda| + 1$ and $\nu \supseteq \lambda$, by Lemma 9.6.1(i). $\square$

Remarkably, the equivariant Chevalley rule determines all structure constants $c_{\lambda\mu}^\nu$ for $H_T^* Gr(d, \mathbb{C}^n)$, and hence also for $H^* Gr(d, \mathbb{C}^n)$. This is far from true of the non-equivariant rule: $H^* Gr(d, \mathbb{C}^n)$ is not generated by the divisor class. A general reason for this phenomenon was given in §7.1. Here we will see an algorithmic proof.

First, we need some more formulas.

Lemma 9.6.3. *We have*

$$(\sigma_\square|_\lambda - \sigma_\square|_\mu)\, c_{\lambda\mu}^\lambda = \sum_{\mu^+} c_{\lambda\mu^+}^\lambda, \tag{i}$$

the sum over μ^+ obtained by adding one box to μ, and

$$(\sigma_\square|_\nu - \sigma_\square|_\lambda)\, c_{\lambda\mu}^\nu = \sum_{\lambda^+} c_{\lambda^+\mu}^\nu - \sum_{\nu^-} c_{\lambda\mu}^{\nu^-}, \tag{ii}$$

the sums over λ^+ obtained by adding one box to λ, and ν^- obtained by removing one box from ν.

Proof For (i), using the formula $c_{\mu\lambda}^\lambda = \sigma_\mu|_\lambda$ together with commutativity, the left-hand side is $(\sigma_\square|_\lambda - \sigma_\square|_\mu)\sigma_\mu|_\lambda$, while the right-hand side is $\sum \sigma_{\mu^+}|_\lambda$. So (i) results from restricting the equivariant Chevalley–Pieri formula to p_λ.

For (ii), we will use associativity. By Chevalley–Pieri,

$$\sigma_\square \cdot (\sigma_\lambda \cdot \sigma_\mu) = \sum_\nu c_{\lambda\mu}^\nu \sigma_\square \cdot \sigma_\nu$$

$$= \sum_{\nu^+} c_{\lambda\mu}^\nu \sigma_{\nu^+} + \sum_\nu c_{\lambda\mu}^\nu (\sigma_\square|_\nu)\sigma_\nu,$$

and

$$(\sigma_\square \cdot \sigma_\lambda) \cdot \sigma_\mu = \sum_{\lambda^+} \sigma_{\lambda^+} \cdot \sigma_\mu + (\sigma_\square|_\lambda)\sigma_\lambda \cdot \sigma_\mu$$

$$= \sum_{\lambda^+} \sum_\nu c_{\lambda^+\mu}^\nu \sigma_\nu + (\sigma_\square|_\lambda) \sum_\nu c_{\lambda\mu}^\nu \sigma_\nu.$$

One obtains (ii) by equating coefficients of σ_ν. $\square$

Theorem 9.6.4 (Molev–Sagan). *The polynomials $c_{\lambda\mu}^\nu$ in $\Lambda^{2(|\lambda|+|\mu|-|\nu|)}$ satisfy and are determined by the following properties:*

$$c_{\lambda\lambda}^\lambda = \sigma_\lambda|_\lambda = \prod_{\substack{i\in I, j\notin I \\ i>j}} (\chi_j - \chi_i), \tag{i}$$

$$(\sigma_\square|_\lambda - \sigma_\square|_\mu)\, c_{\lambda\mu}^\lambda = \sum_{\mu^+} c_{\lambda\mu^+}^\lambda, \tag{ii}$$

and

$$(\sigma_\square|_\nu - \sigma_\square|_\lambda)\, c^\nu_{\lambda\mu} = \sum_{\lambda^+} c^\nu_{\lambda^+\mu} - \sum_{\nu^-} c^{\nu^-}_{\lambda\mu}. \tag{iii}$$

Proof We have seen that (i)–(iii) hold. To prove that they uniquely characterize the coefficients $c^\nu_{\lambda\mu}$, we proceed by induction. Suppose $d^\nu_{\lambda\mu}$ are any polynomials satisfying (i)–(iii). We know $d^\lambda_{\lambda\lambda} = c^\lambda_{\lambda\lambda}$, since this is the explicit formula (i).

Next, $d^\lambda_{\lambda\mu} = \sigma_\mu|_\lambda = c^\lambda_{\lambda\mu}$, by induction on $|\lambda| - |\mu|$: the base case is where $\lambda = \mu$, and is done by (i); for $\lambda \supsetneq \mu$, use formula (ii) and induction. (On the left-hand side of (ii), the factor $(\sigma_\square|_\lambda - \sigma_\square|_\mu)$ is nonzero for $\lambda \neq \mu$. Terms on the right-hand side of (ii) have $|\lambda| - |\mu^+| < |\lambda| - |\mu|$.)

Finally, use (iii) and induction on $|\nu| - |\lambda|$ to get $d^\nu_{\lambda\mu} = c^\nu_{\lambda\mu}$. All terms on the right-hand side of (iii) have $|\nu| - |\lambda^+| < |\nu| - |\lambda|$ and $|\nu^-| - |\lambda| < |\nu| - |\lambda|$. This reduces to the base case $\lambda = \nu$, which was handled previously. □

Remark. By setting $\nu = \mu$ in (iii), one obtains

$$(\sigma_\square|_\mu - \sigma_\square|_\lambda)\, c^\mu_{\lambda\mu} = \sum_{\lambda^+} c^\mu_{\lambda^+\mu}, \tag{ii$'$}$$

since the coefficients $c^{\mu^-}_{\lambda\mu}$ vanish. Using commutativity $(c^\nu_{\lambda\mu} = c^\nu_{\mu\lambda})$ and interchanging λ and μ, one recovers (ii) from (ii$'$). The conditions (i), (ii$'$), and (iii) also characterize $c^\nu_{\lambda\mu}$.

9.7 Grassmann duality

In Chapter 4 we noted the canonical isomorphisms

$$Gr(d, V) = Gr(V, e) = Gr(V^\vee, d) = Gr(e, V^\vee),$$

where $d + e = n = \dim V$, by identifications

$$[F \subseteq V] \leftrightarrow [V \twoheadrightarrow V/F] \leftrightarrow [V^\vee \twoheadrightarrow F^\vee] \leftrightarrow [(V/F)^\vee \subseteq V^\vee].$$

These are equivariant for any group G acting linearly on V, and by the dual representation on $V^\vee$.

To see this in matrices, we fix a basis, so $V \cong \mathbb{C}^n \cong V^\vee$. A point of $Gr(d, V)$ is the image of an embedding $[\mathbb{C}^d \hookrightarrow \mathbb{C}^n]$, so it is represented as the column span of a full-rank matrix A of size $n \times d$. A point of $Gr(V, e)$ is an isomorphism class of quotients $[\mathbb{C}^n \twoheadrightarrow \mathbb{C}^e]$, represented by a full-rank matrix B of size $e \times n$. Dually, a point of $Gr(V^\vee, d)$ is a quotient represented by the transposed matrix $A^\dagger$, and a point of $Gr(e, V^\vee)$ is the column span of $B^\dagger$ (that is, the row span of B).

With this notation, the Grassmann duality isomorphism is

$$\gamma \colon Gr(d, \mathbb{C}^n) \to Gr(e, \mathbb{C}^n),$$
$$F \mapsto \ker(A^\dagger) = \operatorname{im}(B^\dagger).$$

The group GL_n acts on $Gr(d, \mathbb{C}^n)$ by $F \mapsto g \cdot F$, which sends $A \mapsto g \cdot A$ and $B \mapsto B \cdot g^{-1}$. Grassmann duality is therefore equivariant with respect to the group automorphism $\varphi \colon GL_n \to GL_n$, $\varphi(g) = (g^\dagger)^{-1}$. If $T \to GL_n$ is a homomorphism given by characters $\chi_1, \ldots, \chi_n$, then the algebra homomorphism $H_T^* Gr(e, \mathbb{C}^n) \to H_T^* Gr(d, \mathbb{C}^n)$ intertwines the automorphism of $\Lambda_T = \operatorname{Sym}^* M$ induced by $\chi \mapsto -\chi$ for all $\chi \in M$.

Exercise 9.7.1. For λ in the $d \times e$ rectangle, show that $\gamma(\Omega_\lambda) = \widetilde{\Omega}_{\lambda'}$, where λ' is the conjugate partition (i.e., its diagram is the transpose of that of λ).

Duality exchanges the exact sequences

$$0 \to \mathbb{S} \to \mathbb{C}_{Gr}^n \to \mathbb{Q} \to 0$$

and

$$0 \to \mathbb{Q}^\vee \to (\mathbb{C}_{Gr}^n)^\vee \to \mathbb{S}^\vee \to 0.$$

Together with Exercise 9.7.1, this implies a dual Kempf–Laksov formula for Schubert classes:

Corollary 9.7.2. *Let* $x_1, \ldots, x_d$ *be equivariant Chern roots of* $\mathbb{S}^\vee$, *and* $\widetilde{x}_1, \ldots, \widetilde{x}_e$ *equivariant Chern roots of* $\mathbb{Q}$. *For a partition* λ *in the* $d \times e$ *rectangle, we have*

$$\sigma_\lambda = s_\lambda(x|y)$$
$$= s_{\lambda'}(\widetilde{x}|-\widetilde{y}),$$

and

$$\widetilde{\sigma}_\lambda = s_\lambda(x|\widetilde{y})$$
$$= s_{\lambda'}(\widetilde{x}|-y),$$

where T acts on $Gr(d, \mathbb{C}^n)$ by characters $\chi_i = -y_i = -\widetilde{y}_{n+1-i}$.

Expressed as multi-Schur determinants in Chern classes of $\mathbb{S}^\vee$ and $\mathbb{Q}$, these formulas translate into

$$\sigma_\lambda = \Delta_\lambda(c) = \Delta_{\lambda'}(\widetilde{c}')$$

and

$$\widetilde{\sigma}_\lambda = \Delta_\lambda(\widetilde{c}) = \Delta_{\lambda'}(c'),$$

where

$$c(i) = c^T(\mathbb{Q} - E_{e+i-\lambda_i}),$$
$$c'(i) = c^T(\mathbb{S}^\vee - E^\vee_{d+i-\lambda'_i}),$$
$$\widetilde{c}(i) = c^T(\mathbb{Q} - \widetilde{E}_{e+i-\lambda_i}), \quad \text{and}$$
$$\widetilde{c}'(i) = c^T(\mathbb{S}^\vee - \widetilde{E}^\vee_{d+i-\lambda'_i}).$$

(Recall that $\Delta_\lambda(c(E)) = s_{\lambda'}(x_1, \ldots, x_n)$ when $x_1, \ldots, x_n$ are Chern roots of E.)

This lets us prove a refinement of the Cauchy identity used in §4.6.

Corollary 9.7.3. *Let $\delta\colon Gr(d, \mathbb{C}^n) \to Gr(d, \mathbb{C}^n) \times Gr(d, \mathbb{C}^n)$ be the diagonal embedding. Then*

$$\delta_*(1) = \sum_\lambda \Delta_\lambda(c) \times \Delta_{(\lambda^\vee)'}(c').$$

(The partition $(\lambda^\vee)'$ is what we called the *complement* to λ in Chapter 4.)

Proof Use the Kempf–Laksov formulas for Schubert classes, together with the decomposition

$$\delta_*(1) = \sum_\lambda \sigma_\lambda \times \widetilde{\sigma}_{\lambda^\vee}$$

of the diagonal into Poincaré dual classes. $\qquad\square$

The same statement holds, without change, for equivariant Grassmann bundles $\mathbf{Gr}(d, V) \to Y$, so long as the vector bundle $V \to Y$ admits opposite flags $E_\bullet$ and $\widetilde{E}_\bullet$.

Writing $(\mathbb{S}^\vee)^{(1)}$ and $\mathbb{Q}^{(2)}$ for the tautological bundles from the first and second factors of $Gr(d, \mathbb{C}^n) \times Gr(d, \mathbb{C}^n)$, and $x_1, \ldots, x_d$ and $\widetilde{x}_1, \ldots, \widetilde{x}_e$ for their respective Chern roots, the Corollary expresses an equality

$$\prod_{i=1}^{d}\prod_{j=1}^{e}(x_i + \widetilde{x}_j) = \sum_{\lambda} s_\lambda(x|y) \cdot s_{(\lambda^\vee)'}(\widetilde{x}|-y)$$

in $H_T^*(Gr(d, \mathbb{C}^n) \times Gr(d, \mathbb{C}^n))$, or in $H_T^*(\mathbf{Gr}(d, V) \times_Y \mathbf{Gr}(d, V))$.

Exercise 9.7.4. Let $c_{\lambda\mu}^\nu \in \Lambda$ be the coefficient defined by

$$\sigma_\lambda \cdot \sigma_\mu = \sum_\nu c_{\lambda\mu}^\nu \sigma_\nu \qquad \text{in } H_T^* Gr(d, \mathbb{C}^n),$$

as before, and consider similar coefficients defined by

$$\widetilde{\sigma}_\lambda \cdot \widetilde{\sigma}_\mu = \sum_\nu \widetilde{c}_{\lambda\mu}^\nu \widetilde{\sigma}_\nu \qquad \text{in } H_T^* Gr(d, \mathbb{C}^n),$$

and

$$\sigma_{\lambda'} \cdot \sigma_{\mu'} = \sum_{\nu'} c_{\lambda'\mu'}^{\nu'} \sigma_{\nu'} \qquad \text{in } H_T^* Gr(e, \mathbb{C}^n).$$

Show that $c_{\lambda\mu}^\nu$ maps to $\widetilde{c}_{\lambda\mu}^\nu$ under the substitution $\chi_i \mapsto \chi_{n+1-i}$, and $c_{\lambda\mu}^\nu$ maps to $c_{\lambda'\mu'}^{\nu'}$ under $\chi_i \mapsto -\chi_{n+1-i}$.

For example, using $\sigma_\lambda = s_\lambda(x|y)$ and $y_i = -\chi_i$, one computes the product

$$\sigma_{(2)} \cdot \sigma_{(3,1)} = \sigma_{(4,2)} + \sigma_{(3,3)} + (\chi_1 + \chi_3 - \chi_5 - \chi_6)\,\sigma_{(4,1)}$$
$$+ (\chi_1 - \chi_5)\,\sigma_{(3,2)} + (\chi_1 - \chi_5)(\chi_3 - \chi_5)\,\sigma_{(3,1)}$$

in $H_T^* Gr(2, \mathbb{C}^6)$. Compare this with

$$\widetilde{\sigma}_{(2)} \cdot \widetilde{\sigma}_{(3,1)} = \widetilde{\sigma}_{(4,2)} + \widetilde{\sigma}_{(3,3)} + (\chi_6 + \chi_4 - \chi_2 - \chi_1)\,\widetilde{\sigma}_{(4,1)}$$
$$+ (\chi_6 - \chi_2)\,\widetilde{\sigma}_{(3,2)} + (\chi_6 - \chi_2)(\chi_4 - \chi_2)\,\widetilde{\sigma}_{(3,1)}$$

and

$$\sigma_{(1,1)} \cdot \sigma_{(2,1,1)} = \sigma_{(2,2,1,1)} + \sigma_{(2,2,2)} + (\chi_1 + \chi_2 - \chi_4 - \chi_6)\,\sigma_{(2,1,1,1)}$$
$$+ (\chi_2 - \chi_6)\,\sigma_{(2,2,1)} + (\chi_2 - \chi_4)(\chi_2 - \chi_6)\,\sigma_{(2,1,1,1)}$$

in $H_T^* Gr(2, \mathbb{C}^6)$ and $H_T^* Gr(4, \mathbb{C}^6)$, respectively.

9.8   Littlewood–Richardson rules

The ultimate goal is to find a positive formula for the coefficients $c_{\lambda\mu}^{\nu}$. Such a formula is often called a *Littlewood–Richardson rule*. Here we will state several of these rules, without proof.

In the nonequivariant case, the meaning of positivity is clear: $c_{\lambda\mu}^{\nu}$ is a nonnegative integer, and this Littlewood–Richardson rule is classical algebraic combinatorics. With $|\lambda| + |\mu| = |\nu|$, the coefficient $c_{\lambda\mu}^{\nu}$ is the number of ways to fill the boxes of the skew diagram ν/λ with μ_1 1's, μ_2 2's, etc., so that

(a) the filling is weakly increasing along rows;

(b) the filling is strictly increasing down columns; and

(c) when the filling is read from right to left along rows, starting at the top, at each step one has

$$\#(1\text{'s}) \geq \#(2\text{'s}) \geq \cdots .$$

Conditions (a) and (b) say the filling is a semistandard Young tableau on the shape ν/λ. Condition (c), sometimes called the "Yamanouchi word" condition, means that the partition μ grows by reading the filling (in the indicated order), placing a box in the ith row when one reads an entry "i", and each intermediate step is also a partition.

Example 9.8.1.   Let $\lambda = (2, 1, 1)$, $\mu = (3, 2, 1)$, $\nu = (4, 3, 2, 1)$. There are three fillings of ν/λ satisfying the conditions:

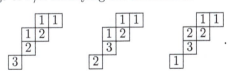

So $c_{\lambda\mu}^{\nu} = 3$. The corresponding reading words—$1\,1\,2\,1\,2\,3$, $1\,1\,2\,1\,3\,2$, and $1\,1\,2\,2\,3\,1$ – satisfy the Yamanouchi condition.

There are many other versions of the Littlewood–Richardson rule. Some have equivariant analogues. In this context, a "positive" formula should express the polynomial $c_{\lambda\mu}^{\nu}$ as a weighted enumeration of some combinatorial set, with weights of the form $\prod(\chi_i - \chi_j)$ for $i < j$. Indeed, the formulas we have seen for special cases have this property, and a general theorem of Graham guarantees that this is always possible. We will return to this in Chapter 19.

Here is one version, due to Kreiman and Molev (working independently). In the statement, reading in *column order* means that entries of a filling of λ are read along columns, from bottom to top, starting at the left.

Theorem 9.8.2. *The structure constants for multiplication in $H_T^* Gr$ $(d, \mathbb{C}^n)$ are given by*

$$c_{\lambda\mu}^{\nu} = \sum_R \sum_{\mathcal{T}} \prod_{(i,j)\in\lambda} (\chi_{e+\mathcal{T}(i,j)-\rho(i,j)_{\mathcal{T}(i,j)}} - \chi_{e+\mathcal{T}(i,j)-(j-i)}),$$

where:

R runs over all sequences

$$\mu = \rho^{(0)} \subset \rho^{(1)} \subset \cdots \subset \rho^{(s)} = \nu$$

such that $\rho^{(i)}$ is obtained by adding one box to $\rho^{(i-1)}$, in row r_i (so $s = |\nu/\mu|$).

$\mathcal{T}$ runs over "reverse barred ν-bounded tableaux" on the shape λ. This means:

- *$\mathcal{T}$ is a filling of the boxes of λ using entries from $\{1,\ldots,d\}$, weakly decreasing along rows and strictly decreasing down columns;*
- *all entries in the jth column of λ are less than or equal to the number of boxes in the jth column of ν, that is, $\mathcal{T}(i,j) \le \nu'_j$;*
- *$s = |\nu/\mu|$ of the entries are marked with a bar. When these entries are read in column order, the resulting word is $\bar{r}_1 \bar{r}_2 \cdots \bar{r}_s$. Thus each barred entry corresponds to a partition $\rho^{(i)}$.*

The product is over all boxes $(i,j) \in \lambda$ such that $\mathcal{T}(i,j)$ is unbarred. If (i,j) is a box with an unbarred entry, $\rho(i,j) = \rho^{(t)}$ is the partition corresponding to the previous barred entry (in column order). If there are no previous barred entries, $\rho(i,j) = \rho^{(0)} = \mu$.

Furthermore, $\rho(i,j)_{\mathcal{T}(i,j)} > j - i$ for all unbarred boxes (i,j).

Example 9.8.3. For $d = 3, n = 6$ (so $e = 6-3 = 3$) and $\lambda = \mu = (2,1)$, $\nu = (3,1,1)$, there are two sequences R:

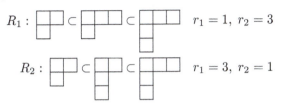

$$R_1: \quad\subset\quad\subset\quad r_1 = 1, \ r_2 = 3$$

$$R_2: \quad\subset\quad\subset\quad r_1 = 3, \ r_2 = 1$$

There is only one tableau for the sequence R_1:

$$\begin{array}{|c|c|}\hline \bar{3} & 1 \\\hline \bar{1} \\\cline{1-1}\end{array} \quad \chi_{3+1-3} - \chi_{3+1+1-2} = \chi_1 - \chi_3.$$

(In this case, $\rho(1,2) = (3,1,1)$, so $\rho(1,2)_{T(1,2)} = 3$.) For R_2, there are two tableaux:

$$\begin{array}{|c|c|}\hline \bar{3} & \bar{1} \\\hline 1 \\\cline{1-1}\end{array} \quad \chi_{3+1-2} - \chi_{3+1+2-1} = \chi_2 - \chi_5 \quad \text{and}$$

$$\begin{array}{|c|c|}\hline \bar{3} & \bar{1} \\\hline 2 \\\cline{1-1}\end{array} \quad \chi_{3+2-1} - \chi_{3+2+2-1} = \chi_4 - \chi_6.$$

(For the first of these, $\rho(2,1) = (2,1)$, so $\rho(2,1)_{T(2,1)} = 2$. For the second, $\rho(2,1) = (2,1)$, and $\rho(2,1)_{T(2,1)} = 1$.) So the rule says $c^\nu_{\lambda\mu} = \chi_4 - \chi_6 + \chi_2 - \chi_5 + \chi_1 - \chi_3$.

Historically, the first positive rule for $c^\nu_{\lambda\mu}$ was given by Knutson and Tao, and involves the combinatorics of *puzzles*. To describe it, we use another encoding of Schubert classes in $Gr(d, \mathbb{C}^n)$. Recall that a partition λ corresponds to a d-element subset $I \subseteq \{1,\ldots,n\}$. The 01-*sequence* corresponding to λ has 1's in positions I, and 0's elsewhere. For example, with $d = 2$ and $n = 5$, the partition $\lambda = (2,0)$ has $I = \{1,4\}$ and 01-sequence $1\,0\,0\,1\,0$.

To compute $c^\nu_{\lambda\mu}$, we label the boundary of an equilateral triangle by 01-sequences corresponding to three partitions λ, μ, and ν, oriented so that the sequence for λ appears along the NW side (from SW to NE), the sequence for μ appears along the NE side (from NW to SE) and the

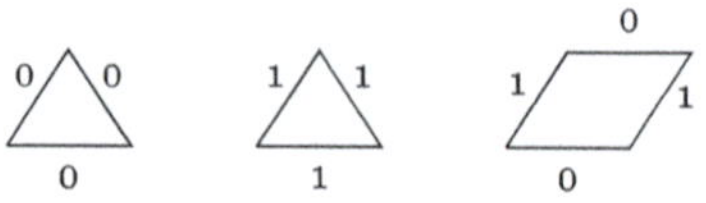

Figure 9.2 Classical puzzle pieces.

Figure 9.3 The equivariant puzzle piece.

sequence for ν appears along the S side (from W to E). A *puzzle* of type $\Delta^\nu_{\lambda\mu}$ is a filling of the triangle by the pieces shown in Figures 9.2 and 9.3. All except the equivariant piece may be rotated; the equivariant piece must appear in its displayed orientation. (See Figure 9.5 for an example.)

Each equivariant piece contributes a factor $\chi_i - \chi_j$, computed from its position as shown in Figure 9.4. The *weight* $\mathrm{wt}(P)$ of a puzzle P is the product of all such factors; it is evidently an element of $\mathbb{Z}_{\geq 0}[\chi_1 - \chi_2, \ldots, \chi_{n-1} - \chi_n]$.

The puzzle rule for computing $c^\nu_{\lambda\mu}$ is this:

Theorem 9.8.4 (Knutson–Tao). *We have*

$$c^\nu_{\lambda\mu} = \sum_{\substack{\text{puzzles } P \\ \text{of type } \Delta^\nu_{\lambda\mu}}} \mathrm{wt}(P).$$

For example, the puzzle in Figure 9.5 contributes $\chi_3 - \chi_5$ to the coefficient of $\sigma_{(3,1)}$ in $\sigma_{(2,1)} \cdot \sigma_{(2)}$. There are two other puzzles, computing $c^{(3,1)}_{(2,1),(2)} = (\chi_1 - \chi_2) + (\chi_2 - \chi_4) + (\chi_3 - \chi_5) = \chi_1 + \chi_3 - \chi_4 - \chi_5$.

The commutativity property $c^\nu_{\lambda\mu} = c^\nu_{\mu\lambda}$ is not immediately obvious from the puzzle rule – in general, there is no bijection between puzzles of types $\Delta^\nu_{\lambda\mu}$ and $\Delta^\nu_{\mu\lambda}$, although the sums of their weights are equal. On the other hand, Grassmann duality (Exercise 9.7.4) is evident: one defines a bijection between puzzles of type $\Delta^\nu_{\lambda\mu}$ and those of type $\Delta^{\nu'}_{\mu'\lambda'}$ by reflecting a puzzle from left to right, and exchanging 0's and 1's.

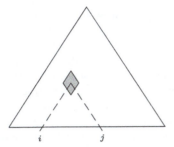

Figure 9.4 An equivariant piece in position (i, j).

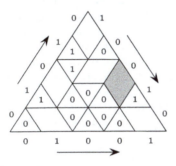

Figure 9.5 A puzzle of type $\Delta_{01010,10010}^{01001}$ and weight $\chi_3 - \chi_5$.

Exercise 9.8.5. Using the puzzle rule, for λ corresponding to a subset I, show that

$$c_{\lambda\lambda}^{\lambda} = \prod_{\substack{i \in I, j \notin I \\ i > j}} (\chi_j - \chi_i)$$

recovering the formula we know for $c_{\lambda\lambda}^{\lambda} = \sigma_\lambda|_\lambda$.

Exercise 9.8.6. Consider $\mathbb{P}^{n-1} = Gr(1, \mathbb{C}^n)$. The Schubert class $y_i \in H_T^*\mathbb{P}^{n-1}$ corresponds to the 01-sequence with a 1 in position $i + 1$, and 0's elsewhere. Use the puzzle rule to recover the formula for c_{ij}^k given in §4.7.

Notes

In the literature dealing with general Lie theory, B^--invariant subvarieties are usually called "opposite." Our conventions are reversed but have the advantage of better stability properties. We will continue this usage through

Chapter 13, and switch to the more standard convention when we discuss general Lie groups in Chapter 15.

The Kempf–Laksov theorem (Theorem 9.2.2) was originally stated in the context of degeneracy loci in Grassmann bundles (Kempf and Laksov, 1974). Its appearance as an equivariant Schubert class represents an early instance of the connection between equivariant geometry and the geometry of fiber bundles, although it was not seen this way at the time.

Studying rank conditions on matrices of homogeneous polynomials, Giambelli (1904) proved the case of Theorem 9.2.2 corresponding to a rectangular partition. We will see more about this in Chapter 11.

In Chapter 4, we saw a basis of Schur determinants $\Delta_\lambda(c^{B^-}(\mathbb{Q}))$, for λ inside the $d \times e$ rectangle. Here we have studied the basis of Schubert classes $[\Omega_\lambda]^{B^-}$, which are expressed (via the Kempf–Laksov formula) as multi-Schur determinants. What is the transition matrix between these two bases? The answer involves certain *flagged Schur polynomials*, which are special cases of the *Schubert polynomials* to be studied in Chapter 10.

The argument for Lemma 9.3.1 comes from Knutson and Tao (2003). The same idea was axiomatized and applied to other settings by Guillemin and Zara (2001) and Tymoczko (2008b). An alternative framework for finding (unique) solutions to such interpolation problems was developed systematically by Fehér and Rimányi (2003).

An excellent reference for double Schur polynomials (and their relatives) is Macdonald's note (1992). In particular, he proves the equivalence of the three characterizations of $s_\lambda(x|y)$ we gave. (We mainly use his "Variation 6".)

Lemma 9.4.1 is due to Okounkov (1996), who shows that these conditions characterize "shifted" Schur functions; see also (Okounkov and Olshanski, 1997) and (Molev and Sagan, 1999).

Proofs of the classical (non-equivariant) case of the Chevalley–Pieri rule, Theorem 9.6.2, can be found in many sources, e.g., (Fulton, 1997, §9.4). A proof of the equivariant version appears in (Knutson and Tao, 2003, Appendix). We will see a complete proof of the analogous formula for general homogeneous spaces G/P in Chapter 19. The characterization theorem (Theorem 9.6.4) is due to Molev and Sagan (1999), and was used extensively by Knutson and Tao (2003). It also has an analogue for G/P, as we will see in Chapter 19.

There are many references for the classical Littlewood–Richardson rule and its variations; see, for example, (Macdonald, 1995; Stanley, 1999; Fulton, 1997). The first complete proof is due to Schützenberger (1977), using a game called *jeu de taquin*. An equivariant *jeu de taquin* rule was given by Thomas and Yong (2018).

A combinatorial rule for the multiplying double Schur polynomials $s_\lambda(x|y)$ was given by Molev and Sagan (1999), but their original formula was not manifestly positive in the variables $y_i - y_j$ (for $i > j$). Molev (2009) later modified this to the positive formula described here; see also (Kreiman, 2010).

Knutson and Tao (2003) gave the first manifestly positive rule for the equivariant structure constants. In fact, they computed $\widetilde{c}_{\lambda\mu}^\nu$, the structure constants for multiplying in the *opposite* Schubert basis $\{\widetilde{\sigma}_\lambda\}$. These are related to $c^\nu_{\lambda\mu}$ by the substitution $\chi_i \mapsto \chi_{n+1-i}$. This is realized by reflecting puzzles left-to-right, which has the effect of exchanging the equivariant and non-equivariant rhombi.

Hints for Exercises

Exercise 9.4.2. Use the bialternant definition. The $(d-p,q)$-entry of the matrix in the numerator of $s_\lambda(y^\mu|y)$ is

$$(y_{\mu_p+d-p+1}|y)^{\lambda_q+d-q} = (y_{\mu_p+d-p+1} - y_1)\cdots(y_{\mu_p+d-p+1} - y_{\lambda_q+d-q}).$$

If $\mu \not\supseteq \lambda$, then some index k has $\mu_k < \lambda_k$ (so also $\mu_p < \lambda_q$ for $q \le k \le p$). But then for all $q \le k \le p$, we have

$$1 \le \mu_p + d - p + 1 \le \lambda_q + d - q,$$

so the above product vanishes, and it follows that the determinant is zero. If $\mu = \lambda$, the matrix is triangular because

$$1 \le \lambda_p + d - p + 1 \le \lambda_q + d - q$$

if $p > q$, so its determinant is the product

$$\prod_{p=1}^{d} \prod_{s=1}^{\lambda_p+d-p} (y_{\lambda_p+d-p+1} - y_s).$$

Dividing by the Vandermonde denominator gives the formula.

Exercise 9.4.4. The key identity is

$$h_p(x|\tau^{1-j}y) = \sum_{a+b=p} h_a(x_1,\ldots,x_d)(-1)^b e_b(y_1,\ldots,y_{d+p-j}),$$

where $y_m = 0$ for $m \le 0$. (This is easy to prove from the tableau definition of $h_p = s_{(p)}$.) Then compare with the degree $p = \lambda_i - i + j$ term of $c^T(\mathbb{Q} - E_{e-p+j}) = c^T(\widetilde{E}_{d+p-j} - \mathbb{S})$.

Exercise 9.7.1. Let $E_\bullet$ and $\widetilde{E}_\bullet$ be the standard and opposite flags in $\mathbb{C}^n$. Under the identification $V \cong V^\vee$ by the chosen basis, we have $E_i \mapsto (V/E_i)^\vee \cong \widetilde{E}_{n-i}$.

Exercise 9.8.5. See (Knutson and Tao, 2003, Proposition 3).

10

Flag Varieties and Schubert Polynomials

In this chapter, we will develop flag variety analogues of what we have just seen for the Grassmannian. The main goal is a "Giambelli" formula for expressing the equivariant class of a Schubert variety as a polynomial in Chern classes of universal bundles. These formulas are comparatively recent: canonical representatives are the *Schubert polynomials* introduced by Lascoux and Schützenberger in the 1980s. This chapter and the next provide a detailed study of the calculations sketched in §1.4.

10.1   Rank functions and Schubert varieties

We generally use "one-line" notation for permutations $w \in S_n$, recording values as $w = w(1)\, w(2) \cdots w(n)$. Let A_w be the permutation matrix having 1's in positions $(w(i), 1)$ and 0's elsewhere. (If e_i is the standard basis vector, this convention means $A_w \cdot e_i = e_{w(i)}$, and $A_{uv} = A_u \cdot A_v$.)

The *rank function* $\boldsymbol{r}_w = \big(r_w(p,q)\big)$ associated to w is defined by

$$r_w(p, q) = \#\{i \leq p \,|\, w(i) \leq q\}$$

for $1 \leq p, q \leq n$. That is, $r_w(p, q)$ is the rank of the upper-left $q \times p$ submatrix of A_w. The same information is encoded in the *dimension function* $\boldsymbol{k}_w = (k_w(p, q))$, defined by

$$k_w(p, q) = \#\{i \leq p \,|\, w(i) > q\}$$

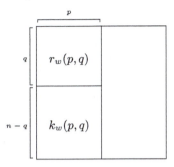

Figure 10.1 r_w and k_w as ranks of submatrices of A_w.

for $1 \leq p, q \leq n$. The number $k_w(p, q)$ is the rank of the lower-left $(n - q) \times p$ submatrix of A_w, so $k_w(p, q) + r_w(p, q) = p$ for all p, q. Figure 10.1 gives a schematic illustration of these ranks of submatrices of A_w.

For an n-dimensional vector space V, consider the complete flag variety $Fl(V) = Fl(1, \ldots, n - 1; V)$ of subspaces $F_1 \subset \cdots \subset F_n = V$. As in the previous chapter, we fix a standard flag $E_\bullet$ and write $E^q = E_{n-q}$, so E_q has dimension q and E^q has codimension q. There are *Schubert cells*

$$\Omega_w^\circ = \Omega_w^\circ(E_\bullet) = \left\{ F_\bullet \mid \mathrm{rk}(F_p \to V/E^q) = r_w(p, q) \text{ for all } 1 \leq p, q \leq n \right\}$$
$$= \left\{ F_\bullet \mid \dim(F_p \cap E^q) = k_w(p, q) \text{ for all } 1 \leq p, q \leq n \right\}$$

and *Schubert varieties*

$$\Omega_w = \Omega_w(E_\bullet) = \left\{ F_\bullet \mid \mathrm{rk}(F_p \to V/E^q) \leq r_w(p, q) \text{ for all } 1 \leq p, q \leq n \right\}$$
$$= \left\{ F_\bullet \mid \dim(F_p \cap E^q) \geq k_w(p, q) \text{ for all } 1 \leq p, q \leq n \right\}$$

in $Fl(V)$. We will soon see that $\Omega_w = \overline{\Omega_w^\circ}$.

First, let us fix conventions for representing flags with matrices. We choose a standard basis $\{e_1, \ldots, e_n\}$, so $V \cong \mathbb{C}^n$. A full-rank matrix $A \in GL_n$ represents a flag $F_\bullet$, where $F_p \subseteq V$ is the span of the first p columns of A. Since initial column-spans are preserved by right multiplication by the Borel subgroup B of upper-triangular matrices, we have $Fl(\mathbb{C}^n) \cong GL_n/B$.

We take the fixed flag $E_\bullet$ to be given by coordinate subspaces $E_q = \langle e_{n-q+1}, \ldots, e_n \rangle \subseteq \mathbb{C}^n$. This flag is fixed by B^-, so Schubert cells and varieties are B^--invariant. In fact, as with the Grassmannian, Schubert cells are B^--orbits, and they decompose the flag variety. For each permutation $w \in S_n$, there is a point $p_w \in Fl(\mathbb{C}^n)$, corresponding to the flag

$$\langle e_{w(1)} \rangle \subset \langle e_{w(1)}, e_{w(2)} \rangle \subset \cdots \subset V,$$

and represented by the permutation matrix A_w. (So our standard flag $E_\bullet$ corresponds to the permutation $n \cdots 2\, 1$.)

Exercise. Show that

$$GL_n = \coprod_{w \in S_n} B^- \cdot A_w \cdot B.$$

(This is the *Bruhat decomposition* of GL_n, an elementary instance of a general phenomenon for algebraic groups.) Then show that

$$\Omega_w^\circ = B^- \cdot p_w,$$

so that $Fl(V) = \coprod_{w \in S_n} \Omega_w^\circ$. Deduce that the classes of Schubert varieties $[\Omega_w]^{B^-}$ form a basis for $H^*_{B^-} Fl(V)$ over Λ_{B^-}.

Given a flagged vector bundle $E_1 \subset \cdots \subset E_n = V$ on a variety Y, with $E^q = E_{n-q}$ having co-rank q in V, there is a Schubert locus $\boldsymbol{\Omega}_w(E_\bullet)$ in the flag bundle $\mathbf{Fl}(V) \to Y$, defined by the conditions

$$\boldsymbol{\Omega}_w(E_\bullet) = \left\{ x \in \mathbf{Fl}(V) \mid \mathrm{rk}(\mathbb{S}_p \to V/E^q) \leq r_w(p, q) \text{ for all } 1 \leq p, q \leq n \right\},$$

where $\mathbb{S}_\bullet$ is the tautological flag of subbundles on $\mathbf{Fl}(V)$.

When $Y = \mathbb{B}T$ (or an approximation space), $V = \mathbb{E}T \times^T \mathbb{C}^n$, and $E_\bullet$ is induced from the standard flag in $\mathbb{C}^n$, we have identifications

$$
\begin{array}{ccc}
\mathbf{Fl}(V) & \!\!=\!\! & \mathbb{E}T \times^T Fl(\mathbb{C}^n) \\
\uparrow & & \uparrow \\
\boldsymbol{\Omega}_w & \!\!=\!\! & \mathbb{E}T \times^T \Omega_w,
\end{array}
$$

so $[\boldsymbol{\Omega}_w] = [\Omega_w]^T$ in $H^* \mathbf{Fl}(V) = H^*_T Fl(\mathbb{C}^n)$.

10.2 Neighborhoods and tangent weights

Now let us consider a torus T acting on $V = \mathbb{C}^n$ with characters $\chi_1, \ldots, \chi_n$, inducing an action on $Fl(V)$. The point p_w is fixed by T, and has a T-invariant affine neighborhood $U \subseteq Fl(V)$. Writing $I_d = \{w(1), \ldots, w(d)\}$ for each $1 \leq d \leq n$, this is the open set defined by the nonvanishing of the $d \times d$ minors on columns $1, \ldots, d$ and rows I_d, for all $d = 1, \ldots, n$. In terms of the natural projections $\pi_d \colon Fl(V) \to Gr(d, V)$, this is

$$U = \bigcap_{d=1}^{n} \pi_d^{-1} U_d,$$

where, for each d, $U_d \subset Gr(d, V)$ is the affine open neighborhood of p_{I_d} in the Grassmannian (see §9.1). More directly, points in U are represented by matrices with free entries in positions $(w(i), j)$ for $i > j$. For $w = 6\ 4\ 2\ 7\ 1\ 3\ 5$, this neighborhood is shown at right. The torus acts on the $(w(i), j)$ entry by the character $\chi_{w(i)} - \chi_{w(j)}$. Just as for Grassmannians,

$$\begin{bmatrix} * & * & * & * & 1 & 0 & 0 \\ * & * & 1 & 0 & 0 & 0 & 0 \\ * & * & * & * & * & 1 & 0 \\ * & 1 & 0 & 0 & 0 & 0 & 0 \\ * & * & * & * & * & * & 1 \\ 1 & 0 & 0 & 0 & 0 & 0 & 0 \\ * & * & * & 1 & 0 & 0 & 0 \end{bmatrix}$$

this invariant neighborhood is an affine space, isomorphic to $\mathbb{C}^{\binom{n}{2}}$, and the tangent space $T_{p_w} Fl(V)$ has torus weights $\{\chi_{w(i)} - \chi_{w(j)} \mid i > j\}$. If all characters χ_i are distinct, these weights are nonzero, and the fixed locus is precisely $Fl(V)^T = \{p_w \mid w \in S_n\}$.

The Schubert cell Ω_w° is the B^--orbit of p_w. In matrices, this means free entries appear only to the left and below the 1's of A_w. For example, with $w = 6\ 4\ 2\ 7\ 1\ 3\ 5$, points in the Schubert cell are represented as

$$\Omega_w^\circ = \begin{bmatrix} 0 & 0 & 0 & 0 & 1 & 0 & 0 \\ 0 & 0 & 1 & 0 & 0 & 0 & 0 \\ 0 & 0 & * & 0 & * & 1 & 0 \\ 0 & 1 & 0 & 0 & 0 & 0 & 0 \\ 0 & * & * & 0 & * & * & 1 \\ 1 & 0 & 0 & 0 & 0 & 0 & 0 \\ * & * & * & 1 & 0 & 0 & 0 \end{bmatrix}.$$

Comparing with the matrix representation of the affine open neighborhood of p_w, this Schubert cell is obtained by setting some of the free entries to zero (shown in bold above). These 0's appear precisely in positions $(w(i), j)$ such that $i > j$ and $w(i) < w(j)$, as shown at right.

$$\begin{bmatrix} 0 & \cdots & 1 \\ & \vdots & \\ 1 & & \end{bmatrix}$$

This picture lets us easily identify the torus weights on the tangent and normal spaces:

$$T_{p_w}\Omega_w^\circ \text{ has weights } \{\chi_{w(i)} - \chi_{w(j)} \mid i > j \text{ and } w(i) > w(j)\};$$
$$N_{p_w}\Omega_w^\circ \text{ has weights } \{\chi_{w(i)} - \chi_{w(j)} \mid i > j \text{ and } w(i) < w(j)\}.$$

The restriction of a Schubert class to a T-fixed point follows:

$$[\Omega_w]^T|_{p_w} = \prod_{\substack{i<j \\ w(i)>w(j)}} (\chi_{w(j)} - \chi_{w(i)}).$$

In fact, computing a little more carefully shows that each Schubert cell is isomorphic (as a variety) to a certain subgroup of B^-. Let $U^- \subseteq B^-$ be the unipotent subgroup of lower-triangular matrices with 1's on the diagonal, and let B be the (opposite) Borel subgroup of upper-triangular matrices. For $w \in S_n$, let $U^-(w) = U^- \cap A_w B^- A_{w^{-1}}$.

Exercise 10.2.1.   Show that the standard T-invariant affine neighborhood is $(A_w B^- A_w^{-1}) \cdot p_w$, and that the map

$$U^-(w) \to \Omega_w^\circ$$
$$u \mapsto u \cdot p_w$$

is an isomorphism of varieties. Furthermore, there is a free entry in position (i, j) of $U^-(w)$ if and only if $i > j$ and $w^{-1}(i) > w^{-1}(j)$.

For example, if $w = 6\ 4\ 2\ 7\ 1\ 3\ 5$, we have

$$U^-(w) = \begin{bmatrix} 1 & 0 & 0 & 0 & 0 & 0 & 0 \\ 0 & 1 & 0 & 0 & 0 & 0 & 0 \\ * & * & 1 & 0 & 0 & 0 & 0 \\ 0 & 0 & 0 & 1 & 0 & 0 & 0 \\ * & * & * & * & 1 & 0 & 0 \\ 0 & 0 & 0 & 0 & 0 & 1 & 0 \\ 0 & * & 0 & * & 0 & * & 1 \end{bmatrix}.$$

10.3 Invariant curves in the flag variety

We have seen formulas for the tangent weights at T-fixed points. Next we will describe T-invariant curves. Let $\tau_{ij} \in S_n$ be the transposition which swaps i and j. For each $w \in S_n$ and $i < j$, there is an invariant curve in $Fl(\mathbb{C}^n)$ through p_w, with tangent weight $\chi_{w(j)} - \chi_{w(i)}$ at this point. The other fixed point is p_v, for $v = w\tau_{ij} = \tau_{kl}w$, where $k = w(i)$ and $l = w(j)$. This is easy to see in matrix coordinates. For example, take $w = 6\ 4\ 2\ 7\ 1\ 3\ 5$, and $v = w\tau_{24} = \tau_{47}w = 6\ 7\ 2\ 4\ 1\ 3\ 5$. The curve connecting these points has matrix representatives in neighborhoods of p_w and p_v as shown below.

$$
\begin{bmatrix}
0 & 0 & 0 & 0 & 1 & 0 & 0 \\
0 & 0 & 1 & 0 & 0 & 0 & 0 \\
0 & 0 & 0 & 0 & 0 & 1 & 0 \\
0 & 1 & 0 & \mathbf{0} & 0 & 0 & 0 \\
0 & 0 & 0 & 0 & 0 & 0 & 1 \\
1 & 0 & 0 & 0 & 0 & 0 & 0 \\
0 & a & 0 & 1 & 0 & 0 & 0
\end{bmatrix}
=
\begin{bmatrix}
0 & 0 & 0 & 0 & 1 & 0 & 0 \\
0 & 0 & 1 & 0 & 0 & 0 & 0 \\
0 & 0 & 0 & 0 & 0 & 1 & 0 \\
0 & \frac{1}{a} & 0 & 1 & 0 & 0 & 0 \\
0 & 0 & 0 & 0 & 0 & 0 & 1 \\
1 & 0 & 0 & 0 & 0 & 0 & 0 \\
0 & 1 & 0 & \mathbf{0} & 0 & 0 & 0
\end{bmatrix}.
$$

In the limit $a \to 0$, one sees p_w, and as $a \to \infty$, one arrives at p_v.

Another way to generate these curves can be useful. There is a subgroup $G_{k,l} \subseteq GL_n$, which is an embedding of GL_2: one places an invertible 2×2 matrix on rows and columns k, l, 1's along the other diagonal entries, and 0's elsewhere. The subgroup $G_{k,l}$ contains the permutation matrix $A_{\tau_{kl}}$. For any $u \in S_n$, the intersection $G_{k,l} \cap uBu^{-1}$ is isomorphic to upper- or lower-triangular matrices in GL_2, so the T-curve connecting p_w and p_v is $G_{k,l} \cdot p_w = G_{k,l} \cdot p_v \cong \mathbb{P}^1$.

In the above example, for $w = 6\ 4\ 2\ 7\ 1\ 3\ 5$, one can generate the T-curve with character $\chi_7 - \chi_4$ by embedding $\begin{bmatrix} 1 & 0 \\ a & 1 \end{bmatrix}$ on rows and columns $4, 7$; the matrix shown on the left-hand side is obtained left multiplication on the permutation matrix A_w.

From our description of Ω_w° in the neighborhood of p_w, we see the curve $G_{k,l} \cdot p_w$ meets Ω_w° either in an affine line (if $k = w(i) < l = w(j)$), or only the point p_w (if $k = w(i) > l = w(j)$).

When the tangent weights $\chi_{w(j)} - \chi_{w(i)}$ are pairwise non-parallel, there are finitely many invariant curves through p_w (§7.2). Assuming this holds at every fixed point p_w, we have a complete description of T-curves in $Fl(\mathbb{C}^n)$.

To summarize: fix w, and $i < j$, with $k = w(i)$ and $l = w(j)$, and $v = w\tau_{ij} = \tau_{kl}w$.

- There is a T-invariant curve $C_{w,v} = G_{kl} \cdot p_w \subseteq Fl(\mathbb{C}^n)$, isomorphic to $\mathbb{P}^1$ and connecting p_w and p_v.

- If $w(i) < w(j)$, the intersection $C_{w,v} \cap \Omega_w^\circ$ is isomorphic to $\mathbb{A}^1$ and dense in $C_{w,v}$. Otherwise, $C_{w,v} \cap \Omega_w^\circ = \{p_w\}$.

- If the characters $\chi_{w(j)} - \chi_{w(i)}$ are pairwise nonparallel (for all $i < j$), then these are all the T-curves through p_w.

10.4   Bruhat order for the symmetric group

Our next goal is to prove that the Schubert variety $\Omega_w \subseteq Fl(V)$, defined by rank inequalities $\mathrm{rk}(F_p \to V/E^q) \leq r_w(p,q)$, is the closure of the Schubert cell Ω_w°, where equality holds. We will need some facts about a partial order on S_n.

Definition 10.4.1.   The *Bruhat order* on S_n is defined as follows. For permutations v, w, we say $v \geq w$ if $r_v(p,q) \leq r_w(p,q)$ for all $1 \leq p, q \leq n$.

From the definition, $v \geq w$ is equivalent to $\Omega_v \subseteq \Omega_w$. There are many other useful characterizations of Bruhat order. Recall (from §9.1) the partial order on p-element subsets of $\{1, \ldots, n\}$: for $J = \{j_1 < \cdots < j_p\}$ and $I = \{i_1 < \cdots < i_p\}$, we say $J \geq I$ if $j_k \geq i_k$ for all $1 \leq k \leq p$.

Exercise 10.4.2.   Show that $v \geq w$ if and only if $\{v(1), \ldots, v(p)\} \geq \{w(1), \ldots, w(p)\}$ for all $1 \leq p \leq n$.

This characterization is sometimes called the *tableau criterion* for comparing elements in Bruhat order. By writing the sets

$$\{w(1), \ldots, w(n)\}, \{w(1), \ldots, w(n-1)\}, \ldots, \{w(1)\}$$

as columns, one represents w by a semistandard tableau of shape $\lambda = (n, n-1, \ldots, 1)$. For instance, one sees $2\,3\,1 > 1\,3\,2$ by comparing two tableaux entry-wise:

$$\begin{array}{|c|c|c|}\hline 1 & 2 & 2 \\\hline 2 & 3 \\\cline{1-2} 3 \\\cline{1-1}\end{array} \qquad \begin{array}{|c|c|c|}\hline 1 & 1 & 1 \\\hline 2 & 3 \\\cline{1-2} 3 \\\cline{1-1}\end{array}\ .$$

The *inversions* of a permutation w are the pairs $i < j$ such that $w(i) > w(j)$. The *length* enumerates inversions:

$$\ell(w) = \#\{i < j \mid w(i) > w(j)\}.$$

There are unique permutations of least and greatest length in S_n. The identity e is evidently the shortest element, with $\ell(e) = 0$. The *longest element* is $w_\circ = n \cdots 2\, 1$, with length $\ell(w_\circ) = \binom{n}{2}$.

The following lemma shows that Bruhat order is graded by length. We will write $v \gtrdot w$ to mean $v > w$ and $\ell(v) = \ell(w) + 1$. As before, $\tau_{ij} \in S_n$ is the transposition swapping i and j.

Lemma 10.4.3. *If $v > w$, then $\ell(v) > \ell(w)$. In fact, there is a permutation w' such that $v \geq w' \gtrdot w$; this w' has the form $w' = w\tau_{ij}$ for some $i < j$ with $w(i) < w(j)$.*

Proof The first statement follows from the second, by induction on Bruhat order, so we focus on the second statement.

We will use the tableau criterion of Exercise 10.4.2. Let i be the smallest index such that $\{v(1), \ldots, v(i)\} \neq \{w(1), \ldots, w(i)\}$ – that is, $v(1) = w(1)$, $v(2) = w(2)$, etc., and $v(i) > w(i)$. Let $r = v(i)$ and $s = w(i)$, so $r > s$. Let $j > i$ be the smallest index such that $w(i) < w(j) \leq r$. (From the choice of i, we know $i < w^{-1}(r)$, so j exists with $j \leq w^{-1}(r)$.) Let $w' = w\tau_{ij}$, where τ_{ij} is the transposition swapping the positions i and j, so

$$w' = w(1) \cdots w(i-1)\, w(j)\, w(i+1) \cdots w(j-1)\, w(i) \cdots w(n).$$

Now $\ell(w') = \ell(w) + 1$. Indeed, by the choice of j, there is no k with $i < k < j$ and $w(i) < w(k) < w(j)$. So w' has all the inversions of w, together with exactly one new inversion $w'(i) > w'(j)$.

Finally, we must check that $v \geq w' \gtrdot w$. If $p < i$ or $p \geq j$, we have

$$\{w'(1), \ldots, w'(p)\} = \{w(1), \ldots, w(p)\}.$$

so in these cases we know $\{v(1), \ldots, v(p)\} \geq \{w'(1), \ldots, w'(p)\}$. When $i \leq p < j$, we have

$$\{w'(1), \ldots, w'(p)\} = (\{w(1), \ldots, w(p)\} \setminus \{w(i)\}) \cup \{w(j)\};$$

that is, the set $\{w'(1), \ldots, w'(p)\}$ is obtained from $\{w(1), \ldots, w(p)\}$ by swapping $w(j)$ for $w(i)$. Since $v(i) \geq w'(i) = w(j) > w(i)$, we see $\{v(1), \ldots, v(p)\} \geq \{w'(1), \ldots, w'(p)\}$ for these p, as well.   □

Now we return to Schubert varieties. From the discussion in the previous section, it follows that Ω_w° – and therefore $\overline{\Omega_w^\circ}$ – has codimension equal to $\ell(w)$ in $Fl(V)$. From the Bruhat decomposition, it follows easily that

$$\Omega_w = \coprod_{v \geq w} \Omega_v^\circ,$$

since the permutation matrix A_v satisfies the conditions defining Ω_w if and only if $v \geq w$. The Schubert cell Ω_v° is the orbit $B^- \cdot p_v$, so this reduces proving $\Omega_w = \overline{\Omega_w^\circ}$ to another characterization of Bruhat order.

Proposition 10.4.4. *For $v, w \in S_n$, we have $v \geq w$ if and only if $p_v \in \overline{\Omega_w^\circ}$. In particular, $\Omega_w = \overline{\Omega_w^\circ}$.*

Proof It is clear from the definitions that $\Omega_w \supseteq \overline{\Omega_w^\circ}$, and we have seen that if $v \not\geq w$, then $p_v \notin \Omega_v$ (since A_v violates one of the rank conditions defining Ω_w). The nontrivial part of the statement is the converse: if $v \geq w$, then $p_v \in \overline{\Omega_w^\circ}$.

We will use descending induction on Bruhat order. The base case is $w = w_\circ$, and the locus $\Omega_{w_\circ}^\circ \subseteq Fl(V)$ is already closed, consisting of just the B^--fixed point: $\Omega_{w_\circ}^\circ = \Omega_{w_\circ} = \{p_{w_\circ}\}$.

It suffices to treat the case $v \gtrdot w$. Indeed, by induction, we know $\Omega_v = \overline{\Omega_v^\circ} = \coprod_{v' \geq v} \Omega_{v'}^\circ$. And by Lemma 10.4.3, for any $v' > w$, one can find v so that $v' \geq v \gtrdot w$. In fact, the lemma says we can find $i < j$ with $w(i) < w(j)$ so that $v = w\tau_{ij} = \tau_{kl}w$, where $k = w(i)$ and $l = w(j)$.

From what we have seen in §10.3, the curve $G_{k,l} \cdot p_w$ has dense intersection with Ω_w° (since $k = w(i) < l = w(j)$). Its other fixed point is p_v, for $v = \tau_{kl}w$. So p_v is a limit of points in Ω_w°, as claimed.   □

Another characterization of Bruhat order is sometimes useful. The adjacent transposition $\tau_{i,i+1}$ is called a *simple transposition* and written s_i. An expression $w = s_{i_1} \cdots s_{i_\ell}$ is *reduced* if ℓ is minimal among all such expressions. It is a basic fact that this minimal ℓ is equal to $\ell(w)$.

Lemma 10.4.5. *Suppose $v \gtrdot w$, so that $v = w\tau_{ab}$ for some transposition τ_{ab} (as in Lemma 10.4.3) and $\ell(w) = \ell(v) - 1$. Let $\ell = \ell(v)$, and let*

$v = s_{i_1} \cdots s_{i_\ell}$ *be a reduced expression for* v. *Then for some* $1 \leq m \leq \ell$, *a minimal expression* $w = s_{i_1} \cdots \widehat{s_{i_m}} \cdots s_{i_\ell}$ *is obtained by omitting one transposition from the expression for* v.

(A proof can be found in references listed in the Notes for this chapter.)

For general permutations $v, w \in S_n$, it follows that $v \geq w$ if and only if there is a reduced expression $v = s_{i_1} \cdots s_{i_\ell}$ so that a *subword* gives a reduced expression $w = s_{j_1} \cdots s_{j_k}$.

To summarize, we have several descriptions of Bruhat order. For $v, w \in S_n$, the following are equivalent:

(i) $v \geq w$.

(ii) $r_v(p, q) \leq r_w(p, q)$ for all $1 \leq p, q \leq n$.

(iii) $\Omega_v \subseteq \Omega_w$.

(iv) $p_v \in \Omega_w$.

(v) $\{v(1), \ldots, v(p)\} \geq \{w(1), \ldots, w(p)\}$ for all $1 \leq p \leq n$.

(vi) There is a chain

$$w = w^{(0)} \to w^{(1)} \to \cdots \to w^{(s)} = v,$$

where each step is of the form $u \to u \cdot \tau_{ij}$ for some transposition so that $\ell(u \cdot \tau_{ij}) = \ell(u) + 1$.

(vii) There is a minimal-length expression $v = s_{i_1} \cdots s_{i_\ell}$ so that w is given by the product of a subsequence of $s_{i_1}, \ldots, s_{i_\ell}$.

(viii) $w_\circ v \leq w_\circ w$.

(ix) $v w_\circ \leq w w_\circ$.

(x) $v^{-1} \geq w^{-1}$.

10.5 Opposite Schubert varieties and Poincaré duality

We have been using the standard flag $E_\bullet$, which in the standard basis is given by $E_q = \langle e_{n+1-q}, \ldots, e_n \rangle$. Just as for Grassmannians, the opposite Schubert cells and opposite Schubert varieties are defined with respect to the opposite flag $\widetilde{E}_\bullet$, which has $\widetilde{E}_q = \langle e_1, \ldots, e_q \rangle$:

$$\tilde{\Omega}^\circ_w = \Omega^\circ_w(\tilde{E}_\bullet) \quad \text{and} \quad \tilde{\Omega}_w = \Omega_w(\tilde{E}_\bullet).$$

Since $\tilde{E}_\bullet$ is preserved by the (opposite) Borel group B of upper-triangular matrices, the opposite Schubert cells and varieties are B-invariant.

Exercise 10.5.1. Assume T acts on $V = \mathbb{C}^n$ by distinct characters $\chi_1, \ldots, \chi_n$. Then $p_{w_\circ w}$ is the only T-fixed point in $\tilde{\Omega}^\circ_w$, and $\tilde{\Omega}^\circ_w = B \cdot p_{w_\circ w}$.

For example, the Schubert cell and opposite Schubert cell for $w = 6\ 4\ 2\ 7\ 1\ 3\ 5$ (so $w_\circ w = 2\ 4\ 6\ 1\ 7\ 5\ 3$) are represented as

$$\Omega^\circ_w = \begin{bmatrix} 0 & 0 & 0 & 0 & 1 & 0 & 0 \\ 0 & 0 & 1 & 0 & 0 & 0 & 0 \\ 0 & 0 & * & 0 & * & 1 & 0 \\ 0 & 1 & 0 & 0 & 0 & 0 & 0 \\ 0 & * & * & 0 & * & * & 1 \\ 1 & 0 & 0 & 0 & 0 & 0 & 0 \\ * & * & * & 1 & 0 & 0 & 0 \end{bmatrix} \quad \text{and} \quad \tilde{\Omega}^\circ_{w_\circ w} = \begin{bmatrix} * & * & * & * & 1 & 0 & 0 \\ * & * & 1 & 0 & 0 & 0 & 0 \\ * & * & 0 & * & 0 & 1 & 0 \\ * & 1 & 0 & 0 & 0 & 0 & 0 \\ * & 0 & 0 & * & 0 & 0 & 1 \\ 1 & 0 & 0 & 0 & 0 & 0 & 0 \\ 0 & 0 & 0 & 1 & 0 & 0 & 0 \end{bmatrix}.$$

The opposite Schubert varieties also give a cell decomposition of $Fl(V)$, so their classes form a basis for $H^*_B Fl(V)$. Furthermore,

$$\tilde{\Omega}_w = \coprod_{v \geq w} \tilde{\Omega}^\circ_v,$$

by the same argument as in the previous section. Equivalently, $p_{w_\circ v}$ lies in $\tilde{\Omega}_w$ if and only if $v \geq w$ in Bruhat order. We also have an analogous computation of the weights at fixed points:

$$T_{p_w} \tilde{\Omega}^\circ_{w_\circ w} \text{ has weights } \{\chi_{w(i)} - \chi_{w(j)} \mid i > j \text{ and } w(i) < w(j)\};$$
$$N_{p_w} \tilde{\Omega}^\circ_{w_\circ w} \text{ has weights } \{\chi_{w(i)} - \chi_{w(j)} \mid i > j \text{ and } w(i) > w(j)\}.$$

These computations show that the intersection $\Omega^\circ_w \cap \tilde{\Omega}^\circ_{w_\circ w}$ is transverse. As before, we also obtain a simple formula for the restriction to a fixed point:

$$[\tilde{\Omega}_{w_\circ w}]^T |_{p_w} = \prod_{\substack{i<j \\ w(i)<w(j)}} (\chi_{w(j)} - \chi_{w(i)}).$$

Since both Schubert varieties and opposite Schubert varieties are T-equivariant, we have two bases for $H^*_T Fl(V)$. In fact, they are Poincaré dual.

Proposition 10.5.2. *Let* $\sigma_w = [\Omega_w]^T$ *and* $\widetilde{\sigma}_w = [\widetilde{\Omega}_w]^T$. *Then* $\{\sigma_w\}$ *and* $\{\widetilde{\sigma}_{w_\circ w}\}$ *form Poincaré dual bases for* $H^*_T Fl(V)$ *as a* Λ_T-*module. That is,*

$$\rho_*(\sigma_u \cdot \widetilde{\sigma}_v) = \delta_{u,w_\circ v}$$

in Λ_T.

The proof is analogous to that of the corresponding fact for Grassmannians (Proposition 9.5.2).

On the other hand, with a more careful argument, we can prove something stronger. We will need a simple fact about Schubert varieties.

Exercise 10.5.3. For two flags $F_\bullet, G_\bullet \in Fl(V)$, show that $F_\bullet \in \Omega_w(G_\bullet)$ if and only if $G_\bullet \in \Omega_{w^{-1}}(F_\bullet)$.

Proposition 10.5.4. *If* $u \leq w_\circ v$, *then the intersection* $\Omega_u \cap \widetilde{\Omega}_v$ *is reduced and irreducible of dimension* $\ell(w_\circ v) - \ell(u)$. *Otherwise, this intersection is empty.*

Proof Consider a double Schubert variety in $Fl(V) \times Fl(V)$, defined by

$$\boldsymbol{\Omega}_v = \Big\{ (G_\bullet, F_\bullet) \,\big|\, F_\bullet \in \Omega_v(G_\bullet) \Big\}$$
$$= \Big\{ (G_\bullet, F_\bullet) \,\big|\, G_\bullet \in \Omega_{v^{-1}}(F_\bullet) \Big\}.$$

The equivalence of the two descriptions follows from Exercise 10.5.3. Both projections $pr_i \colon \boldsymbol{\Omega}_v \to Fl(V)$ $(i = 1, 2)$ are locally trivial fiber bundles, with Schubert varieties as fibers. More precisely, the fibers are $pr_1^{-1}(G_\bullet) = \Omega_v(G_\bullet)$ and $pr_2^{-1}(F_\bullet) = \Omega_{v^{-1}}(F_\bullet)$. In particular, $\boldsymbol{\Omega}_v$ is reduced and irreducible.

Let $Z = \boldsymbol{\Omega}_v \cap (Fl(V) \times \Omega_u)$, so there is a diagram

$$
\begin{array}{ccc}
& Z & \longrightarrow \Omega_u \\
{\scriptstyle f}\swarrow & \downarrow & \downarrow \\
Fl(V) \longleftarrow & \boldsymbol{\Omega}_v & \longrightarrow Fl(V),
\end{array}
$$

where the square on the right is a fiber square, and all maps are B^--equivariant. The second projection makes $Z \to \Omega_u$ a locally trivial fiber bundle, whose fiber over $F_\bullet$ is the Schubert variety $\Omega_{v^{-1}}(F_\bullet)$. So Z is also reduced and irreducible, of dimension

$$\dim Z = \dim \Omega_u + \dim \Omega_{v^{-1}}$$
$$= \dim Fl(V) - \ell(u) + \ell(w_\circ v),$$

using $\ell(v^{-1}) = \ell(v)$.

Under the first projection $pr_1\colon \Omega_v \to Fl(V)$, the fiber over p_e is $\Omega_v(\widetilde{E}_\bullet) = \widetilde{\Omega}_v$. So the fiber of $f\colon Z \to Fl(V)$ is $f^{-1}(p_e) = \Omega_u \cap \widetilde{\Omega}_v$. On the other hand, by B^--equivariance, for every $x = b^- \cdot p_e$ in the dense open set $B^- \cdot p_e$, the fiber $f^{-1}(x)$ is isomorphic to $\Omega_u \cap \widetilde{\Omega}_v$. By analyzing T-fixed points, we have already seen that $\Omega_u \cap \widetilde{\Omega}_v$ is empty if and only if $u \not\leq w_\circ v$. Assuming $u \leq w_\circ v$, the asserted dimension follows from the formula for $\dim Z$.

There is a section of $f\colon Z \to Fl(V)$ over the open set $B^- \cdot p_e$, sending $b^- \cdot p_e$ to $(b^- \cdot p_e, b^- \cdot p_v) \in f^{-1}(B^- \cdot p_e)$. (This is well defined!) It follows that $f^{-1}(B^- \cdot p_e) \cong (B^- \cdot p_e) \times (\Omega_u \cap \widetilde{\Omega}_v)$. Since Z is reduced and irreducible, and $B^- \cdot p_e$ is an affine space, we conclude that $\Omega_u \cap \widetilde{\Omega}_v$ is reduced and irreducible. $\qquad\square$

In fact, the proof shows much more.

Corollary 10.5.5. *Let P be a property of schemes such that*

(1) all Schubert varieties have P;

(2) X has P if and only if every nonempty open $U \subseteq X$ has P; and

(3) X has P if and only if $X \times \mathbb{A}^n$ has P.

Then the intersection $\Omega_u \cap \widetilde{\Omega}_v$ has P.

For example, this shows $\Omega_u \cap \widetilde{\Omega}_v$ is Cohen–Macaulay, since Schubert varieties have this property.

10.6   Schubert polynomials

Now we can address the Giambelli problem, expressing the Schubert class $\sigma_w = [\Omega_w]^T$ as a polynomial in Chern classes. In Chapter 4 we saw an isomorphism

$$H_T^* Fl(V) = \Lambda[x_1, \dots, x_n]/I,$$

where $x_i = -c_1^T(\mathbb{S}_i/\mathbb{S}_{i-1})$, the torus acts on $V = \mathbb{C}^n$ by characters $\chi_i = -y_i$, and I is generated by $e_k(x) - e_k(y)$ for $k = 1, \ldots, n$. We seek polynomials in $\Lambda[x]$ which map to Schubert classes.

The answer is given by Schubert polynomials. These are defined inductively, for each $w \in S_n$, starting from the longest element $w_\circ$ and moving down in Bruhat order by passing from w to ws_k, where s_k is the simple transposition swapping positions k and $k+1$.

The *divided difference operator* ∂_k acts on a polynomial ring $R[x] = R[x_1, x_2, \ldots]$ by

$$\partial_k(f) = \frac{f(\ldots, x_k, x_{k+1}, \ldots) - f(\ldots, x_{k+1}, x_k, \ldots)}{x_k - x_{k+1}} = \frac{f - s_k f}{x_k - x_{k+1}}.$$

Here a permutation w acts on a polynomial f by

$$(wf)(x_1, x_2, \ldots) = f(x_{w^{-1}(1)}, x_{w^{-1}(2)}, \ldots).$$

If f is homogeneous of degree d, then $\partial_k(f)$ is homogeneous of degree $d - 1$.

Definition 10.6.1. Taking $R = \mathbb{Z}[y] = \mathbb{Z}[y_1, \ldots, y_n]$, the *double Schubert polynomials* are defined by

(i) $\mathfrak{S}_{w_\circ}(x; y) = \displaystyle\prod_{i+j \leq n} (x_i - y_j);$

(ii) $\mathfrak{S}_{ws_k}(x; y) = \partial_k \mathfrak{S}_w(x; y)$ if $\ell(ws_k) < \ell(w)$.

Setting the y variables to zero defines the *single Schubert polynomials*, $\mathfrak{S}_w(x) = \mathfrak{S}_w(x; 0)$.

Since every $w \in S_n$ is obtained from $w_\circ$ by successive transpositions s_k, this defines all $\mathfrak{S}_w$. Concretely, for any $w \in S_n$, writing $w = w_\circ s_{i_1} \cdots s_{i_\ell}$ with ℓ minimal (so $\ell = \ell(w_\circ) - \ell(w)$), we have

$$\mathfrak{S}_w = \partial_{i_\ell} \circ \cdots \circ \partial_{i_1} \mathfrak{S}_{w_\circ}.$$

It follows that $\mathfrak{S}_w$ is homogeneous of degree $\ell(w)$. (To obtain such an expression for w, successively swap adjacent entries of w until one reaches $w_\circ$. For example,

$$w = 3\,5\,1\,2\,4 \xrightarrow{\cdot s_1} 5\,3\,1\,2\,4 \xrightarrow{\cdot s_3} 5\,3\,2\,1\,4$$
$$\xrightarrow{\cdot s_4} 5\,3\,2\,4\,1 \xrightarrow{\cdot s_3} 5\,3\,4\,2\,1 \xrightarrow{\cdot s_2} 5\,4\,3\,2\,1 = w_\circ$$

shows that $w\,s_1\,s_3\,s_4\,s_3\,s_2 = w_\circ$, so $w = w_\circ\,s_2\,s_3\,s_4\,s_3\,s_1$.) In general, there are many ways to obtain w from $w_\circ$ by simple transpositions, and it is not immediately obvious that this definition of $\mathfrak{S}_w$ is independent choices. One can prove it algebraically, by showing that difference operators satisfy the braid relation $\partial_k \partial_{k+1} \partial_k = \partial_{k+1} \partial_k \partial_{k+1}$. Later we will see geometric reasons for this and other properties of Schubert polynomials.

Exercise 10.6.2. Show that $\partial_k \partial_{k+1} \partial_k = \partial_{k+1} \partial_k \partial_{k+1}$ as operators on polynomials. Conclude that for any permutation w, choosing a reduced expression $w = s_{i_1} \cdots s_{i_\ell}$ defines an operator $\partial_w = \partial_{i_1} \circ \cdots \circ \partial_{i_\ell}$ which is independent of the choice of reduced expression.

Several easy properties of difference operators are particularly useful in computing:

- $\partial_k(f) = 0$ if and only if f is symmetric in x_k and x_{k+1}; that is, whenever $f = s_k f$.

- $\partial_k(f)$ is symmetric in x_k and x_{k+1} (so $\partial_k^2 = 0$).

- (Leibniz rule) $\partial_k(f \cdot g) = \partial_k(f) \cdot g + (s_k f) \cdot \partial_k(g)$.

It follows that if a polynomial f is symmetric in x_k and x_{k+1}, it acts as a scalar with respect to ∂_k; that is, $\partial_k(f \cdot g) = f \cdot \partial_k(g)$. This means that if $I \subseteq R[x]$ is any ideal whose generators are symmetric in x_k and x_{k+1}, then ∂_k descends to an operator on $R[x]/I$. In particular, all divided difference operators act on $H_T^* Fl(V) = \Lambda[x_1, \ldots, x_n]/I$, using the presentation from Chapter 4.

Example 10.6.3. The double Schubert polynomials for $n = 3$ are as follows:

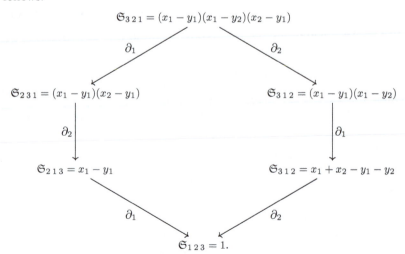

$$\mathfrak{S}_{321} = (x_1 - y_1)(x_1 - y_2)(x_2 - y_1)$$

$\partial_1 \qquad\qquad \partial_2$

$$\mathfrak{S}_{231} = (x_1 - y_1)(x_2 - y_1) \qquad\qquad \mathfrak{S}_{312} = (x_1 - y_1)(x_1 - y_2)$$

$\partial_2 \qquad\qquad\qquad\qquad\qquad\qquad \partial_1$

$$\mathfrak{S}_{213} = x_1 - y_1 \qquad\qquad\qquad \mathfrak{S}_{312} = x_1 + x_2 - y_1 - y_2$$

$\partial_1 \qquad\qquad \partial_2$

$$\mathfrak{S}_{123} = 1.$$

We will use divided difference operators to prove the Giambelli formula. Let $E_\bullet$ be the standard flag as before, with quotients V/E^q, and let $\mathbb{S}_p \subset V$ be the tautological subbundle on $Fl(V)$. Let

$$x_i = -c_1^T(\mathbb{S}_i/\mathbb{S}_{i-1}) \quad \text{and} \quad y_i = -c_1^T(E^{i-1}/E^i),$$

so $y_i = -\chi_i$, where T acts on V with characters $\chi_1, \ldots, \chi_n$. (In Chapter 4, the variables x_i had the opposite sign!)

Theorem 10.6.4. *We have* $\sigma_w = \mathfrak{S}_w(x; y)$ *in* $H_T^* Fl(V)$.

To prove this, we need a basic construction. Let $Fl = Fl(V)$, and let

$$Fl(\widehat{k}) = Fl(1, \ldots, k-1, k+1, \ldots, n-1; V)$$

be the partial flag variety omitting the k-dimensional subspace. There is a diagram

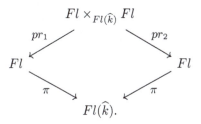

The fiber product parametrizes pairs $(F_\bullet^{(1)}, F_\bullet^{(2)})$ of flags such that $F_i^{(1)} = F_i^{(2)}$ for all $i \neq k$. This identifies the projection pr_2 as a $\mathbb{P}^1$-bundle

$$Fl \times_{Fl(\widehat{k})} Fl \cong \mathbb{P}(\mathbb{S}_{k+1}/\mathbb{S}_{k-1}) \xrightarrow{pr_2} Fl.$$

The corresponding tautological subbundle is

$$\mathbb{L} = pr_1^*(\mathbb{S}_k/\mathbb{S}_{k-1}) \subseteq \mathbb{S}_{k+1}/\mathbb{S}_{k-1},$$

so $c_1^T(\mathbb{L}^\vee) = pr_1^* x_k$. The homomorphism $pr_{2*} pr_1^* \colon H_T^* Fl \to H_T^* Fl$ reduces degrees by 2.

Lemma 10.6.5. *Under the isomorphism* $H_T^* Fl(V) \cong \Lambda[x_1, \ldots, x_n]/I$, *we have* $pr_{2*} pr_1^* = \partial_k$.

Proof We need to compute the pushforward pr_{2*}. Write any class $\alpha \in H_T^* Fl$ as $\alpha = a + bx_k$, where a, b are symmetric in x_k and x_{k+1}. (To do this, recall from Chapter 4 that $\{x_1^{i_1} \cdots x_n^{i_n} \mid 0 \leq i_j \leq n - j\}$ is a Λ-basis for $H_T^* Fl$, as is any permutation of the x-variables. Choosing a permutation so that $k \mapsto n-1$ and $k+1 \mapsto n$ gives a basis where α can be written in the desired form.)

Now $pr_1^* \alpha = a + bc_1^T(\mathbb{L}^\vee)$, so by basic properties of Chern classes, we have $pr_{2*}pr_1^*\alpha = b$. (See Appendix A, §A.5.) But this agrees with $\partial_k(a + bx_k)$. □

This shows that $pr_{2*}pr_1^*$ acts as ∂_k. Next, we check that it operates on Schubert classes just as ∂_k operates on Schubert polynomials.

Lemma 10.6.6. *For $w \in S_n$,*

$$pr_{2*}pr_1^*\sigma_w = \begin{cases} \sigma_{ws_k} & \text{if } \ell(ws_k) < \ell(w); \\ 0 & \text{otherwise.} \end{cases}$$

Proof First we will show that $pr_2(pr_1^{-1}\Omega_w^\circ) = \Omega_w^\circ \cup \Omega_{ws_k}^\circ$, for any w and k. To do this, it suffices to keep track of B^--orbits, since both projections are B^--equivariant. The T-fixed points in $pr_1^{-1}(p_w)$ are (p_w, p_w) and (p_w, p_{ws_k}), because this fiber is naturally identified with $\mathbb{P}^1 = \mathbb{P}(\mathbb{C}e_{w(k)} \oplus \mathbb{C}e_{w(k+1)})$. (It is illustrative to check this with matrix representatives.) Projecting by pr_2 and taking B^--orbits, we see the image is $\Omega_w^\circ \cup \Omega_{ws_k}^\circ$, as claimed.

Next, we observe that

$$\dim pr_1^{-1}\Omega_w = \dim \Omega_w + 1 = \begin{cases} \dim \Omega_{ws_k} & \text{if } \ell(ws_k) < \ell(w); \\ \dim \Omega_{ws_k} + 2 & \text{otherwise.} \end{cases}$$

So $\dim pr_2(pr_1^{-1}\Omega_w) < \dim pr_1^{-1}\Omega_w$ if $\ell(ws_k) > \ell(w)$, and we see that $pr_{2*}pr_1^*\sigma_w = 0$ in this case.

Finally, in the case $\ell(ws_k) < \ell(w)$, we claim that the projection pr_2 maps the open set $B^- \cdot (p_w, p_{ws_k}) \subseteq pr_1^{-1}\Omega_w$ isomorphically onto $\Omega_{ws_k}^\circ$. The explicit description of $U^-(w)$ found in Exercise 10.2.1 implies that $U^-(w) \subseteq U^-(ws_k)$ if and only if $\ell(ws_k) < \ell(w)$. So we have $B^- \cdot (p_w, p_{ws_k}) = U^-(ws_k) \cdot (p_w, p_{ws_k})$, and this orbit is a principal homogeneous space for $U^-(ws_k)$, as is $\Omega_{ws_k}^\circ$. Any equivariant map between principal homogeneous spaces is an isomorphism. It follows that $pr_{2*}pr_1^*\sigma_w = \sigma_{ws_k}$ in this case. □

To complete the proof of the theorem, we must check the formula in the base case $w = w_\circ$. This is the easiest part: $\Omega_{w_\circ}$ is defined by conditions

$\mathrm{rk}(\mathbb{S}_i \to V/E^i) = 0$ for all $i \quad \Leftrightarrow \quad \mathbb{S}_i/\mathbb{S}_{i-1} \to V/E^i$ is 0 for all i.

That is, $\Omega_{w_\circ}$ is the zeroes of a section of

$$\bigoplus_{i=1}^{n-1} \mathrm{Hom}(\mathbb{S}_i/\mathbb{S}_{i-1}, V/E^i) = \bigoplus_{i+j\leq n} (\mathbb{S}_i/\mathbb{S}_{i-1})^\vee \otimes E^{j-1}/E^j).$$

Taking the top Chern class, we obtain

$$\sigma_{w_\circ} = \prod_{i+j\leq n} (x_i - y_j),$$

as needed. $\qquad\qquad\qquad\qquad\qquad\qquad\qquad\qquad\qquad\qquad\qquad\qquad$ $\square$

Exchanging y_i with $\widetilde{y}_i = y_{n+1-i} = -\chi_{n+1-i}$, the same argument shows $\widetilde{\sigma}_w = \mathfrak{S}_w(x; \widetilde{y})$ in $H_T^* Fl(V)$.

10.7 Multiplying Schubert classes

A positive combinatorial formula for the structure constants c_{uv}^w for multiplying Schubert classes in $Fl(V)$—or equivalently, for multiplying Schubert polynomials—is unknown, in general. However, there are formulas in special cases, as well as a characterization result similar to what we saw for Grassmannians.

These structure constants satisfy properties analogous to those for the Grassmannian. As before, we assume T acts on $V = \mathbb{C}^n$ by distinct characters $\chi_1, \ldots, \chi_n$.

Lemma 10.7.1. *The coefficients c_{uv}^w satisfy the following properties:*

(i) $c_{uv}^w = 0$ unless $u \leq w$ and $v \leq w$.

(ii) $c_{uv}^v = \sigma_u|_v$.

(iii) $c_{uu}^u = \displaystyle\prod_{\substack{i<j \\ u(i)>u(j)}} (\chi_{u(j)} - \chi_{u(i)})$.

The proof is essentially the same as for the Grassmannian (Lemma 9.6.1).

Next we will see a "Chevalley–Monk" formula for multiplying by divisor classes. We need a formula for restricting divisors, corresponding to simple transpositions s_k.

Exercise 10.7.2.   Show that $\mathfrak{S}_{s_k}(x;y) = x_1 + \cdots + x_k - y_1 - \cdots - y_k$. Setting $y_i = -\chi_i$, conclude that

$$\sigma_{s_k}|_w = \sum_{i=1}^{k} \chi_i - \sum_{i=1}^{k} \chi_{w(i)}$$

for each permutation w. In particular, for any $w \neq v$, there is a k such that $\sigma_{s_k}|_w \neq \sigma_{s_k}|_v$.

To state the formula, we need some notation. Recall that $v \gtrdot w$ if and only if there is a transposition τ_{ij} so that $v = w\tau_{ij}$ and $\ell(v) = \ell(w) + 1$. We will write $v \gtrdot_k w$ if $v = w\tau_{ij}$ with $i \leq k < j$.

Theorem 10.7.3 (Equivariant Monk formula).   *We have*

$$\sigma_{s_k} \cdot \sigma_w = \sum_{w^+ \gtrdot_k w} \sigma_{w^+} + \left(\sum_{i=1}^{k} \chi_i - \sum_{i=1}^{k} \chi_{w(i)} \right) \sigma_w.$$

For example, using a vertical bar to indicate the position k, we compute

$$\sigma_{s_3} \cdot \sigma_{21534} = \sigma_{315|24} + \sigma_{235|14} + (\chi_3 - \chi_5)\sigma_{21534}$$

and

$$\sigma_{s_2} \cdot \sigma_{21534} = \sigma_{51|234} + \sigma_{31|524} + \sigma_{25|134} + \sigma_{23|514}.$$

In the theorem, the sum over w^+ is the classical Monk formula. Combinatorial and algebraic proofs using Schubert polynomials can be found in the Notes. (The reader may find it a pleasant challenge – note that one can work with single Schubert polynomials $\mathfrak{S}_w(x)$.) The equivariant coefficient is $\sigma_{s_k}|_w$, and it appears for the same reason as in the Grassmannian case (Theorem 9.6.2).

Finally, we have a characterization result analogous to the one for Grassmannians.

Theorem 10.7.4.   *The coefficients c_{uv}^w are the unique homogeneous polynomials (of degree $\ell(u)+\ell(v)-\ell(w)$) in Λ which satisfy the following properties, for all $1 \leq k < n$:*

(i) $c_{uu}^u = \sigma_u|_u = \displaystyle\prod_{\substack{i<j \\ u(i)>u(j)}} (\chi_{u(j)} - \chi_{u(i)}),$

(ii) $(\sigma_{s_k}|_u - \sigma_{s_k}|_v)\, c_{uv}^u = \sum_{v^+ \,\gtrdot_k\, v} c_{uv^+}^u,$ *and*

(iii) $(\sigma_{s_k}|_w - \sigma_{s_k}|_u)\, c_{uv}^w = \sum_{u^+ \,\gtrdot_k\, u} c_{u^+ v}^w - \sum_{w^- \,\lessdot_k\, w} c_{uv}^{w^-}.$

Again, the proof is almost the same as in the Grassmannian case. In showing these properties uniquely characterize c_{uv}^w, one needs to choose k so that the factors $(\sigma_{s_k}|_w - \sigma_{s_k}|_u)$ are nonzero; this can be done, thanks to Exercise 10.7.2.

10.8 Partial flag varieties

In Chapter 4, we saw presentations and bases for the equivariant cohomology of partial flag varieties $Fl(\mathbf{d}, V)$, where the dimensions of subspaces are indicated by $\mathbf{d} = (0 < d_1 < \cdots < d_r < n)$. Partial flag varieties have T-equivariant decompositions into Schubert cells, and their equivariant geometry is similar to what we have seen for Grassmannians and complete flag varieties. Here we state the facts; working out examples and proofs make useful exercises.

There is a *Young subgroup*

$$S_{\mathbf{d}} = S_{d_1} \times S_{d_2 - d_1} \times \cdots \times S_{n - d_r} \subseteq S_n,$$

and Schubert varieties in $Fl(\mathbf{d}, V)$ are indexed by cosets $[w] \in S_n/S_{\mathbf{d}}$. Each coset has a **minimal representative** $w^{\min}$, characterized by requiring $w(i) < w(i+1)$ whenever i is not among the d_p. Similarly, there is a **maximal representative** $w^{\max}$, with $w(i) > w(i+1)$ whenever i is not among the d_p.

For any two cosets $[u], [v] \in S_n/S_{\mathbf{d}}$, we have $u^{\min} \leq v^{\min}$ if and only if $u^{\max} \leq v^{\max}$ in Bruhat order on S_n. This induces partial order on $S_n/S_{\mathbf{d}}$, also called Bruhat order, by

$$[u] \leq [v] \quad \text{iff} \quad u^{\min} \leq v^{\min} \quad \text{iff} \quad u^{\max} \leq v^{\max}.$$

For each coset $[w]$, and any representative $w \in [w]$, there is a Schubert variety

$$\Omega_{[w]} = \Big\{ F_\bullet \mid \mathrm{rk}(F_{d_p} \to V/E^q) \leq r_w(d_p, q) \text{ for } 1 \leq p \leq r,\, 1 \leq q \leq n \Big\}$$
$$= \Big\{ F_\bullet \mid \dim(F_{d_p} \cap E^q) \geq k_w(d_p, q) \text{ for } 1 \leq p \leq r,\, 1 \leq q \leq n \Big\}$$

in $Fl(\mathbf{d}, V)$. Its codimension is equal to $\ell(w^{\min})$. Schubert cells $\Omega^\circ_{[w]}$ are defined similarly, with equalities on ranks and dimensions in place of inequalities; they are B^--orbits of fixed points $p_{[w]}$.

Containment among Schubert varieties is described by Bruhat order on cosets. Equivalently, one has $p_{[v]} \in \Omega_{[w]}$ if and only if $[v] \geq [w]$.

Points in $Fl(\mathbf{d}, V)$ can be represented in matrices, so that the space F_{d_p} is the span of the first d_p columns. Schubert cells can be written by taking a minimal representative $w = w^{\min}$ and using column echelon form having the 1's in the permutation matrix A_w as pivots. For example, in $Fl(1, 4; \mathbb{C}^6)$, with $w = 2\,1\,4\,6\,3\,5$, the Schubert cell is

$$\Omega^\circ_{[w]} = \begin{bmatrix} 0 & 1 & 0 & 0 & 0 & 0 \\ 1 & 0 & 0 & 0 & 0 & 0 \\ * & * & 0 & 0 & 1 & 0 \\ * & 0 & 1 & 0 & 0 & 0 \\ * & * & * & 0 & 0 & 1 \\ * & 0 & 0 & 1 & 0 & 0 \end{bmatrix}.$$

Since the cells $\Omega^\circ_{[w]}$ decompose the partial flag variety, the Schubert classes $\sigma_{[w]} = [\Omega_{[w]}]^T$ form a basis for $H^*_T Fl(\mathbf{d}, V)$ over Λ, as $[w]$ ranges over cosets $S_n/S_{\mathbf{d}}$, or equivalently, as w ranges over minimal representatives.

Opposite Schubert cells and varieties are defined in the same way, using the opposite flag $\tilde{E}^\bullet$ in place of $E^\bullet$. The opposite Schubert cell $\tilde{\Omega}^\circ_{[w]}$ is the B-orbit of $p_{[w_\circ w]}$. The fixed points of $\tilde{\Omega}_{[w_\circ w]}$ are those $p_{[v]}$ such that $[v] \leq [w]$. In particular,

$$\Omega_{[u]} \cap \tilde{\Omega}_{[w_\circ w]} = \emptyset \quad \text{unless} \quad [u] \leq [w].$$

The opposite Schubert classes $\tilde{\sigma}_{[w_\circ w]} = [\tilde{\Omega}_{[w_\circ w]}]^T$ form the Poincaré dual basis to the basis of Schubert classes $\sigma_{[w]}$.

The pullback by the projection $\pi\colon Fl(V) \twoheadrightarrow Fl(\mathbf{d}, V)$ embeds $H^*_T Fl(\mathbf{d}, V)$ in $H^*_T Fl(V) = \Lambda[x]/I$ as the subring of invariants for the natural action of $S_{\mathbf{d}}$ on the x variables. It follows that $\pi^* \sigma_{[w]} = \sigma_{w^{\min}}$, and therefore:

The class $\sigma_{[w]}$ is represented by the Schubert polynomial $\mathfrak{S}_{w^{\min}}(x; y)$.

(It is easy to see that $\mathfrak{S}_{w^{\min}}(x; y)$ is invariant under $S_{\mathbf{d}}$, since $w^{\min} s_k > w^{\min}$ unless k is among the d_p.)

There are characterizations of the structure constants $c_{[u],[v]}^{[w]}$ similar to what we saw in §10.7, as well as formulas for multiplying by a divisor class. We will see generalizations to other projective homogeneous varieties in later chapters.

10.9 Stability

A key property of Schubert polynomials is that they are *stable* representatives of Schubert classes, with respect to natural embeddings of symmetric groups and maps between flag varieties. That is, for $w \in S_n$, and the standard embedding $\iota\colon S_n \hookrightarrow S_m$ (for $m > n$), we have

$$\mathfrak{S}_{\iota(w)}(x; y) = \mathfrak{S}_w(x; y).$$

From the definition of Schubert polynomials, it suffices to prove this for the longest element $w = w_{\circ}^{(n)}$ in S_n, so

$$\iota(w_{\circ}^{(n)}) = n \cdots 2\, 1\, n+1 \cdots m$$

in S_m. It also suffices to treat the case $m = n+1$. This is rather straightforward algebraically:

Exercise 10.9.1. Assume $m = n + 1$, and show

$$\mathfrak{S}_{\iota(w_{\circ}^{(n)})}(x; y) = \partial_n \cdots \partial_2 \partial_1 \mathfrak{S}_{w_{\circ}^{(n+1)}}(x; y)$$

by direct computation.

A more precise statement comes from the geometry of certain partial flag varieties. Consider

$$Fl^{(n)}(\mathbb{C}^m) = Fl(1, \ldots, n; \mathbb{C}^m)$$

for $n \le m$. For $m \gg 0$, and a torus T acting on $\mathbb{C}^m$, there are no relations in low degree among the Chern classes $x_i = -c_1^T(\mathbb{S}_i/\mathbb{S}_{i-1})$, so

$$H_T^* Fl^{(n)}(\mathbb{C}^m) = \Lambda[x_1, \ldots, x_n] \text{ (modulo relations in degree } > m - n).$$

(We saw this for ordinary cohomology when describing the classifying space $\mathbb{B}B$ in Chapter 2; the equivariant statement holds for the same reasons.)

Let $S_m^{(n)} \subset S_m$ be the subset

$$S_m^{(n)} = \left\{ w \in S_m \,\middle|\, w(i) < w(i+1) \text{ for } i \geq n \right\}.$$

These are minimal representatives for the cosets parametrizing fixed points and Schubert varieties in $Fl^{(n)}(\mathbb{C}^m)$, as we saw in the previous section. So for each $w \in S_m^{(n)}$, there is a Schubert variety

$$\Omega_{[w]} = \left\{ F_\bullet \,\middle|\, \dim(F_p \cap E^q) \geq k_w(p,q) \text{ for } 1 \leq p \leq n, \, 1 \leq q \leq m \right\}$$

in $Fl^{(n)}(\mathbb{C}^m)$. The classes of these Schubert varieties form a basis for $H_T^* Fl^{(n)}(\mathbb{C}^m)$.

Schubert varieties and their classes are compatible with standard embeddings of partial flag varieties. For $m \leq m'$, we have an inclusion $\mathbb{C}^m \subseteq \mathbb{C}^{m'}$ as the span of the first m standard basis vectors, and a corresponding embedding $\iota\colon Fl^{(n)}(\mathbb{C}^m) \hookrightarrow Fl^{(n)}(\mathbb{C}^{m'})$. Echoing this notation, let $\iota\colon S_m \hookrightarrow S_{m'}$ be the standard embedding of symmetric groups, so $\iota(w)(k) = k$ for $k > m$.

Exercise 10.9.2.   Show that

$$\iota(Fl^{(n)}(\mathbb{C}^m)) = \widetilde{\Omega}_{[w_\circ^{(m')} \cdot \iota(w_\circ^{(m)})]}$$

inside $Fl^{(n)}(\mathbb{C}^{m'})$, where $w_\circ^{(m)}$ is the longest element in S_m. Also, show that $\iota^{-1}\Omega_{[\iota(w)]} = \Omega_{[w]}$ for any w.

Conclude that

$$\iota^* \sigma_{[\iota(w)]} = \sigma_{[w]} \quad \text{in} \quad H_T^* Fl^{(n)}(\mathbb{C}^m),$$

and $\iota^* \sigma_{[u]} = 0$ unless $[u] \leq [w_\circ^{(m)}]$. For $m = n$, this says $\iota^* \sigma_{[u]} = 0$ unless $u \in S_n$.

Using the standard embeddings of symmetric groups, consider the group $S_\infty = \bigcup_m S_m$ (permutations of $\{1, 2, \ldots\}$ which fix all but finitely many integers), and the subset $S_\infty^{(n)} = \bigcup_m S_m^{(n)}$ (the elements w such that $w(i) < w(i+1)$ whenever $i \geq n$). By the above exercise, one can define $\sigma_{[w]}$ unambiguously for $w \in S_\infty^{(n)}$.

In the previous section, we saw that $\sigma_{[w]} = \mathfrak{S}_w(x; y)$ for $w \in S_m^{(n)}$. If m is large enough relative to the degree of $\sigma_{[w]}$ so that $H_T^* Fl^{(n)}(\mathbb{C}^m)$ has no relations in this degree, then $\mathfrak{S}_w(x; y)$ is the unique polynomial representative for $\sigma_{[w]}$. (Taking $m > \ell(w) + n$ suffices.)

These observations are summarized by the following theorem:

Theorem 10.9.3. *For $w \in S_\infty^{(n)}$, $\mathfrak{S}_w(x; y)$ is the unique polynomial in $\Lambda[x_1, \ldots, x_n]$ mapping to $\sigma_w \in H_T^* Fl^{(n)}(\mathbb{C}^m) = \Lambda[x_1, \ldots, x_n]/I$ for all $m \geq n$.*

In other words, the Schubert polynomials $\mathfrak{S}_w(x; y)$ are the unique polynomials in $\Lambda[x_1, \ldots, x_n]$ stably representing Schubert classes. (The theorem follows from three facts: (1) Schubert polynomials represent Schubert classes; (2) for fixed d and sufficiently large m, the graded rings $H_T^* Fl^{(n)}(\mathbb{C}^m)$ and $\Lambda[x_1, \ldots, x_n]$ are isomorphic in degree at most d; and (3) Schubert polynomials are stable.)

Stability may also be expressed by considering formulas in an infinite flag manifold, as we will see in Chapter 12.

10.10 Properties of Schubert polynomials

Schubert polynomials have remarkable algebraic and combinatorial properties, many of which can be proved easily with geometry. We will repeatedly use the fact that identities of polynomials can be deduced from identities of classes in $H_T^* Fl^{(n)}(\mathbb{C}^m)$ by taking sufficiently large n and m.

10.10.1 Schubert Classes. From the definition, $\mathfrak{S}_w(x, y)$ is a homogeneous polynomial of degree $\ell(w)$ in $\mathbb{Z}[x_1, \ldots, x_n, y_1, \ldots, y_n]$. Perhaps its most fundamental property is the one we proved in Theorem 10.6.4. Let T act on $Fl(\mathbb{C}^n)$ by characters $\chi_1, \ldots, \chi_n$.

> *For $w \in S_n$, the polynomial $\mathfrak{S}_w(x; y)$ maps to $\sigma_w = [\Omega_w]^T$ under the evaluation*
>
> $$x_i = -c_1^T(\mathbb{S}_i/\mathbb{S}_{i-1}), \quad y_i = -\chi_i.$$
>
> *Similarly, $\mathfrak{S}_w(x; \widetilde{y})$ maps to $\widetilde{\sigma}_w = [\widetilde{\Omega}_w]^T$ under the evaluation*
>
> $$x_i = -c_1^T(\mathbb{S}_i/\mathbb{S}_{i-1}), \quad \widetilde{y}_i = -\chi_{n+1-i}.$$

Here $\mathbb{S}_i$ is the tautological bundle on $Fl(\mathbb{C}^n)$.

The classes σ_w have better stability properties, so from now on we focus on them and the corresponding polynomials $\mathfrak{S}_w(x; y)$. (To recover statements for B-invariant Schubert varieties, one can substitute $\tilde{y}_i = y_{n+1-i}$ in any formula.)

10.10.2 Stability. Consider the standard embedding $\iota\colon S_n \hookrightarrow S_{n'}$ for any $n' > n$, so $\iota(w)(i) = i$ for $n < i \leq n'$. Another fundamental property of Schubert polynomials is stability with respect to such embeddings, as we saw in the previous section:

We have $\mathfrak{S}_{\iota(w)}(x; y) = \mathfrak{S}_w(x; y)$.

The stability property lets us unambiguously regard a permutation $w \in S_n$ as one in S_m, for any $m \geq n$. In fact, let $S_\infty = \bigcup S_n$ be the infinite symmetric group, that is, the group of all bijections $w\colon \mathbb{Z}_{>0} \to \mathbb{Z}_{>0}$ such that $w(i) = i$ for all but finitely many i. Let $x = (x_1, x_2, \ldots)$ and $y = (y_1, y_2, \ldots)$ be infinite sets of variables, and let $\mathbb{Z}[x, y]$ be the polynomial ring in these variables. By stability, $\mathfrak{S}_w(x; y)$ is a well-defined element of $\mathbb{Z}[x, y]$ for any $w \in S_\infty$, so we usually suppress the notation $\iota(w)$.

The Schubert class and stability properties characterize Schubert polynomials, as we saw in Theorem 10.9.3. That is, $\mathfrak{S}_w$ is the unique polynomial which represents σ_w in $H_T^* Fl(\mathbb{C}^m)$ for all sufficiently large m.

10.10.3 Basis. Schubert polynomials form a linear basis for all polynomials:

The set $\{\mathfrak{S}_w \mid w \in S_\infty\}$ is a basis for $\mathbb{Z}[x, y]$ as a module over $\mathbb{Z}[y]$.

We can be more precise:

Let $S_\infty^{(n)} \subseteq S_\infty$ be the subset of w such that $w(i) < w(i+1)$ for all $i > n$. Then

(1) $\mathfrak{S}_w \in \mathbb{Z}[y][x_1, \ldots, x_n]$ if and only if $w \in S_\infty^{(n)}$, and

(2) $\{\mathfrak{S}_w \mid w \in S_\infty^{(n)}\}$ is a basis for $\mathbb{Z}[y][x_1, \ldots, x_n]$ as a module over $\mathbb{Z}[y]$.

For (1), the condition $w \in S_\infty^{(n)}$ is equivalent to $ws_k > w$ for any $k > n$, which in turn is equivalent to $\partial_k \mathfrak{S}_w = 0$. This means that $\mathfrak{S}_w$

is symmetric in the variables $x_{n+1}, x_{n+2}, \ldots$. Since only finitely many variables appear in $\mathfrak{S}_w$, it is the same to say only $x_1, \ldots, x_n$ appear.

The proof of (2) follows from results of §10.9. Consider the partial flag variety $Fl^{(n)}(\mathbb{C}^m)$, with the projection $\pi \colon Fl(\mathbb{C}^m) \to Fl^{(n)}(\mathbb{C}^m)$, for $m \gg 0$. As w ranges over $S_m^{(n)}$, the Schubert classes $\sigma_{[w]}$ form a basis for $H_T^* Fl^{(n)}(\mathbb{C}^m)$ over $\Lambda = \mathbb{Z}[y_1, \ldots, y_m]$. For sufficiently large m, the Schubert polynomial $\mathfrak{S}_w$ is the unique representative for $\sigma_{[w]}$ (Theorem 10.9.3). $\qquad\square$

Schubert polynomials also form a basis for the ideal I in the presentation $H_T^* Fl(\mathbb{C}^n) = \Lambda[x_1, \ldots, x_n]/I$. For this, we restrict to finitely many y variables by setting $y_i = 0$ for $i > n$, so $\Lambda = \mathbb{Z}[y_1, \ldots, y_n]$.

The ideal I has a basis $\{\mathfrak{S}_w \mid w \in S_\infty^{(n)}, w \notin S_n\}$ as a Λ-module.

Consider the embedding $\iota \colon Fl(\mathbb{C}^n) \hookrightarrow Fl^{(n)}(\mathbb{C}^m)$ defined by regarding $\mathbb{C}^n = \widetilde{E}_n \subset \mathbb{C}^m$ as part of the opposite flag $\widetilde{E}_\bullet$ in $\mathbb{C}^m$. By Exercise 10.9.2, for $w \in S_m^{(n)}$ we have

$$\iota^* \sigma_w = \begin{cases} \sigma_w & \text{if } w \in S_n; \\ 0 & \text{if } w \notin S_n, \end{cases}$$

and the claim follows. $\qquad\square$

10.10.4 Multiplication. Schubert polynomials multiply exactly as Schubert classes in $H_T^* Fl(\mathbb{C}^n)$, when $n \gg 0$. More precisely, by the basis property, we can write

$$\mathfrak{S}_u \cdot \mathfrak{S}_v = \sum_{w \in S_\infty} c_{uv}^w \, \mathfrak{S}_w,$$

for some polynomials $c_{uv}^w \in \mathbb{Z}[y]$.

We have

$$\sigma_u \cdot \sigma_v = \sum_{w \in S_n} c_{uv}^w \, \sigma_w$$

in $H_T^* Fl(\mathbb{C}^n)$.

That is, the product of Schubert classes in $H_T^* Fl(\mathbb{C}^n)$ is obtained by multiplying the corresponding Schubert polynomials in $\mathbb{Z}[x, y]$ and discarding terms $\mathfrak{S}_w$ for $w \notin S_n$. This follows from the description of the kernel of $\mathbb{Z}[x, y] \to H_T^* Fl(\mathbb{C}^n)$.   □

10.10.5 Localization and Interpolation. Like double Schur polynomials, the double Schubert polynomials satisfy and are characterized by interpolation properties.

For $w \in S_n$, the double Schubert polynomial $\mathfrak{S}_w$ specializes as

$$\mathfrak{S}_w(y_{w(1)}, \ldots, y_{w(n)}; y_1, \ldots, y_n) = \prod_{\substack{i<j \\ w(i)>w(j)}} (y_{w(i)} - y_{w(j)}) \quad (*)$$

and

$$\mathfrak{S}_w(y_{v(1)}, \ldots, y_{v(n)}; y_1, \ldots, y_n) = 0 \quad \text{if } v \not\geq w, \quad (**)$$

and it is the unique homogeneous polynomial of degree $\ell(w)$ satisfying $(*)$ and $(**)$.

The formulas $(*)$ and $(**)$ follow by restricting the corresponding Schubert classes: letting T act on $\mathbb{C}^n$ by characters $\chi_i = -y_i$, we have seen $\sigma_w|_w = \prod(y_{w(i)} - y_{w(j)})$ in §10.2, and for $v \not\geq w$ we have $\sigma_w|_v = 0$ since $p_v \notin \Omega_w$.

These properties characterize Schubert classes, by the same argument as for Grassmannians (Lemma 9.3.1). So they also characterize Schubert polynomials.   □

10.10.6 Grassmannian Permutations. Consider a partition λ in the $d \times (n-d)$ rectangle, with corresponding subsets $I = \{i_1 < \cdots < i_d\}$ and $J = \{j_1 < \cdots < j_{n-d}\}$ of $\{1, \ldots, n\}$, as in Chapter 9. We define a permutation in S_n by

$$w(\lambda) = i_1 \cdots i_d \, j_1 \cdots j_{n-d}.$$

This is a *Grassmannian permutation*, meaning it has a single descent, $w(d) > w(d+1)$. (All Grassmannian permutations arise this way, for some d.)

We have

$$\mathfrak{S}_{w(\lambda)}(x; y) = s_\lambda(x|y),$$

where $x = (x_1, \ldots, x_d)$. *In particular, Schubert polynomials for Grassmannian permutations have determinantal expressions.*

To prove this, we compare opposite Schubert classes in the Grassmannian and flag variety via the projection $\pi\colon Fl(\mathbb{C}^n) \to Gr(d, \mathbb{C}^n)$. We have $\pi^{-1}\Omega_\lambda = \Omega_{w(\lambda)}$, so $\pi^*\sigma_\lambda = \sigma_{w(\lambda)}$. Since $\mathfrak{S}_{w(\lambda)}(x; y) = \sigma_{w(\lambda)}$ and $s_\lambda(x|y) = \sigma_\lambda$ (see Corollary 9.7.2), the identity of polynomials follows, by taking n large enough so that there are no relations among the participating variables. $\qquad\square$

For example, take $n = 3$, $d = 1$, and $\lambda = (2)$. We have $I = \{3\}$ and $J = \{1, 2\}$, so $w(\lambda) = 3\,1\,2$. Comparing polynomials, we see

$$s_{\square}(x|y) = (x_1 - y_1)(x_1 - y_2) = \mathfrak{S}_{3\,1\,2}(x; y),$$

as claimed.

Notes

The "Bruhat order" on the symmetric group was first discussed by Ehresmann (1934) in the context of the (Schubert) cell decomposition of flag varieties and some other homogeneous spaces. Up to re-indexing, he uses the tableau criterion to describe a partial order on cells. A proof of Lemma 10.4.5, in the context of general Coxeter groups, can be found in (Humphreys, 1990, §5.10). Other criteria for Bruhat order can be found in the notes by Macdonald (1991) or the book by Björner and Brenti (2005).

Bernstein, Gelfand, and Gelfand (1973) and Demazure (1974) used divided difference operators to compute the cohomology classes of Schubert varieties, as in Lemma 10.6.6. The operators themselves go back at least to Newton's interpolation formulas, and can also be seen in some of Giambelli's reduction formulas (Giambelli, 1904).

Schubert polynomials were introduced by Lascoux and Schützenberger, first in their "single" versions $\mathfrak{S}_w(x) = \mathfrak{S}_w(x; 0)$ as canonical representatives for Schubert classes, with the double Schubert polynomials appearing shortly afterward in the context of interpolation (Lascoux and Schützenberger, 1982, 1985). The geometric significance of double Schubert polynomials seems to have been noticed somewhat later (Fulton, 1992; Fehér and Rimányi, 2002, 2003; Knutson and Miller, 2005). Algebraic proofs of many of their properties can be found in (Macdonald, 1991).

The intersections $\Omega_u \cap \widetilde{\Omega}_v$ are called *Richardson varieties*, and they appear frequently in Schubert calculus. Irreducibility was proved by Richardson (1992), extending a corresponding result of Deodhar (1985) for the intersections of cells $\Omega_u^\circ \cap \widetilde{\Omega}_v^\circ$. Our proof of Proposition 10.5.4 is essentially that of Brion and Lakshmibai (2003), who also observe that Richardson varieties have local properties of Schubert varieties (as stated in Corollary 10.5.5). The fact that Schubert varieties are Cohen–Macaulay is due to many authors; a particularly elegant argument using Frobenius splitting was given by Ramanathan (1985). A more detailed study of the local properties of Richardson varieties can be found in (Knutson et al., 2013).

Monk's formula appears in (Monk, 1959). An algebraic proof using Schubert polynomials is in (Macdonald, 1991, (4.15″)), and a combinatorial one was given by Bergeron and Billey (1993). Chevalley (1994) gave a general formula for multiplying by a divisor in a generalized flag variety G/P. We will see a proof of this formula in Chapter 19.

Hints for Exercises

Exercise 10.4.2. For each p, the number of elements of $\{v(1), \ldots, v(p)\}$ which are $\leq q$ is equal to $r_v(p, q)$. In fact, it suffices to consider only those p such that $w(p) > w(p+1)$, i.e., the descents of w; see (Björner and Brenti, 2005, §2.6).

Exercise 10.5.1. Note that $\widetilde{\Omega}_w^\circ = w_\circ \cdot \Omega_w^\circ$.

Exercise 10.6.2. To prove the braid identity, it suffices to consider the case $k = 1$. Writing $\partial_k = \frac{1}{x_k - x_{k+1}}(1 - s_k)$, one can expand the operator $\partial_1 \partial_2 \partial_1$ as

$$\frac{1}{x_1 - x_2}(1 - s_1)\frac{1}{x_2 - x_3}(1 - s_2)\frac{1}{x_1 - x_2}(1 - s_1)$$
$$= \frac{1}{(x_1 - x_2)(x_1 - x_3)(x_2 - x_3)}\sum_{w \in S_3}(-1)^{\ell(w)}w,$$

and obtain the same expansion for $\partial_2 \partial_1 \partial_2$.

Exercise 10.7.2. There are several easy ways to prove the formula. One is to notice Ω_{s_k} is the locus where $\mathrm{rk}(\mathbb{S}_k \to V/E^k) \leq k - 1$, so its class is $c_1^T(V/E_k - \mathbb{S}_k)$ (cf. §9.6). Another is algebraic, using the interpolation condition of §10.10.5. A third is to compare with Schur polynomials: as noted in §10.10.6, $\mathfrak{S}_{s_k}(x; y) = s_\square(x_1, \ldots, x_k | y)$. Finally, to see $\sigma_{s_k}|_w \neq \sigma_{s_k}|_v$ if $w \neq v$, take k to be the first position where $w(k) \neq v(k)$.

Exercise 10.9.2. The opposite Schubert cell $\widetilde{\Omega}_{[w_\circ^{(m')} \cdot \iota(w(n,m))]} \subseteq Fl^{(n)}$ $(\mathbb{C}^{m'})$ is the B-orbit of $p_{[\iota(w_\circ^{(m)})]}$. In matrices, this is

$$
\begin{bmatrix}
 & & & & & 1 & & & \\
 & * & \cdot & & \cdot\cdot^{\cdot} & & & 0 & \\
 & & & 1 & & & & & \\
\hline
* & * & \cdots & 1 & & & & & \\
* & \cdots & 1 & & & 0 & & 0 & \\
\vdots & & \cdot\cdot^{\cdot} & & & & & & \\
1 & & & & & & & & \\
\hline
 & & & & & & 1 & & \\
 & 0 & & & 0 & & & \cdot\cdot_{\cdot} & \\
 & & & & & & & & 1
\end{bmatrix},
$$

which identifies with the open Schubert cell in $Fl^{(n)}(\mathbb{C}^m)$ under the embedding ι.

11

Degeneracy Loci

There is a close connection between degeneracy loci for maps of vector bundles and equivariant cohomology classes of Schubert varieties. In fact, equivariant classes of Schubert varieties give universal formulas for the classes of degeneracy loci. In this chapter, we will make this connection precise. Along the way, we will see how Schubert polynomials arise naturally in the equivariant cohomology of the space of matrices, and deduce some further properties of these polynomials from the geometry of degeneracy loci.

11.1 The Cayley–Giambelli–Thom–Porteous formula

Let E and F be vector spaces of dimensions n and m, respectively, and set $M = \mathrm{Hom}(F, E)$. The group $G = GL(E) \times GL(F)$ acts on M, by

$$((g, h) \cdot \varphi)(v) = g \cdot \varphi(h^{-1} \cdot v).$$

Choosing bases, this is $M \cong M_{n,m}$, the space of $n \times m$ matrices, with $GL(E)$ acting by left multiplication (row operations) and $GL(F)$ acting by right multiplication (column operations). Since M is contractible, we have

$$H_G^* M = \Lambda_G = \Lambda_{GL(E)} \otimes \Lambda_{GL(F)}$$
$$= \mathbb{Z}[a_1, \ldots, a_n, b_1, \ldots, b_m],$$

where $a_i = c_i^{GL(E)}(E)$ and $b_i = c_i^{GL(F)}(F)$. (See Exercise 3.2.4.)

The G-orbits in M are precisely the sets

$$D_r^\circ = \{\varphi \in M \mid \text{rk}(\varphi) = r\},$$

for $r = 0, \ldots, \min\{m, n\}$, so the (irreducible) G-invariant subvarieties are their closures

$$D_r = \overline{D_r^\circ} = \{\varphi \mid \text{rk}(\varphi) \leq r\}.$$

The class $[D_r]^G$ is a canonical polynomial in $\Lambda_G = \mathbb{Z}[a, b]$. What is it?

The answer is given by the *(Cayley–)Giambelli–Thom–Porteous formula*:

Proposition 11.1.1. *We have*

$$[D_r]^G = \left| c_{n-r+j-i} \right|_{1 \leq i, j \leq m-r},$$

where

$$c = \frac{1 + a_1 + a_2 + \cdots}{1 + b_1 + b_2 + \cdots} = 1 + (a_1 - b_1) + (a_2 - a_1 b_1 + b_1^2 - b_2) + \cdots$$

is the total equivariant Chern class $c = c^G(E - F)$.

The right-hand side of the formula is the Schur determinant $\Delta_\lambda(c)$, where $\lambda = (m - r)^{n-r}$ is the $(m - r) \times (n - r)$ rectangle.

Proof This can be deduced from formulas we already know for the Grassmannian, using a graph construction. With the group $G = GL(E) \times GL(F)$ acting on $Gr = Gr(m, E \oplus F)$ in the evident way via the block-diagonal inclusion in $GL(E \oplus F)$, there is a G-equivariant map

$$f \colon M \to Gr, \qquad \varphi \mapsto \text{Graph}(\varphi)$$

which sends each homomorphism φ to its *graph*

$$\text{Graph}(\varphi) = \{(\varphi(v), v)\} \subseteq E \oplus F.$$

In fact, f embeds M as an open set in Gr, identifying it with the tangent space at the point $0 \oplus F$.

Let $\Omega \subseteq Gr$ be

$$\Omega = \left\{ L \subseteq E \oplus F \mid \dim(L \cap (0 \oplus F)) \geq m - r \right\}.$$

This is a Schubert variety $\Omega_\lambda(E_\bullet)$ with respect to any flag $E_\bullet$ such that $E_m = 0 \oplus F$, where $\lambda = (n - r, \ldots, n - r)$ is the $(m - r) \times (n - r)$

rectangle, as in the statement of the proposition. Observing that $\ker(\varphi) = \mathrm{Graph}(\varphi) \cap (0 \oplus F)$, we see that φ has rank at most r if and only if $\dim(\mathrm{Graph}(\varphi) \cap (0 \oplus F)) \geq m - r$. So $D_r = f^{-1}\Omega$, and since f is an open embedding, it follows that $[D_r]^G = f^*[\Omega]^G$. Since $F \cong f^*E_m$ and $E \cong f^*Q$, we get our formula for $D_r \subseteq M$ from one we have seen for $\Omega \subseteq Gr$, namely the determinantal Kempf–Laksov formula $\Delta_\lambda(c^T(Q - E_m))$ (Theorem 9.2.2). $\qquad\square$

The torus $T = (\mathbb{C}^*)^n \times (\mathbb{C}^*)^m \subseteq G = GL(E) \times GL(F)$, acts on matrices by $(g, h) \cdot A = gAh^{-1}$. If $x_1, \ldots, x_n$ and $y_1, \ldots, y_m$ are the standard characters on T, this action is by the character $x_i - y_j$ on the (i, j) entry of a matrix. The formula restricts to

$$[D_r]^T = s_{\lambda'}(x|y),$$

where λ' is the transposed partition – that is, the $(n - r) \times (m - r)$ rectangle. (This is Corollary 9.7.2, using x in place of $\widetilde{x}$ and y in place of $-\widetilde{y}$.)

Example 11.1.2. Consider $T = \mathbb{C}^*$ acting on $M = M_{n,m}$ by scaling all entries. (Writing z for the identity character of $\mathbb{C}^*$, this is the case $x_1 = \cdots = x_n = z$, $y_1 = \cdots = y_m = 0$.) Then $D_1 \subseteq M$ is the cone over the Segre variety $\mathbb{P}^{n-1} \times \mathbb{P}^{m-1} \subseteq \mathbb{P}^{nm-1}$, so $[D_1]^T = d \cdot z^{(m-1)(n-1)}$, where d is the degree of the Segre variety. The Cayley–Giambelli–Thom–Porteous formula computes this degree as

$$\mathrm{degree}(\mathbb{P}^{n-1} \times \mathbb{P}^{m-1}) = \left| \binom{n}{n - 1 + j - i} \right|_{1 \leq i, j \leq m-1}$$

$$= s_{(m-1)^{n-1}}(\underbrace{1, \ldots, 1}_{n})$$

$$= \# \left(\begin{array}{c} \text{SSYT of shape } (n - 1) \times (m - 1) \\ \text{with entries } 1, \ldots, n \end{array} \right)$$

$$= \prod_{i=1}^{n-1} \prod_{j=1}^{m-1} \frac{m + i - j}{i + j - 1}.$$

For instance, the degree of $\mathbb{P}^1 \times \mathbb{P}^{m-1}$ is m.

11.2 Flagged degeneracy loci

Next we turn to degeneracy loci for maps between flagged vector spaces. Suppose we have

$$F_1 \subset \cdots \subset F_m = F \xrightarrow{\varphi} E = E_n \twoheadrightarrow \cdots \twoheadrightarrow E_1,$$

where subscripts indicate dimension. Imposing rank conditions on $\varphi_{pq} \colon F_q \to E_p$ defines a subscheme of $M = \operatorname{Hom}(F, E)$. Such conditions will be B-invariant, where $B = B(E_\bullet) \times B(F_\bullet) \subseteq GL(E) \times GL(F)$ is the subgroup preserving both flags. Thus, given an $n \times m$ *rank matrix* of integers $\boldsymbol{r} = \big(r(p, q)\big)$, we have a B-invariant degeneracy locus

$$D_{\boldsymbol{r}} = \{\varphi \mid \operatorname{rk}(\varphi_{pq}) \leq r(p, q) \text{ for all } p, q\} \subseteq M.$$

Two basic questions arise:

(1) Which rank matrices $\boldsymbol{r}$ define irreducible loci $D_{\boldsymbol{r}}$?

(2) When $D_{\boldsymbol{r}}$ is irreducible, what is a formula for its class in $H_B^* M$?

To address these questions, let us fix bases $e_1, \ldots, e_n$ for E and $f_1, \ldots, f_m$ for F, so that E_p is the span of $\{e_1, \ldots, e_p\}$ and F_q is the span of $\{f_1, \ldots, f_q\}$. Then $M = M_{n,m}$ is the space of $n \times m$ matrices, and the map φ_{pq} is represented as the upper-left $p \times q$ submatrix of φ:

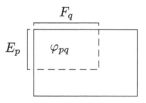

With respect to these bases, we have $B = B_n^- \times B_m^+ \subseteq GL_n \times GL_m$, where B^- and B^+ denote lower- and upper-triangular matrices, respectively: rank conditions on the submatrices φ_{pq} are preserved by lower-triangular row operations and upper-triangular column operations. (To see why $B(E_\bullet) = B_n^-$, note that $\ker(E \to E_p) = \operatorname{span}\{e_n, \ldots, e_{p+1}\}$.)

We take the maximal torus $T = (\mathbb{C}^*)^n \times (\mathbb{C}^*)^m$ to be diagonalized in these bases, and we denote the standard characters for the actions on E and F by $x_1, \ldots, x_n$ and $y_1, \ldots, y_m$, respectively. Since T acts on M by

$(g, h) \cdot A = gAh^{-1}$, the affine space M has weights $x_i - y_j$, for $1 \leq i \leq n$, $1 \leq j \leq m$. As before, the class $[D_r]^B = [D_r]^T$ we seek is a canonical polynomial in

$$H_B^* M = H_T^* M = \mathbb{Z}[x_1, \ldots, x_n, y_1, \ldots, y_m].$$

We will see that the answer is given by the Schubert polynomials encountered in the previous chapter. (This is how Schubert polynomials could have been discovered in the 1960s!)

11.3   Irreducibility

A *partial permutation matrix* is an $n \times m$ matrix of 0's and 1's, with at most one 1 in each row and column. Often it is convenient to represent such a matrix by replacing 1's by dots and 0's by blank spaces. The associated rank $r(p, q)$ of the upper-left $p \times q$ submatrix is then simply the number of dots in this region. Here is an example, with a partial permutation matrix next to its rank matrix $r = (r(p, q))$ (with the 1's represented by dots in the latter):

0	0	0	0	0
0	0	0	1	0
0	1	0	0	0
0	0	0	0	0

0	0	0	0	0
0	0	0	1●	1
0	1●	1	2	2
0	1	1	2	2

Any $n \times m$ partial permutation can be viewed as an injective map $\widehat{w} \colon I \to \{1, \ldots, m\}$ where $I \subseteq \{1, \ldots, n\}$ is the subset of rows which are occupied: $\widehat{w}(i) = j$ if there is a 1 in position (i, j). This can be extended canonically to a permutation $w \in S_N$, for sufficiently large N. (The minimal possibility is $N = n + m - \#(\text{dots})$.) The permutation w is the minimal one in Bruhat order such that $w(i) = j$ for $i \in I$, and $w(p) > m$ for $p \in \{1, \ldots, n\} \smallsetminus I$.

This is done algorithmically in three steps, as follows. Start by writing the values $w(i) = j$ in position i, for $i \in I$. For the empty positions $\{1, \ldots, n\} \smallsetminus I$, fill in values $m+1, m+2, \ldots$ in increasing order. Finally, write any unused values from $\{1, \ldots, N\}$ in increasing order, starting at position $n + 1$.

Continuing the above example, so $n = 4$ and $m = 5$, we start with the partial permutation given by $\widehat{w}(2) = 4$ and $\widehat{w}(3) = 2$. Taking $N = 7$, we obtain a permutation $w = 6\ \mathbf{4}\ \mathbf{2}\ 7\ 1\ 3\ 5$. Writing out the three steps, this is

$$\cdot\ 4\ 2\ \cdot$$

$$6\ 4\ 2\ 7$$

$$6\ 4\ 2\ 7\ 1\ 3\ 5.$$

The transposed permutation matrix $A_w^\dagger$ contains the partial permutation matrix of $\widehat{w}$ as its upper-left $n \times m$ submatrix, as shown below.

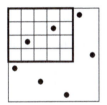

In case the construction produces $w \in S_N$ for $N < m+n$, one can always extend to S_{m+n} by appending values $w(i) = i$. For instance, the above example leads to $w = 6\ 4\ 2\ 7\ 1\ 3\ 5\ 8\ 9$.

The *length* of a partial permutation $\widehat{w}$ is

$$\ell(\widehat{w}) = \#\{i < j \mid i, j \in I \text{ and } \widehat{w}(i) > \widehat{w}(j)\}.$$

If w is a (complete) permutation minimally extending $\widehat{w}$, then $\ell(\widehat{w}) = \ell(w)$ is the usual length of w.

All feasible rank functions $\boldsymbol{r} = \big(r(p, q)\big)$ arise from partial permutations this way. That is, if there exists an $n \times m$ matrix A whose upper-left ranks are given by an array $\boldsymbol{r}$, then $\boldsymbol{r}$ is the rank function of some partial permutation. (To see this, find a unique partial permutation matrix in each B-orbit on M.) In fact, the set

$$D_r^\circ = \Big\{\varphi \mid \mathrm{rk}(\varphi_{pq}) = r(p, q) \text{ for all } p, q\Big\} \subseteq M$$

is the orbit of a partial permutation matrix $\widehat{w}$ corresponding to $\boldsymbol{r}$, or empty if $\boldsymbol{r}$ does not come from a partial permutation. When $\boldsymbol{r}$ corresponds to a partial permutation $\widehat{w}$, the locus $D_{\boldsymbol{r}}^{\circ}$ has codimension $\ell(\widehat{w})$ in M. For this reason we use the notation

$$\ell(\boldsymbol{r}) = \ell(\widehat{w}) = \ell(w),$$

where $\boldsymbol{r}$ is the rank function of a partial permutation $\widehat{w}$, and w is the minimal extension of $\widehat{w}$ to a complete permutation.

There is a partial order on $n \times m$ rank functions, where $\boldsymbol{r}' \leq \boldsymbol{r}$ if $r'(p, q) \leq r(p, q)$ for all p, q. This induces an order on partial permutations, by declaring $\widehat{w}' \geq \widehat{w}$ if the corresponding rank functions satisfy $\boldsymbol{r}' \leq \boldsymbol{r}$ (note the reversal). Equivalently, $\widehat{w}' \geq \widehat{w}$ if the corresponding complete permutations satisfy $w' \geq w$ in usual Bruhat order on S_{m+n}.

These considerations show that M decomposes into finitely many B-orbits indexed by partial permutations:

$$M = \coprod D_{\boldsymbol{r}}^{\circ},$$

and more generally, for any rank function $\boldsymbol{r}$,

$$D_{\boldsymbol{r}} = \coprod_{\boldsymbol{r}' \leq \boldsymbol{r}} D_{\boldsymbol{r}'}^{\circ},$$

noting as before that $D_{\boldsymbol{r}'}^{\circ}$ is nonempty if and only if $\boldsymbol{r}'$ comes from a partial permutation. This generalizes the Bruhat decomposition of GL_n. In the case $n = m$ and $\boldsymbol{r}$ comes from a permutation $w \in S_n$ – so $r(p, q) = r_w(p, q)$ is the rank of the upper-left $p \times q$ submatrix of $A_w^\dagger$ – we will usually write D_w and D_w° for $D_{\boldsymbol{r}}$ and $D_{\boldsymbol{r}}^{\circ}$. In this notation, the Bruhat decomposition says $GL_n = \coprod_{w \in S_n} D_w^{\circ}$.

Exercise 11.3.1. The following conditions on $r(p, q)$ (for all p, q) are necessary and sufficient for a matrix of nonnegative integers $\boldsymbol{r} = (r(p, q))$ to be the rank matrix of a partial permutation:

(1) $r(p, q) \leq \min\{p, q\}$;

(2) $r(p + 1, q) = r(p, q)$ or $r(p, q) + 1$;

(3) $r(p, q + 1) = r(p, q)$ or $r(p, q) + 1$; and

(4) $r(p + 1, q + 1) + r(p, q) - r(p + 1, q) - r(p, q + 1) = 0$ or 1.

Remark. The first three conditions in the exercise are intuitive: they say that a $p \times q$ submatrix has rank at most $\min\{p, q\}$, and that the rank can increase by at most 1 when adding a row or column. Requiring only these three conditions leads to rank matrices associated to *alternating sign matrices*.

Now we can identify the irreducible loci.

Proposition 11.3.2. *If r is a rank function on $n \times m$ matrices coming from a partial permutation $\widehat{w}$, then $D_r = \overline{D_r^\circ}$. These are precisely the irreducible loci: D_r is irreducible if and only if r comes from a partial permutation.*

Using the fact that M decomposes into D_r°, as r ranges over partial-permutation rank functions, the second statement follows from the first. The claim $D_r = \overline{D_r^\circ}$ can be proved in the same way as the analogous claim for Schubert cells and varieties.

Exercise 11.3.3. Let r be a rank function corresponding to a partial permutation $\widehat{w}$. Imitating the argument for Proposition 10.4.4, show that a partial permutation matrix $\widehat{w}'$ lies in $\overline{D_r^\circ}$ if and only if $\widehat{w}' \geq \widehat{w}$. Deduce Proposition 11.3.2.

In the next section, we will see that D_r may be identified with an open subset of a certain Schubert variety, which gives another proof of Proposition 11.3.2.

11.4 The class of a degeneracy locus

Before writing down the formula for $[D_r]^T$, we record a duality property which it must satisfy. When $n = m$ and r comes from a (complete) permutation $w \in S_n$, we continue to write $D_r = D_w \subseteq M$.

Lemma 11.4.1. *Assume $m = n$, and let $T = (\mathbb{C}^*)^n \times (\mathbb{C}^*)^n$ act on $M = \mathrm{Hom}(F, E)$ as usual. Identifying $H_T^* M$ with the polynomial ring $\mathbb{Z}[x_1, \ldots, x_n; y_1, \ldots, y_n]$, write $[D_w]^T = P_w(x; y)$, for each $w \in S_n$. Then $P_{w^{-1}}(x; y) = P_w(-y; -x)$.*

Proof We use the "transpose" automorphism $\tau \colon M \to M$, coming from the identification

$$M = \mathrm{Hom}(F, E) \cong \mathrm{Hom}(E^\vee, F^\vee).$$

This is equivariant with respect to the automorphism of $B^- \times B^+$ sending $(g, h) \mapsto ((h^{-1})^\dagger, (g^{-1})^\dagger)$. Restricting to the torus, the pullback homomorphism $\tau^* \colon H_T^* M \to H_T^* M$ is given by $x_i \mapsto -y_i$, $y_i \mapsto -x_i$.

On the other hand, τ maps $A_w^\dagger$ to $A_w = A_{w^{-1}}^\dagger$, so $\tau^{-1}(D_w) = D_{w^{-1}}$ and therefore

$$P_w(-y; -x) = \tau^* [D_w]^T = [D_{w^{-1}}]^T = P_{w^{-1}}(x; y),$$

as claimed. $\qquad\qquad\qquad\qquad\qquad\qquad\qquad\qquad\qquad\qquad\qquad\qquad\qquad\qquad\square$

Next, we use another graph construction to equivariantly embed $M = \mathrm{Hom}(F, E)$ into $Fl(E \oplus F)$. The map is given by

$$\varphi \mapsto \big(\mathrm{Graph}(\varphi|_{F_1}) \subset \cdots \subset \mathrm{Graph}(\varphi|_{F_m}) = \mathrm{Graph}(\varphi)$$
$$\subset E_1 + \mathrm{Graph}(\varphi) \subset \cdots \subset E_n + \mathrm{Graph}(\varphi) = E \oplus F\big),$$

where $E_\bullet$ and $F_\bullet$ are the standard flags. Using the standard basis $e_1, \ldots,$ e_n, $f_1, \ldots, f_m$, this map is represented in matrices by

$$\varphi \mapsto \left[\begin{array}{c|c} \begin{matrix} & & 1 \\ & \varphi & & \ddots \\ & & & & 1 \end{matrix} & \\ \hline \begin{matrix} 1 & & \\ & \ddots & \\ & & 1 \end{matrix} & \;0\; \end{array} \right]$$

(So the first column is $(\varphi(f_1), f_1)$, the second is $(\varphi(f_2), f_2)$, etc.)

It is easy to identify the image of this graph embedding. Let $v \in S_{n+m}$ be the permutation

$$v = [n + 1, \ldots, n + m, 1, \ldots, m].$$

(So $\varphi = 0$ maps to the permutation matrix A_v.) Then the embedding is a T-equivariant isomorphism

$$M \cong \widetilde{\Omega}^\circ$$

onto the opposite Schubert cell $\widetilde{\Omega}^\circ = \widetilde{\Omega}^\circ_{w_\circ v}$ containing the fixed point p_v.

Using the isomorphism $M \cong \widetilde{\Omega}^\circ$, we find a useful realization of the degeneracy locus D_r.

Lemma 11.4.2. *Let r be the rank function coming from an $n \times m$ partial permutation $\widehat{w}$, and let $w \in S_{n+m}$ be the corresponding permutation. Under the identification $M \cong \widetilde{\Omega}^\circ \subseteq Fl(E \oplus F)$, we have $D_r = \Omega_{w^{-1}} \cap \widetilde{\Omega}^\circ$.*

Proof Let $E^\bullet$ be the flag in $E \oplus F$ corresponding to the ordered basis $\{e_n, \ldots, e_1, f_m, \ldots, f_1\}$, indexed so that $E^p \subseteq E \oplus F$ has codimension p. So if $p \leq n$, then $E^p = \ker(E \twoheadrightarrow E_p) \oplus F$. This means $\mathrm{Graph}(\varphi|_{F_q}) \cap E^p = \ker \varphi_{pq}$ for $1 \leq p \leq n$ and $1 \leq q \leq m$.

The Schubert variety $\Omega_{w^{-1}}$ is defined as the locus of flags $G_\bullet$ satisfying conditions

$$\mathrm{rk}(G_q \to (E \oplus F)/E^p) \leq r_{w^{-1}}(q, p) = r_w(p, q)$$

for all p, q. Our choice of w – as the minimal permutation so that $r_w(p, q) = r(p, q)$ whenever $p \leq n$ and $q \leq m$ – means that all the above conditions on $\mathrm{rk}(G_q \to (E \oplus F)/E^p)$ follow from those where $p \leq n$ and $q \leq m$. Restricting to $M \cong \widetilde{\Omega}^\circ$, these conditions become

$$\mathrm{rk}(F_q \to E_p) \leq r(p, q),$$

as claimed. $\qquad\qquad\qquad\qquad\qquad\qquad\qquad\qquad\qquad\qquad\qquad\square$

Together with two results from Chapter 10 (Proposition 10.5.4 and Corollary 10.5.5), the above lemma implies that D_r is irreducible and Cohen–Macaulay, of codimension $\ell(w) = \ell(\widehat{w})$ in M.

Theorem 11.4.3. *Let r be the rank matrix of an $n \times m$ partial permutation $\widehat{w}$, which extends to a permutation $w \in S_{n+m}$. Then*

$$[D_r]^T = \mathfrak{S}_w(x; y)$$

in $H_T^* M = \mathbb{Z}[x_1, \ldots, x_n; y_1, \ldots, y_m]$, where $x_i = c_1^T(\ker(E_i \twoheadrightarrow E_{i-1}))$ and $y_j = c_1^T(F_j/F_{j-1})$, so the weight of T on the (i, j) entry of M is $x_i - y_j$.

Proof Let $\iota \colon M \hookrightarrow Fl(E \oplus F)$ be the graph embedding, identifying $M \cong \widetilde{\Omega}^\circ$ as above, where $\widetilde{\Omega}^\circ = \widetilde{\Omega}^\circ_{w_\circ v}$. By Lemma 11.4.2, $D_r = \Omega_{w^{-1}} \cap \widetilde{\Omega}^\circ$. Since $\widetilde{\Omega}^\circ$ is an affine space containing the fixed point p_v, the pullback $\iota^* \colon H_T^* Fl(E \oplus F) \to H_T^* M$ is the same as the restriction homomorphism $H_T^* Fl(E \oplus F) \to H_T^*(p_v)$. So $[D_r]^T = \iota^*[\Omega_{w^{-1}}]^T$.

In $H_T^* Fl(E \oplus F)$, let $z_i = -c_1^T(\mathbb{S}_i/\mathbb{S}_{i-1})$ and write $t_1, \ldots, t_{n+m}$ for the characters $-x_1, \ldots, -x_n, -y_1, \ldots, -y_m$. We have

$$[D_r]^T = \mathfrak{S}_{w^{-1}}(z;t)|_v$$
$$= \mathfrak{S}_{w^{-1}}(t_{n+1},\ldots,t_{n+m},t_1,\ldots,t_n;t_1,\ldots,t_{n+m})$$
$$= \mathfrak{S}_{w^{-1}}(-y,-x;-x,-y).$$

The theorem then follows from an algebraic identity of Schubert polynomials, which we will prove using geometry.

Consider the case where $n = m$ and r is the rank function of a permutation $w \in S_n$, so $D_r = D_w$. Regarding w as an element of S_{2n} by the standard embedding $S_n \hookrightarrow S_{2n}$, the above argument shows

$$[D_w]^T = \mathfrak{S}_{w^{-1}}(z;t)|_v.$$

By the stability property of Schubert polynomials, $\mathfrak{S}_{w^{-1}}(z;t)$ depends only on $z_1,\ldots,z_n$ and $t_1,\ldots,t_n$, since $w^{-1} \in S_n$. So this specialization is

$$\mathfrak{S}_{w^{-1}}(z;t)|_v = \mathfrak{S}_{w^{-1}}(-y_1,\ldots,-y_n;-x_1,\ldots,-x_n).$$

Applying Lemma 11.4.1, we have

$$\mathfrak{S}_{w^{-1}}(-y;-x) = \mathfrak{S}_w(x;y),$$

and the theorem is proved in this case.

Now we turn back to the general case. When $w(i) < w(i+1)$ for all $i \geq n$, the polynomial $\mathfrak{S}_w(x;y)$ depends only on the variables $x_1,\ldots,x_n$ (see §10.10.3). From the identity $\mathfrak{S}_w(x;y) = \mathfrak{S}_{w^{-1}}(-y;-x)$, it follows that $\mathfrak{S}_w(x;y)$ depends only on $y_1,\ldots,y_m$ if $w^{-1}(i) < w^{-1}(i+1)$ for all $i \geq m$. Both of these conditions hold when w comes from an $n \times m$ partial permutation. Applying these conclusions to the above formula $[D_r]^T = \mathfrak{S}_{w^{-1}}(-y,-x;-x,-y)$, we obtain

$$[D_r]^T = \mathfrak{S}_w(x_1,\ldots,x_n;y_1,\ldots,y_m),$$

as claimed. $\square$

In the course of the proof, we identified several useful properties of Schubert polynomials:

Corollary 11.4.4. *(1) For any permutation w, there is a duality identity,*

$$\mathfrak{S}_{w^{-1}}(x;y) = \mathfrak{S}_w(-y;-x).$$

*(2) Double Schubert polynomials may be computed recursively by divided
difference operators ∂_k^y acting on the y-variables:*

$$\partial_k^y \mathfrak{S}_w(x;y) = \begin{cases} -\mathfrak{S}_{s_k w}(x;y) & \text{if } s_k w < w; \\ 0 & \text{otherwise.} \end{cases}$$

(3) We have

$$\mathfrak{S}_w(x;y) \in \mathbb{Z}[x_1, \ldots, x_n; y_1, \ldots, y_m]$$

*if and only if $w(i) < w(i+1)$ and $w^{-1}(j) < w^{-1}(j+1)$ for all $i \geq n$
and all $j \geq m$.*

11.5   Essential sets

If r is a rank function coming from an $n \times m$ partial permutation matrix,
the nm rank conditions $\text{rk}(F_q \to E_p) \leq r(p,q)$ defining D_r are highly
redundant. A more efficient list of conditions is given by the *essential
set*.

First, we need to construct the *diagram* of a (partial) permutation.
This is the collection of boxes which remain after crossing out all boxes
to the right or below a dot in the matrix. It is not hard to see that $\ell(\widehat{w})$ is
the number of boxes in the diagram. Here is an example, continuing the
one above. The crossed-out boxes are shaded, so the diagram consists of
the unshaded boxes; we have $\ell(\widehat{w}) = 12$.

The essential set is the set of conditions coming from boxes (p,q) in the
southeast corners of the diagram. In our running example, the essential
set consists of the five conditions

$$\text{rk}(F_5 \to E_1) \leq 0, \qquad \text{rk}(F_3 \to E_2) \leq 0,$$
$$\text{rk}(F_1 \to E_4) \leq 0, \qquad \text{rk}(F_3 \to E_4) \leq 1,$$
$$\text{rk}(F_5 \to E_4) \leq 2.$$

These suffice to define D_r. That is, each condition $\mathrm{rk}(F_q \to E_p) \leq r$ means that all the size $r + 1$ minors of the upper-left $p \times q$ submatrix vanish; as (p, q) ranges over the essential set, the determinantal ideal generated by these minors defines D_r as a subscheme of the affine space M.

By construction, if $w \in S_N$ is the permutation associated to $\widehat{w}$, its diagram – and hence its essential set – is the same as that of $\widehat{w}$. Continuing with $w = 6\ 4\ \mathbf{2}\ 7\ 1\ 3\ 5$, here is an example.

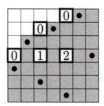

Proofs of these facts about essential sets and the schemes D_r can be found in the references listed at the end of the chapter. Lemma 11.4.2 is one useful consequence – it follows from the observation that the essential set of w is that same as that of the partial permutation matrix it comes from.

Exercise. For a permutation $\widehat{w} = w$, verify the above claim that the number of boxes in the diagram of w is equal to the number of inversions, $\ell(w) = \#\{i < j \mid w(i) > w(j)\}$. Observing that the matrix, and hence the diagram, of w^{-1} is obtained by transposing that of w, conclude that $\ell(w^{-1}) = \ell(w)$.

11.6 Degeneracy loci for maps of vector bundles

The degeneracy locus $D_r \subseteq M$, described above for maps of vector spaces, globalizes to the degeneracy of maps of vector bundles on a variety X. This is the setting for the main theorems of this chapter, and it leads to a central theme of this book.

Here is the setup. We have a morphism $\varphi \colon F \to E$ of vector bundles on a nonsingular variety X, along with complete flags of sub- and quotient bundles, so

$$F_1 \subset \cdots \subset F_m = F \xrightarrow{\varphi} E = E_n \twoheadrightarrow \cdots \twoheadrightarrow E_1,$$

with subscripts indicating ranks, as before; we often abbreviate this by writing $\varphi \colon F_\bullet \to E_\bullet$. Given a rank function $\boldsymbol{r} = (r(p,q))$, the degeneracy locus

$$D_{\boldsymbol{r}}(\varphi) = \left\{ x \in X \mid \mathrm{rk}(F_q \xrightarrow{\varphi_{pq}} E_p) \leq r(p,q) \text{ for all } p, q \right\}$$

is defined scheme-theoretically by the evident determinantal equations.

From now on, we will assume $\boldsymbol{r}$ comes from an $n \times m$ partial permutation, which extends minimally to a permutation $w \in S_{n+m}$, so the matrix locus $D_{\boldsymbol{r}} \subseteq M$ is irreducible. Recall that $\ell(\boldsymbol{r}) = \ell(w)$ denotes the length, which is also the codimension of $D_{\boldsymbol{r}}$ in the space of $n \times m$ matrices M.

A key feature of our degeneracy locus formulas is that they produce classes which are supported on the locus, regardless of any genericity assumption on codimension. A cohomology class $\alpha \in H^k X$ is said to be *supported on a subvariety* $Y \subseteq X$ if it lies in the image of the canonical homomorphism

$$H_{2 \dim X - k} Y = H^k(X, X \smallsetminus Y) \to H^k X.$$

Here is the first main theorem.

Theorem 11.6.1. *Let $\varphi \colon F_\bullet \to E_\bullet$ be as above. There is a unique class $\mathbb{D}_{\boldsymbol{r}} \in H^{2\ell(\boldsymbol{r})} X$ with the following properties.*

(1) Whenever $D_r(\varphi) \subseteq X$ has codimension equal to $\ell(\boldsymbol{r})$, we have $\mathbb{D}_{\boldsymbol{r}} = [D_{\boldsymbol{r}}(\varphi)]$.

(2) Given $f \colon X' \to X$, with pullback morphism $\varphi' \colon F'_\bullet \to E'_\bullet$, the corresponding class $\mathbb{D}'_{\boldsymbol{r}} \in H^{2\ell(\boldsymbol{r})} X'$ satisfies $\mathbb{D}'_{\boldsymbol{r}} = f^* \mathbb{D}_{\boldsymbol{r}}$. (The class $\mathbb{D}_{\boldsymbol{r}}$ is "stable under pullbacks.")

Furthermore, for any φ, the class $\mathbb{D}_{\boldsymbol{r}}$ is supported on $D_{\boldsymbol{r}}(\varphi)$.

Two aspects of this theorem are worth emphasizing. First, the class $\mathbb{D}_{\boldsymbol{r}}$ is independent of the particular morphism φ; it depends only on the vector bundles $E_\bullet$ and $F_\bullet$. Second, the fact that $\mathbb{D}_{\boldsymbol{r}}$ is supported on $D_{\boldsymbol{r}}(\varphi)$ is often significant – for example, this means that the locus $D_{\boldsymbol{r}}(\varphi)$ is nonempty whenever the class $\mathbb{D}_{\boldsymbol{r}}$ is nonzero.

The second main theorem gives a formula for the class $\mathbb{D}_{\boldsymbol{r}}$.

Theorem 11.6.2. *With notation as in Theorem 11.6.1, we have*

$$\mathbb{D}_{\boldsymbol{r}} = \mathfrak{S}_w(x; y),$$

where $w \in S_{n+m}$ is the permutation associated to the rank matrix $\boldsymbol{r}$, and the variables are evaluated as $x_i = c_1(\ker(E_i \twoheadrightarrow E_{i-1}))$ and $y_j = c_1(F_j/F_{j-1})$.

We quote a basic fact from intersection theory.

Lemma 11.6.3. *Let $f: X \to Y$ be a morphism of nonsingular varieties. Let $W \subseteq Y$ be a Cohen–Macaulay subvariety of codimension c, and let $Z = f^{-1}W \subseteq X$ be the (scheme-theoretic) inverse image.*

(1) If $Z \subseteq X$ also has codimension c, then it is Cohen–Macaulay and

$$[Z] = f^*[W]$$

in $H^{2c}X$.

(2) In general, we have $\operatorname{codim}_X Z \leq c$, and the class $f^*[W] \in H^{2c}X$ lies in the image of

$$\overline{H}_{2d}Z = H^{2c}(X, X \smallsetminus Z) \to H^{2c}X,$$

where $d = \dim X - c$.

Now we can prove the first theorem.

Proof of Theorem 11.6.1 Consider the bundle $\operatorname{Hom} = \operatorname{Hom}(F, E)$ on X, along with the tautological homomorphism $\Phi: F_\bullet \to E_\bullet$ of bundles pulled back to Hom, and the corresponding degeneracy locus $D_{\boldsymbol{r}}(\Phi) \subseteq$ Hom. Over a point of X, the fibers are isomorphic to the matrix loci $D_{\boldsymbol{r}} \subseteq M$ considered in §11.2. In particular, $D_{\boldsymbol{r}}(\Phi)$ is irreducible and Cohen–Macaulay, of codimension $\ell(\boldsymbol{r})$ in Hom. Replacing X by Hom, we must have $\mathbb{D}_{\boldsymbol{r}} = [D_{\boldsymbol{r}}(\Phi)]$ in H^* Hom, by the first property of $\mathbb{D}_{\boldsymbol{r}}$.

The morphism $\varphi: F_\bullet \to E_\bullet$ of bundles on X corresponds to a section $s_\varphi: X \to \operatorname{Hom}$, with $D_{\boldsymbol{r}}(\varphi) = s_\varphi^{-1}D_{\boldsymbol{r}}(\Phi)$. Since Hom $\to X$ is a vector bundle, the pullback homomorphism s_φ^* is a canonical isomorphism $H^* \operatorname{Hom} = H^*X$. So by the second property, we must have $\mathbb{D}_{\boldsymbol{r}} = s_\varphi^*[D_{\boldsymbol{r}}(\Phi)] = [D_{\boldsymbol{r}}(\Phi)]$ in $H^*X = H^* \operatorname{Hom}$. (If $D_{\boldsymbol{r}}(\varphi) \subseteq X$ has codimension $\ell(\boldsymbol{r})$, then $s_\varphi^*[D_{\boldsymbol{r}}(\Phi)] = [D_{\boldsymbol{r}}(\varphi)]$ by Lemma 11.6.3, so the two properties are compatible.)

This shows that the class $\mathbb{D}_r$ is determined by its properties, and constructs the class as $s_\varphi^*[D_r(\Phi)]$. By functoriality of pullbacks, it also shows that $\mathbb{D}_r$ comes from $H^*(X, X \smallsetminus D_r(\varphi))$, as claimed. $\qquad\square$

To prove the second theorem, we will reduce to the equivariant class of $D_r \subseteq M$, considered above. Combined with Lemma 11.6.3, the next lemma shows that these equivariant classes are universal cases of $\mathbb{D}_r$. It is proved in Appendix E.

Lemma 11.6.4. *Let E be a vector bundle of rank n on a variety X. Then there is an approximation space $\mathbb{B} = \mathbb{B}_N GL_n$, together with morphisms $p\colon X' \to X$ and $f\colon X' \to \mathbb{B}$ such that $p^*\colon H^*X \to H^*X'$ is injective and $p^*E \cong f^*\mathcal{E}$, where $\mathcal{E} \to \mathbb{B}$ is the universal vector bundle.*

If E has a complete flag of sub- or quotient bundles $E_\bullet$, then the above holds with $\mathbb{B}$ replaced by an approximation space for $B \subseteq GL_n$, a Borel subgroup of upper or lower triangular matrices, and with $\mathcal{E}$ replaced by $\mathcal{E}_\bullet$, the universal flag on $\mathbb{B}$.

Proof of Theorem 11.6.2 We apply Lemma 11.6.4 to the flagged bundles $E_\bullet$ and $F_\bullet$, with $B = B^- \times B^+$. Replacing X by X' if necessary, we obtain a morphism $X \to \mathbb{B}_N B$ by which Hom is pulled back from a universal bundle. That is, we have a diagram

$$
\begin{array}{ccc}
D_r(\Phi) & \longrightarrow & \mathbb{E} \times^B D_r \\
\uparrow & & \uparrow \\
\mathrm{Hom} & \overset{g}{\longrightarrow} & \mathbb{E} \times^B M \\
\downarrow & & \downarrow \\
X & \overset{f}{\longrightarrow} & \mathbb{B}.
\end{array}
$$

Writing $E_\bullet'$ and $F_\bullet'$ for the flagged vector spaces which define $D_r \subseteq M$, the universal bundles on $\mathbb{B}$ are $\mathcal{E}_\bullet = \mathbb{E} \times^B E_\bullet'$ and $\mathcal{F}_\bullet = \mathbb{E} \times^B F_\bullet'$. These have Chern classes

$$
x_i' = c_1^T(\ker(E_i' \twoheadrightarrow E_{i-1}')) = c_1(\ker(\mathcal{E}_i \twoheadrightarrow \mathcal{E}_{i-1}))
$$

and

$$
y_j' = c_1^T(F_j'/F_{j-1}') = c_1(\mathcal{F}_j/\mathcal{F}_{j-1}),
$$

so under the pullback map $f^*\colon H_T^*(\mathrm{pt}) \to H^*X$, which is the same as $g^*\colon H_T^*M \to H^*\,\mathrm{Hom}$, we have $\mathfrak{S}_w(x';y') \mapsto \mathfrak{S}_w(x;y)$.

The locally trivial fiber bundle $\mathbb{E} \times^B D_{\boldsymbol{r}} \to \mathbb{B}$ has Cohen–Macaulay fibers $D_{\boldsymbol{r}}$, so the variety $\mathbb{E} \times^B D_{\boldsymbol{r}}$ is also Cohen–Macaulay. Therefore we may apply Lemma 11.6.3 to conclude that $[D_{\boldsymbol{r}}(\Phi)] = g^*[\mathbb{E} \times^B D_{\boldsymbol{r}}]$ in $H^*\,\mathrm{Hom}$.

Since $[\mathbb{E} \times^B D_{\boldsymbol{r}}] = [D_{\boldsymbol{r}}]^B = [D_{\boldsymbol{r}}]^T$, the theorem follows from the formula for $[D_{\boldsymbol{r}}]^T$ (Theorem 11.4.3). $\qquad\square$

The theorems immediately imply their equivariant analogues. This is most often used for torus actions, but in fact the statement is the same for any (linear algebraic) group G. In the equivariant setting, the main results can be summarized as follows.

Corollary 11.6.5. *Suppose G acts on a nonsingular variety X, with a G-equivariant homomorphism of flagged vector bundles $\varphi\colon F_\bullet \to E_\bullet$, and consider the degeneracy locus $D_{\boldsymbol{r}}(\varphi)$, a G-invariant subscheme of X.*

(1) There is a class $\mathbb{D}_{\boldsymbol{r}}^G \in H_G^{2\ell(\boldsymbol{r})}X$ with the properties specified in Theorem 11.6.1.

(2) Let $x_i = c_1^G(\ker(E_i \to E_{i-1}))$ and $y_j = c_1^G(F_j/F_{j-1})$ in H_G^*X. Then

$$\mathbb{D}_{\boldsymbol{r}}^G = \mathfrak{S}_w(x;y).$$

(3) For every φ, the locus $D_{\boldsymbol{r}}(\varphi)$ is either empty, or has codimension at most $\ell(\boldsymbol{r})$ in X. If its codimension is equal to $\ell(\boldsymbol{r})$, then it is Cohen–Macaulay.

(4) With variables x and y as above, if $\mathfrak{S}_w(x;y)$ is nonzero in H_G^*X, then for every φ, the locus $D_{\boldsymbol{r}}(\varphi)$ is nonempty.

(Taking a principal bundle $\mathbb{E} \to \mathbb{B}$ for G, one applies Theorems 11.6.1 and 11.6.2 to the case where X is replaced by $\mathbb{E} \times^G X$, and the vector bundles E and F are replaced by $\mathbb{E} \times^G E$ and $\mathbb{E} \times^G F$, respectively.)

Example 11.6.6. Let V be a G-equivariant vector bundle on Y, and suppose there is a complete flag of subbundles $E^\bullet$, indexed by codimension. Let us write $E'_\bullet$ for the quotient flag, so $E_p = V/E^p$. Consider the flag bundle $X = \mathbf{Fl}(V) \to Y$, with tautological subbundles $F_q = \mathbb{S}_q$. Then we have

$$D_{r_w}(\mathbb{S}_\bullet \to E'_\bullet) = \Omega_{w^{-1}}(E^\bullet) \subseteq X,$$

where r_w is the rank function corresponding to a permutation w. (Comparing with §10.1, we have swapped p and q, and used $r_w(p,q) = r_{w^{-1}}(q,p)$.)

In the particular case where $Y = \mathbb{B}$ is an approximation space for T, the vector bundle is $V = \mathbb{E} \times^T \mathbb{C}^n$, and $E^\bullet$ is the standard flag, we have $X = \mathbb{E} \times^T Fl(\mathbb{C}^n)$ and

$$D_{r_w}(\mathbb{S}_\bullet \to E'_\bullet) = \mathbb{E} \times^T \Omega_{w^{-1}}.$$

Evaluating the variables as $x_i \mapsto c_1^T(\ker(E'_i \to E'_{i-1})) = c_1^T(E^{i-1}/E^i)$ and $y_i \mapsto -c_1^T((\mathbb{S}_i/\mathbb{S}_{i-1})^*)$, the identity between the corresponding formulas

$$[\Omega_{w^{-1}}]^T = \mathfrak{S}_{w^{-1}}(-y; -x) \quad \text{and} \quad [D_{r_w}(\mathbb{S}_\bullet \to E'_\bullet)] = \mathfrak{S}_w(x; y)$$

is an instance of the duality property of Schubert polynomials (Corollary 11.4.4(i)).

11.7 Universal properties of Schubert polynomials

For the rest of the chapter, we only consider degeneracy loci associated to a permutation, writing $D_r = D_w$. The key results of the previous section may be understood as a universal property of Schubert polynomials. Here we spell this out in two (equivalent) ways.

First, continuing our earlier setup, we have a map of flagged vector bundles on X,

$$F_1 \hookrightarrow \cdots \hookrightarrow F_n \xrightarrow{\varphi} E_n \twoheadrightarrow \cdots \twoheadrightarrow E_1,$$

and a degeneracy locus $D_w(\varphi) = D_w(F_\bullet \to E_\bullet) \subseteq X$ defined by placing conditions on the ranks of the maps $\varphi_{pq} \colon F_q \to E_p$. In this situation, the universal property is a consequence of results from §11.6.

Proposition 11.7.1. *The Schubert polynomial $\mathfrak{S}_w(x; y)$ is the unique polynomial in $\mathbb{Z}[x; y]$ which maps to $\mathbb{D}_w \in H^{2\ell(w)}X$ under the evaluation*

$$x_i \mapsto c_1(\ker(E_i \twoheadrightarrow E_{i-1})), \quad y_j \mapsto c_1(F_j/F_{j-1}),$$

for all degeneracy loci $D_w(F_\bullet \to E_\bullet)$ as above.

(The formula holds by Theorem 11.6.2, and uniqueness follows from the case where X is an approximation to $\mathbb{E}T \times^T M$, because H^*X agrees with $\mathbb{Z}[x, y]$ in relevant degrees.)

In the second situation, we have a vector bundle V on X, along with two flags of subbundles

$$F_1 \subset F_2 \subset \cdots \subset V \supset E^1 \supset E^2 \supset \cdots ,$$

where F_p has rank p, and E^q has co-rank q in V, so $E'_q = V/E^q$ has rank q. Here a locus is defined by placing conditions on the intersections $F_p \cap E^q$. Specifically,

$$D_w(F_\bullet \cap E^\bullet) := \Big\{x \in X \mid \dim(F_p \cap E^q) \geq k_w(p, q) \text{ for all } p, q\Big\}.$$

By a construction analogous to the proof of Theorem 11.6.1, one has $D_w(F_\bullet \cap E^\bullet) = f^{-1}(\mathbb{E} \times^{B^-} \Omega_w)$ for a map $f\colon X \to \mathbb{E} \times^{B^-} Fl(\mathbb{C}^n)$, and this defines the scheme structure. (As explained before, it may be necessary to replace X by an affine bundle.)

Setting $E'_q = V/E^q$, this reduces to an instance of the first situation. As in Example 11.6.6, we have

$$D_w(F_\bullet \cap E^\bullet) = D_{w^{-1}}(F_\bullet \to E'_\bullet),$$

using the compositions $F_p \hookrightarrow V \twoheadrightarrow E'_q$.

Proposition 11.7.2. *The Schubert polynomial $\mathfrak{S}_w(x; y)$ is the unique polynomial in $\mathbb{Z}[x; y]$ which maps to $\mathbb{D}_{w^{-1}}$ under*

$$x_i \mapsto -c_1(F_i/F_{i-1}), \quad y_j \mapsto -c_1(E^{j-1}/E^j),$$

for all degeneracy loci $D_w(F_\bullet \cap E^\bullet)$ as above.

(Use the identity $\mathfrak{S}_{w^{-1}}(-y; -x) = \mathfrak{S}_w(x; y)$.)

Conversely, the locus $D_w(\varphi\colon F_\bullet \to E_\bullet)$ is an instance of the second situation, via a graph construction. This is similar to the proof of Theorem 11.4.3. Set $V = E_n \oplus F_n$, and use flags of subbundles $F'_\bullet$ and $(E')^\bullet$, where $F'_p \subseteq E_n \oplus F_n$ is the graph of $\varphi|_{F_p}$ and $(E')^q$ is the kernel of $V \twoheadrightarrow E_q \oplus 0$. Then $D_w(\varphi) = D_{w^{-1}}(F'_\bullet \cap (E')^\bullet)$.

As in §11.6, these results extend directly to the equivariant situation, where a linear algebraic group G acts on X, with an equivariant morphism of equivariant vector bundles $\varphi \colon F_\bullet \to E_\bullet$ in the first setting (Proposition 11.7.1), or equivariant subbundles $F_\bullet, E^\bullet$ of V in the second setting (Proposition 11.7.2). The equivariant class $\mathbb{D}_w$ lives in $H_G^{2\ell(w)} X$, and the formulas are given by evaluating Schubert polynomials at the corresponding G-equivariant Chern classes.

11.8 Further properties of Schubert polynomials

Many other properties of Schubert polynomials can be deduced from the geometry of degeneracy loci. Here we give a few more examples.

Duality. We have already seen one form of duality for Schubert polynomials:

$$\mathfrak{S}_{w^{-1}}(-y; -x) = \mathfrak{S}_w(x; y). \tag{11.1}$$

As shown in Lemma 11.4.1, this identity follows by applying the transpose isomorphism $\tau \colon \mathrm{Hom}(F, E) \xrightarrow{\sim} \mathrm{Hom}(E^\vee, F^\vee)$. This is equivariant with respect to the homomorphism

$$GL(E) \times GL(F) \to GL(F^\vee) \times GL(E^\vee), \quad (g, h) \mapsto ((h^\dagger)^{-1}, (g^\dagger)^{-1}),$$

and it maps the locus D_w to $D_{w^{-1}}$.

In terms of general degeneracy loci for vector bundles $F_\bullet \xrightarrow{\varphi} E_\bullet$ on a variety X, the reason for identity (11.1) is even simpler. Since $r_w(p, q) = r_{w^{-1}}(q, p)$, the conditions

$$\mathrm{rk}(F_q \to E_p) \leq r_w(p, q) \quad \text{and} \quad \mathrm{rk}(E_p^\vee \to F_q^\vee) \leq r_{w^{-1}}(q, p)$$

are equivalent, so $D_w(F_\bullet \to E_\bullet) = D_{w^{-1}}(E_\bullet^\vee \to F_\bullet^\vee)$ as subschemes of X. The effect of exchanging $F_\bullet$ with $E_\bullet^\vee$ and $E_\bullet$ with $F_\bullet^\vee$ is to swap x_i and $-y_i$, establishing the identity.

There is another form of duality, coming from the exchange of flags of vector bundles on a variety X,

$$F_1 \subset \cdots \subset F_n = V = E_n \twoheadrightarrow \cdots \twoheadrightarrow E_1,$$

with the flags

$$E_1' \subset \cdots \subset E_n' = V = F_n' \twoheadrightarrow \cdots \twoheadrightarrow F_1',$$

where $E_p' = \ker(V \twoheadrightarrow E_{n-p})$ and $F_q' = V/F_{n-q}$.

Combinatorially, this is related to the involution on permutations which takes $w \in S_n$ to $w' = w_\circ w w_\circ$. This is w read "opposite and backwards," so $w'(i) = n + 1 - w(n + 1 - i)$. For example, $(2\ 1\ 6\ 3\ 5\ 4)' = 3\ 2\ 4\ 1\ 6\ 5$.

Exercise 11.8.1. With notation as above, let $D_w = D_w(F_\bullet \to E_\bullet)$ and $D_w' = D_w(E_\bullet' \to F_\bullet')$. Then $D_w' = D_{(w')^{-1}} \subseteq X$.

Example 11.8.2. Fix a flag $E_\bullet$ in a vector space V, with $E^q = E_{n-q}$. Recall that the Schubert variety $\Omega_w(E_\bullet) \subseteq Fl(V)$ is defined by conditions $\mathrm{rk}(F_p \to V/E^q) \le r_w(p, q)$ for all $1 \le p, q \le n$. Writing $F_p' = V/F_{n-p}$ for the quotient flag as above, we have

$$\Omega_w(E_\bullet) = \left\{ F_\bullet' \mid \mathrm{rk}(E_q \to F_p') \le r_{w'}(p, q) \text{ for all } 1 \le p, q \le n \right\}.$$

Now let

$$x_i = c_1(\ker(E_i \to E_{i-1})) \qquad y_i = c_1(F_i/F_{i-1}),$$
$$x_i' = c_1(E_i'/E_{i-1}') \qquad y_i' = c_1(\ker(F_i' \to F_{i-1}')),$$

so $x_i' = x_{n+1-i}$ and $y_i' = y_{n+1-i}$. Since $E_n = V = F_n$ – that is, the flags $E_\bullet$ and $F_\bullet$ are in the same vector bundle – there are necessarily relations between the x and y variables:

$$e_k(x_1, \ldots, x_n) = e_k(y_1, \ldots, y_n) = c_k(V).$$

So identities proved using this geometry are valid modulo the ideal generated by differences $e_k(x) - e_k(y)$ of elementary symmetric polynomials (for $1 \le k \le n$). Let $I \subseteq \mathbb{Z}[x; y]$ be this ideal.

Exchanging $E_\bullet \leftrightarrow F_\bullet'$ and $F_\bullet \leftrightarrow E_\bullet'$ swaps x_i with $y_i' = y_{n+1-i}$, so Exercise 11.8.1 implies

$$\mathfrak{S}_w(x; y) \equiv \mathfrak{S}_{w_\circ w^{-1} w_\circ}(y'; x') \pmod{I}.$$

Applying the previous duality identity (11.1), we obtain

$$\mathfrak{S}_{w_\circ w w_\circ}(-x_n, \ldots, -x_1; -y_n, \ldots, -y_1)$$
$$= \mathfrak{S}_{w_\circ w^{-1} w_\circ}(y_n, \ldots, y_1; x_n, \ldots, x_1)$$
$$\equiv \mathfrak{S}_w(x; y) \pmod{I}. \tag{11.2}$$

For example, we know $\mathfrak{S}_{s_k}(x;y) = x_1 + \cdots + x_k - y_1 - \cdots - y_k$. Since $w_\circ s_k w_\circ = s_{n-k}$, this Schubert polynomial is congruent to

$$\mathfrak{S}_{s_{n-k}}(-x_n, \ldots, -x_1; -y_n, \ldots, -y_1) = -x_n - \cdots - x_{k+1} + y_n + \cdots + y_{k+1}$$

modulo I. (Since $\mathfrak{S}_{s_k}(x;y) = \mathfrak{S}_{s_{n-k}}(-x'; -y') + e_1(x) - e_1(y)$, this is easy to see directly.) Similarly, for $n = 3$ and $w = 2\ 3\ 1$, we have $w_\circ w w_\circ = 3\ 1\ 2$, and the polynomials

$$\mathfrak{S}_{2\,3\,1}(x_1, x_2, x_3; y_1, y_2, y_3) = (x_1 - y_1)(x_2 - y_1)$$

and

$$\mathfrak{S}_{3\,1\,2}(-x_3, -x_2, -x_1; -y_3, -y_2, -y_1) = (-x_3 + y_3)(-x_1 + y_2)$$

are congruent modulo I.

Example 11.8.3.   This duality generalizes one we saw for Grassmannians. For an n-dimensional vector space V, there is a canonical isomorphism $Fl(V) \xrightarrow{\sim} Fl(V^\vee)$, defined by sending a flag $F_\bullet$ in V to the flag $(V/F_{n-1})^\vee \subset \cdots \subset (V/F_1)^\vee \subset V^\vee$. This isomorphism is equivariant for any G acting linearly on V and by the dual representation on $V^\vee$.

Now fix a basis $e_1, \ldots, e_n$ for V, with dual basis $e_1^*, \ldots, e_n^*$. Then $e_i^* \mapsto e_i$ determines an isomorphism $V^\vee \to V$. This is equivariant with respect to the involution $g \mapsto (g^\dagger)^{-1}$ of $GL(V)$. Composing with the above isomorphism of flag varieties, one obtains an involution $Fl(V) \to Fl(V)$, defined by sending $F_\bullet$ to the flag

$$(V/F_{n-1})^\vee \subset \cdots \subset (V/F_1)^\vee \subset V^\vee \cong V.$$

This involution of $Fl(V)$ is equivariant with respect to $g \mapsto (g^\dagger)^{-1}$.

Taking the standard flag in V, so E^q is spanned by $e_{q+1}, \ldots, e_n$, this involution takes $E^\bullet$ to the opposite flag $\widetilde{E}^\bullet$, where $\widetilde{E}^q$ is spanned by $e_1, \ldots, e_{n-q}$. Similarly, it sends the quotient flag, $E'_q = V/E^q$, to the opposite quotient flag, $\widetilde{E}'_q = V/\widetilde{E}^q$.

The involution on $Fl(V)$ acts on Schubert varieties by sending

$$\Omega_w = D_{w^{-1}}(\mathbb{S}_\bullet \to E'_\bullet) \quad \text{to} \quad \widetilde{\Omega}_{w'} = D_{(w')^{-1}}(\mathbb{S}_\bullet \to \widetilde{E}'_\bullet).$$

So under pullback, the substitution $x_i \mapsto -x_{n+1-i}$, $y_i \mapsto -y_{n+1-i}$ sends the equivariant class $\sigma_w = [\Omega_w]^T$ to $\widetilde{\sigma}_{w'} = [\widetilde{\Omega}_{w'}]^T$.

One consequence is this. Recall the coefficients $c_{uv}^w \in \Lambda_T$ defined by

$$\sigma_u \cdot \sigma_v = \sum_w c_{uv}^w \sigma_w$$

in $H_T^* Fl(\mathbb{C}^n)$.

Corollary 11.8.4. *The substitution $y_i \mapsto -y_{n+1-i}$ sends c_{uv}^w to $c_{u'v'}^{w'}$.*

This generalizes the substitution which relates $c_{\lambda\mu}^\nu$ to $c_{\lambda'\mu'}^{\nu'}$ (Exercise 9.7.4).

Exercise 11.8.5. For $n = d + e$, let λ be a partition in the $d \times e$ rectangle, and let λ' be its conjugate partition (in the $e \times d$ rectangle). Show that the corresponding Grassmannian permutations are related by $w(\lambda') = w(\lambda)'$.

Cauchy Formula. The *Cauchy formula* for Schubert polynomials is a very useful identity:

$$\mathfrak{S}_w(x; y) = \sum_{vu \doteq w} \mathfrak{S}_u(x; t)\, \mathfrak{S}_v(t; y)$$

$$= \sum_{v^{-1}u \doteq w} \mathfrak{S}_u(x; t)\, \mathfrak{S}_v(-y; -t), \qquad (11.3)$$

where the notation $vu \doteq w$ means $vu = w$ and $\ell(u) + \ell(v) = \ell(w)$. No t variables appear on the left-hand side – so the right-hand side is also independent of t!

As usual, by choosing $n \gg 0$, this is equivalent to a formula in cohomology. Consider the locus

$$\Omega_w = D_w(\mathbb{S}_\bullet^{(2)} \to \mathbb{Q}_\bullet^{(1)}) \subseteq Fl(\mathbb{C}^n) \times Fl(\mathbb{C}^n),$$

where $\mathbb{S}_\bullet^{(2)}$ is the tautological subspace flag pulled back by the second projection pr_2, and $\mathbb{Q}_\bullet^{(1)}$ is the tautological quotient flag pulled back by the first projection pr_1. This is a "double Schubert variety" of the same type used in Proposition 10.5.4. In particular, the fiber of pr_1 over the opposite quotient flag $\widetilde{E}_\bullet'$ is

$$pr_1^{-1}(\widetilde{E}_\bullet') = D_w(\mathbb{S}_\bullet \to \widetilde{E}_\bullet') = \widetilde{\Omega}_{w^{-1}}.$$

Using Exercise 11.8.1, the fiber of the second projection is

$$pr_2^{-1}(E_\bullet) = D_w(E_\bullet \to \mathbb{Q}_\bullet) = D_{(w')^{-1}}(\mathbb{S}_\bullet \to E_\bullet') = \Omega_{w'},$$

where $w' = w_\circ w w_\circ$. The formula (11.3) becomes

$$[\boldsymbol{\Omega}_w]^T = \sum_{v^{-1}u \doteq w} [\Omega_{w'}]^T \times [\tilde{\Omega}_v]^T \tag{11.4}$$

in $H_T^*(Fl(\mathbb{C}^n) \times Fl(\mathbb{C}^n))$, with respect to the diagonal action of the torus. Here Ω_w and $\tilde{\Omega}_w$ are the usual Schubert varieties and opposite Schubert varieties in $Fl(\mathbb{C}^n)$, and in the degeneracy locus formula, the variables are specialized as $x_i = c_1^T(\ker(\mathbb{Q}_i^{(1)} \twoheadrightarrow \mathbb{Q}_{i-1}^{(1)}))$, $y_i = c_1^T(\mathbb{S}_i^{(2)}/\mathbb{S}_{i-1}^{(2)})$, and $t_i = c_1^T(E_i/E_{i-1})$.

Proof The general case of Equation (11.3) can be deduced easily from the case $w = w_\circ$ by applying difference operators. On the other hand, $\boldsymbol{\Omega}_{w_\circ} \subseteq Fl(\mathbb{C}^n) \times Fl(\mathbb{C}^n)$ is the diagonal, and the case $w = w_\circ$ of (11.4) becomes

$$[\boldsymbol{\Omega}_{w_\circ}]^T = \sum_u [\Omega_{w_\circ u}]^T \times [\tilde{\Omega}_u]^T,$$

which is precisely the (equivariant) Künneth decomposition. (See §3.7 and §4.6.) $\qquad\square$

More Products of Schubert Polynomials. One can consider more general products of Schubert polynomials, of the form

$$\mathfrak{S}_u(x;s) \cdot \mathfrak{S}_v(x;t) = \sum_w c_{uv}^w(s,t)\, \mathfrak{S}_w(x;t),$$

where $c_{uv}^w(s,t)$ is homogeneous of degree $\ell(u) + \ell(v) - \ell(w)$ in variables $s = (s_1, s_2, \dots)$ and $t = (t_1, t_2, \dots)$. These polynomials specialize to the equivariant coefficients $c_{uv}^w = c_{uv}^w(t,t)$. They also satisfy a vanishing property:

We have $c_{uv}^w(s,t) = 0$ unless $v \leq w$ in Bruhat order.

(Note that $c_{uv}^w(s,t)$ need not vanish when $u \not\leq w$!)

Proof On a variety X, consider a vector bundle E of rank $n \gg 0$, equipped with two general flags of subbundles $S_1 \subset \cdots \subset S_n = E$ and $T_1 \subset \cdots \subset T_n = E$. Let $s_i = c_1(S_i/S_{i-1})$ and $t_i = c_1(T_i/T_{i-1})$. Consider the flag bundle $\mathbf{Fl}(E) \to X$, with tautological quotient flag $\mathbb{Q}_\bullet$, and $x_i = c_1(\ker(\mathbb{Q}_i \to \mathbb{Q}_{i-1}))$. Then the degeneracy locus formula gives

$$\mathfrak{S}_u(x;s) = [D_u(S_\bullet \to \mathbb{Q}_\bullet)] \quad \text{and} \quad \mathfrak{S}_v(x;t) = [D_v(T_\bullet \to \mathbb{Q}_\bullet)].$$

The classes $\mathfrak{S}_w(x;t)$ form a basis for $H^*\mathbf{Fl}(E)$ over H^*X, so one can write

$$\mathfrak{S}_u(x;s)\,\mathfrak{S}_v(x;t) = \sum_w c_{uv}^w(s,t)\,\mathfrak{S}_w(x;t)$$

modulo relations among these variables – but as usual, by taking n sufficiently large we may assume there are no relations in relevant degrees, so this is an identity of polynomials.

The class $\mathfrak{S}_u(x;s) \cdot \mathfrak{S}_v(x;t)$ is supported on $D_v = D_v(T_\bullet \to \mathbb{Q}_\bullet)$, so it comes from a refined class in $H^*(\mathbf{Fl}(E), \mathbf{Fl}(E) \smallsetminus D_v)$. As an H^*X-module, $H^*(\mathbf{Fl}(E) \smallsetminus D_v)$ has a basis of classes $[D_w(T_\bullet \to \mathbb{Q}_\bullet)]$, ranging over w such that $v \not\leq w$, so the vanishing follows from the exact sequence for the pair $(\mathbf{Fl}(E), \mathbf{Fl}(E) \smallsetminus D_v)$. $\qquad\square$

Notes

The case of Proposition 11.1.1 where $m = n + 1$ and $r = n - 1$ was treated by Cayley (1849). Salmon and Roberts gave the answer for $n \times m$ matrices of sub-maximal rank; see (Salmon, 1852, pp. 285–300). The general case of the proposition was proved by Giambelli (1904).

The diagram of a permutation is sometimes called the *Rothe diagram*. It was invented by (Rothe, 1800), who used it to show $\ell(w) = \ell(w^{-1})$, as in the Exercise from §11.5. The essential set was defined in (Fulton, 1992).

The loci D_w are often called *matrix Schubert varieties*, and were studied systematically in (Fulton, 1992), where the degeneracy locus formula (Theorem 11.6.2) was proved. The connection with equivariant cohomology was made by Graham (1997), Fehér and Rimányi (2002, 2003), and Knutson and Miller (2005).

Theorem 11.6.1 appears in (Fulton, 1992, Theorem 8.2).

Lemma 11.6.3 is proved in (Fulton and Pragacz, 1998, Appendix A). The version stated there weakens the hypothesis that X be nonsingular, requiring only that X be Cohen–Macaulay. The fact that the matrix Schubert varieties D_w are Cohen–Macaulay appears first in (Fulton, 1992), where it is deduced from the corresponding fact about Schubert varieties in the flag variety. The role of the Cohen–Macaulay condition was emphasized by Kempf and Laksov (1974); the application of degeneracy locus formulas to nonemptiness was used by Kleiman and Laksov (1974) to establish the existence of Brill–Noether special divisors on curves. Further history and applications may be found in (Fulton and Pragacz, 1998).

The Cauchy formula for double Schubert polynomials (11.3) appears in (Fomin and Kirillov, 1996a, Theorem 8.1). One can give a direct proof of the geometric Cauchy formula (11.4) by a transversality argument; see (Anderson, 2007a).

For the Grassmannian—so $u = w(\lambda)$, etc.—the coefficients $c^\nu_{\lambda\mu}(s, t)$ were studied by Molev and Sagan (1999).

Hints for Exercises

Exercise 11.3.1. One recovers the partial permutation matrix $A = (a_{pq})$ from r by setting $a_{pq} = r(p, q) + r(p - 1, q - 1) - r(p, q - 1) - r(p - 1, q)$, with the convention $r(p, 0) = r(0, q) = r(0, 0) = 0$.

Exercise 11.8.1. Check that $r_{w'}(n - p, n - q) = n - p - q + r_w(p, q)$.

12

Infinite-Dimensional Flag Varieties

In the last chapter, we saw how Schubert polynomials arise naturally from degeneracy loci. There we had incidence conditions between sub-bundles of given ranks or co-ranks in an ambient vector bundle V. Another way of setting up the degeneracy locus problem is to impose conditions on subbundles whose ranks are near half the rank of V. This arises from a different notion of stability, and leads to a richer theory of Schubert polynomials, which will provide a helpful link to the symplectic story in the next chapter. As a warmup, we start by rephrasing the stability property from Chapter 10 in terms of classes in infinite-dimensional flag varieties.

12.1 Stability revisited

In Chapter 10, we saw that Schubert polynomials are characterized by a certain type of stability. Infinite flag manifolds provide an elegant framework for expressing this property. The variety $Fl(\mathbb{C}^\infty)$ parametrizes all flags of finite-dimensional subspaces of $\mathbb{C}^\infty$:

$$Fl(\mathbb{C}^\infty) = \left\{ (F_1 \subset F_2 \subset \cdots \subset \mathbb{C}^\infty) \,\middle|\, \dim F_i = i \right\}.$$

By now, we are used to working with infinite-dimensional spaces via finite-dimensional approximations, and this case will be similar.

To start, we consider the partial flag varieties

$$Fl^{(n)}(\mathbb{C}^m) = Fl(1, \ldots, n; \mathbb{C}^m),$$

for $n \leq m$. There are two evident systems of maps among these varieties. We will construct $Fl(\mathbb{C}^\infty)$ as a certain limit, and see that Schubert polynomials are uniquely characterized by stability with respect to these maps.

First, we have the standard embeddings: for $m \leq m'$, we have the inclusion $\mathbb{C}^m \subseteq \mathbb{C}^{m'}$ as the span of the first m standard basis vectors, and a corresponding embedding $\iota\colon Fl^{(n)}(\mathbb{C}^m) \hookrightarrow Fl^{(n)}(\mathbb{C}^{m'})$. Second, we have the canonical projections $\pi\colon Fl^{(n')}(\mathbb{C}^m) \to Fl^{(n)}(\mathbb{C}^m)$, for $n' \geq n$.

The union over the embeddings $\iota\colon Fl^{(n)}(\mathbb{C}^m) \hookrightarrow Fl^{(n)}(\mathbb{C}^{m'})$ is

$$Fl^{(n)}(\mathbb{C}^\infty) = \left\{ (F_1 \subset \cdots \subset F_n \subset \mathbb{C}^\infty) \,\middle|\, \dim F_i = i \right\}.$$

These embeddings were used in §10.9, and as noted there, the union is a model for the classifying space $\mathbb{B}B$. The projections π are compatible with this union, so we obtain projections $\pi\colon Fl^{(n')}(\mathbb{C}^\infty) \to Fl^{(n)}(\mathbb{C}^\infty)$. The infinite flag variety $Fl(\mathbb{C}^\infty)$ is the inverse limit of the spaces $Fl^{(n)}(\mathbb{C}^\infty)$ with respect to these projections. Thus $Fl(\mathbb{C}^\infty)$ is the inverse limit of a direct limit of finite-dimensional varieties—a "pro-ind-variety."

Next we consider group actions. A torus T acts on $Fl^{(n)}(\mathbb{C}^m)$ via actions on $\mathbb{C}^m$ for all m, by characters $y_1, y_2, \ldots$. One may regard these as an infinite sequence of distinct characters of an arbitrary torus, or take them to be a basis of characters for the infinite torus $T = \prod_{i \geq 1} \mathbb{C}^*$. In the latter case, one has $\Lambda = \Lambda_T = \mathbb{Z}[y_1, y_2, \ldots]$, the polynomial ring in countably many independent variables. In fact, $\mathbb{B}T = \prod_{k \geq 1} \mathbb{P}^\infty$ is the countable product of projective spaces. (See Appendix A, §A.8 for a computation of this cohomology ring.)

The embeddings $\iota\colon Fl^{(n)}(\mathbb{C}^m) \hookrightarrow Fl^{(n)}(\mathbb{C}^{m'})$ are equivariant with respect to the inclusion of GL_m in the parabolic subgroup of $GL_{m'}$ which preserves $\mathbb{C}^m \subseteq \mathbb{C}^{m'}$. So they are equivariant with respect to inclusions of Borel subgroups B, but not with respect to B^-.

The projections $\pi\colon Fl^{(n')}(\mathbb{C}^m) \to Fl^{(n)}(\mathbb{C}^m)$, on the other hand, are equivariant with respect to GL_m.

All maps are equivariant with respect to T, and we may compute the equivariant cohomology via limits. Naively, the cohomology of a direct limit of spaces should be the inverse limit of cohomology rings; the cohomology of an inverse limit should the direct limit of cohomology rings. (Such limits are taken in the category of graded rings. Justification for these naive expectations may be found in Appendix A, §A.8.) First, we compute

$$H_T^* Fl^{(n)}(\mathbb{C}^\infty) = \varprojlim_m H_T^* Fl^{(n)}(\mathbb{C}^m)$$

$$= \Lambda[x_1, \ldots, x_n].$$

Then we have

$$H_T^* Fl(\mathbb{C}^\infty) = \varinjlim_n H_T^* Fl^{(n)}(\mathbb{C}^\infty)$$

$$= \Lambda[x_1, x_2, \ldots]$$

$$= \Lambda[x],$$

the polynomial ring over in countably many variables x with coefficients in Λ. As usual, any finite computation in these rings may be carried out by taking m and n finite but sufficiently large.

In §10.9, we saw Schubert varieties $\Omega_{[w]} \subseteq Fl^{(n)}(\mathbb{C}^m)$, indexed by $w \in S_m^{(n)}$. Schubert varieties in $Fl(\mathbb{C}^\infty)$ are indexed by permutations in S_∞, and are defined similarly, by

$$\Omega_w = \Big\{ F_\bullet \mid \dim(F_p \cap E^q) \geq k_w(p,q) \text{ for all } p, q > 0 \Big\},$$

where $E^q \subset \mathbb{C}^\infty$ is the codimension q subspace where the first q coordinates vanish.

The embeddings ι let us define Schubert classes in $H_T^* Fl^{(n)}(\mathbb{C}^\infty) = \Lambda[x_1, \ldots, x_n]$; see Exercise 10.9.2. The following exercise establishes the analogous compatibility with respect to the projections π.

Exercise. For $w \in S_m^{(n)}$, and $n \leq n' \leq m$, show that $\pi^{-1}\Omega_{[w]} = \Omega_{[w]} \subseteq Fl^{(n')}(\mathbb{C}^m)$, so

$$\pi^* \sigma_{[w]} = \sigma_{[w]} \quad \text{in} \quad H_T^* Fl^{(n')}(\mathbb{C}^m).$$

In particular, taking $n' = m$, we have $\pi^* \sigma_{[w]} = \sigma_w$ in $H_T^* Fl(\mathbb{C}^m)$.

By stability with respect to the system of inclusions (Exercise 10.9.2), for each $w \in S_\infty^{(n)} = \bigcup_m S_m^{(n)}$, there is a stable Schubert class σ_w in $H_T^* Fl^{(n)}(\mathbb{C}^\infty) = \Lambda[x_1, \ldots, x_n]$. By stability with respect to the system of projections (the exercise above), for any permutation $w \in S_\infty$ there is a stable Schubert class σ_w in $H_T^* Fl(\mathbb{C}^\infty) = \Lambda[x]$. Such classes are therefore uniquely represented by a polynomial in $\Lambda[x]$, and this determines the Schubert polynomials.

12.2 Infinite Grassmannians and flag varieties

Consider a vector space $\mathbb{C}^{2m} = \mathbb{C}^m_- \oplus \mathbb{C}^m_+$, with standard basis

$$e_{-m}, \ldots, e_{-1}, e_1, \ldots, e_m,$$

so that $\mathbb{C}^m_-$ is the span of the negative standard basis vectors and $\mathbb{C}^m_+$ is the span of the positive ones. An embedding $\mathbb{C}^{2m} \hookrightarrow \mathbb{C}^{2m+2}$ is defined in the evident way, by identifying $\mathbb{C}^{2m+2}$ with $\mathbb{C}e_{-m-1} \oplus \mathbb{C}^{2m} \oplus \mathbb{C}e_{m+1}$.

There are corresponding embeddings of Grassmannians, defined by

$$Gr(m, \mathbb{C}^{2m}) \hookrightarrow Gr(m+1, \mathbb{C}^{2m+2}),$$

$$(F \subset \mathbb{C}^{2m}) \mapsto (\mathbb{C}e_{-m-1} \oplus F \subset \mathbb{C}^{2m+2}).$$

Taking the union, we obtain an infinite Grassmannian,

$$Gr^\infty = \bigcup_m Gr(m, \mathbb{C}^{2m}).$$

This infinite Grassmannian parametrizes subspaces of both infinite dimension and infinite codimension in a countable-dimensional vector space, and some care is required in its interpretation. The vector space is $\mathbb{C}^\infty_- \oplus \mathbb{C}^\infty_+$, where $\mathbb{C}^\infty_- = \operatorname{span}\{e_{-1}, e_{-2}, \ldots\}$ and $\mathbb{C}^\infty_+ = \operatorname{span}\{e_1, e_2, \ldots\}$, and one can use this splitting to give a more intrinsic description of Gr^∞, which is sometimes also called the *Sato Grassmannian*. (See the Notes for details and references.)

For our purposes, it is usually simplest to regard any statement about Gr^∞ as shorthand for one which takes place on $Gr(m, \mathbb{C}^{2m})$ for all sufficiently large m and compatibly with respect to these embeddings.

As in §12.1, an infinite torus T_+ acts on $\mathbb{C}^\infty_+$. There are similar actions of T_- on $\mathbb{C}^\infty_-$ and of $T = T_- \times T_+$ on $\mathbb{C}^\infty_- \oplus \mathbb{C}^\infty_+$. In the basis $\{\ldots, e_{-2}, e_{-1}, e_1, e_2, \ldots\}$ for $\mathbb{C}^\infty_- \oplus \mathbb{C}^\infty_+$, we will write

$$\ldots, b_2, b_1, y_1, y_2, \ldots$$

for the characters of this T action. That is, for $i \geq 1$, T acts on e_i by the character y_i and on e_{-i} by the character b_i. So T acts on Gr^∞, and we wish to compute its equivariant cohomology in terms of certain Chern classes.

The cohomology rings $H_T^* Gr(m, \mathbb{C}^{2m})$ were computed in Chapter 4, but here we want a presentation which is compatible with our embeddings of Grassmannians. This is straightforward, using the presentations we already know.

Exercise 12.2.1. Show that sending

$$c_i \mapsto c_i^T(\mathbb{C}_-^m - \mathbb{S})$$

defines a surjective homomorphism $\Lambda[c_1, \ldots, c_m] \to H_T^* Gr(m, \mathbb{C}^{2m})$, with kernel generated by relations

$$\sum_{i=0}^m c_i \cdot e_{k-i}(y_1, \ldots, y_m) = 0$$

for $k = m+1, \ldots, 2m$.

The total Chern class $c^T(\mathbb{C}_-^{m+1} - \mathbb{S})$ restricts to $c^T(\mathbb{C}_-^m - \mathbb{S})$ under the embedding $Gr(m, \mathbb{C}^{2m}) \hookrightarrow Gr(m+1, \mathbb{C}^{2m+2})$, so the exercise gives a stable presentation. It follows that the cohomology ring of the infinite Grassmannian is

$$H_T^* Gr^\infty = \varprojlim H_T^* Gr(m, \mathbb{C}^{2m}) = \Lambda[c_1, c_2, \ldots],$$

where c restricts to $c^T(\mathbb{C}_-^m - \mathbb{S})$ on each finite-dimensional Grassmannian $Gr(m, \mathbb{C}^{2m})$. (See Appendix A, §A.8.)

The polynomial ring $\Lambda[c] = \Lambda[c_1, c_2, \ldots]$ may be regarded as a ring of symmetric functions in new variables $a_1, a_2, \ldots$, with coefficients in $\Lambda = \mathbb{Z}[\ldots, b_2, b_1, y_1, y_2, \ldots]$. A natural way to do this is by writing

$$c = c^T(\mathbb{C}_-^\infty - \mathbb{S}) = \prod_{i \geq 1} \frac{1 + b_i}{1 - a_i},$$

so the a_i are Chern roots for $\mathbb{S}^\vee$. Viewed this way, each c_k is a *super-symmetric function* in the a and b variables. By definition, a function $f(a; b)$ in two sets of variables is super-symmetric if it is separately symmetric in a and in b, and satisfies a cancellation property: the evaluation $f(a; b)|_{a_1 = -b_1 = t}$ is independent of t.

Exercise 12.2.2. For a commutative ring R, show that the homomorphism

$$R[c_1, \ldots, c_n] \to R[h_1, \ldots, h_m, e_1, \ldots, e_p]$$

defined by

$$c_k \mapsto h_k + h_{k-1} e_1 + \cdots + h_1 e_{k-1} + e_k$$

is injective whenever $n \leq m + p$.

Exercise 12.2.3. Let $\text{Symm}(x)$ denote the ring of symmetric functions in infinitely many variables $x = (x_1, x_2, \ldots)$, with coefficients in a ring R, so $\text{Symm}(x) \cong R[h_1, h_2, \ldots]$, where $h_k = h_k(x)$ is the complete homogeneous symmetric function. Consider the homomorphism

$$\text{Symm}(x) \to \text{Symm}(a) \otimes_R \text{Symm}(b)$$

defined by sending $h_k(x)$ to

$$c_k(a|b) = h_k(a) + h_{k-1}(a) e_1(b) + h_{k-2}(a) e_2(b) + \cdots + e_k(b),$$

where $h_k(a)$ and $e_k(b)$ are the complete homogeneous and elementary symmetric functions in a and b variables, respectively. Show that this is an isomorphism onto the subring of super-symmetric functions, so the elements c_k are algebraically independent generators of this subring.

Analogously to the construction of $Fl(\mathbb{C}^\infty)$ in §12.1, the infinite flag variety is a limit of partial flag varieties,

$$Fl^\infty = \varprojlim_n \bigcup_m Fl(m, m+1, \ldots, m+n; \mathbb{C}^{2m}).$$

The projections $Fl(m, m+1, \ldots, m+n; \mathbb{C}^{2m}) \to Gr(m, \mathbb{C}^{2m})$ are compatible, making $Fl^\infty \to Gr^\infty$ a fiber bundle with fibers isomorphic to $Fl(\mathbb{C}^\infty)$. The stable way to index flags in $Fl(m, m+1, \ldots, m+n; \mathbb{C}^{2m})$ (or equivalently, in Fl^∞) is by writing

$$F = F_0 \subset F_1 \subset F_2 \subset \cdots,$$

so that at any finite stage, $\dim F_i = m + i$. We use the same indexing for tautological bundles $\mathbb{S}_\bullet$ on $Fl(m, m+1, \ldots, m+n; \mathbb{C}^{2m})$ (or Fl^∞), so

$$\mathbb{S} = \mathbb{S}_0 \subset \mathbb{S}_1 \subset \mathbb{S}_2 \subset \cdots.$$

The computation of the equivariant cohomology of Fl^∞ is similar to that of the infinite Grassmannian: we have

$$H_T^* Fl^\infty = \varinjlim_n \left(\varprojlim_m H_T^* Fl(m, m+1, \ldots, m+n; \mathbb{C}^{2m}) \right)$$

$$= \Lambda[c_1, c_2, \ldots][x_1, x_2, \ldots],$$

where on any finite partial flag variety $Fl(m, m+1, \ldots, m+n; \mathbb{C}^{2m})$, the variable c_k restricts to $c_k^T(\mathbb{C}_-^m - \mathbb{S})$ and x_i restricts to $c_1^T((\mathbb{S}_i/\mathbb{S}_{i-1})^\vee)$.

12.3   Schubert varieties and Schubert polynomials

Schubert varieties in Fl^∞ are defined with respect to the descending flag of subspaces $E^q = \mathrm{span}\{e_i \,|\, i > q\}$, for $q = 0, 1, 2, \ldots$, by setting

$$\Omega_w^\infty = \left\{ F_\bullet \,\middle|\, \dim(F_p \cap E^q) \geq k_w(p, q) \text{ for all } p \geq 1, q \geq 0 \right\}$$

for each permutation $w \in S_\infty$, where as usual

$$k_w(p, q) = \#\{i \leq p \,|\, w(i) > q\}.$$

These are compatible with restrictions to each finite flag variety. In fact, each Ω_w^∞ comes from an ordinary Schubert variety Ω_v in $Fl(m, m+1, \ldots, m+n; \mathbb{C}^{2m})$. The distinction arises when considering stability.

In §10.9, we saw that Schubert varieties satisfy a stability property with respect to the inclusion $\iota\colon S_n \hookrightarrow S_{n+m}$, defined by $\iota(w)(p) = p$ if $k > n$. There is another natural embedding, defined by

$$\iota'(w)(p) = \begin{cases} p & \text{if } 1 \leq p \leq m; \\ w(p - m) + m & \text{if } i > m. \end{cases}$$

This is written $w \mapsto 1^m \times w$, so in permutation matrices, we have

$$A_{1^m \times w} = \left[\begin{array}{c|c} \mathrm{id}_m & 0 \\ \hline 0 & A_w \end{array} \right].$$

Using this notation, we can say more precisely how Ω_w^∞ is determined by Schubert varieties in the finite-dimensional flag variety.

Lemma 12.3.1.   *The conditions defining Ω_w^∞ are the same as those defining $\Omega_{1^m \times w}$ in the finite-dimensional flag variety $Fl(\mathbb{C}^{2m})$.*

Proof Let us examine the diagram of $1^m \times w$, as shown below.

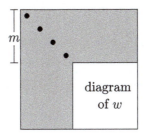

The essential set of $1^m \times w$ lies in the lower-right corner, and – up to a shift in indexing – coincides with that of w. In particular, using this indexing shift, the conditions $\dim(F_p \cap E^q) \geq k_w(p, q)$ define $\Omega_{1^m \times w} \subseteq Fl(\mathbb{C}^{2m})$. □

For instance, with $w = 2\,3\,1$ and $m = 3$, so $1^m \times w = 1\,2\,3\,5\,6\,4$, the Schubert variety is the closure of a cell

$$\Omega^\circ_{1^m \times w} = \left[\begin{array}{ccc|ccc} 1 & 0 & 0 & 0 & 0 & 0 \\ * & 1 & 0 & 0 & 0 & 0 \\ * & * & 1 & 0 & 0 & 0 \\ \hline * & * & * & 0 & 0 & 1 \\ * & * & * & 1 & 0 & 0 \\ * & * & * & * & 1 & 0 \end{array} \right].$$

When writing matrix representatives like this, we use standard basis vectors $e_{-m}, \ldots, e_{-1}, e_1, \ldots, e_m$, and the first $m + p$ columns span F_p. In our example, the essential condition is $\dim(F_2 \cap E^1) \geq 2$.

By stability, the Schubert variety Ω^∞_w determines a class σ^∞_w in $H^*_T Fl^\infty$. Indeed, the projection and inclusion maps

$$\pi_n \colon Fl^\infty \to \bigcup_m Fl(m, \ldots, m + n; \mathbb{C}^{2m})$$

and

$$\iota_m \colon Fl(m, \ldots, m + n; \mathbb{C}^{2m}) \hookrightarrow \bigcup_{m'} Fl(m', \ldots, m' + n; \mathbb{C}^{2m'}),$$

along with the subvarieties $\Omega_{1^m \times w} \subseteq Fl(m, \ldots, m + n; \mathbb{C}^{2m})$, satisfy the properties needed to define such a class, as described in Appendix A, §A.8. That is, if $w \in S_n$, there is a class $\sigma^{\infty, n}_w$ in the cohomology of

$\bigcup_{m'} Fl(m', \ldots, m'+n; \mathbb{C}^{2m'})$, such that $\iota_m^* \sigma_w^{\infty,n} = \sigma_{1^m \times w}$ for all m; and then σ_w^∞ is defined to be $\pi_n^* \sigma_w^{\infty,n}$.

Definition 12.3.2. The *(enriched) Schubert polynomial* is the polynomial

$$\mathrm{S}_w(c; x; y) \in \Lambda_T[c; x] = \mathbb{Z}[c; x; y; b]$$

corresponding to the class σ_w^∞ under the identification of $H_T^* Fl^\infty$ with $\Lambda_T[c; x]$ given in §12.2.

Recall that the torus acts by characters b_i on e_{-i} and y_i on e_i, and the isomorphism $\Lambda_T[c; x] = H_T^* Fl^\infty$ sends $x_i \mapsto c_1^T((\mathbb{S}_i/\mathbb{S}_{i-1})^\vee)$ and $c_k \mapsto c_k^T(\mathbb{C}_-^m - \mathbb{S})$ (stably, for any m).

Knowing how Ω_w^∞ restricts to each finite-dimensional flag variety will let us relate the polynomial $\mathrm{S}_w(c; x; y)$ to the ordinary (Lascoux–Schützenberger) double Schubert polynomial $\mathfrak{S}_w(x; y)$. First we need another stability property of the latter polynomials.

Lemma 12.3.3. *For $v \in S_\infty$, we have*

$$\mathfrak{S}_v(b_1, \ldots, b_m, x_1, x_2, \ldots; b_1, \ldots, b_m, y_1, y_2, \ldots)$$
$$= \begin{cases} \mathfrak{S}_w(x_1, x_2, \ldots; y_1, y_2, \ldots) & \text{if } v = 1^m \times w \text{ for some } w \in S_\infty; \\ 0 & \text{otherwise.} \end{cases}$$

Proof We may assume $v \in S_{m+n}$ for some n, and then $w \in S_n$ in the first case of the asserted formula. Consider a morphism of flagged vector bundles $F_\bullet \xrightarrow{\varphi} E_\bullet$ on a variety X, along with general line bundles $L_1, \ldots, L_m$, all chosen so that there are no relations among $x_i = c_1(\ker(E_i \to E_{i-1}))$, $y_i = c_1(F_i/F_{i-1})$, and $b_i = c_1(L_i)$. Assuming the situation is sufficiently generic, the degeneracy locus formula of Chapter 11 says $[D_w(\varphi)] = \mathfrak{S}_w(x; y)$. Let $H_i = L_1 \oplus \cdots \oplus L_i$. Then the homomorphism $\mathrm{id} \oplus \varphi \colon H_m \oplus F_n \to H_m \oplus E_n$ determines a map of flags $(H \oplus F)_\bullet \to (H \oplus E)_\bullet$. (More precisely, $(H \oplus F)_\bullet$ is the flag defined by

$$(H \oplus F)_i = \begin{cases} H_i \oplus 0 & \text{if } i \leq m; \\ H_m \oplus F_{i-m} & \text{if } i > m, \end{cases}$$

and $(H \oplus E)_\bullet$ is defined analogously.) The locus $D_v(\mathrm{id} \oplus \varphi)$ is empty if $v(i) \neq i$ for some $i \leq m$, since this forces the composition

$$H_m \hookrightarrow H_m \oplus F_n \to H_m \oplus E_n \twoheadrightarrow H_m$$

to have rank strictly less than m, when it should be the identity. So the locus is nonempty only if $v = 1^m \times w$ for some $w \in S_n$. In this case, we have $D_{1^m \times w}(\mathrm{id} \oplus \varphi) = D_w(\varphi)$, as one sees by examining the diagram of $1^m \times w$, as in the proof of Lemma 12.3.1 above. $\qquad \square$

The next theorem explains how $\mathfrak{S}_w$ specializes to $\mathfrak{S}_w$.

Theorem 12.3.4. *Under the evaluation*

$$c \mapsto c^{(m)} := \prod_{i=1}^{m} \frac{1 + b_i}{1 - a_i},$$

we have

$$\mathfrak{S}_w(c^{(m)}; x; y)$$
$$= \mathfrak{S}_{1^m \times w}(a_1, \dots, a_m, x_1, x_2, \dots; -b_1, \dots, -b_m, -y_1, -y_2, \dots).$$

Proof Assume $w \in S_n$. From the definitions,

$$\mathfrak{S}_w(c; x; y) = \sigma_w^{\infty} := \pi_n^* \sigma_w^{\infty, n} \quad \text{in } \Lambda_T[c; x] = H_T^* Fl^{\infty},$$

and $\iota_m^* \sigma_w^{\infty, n} = \sigma_{1^m \times w}$ in $H_T^* Fl(m, \dots, m+n; \mathbb{C}^{2m})$, where

$$\pi_n \colon Fl^{\infty} \to \bigcup_m Fl(m, \dots, m+n; \mathbb{C}^{2m})$$

and

$$\iota_m \colon Fl(m, \dots, m+n; \mathbb{C}^{2m}) \hookrightarrow \bigcup_{m'} Fl(m', \dots, m'+n; \mathbb{C}^{2m'})$$

are the projections and inclusions. The homomorphism π_n^* is injective. The homomorphism ι_m^* sends c to $c^T(\mathbb{C}_-^m - \mathbb{S})$, which evaluates as asserted, when $a_1, \dots, a_m$ are Chern roots of $\mathbb{S}^{\vee}$ and $b_1, \dots, b_m$ are the characters on $\mathbb{C}_-^m$. For any permutation v, we have seen that $\sigma_v = \mathfrak{S}_v(z; t)$, where $-z_1, \dots, -z_i$ are Chern roots of the rank i tautological subbundle, and the t variables are appropriate characters (§10.6). Evaluating the z and t variables in the present situation, this says $\sigma_{1^m \times w} = \mathfrak{S}_{1^m \times w}(a, x; -b, -y)$.

This argument shows that $\mathfrak{S}_w(c^{(m)}; x; y) = \mathfrak{S}_{1^m \times w}(a, x; -b, -y)$ modulo the ideal defining $H_T^* Fl(m, \dots, m+n; \mathbb{C}^{2m})$. When m is sufficiently large (relative to $\ell(w)$, the degree of these polynomials), there are no relations among the variables in relevant degrees, so this is an identity of polynomials. The identity for general m follows from this together with Lemma 12.3.3.   □

Evaluating at $c = 1$ recovers the ordinary Schubert polynomials:

$$S_w(1; x; y) = \mathfrak{S}_w(x; -y).$$

(This follows from the $m = 0$ case of the theorem, or from Lemma 12.3.3.) Since the ordinary Schubert polynomials $\mathfrak{S}_w(x; y)$ form a basis for $\mathbb{Z}[x, y]$ over $\mathbb{Z}[y]$ (§10.10.3), it follows that the enriched Schubert polynomials also form a basis.

Corollary 12.3.5. *The polynomials $S_w(c; x; y)$, for $w \in S_\infty$, form a basis for $\mathbb{Z}[c, x, y]$ over $\mathbb{Z}[c, y]$.*

Theorem 12.3.4 characterizes the polynomial $S_w(c; x; y)$, because one can find a sufficiently large m so that there are no relations among the $c_k^{(m)}$ in any given degree. To write out the polynomial $S_w(c; x; y)$, first observe that for any $1 \leq i \leq m$, both $(1^m \times w) \cdot s_i$ and $s_i \cdot (1^m \times w)$ are greater than $(1^m \times w)$ in Bruhat order, so

$$\partial_i \mathfrak{S}_{1^m \times w} = \partial_i^y \mathfrak{S}_{1^m \times w} = 0$$

for $1 \leq i \leq m$. It follows that the polynomial

$$\mathfrak{S}_{1^m \times w}(a_1, \ldots, a_m, x_1, x_2, \ldots; b_1, \ldots, b_m, y_1, y_2, \ldots)$$

is separately symmetric in the a and b variables. Lemma 12.3.3 then shows it is super-symmetric in a and b. It follows that this Schubert polynomial may be written as a polynomial in the variables $c^{(m)}$, x, and y, since the elements $c_k^{(m)}$ generate the ring of super-symmetric polynomials (Exercise 12.2.3). In fact, using Exercise 12.2.2, such an expression is determined by evaluating at any m with $2m \geq \ell(w)$.

Example 12.3.6. For $w = 2\,3\,1$, we have

$$S_{2\,3\,1}(c; x; y) = (x_1 + y_1)(x_2 + y_1) + (x_1 + x_2 + y_1)c_1 + c_1^2 - c_2.$$

This can be checked by evaluating at $m = 1$, since there are no relations among $c_1^{(1)}$ and $c_2^{(1)}$. One computes

$$\mathfrak{S}_{12453} = (x_1 + y_1)(x_2 + y_1) + (x_1 + x_2 + y_1)(a_1 + b_1)$$
$$+ a_1 b_1 + b_1^2,$$

which agrees with the asserted formula for $S_{231}(c^{(1)}; x; y)$.

12.4 Degeneracy loci

The general degeneracy locus setting is this. On a nonsingular variety X, we have a vector bundle V of rank $2m$, with two flags of subbundles $E^\bullet$ and $F_\bullet$, arranged so that

$$F = F_0 \subset F_1 \subset F_2 \subset \cdots \subset V \supset E = E^0 \supset E^1 \supset \cdots,$$

with $\operatorname{rk} F = \operatorname{rk} E = m$. For a permutation w, we define a degeneracy locus in X by

$$D_w^\infty(F_\bullet \cap E^\bullet) = \left\{ x \in X \mid \dim(F_p \cap E^q) \geq k_w(p, q) \text{ for all } p, q \right\}.$$

Its class is computed by evaluating the polynomials $S_w(c; x; y)$.

Proposition 12.4.1. *There is a canonical class $\mathbb{D}_w^\infty \in H^{2\ell(w)}X$, supported on $D_w^\infty(F_\bullet \cap E^\bullet)$, compatible with pullbacks, and equal to $[D_w^\infty(F_\bullet \cap E^\bullet)]$ when the locus has expected codimension $\ell(w)$ in X.*

The polynomial $S_w(c; x; y)$ is the unique polynomial in $\mathbb{Z}[c; x; y]$ which maps to $\mathbb{D}_w^\infty$ under the evaluation

$$c_k \mapsto c_k(V - E - F), \quad x_i \mapsto -c_1(F_i/F_{i-1}), \quad y_i \mapsto c_1(E^{i-1}/E^i)$$

for all degeneracy loci $D_w^\infty(F_\bullet \cap E^\bullet) \subseteq X$ as above.

Proof The Schubert varieties $\Omega_{1^m \times w} \subseteq Fl(\mathbb{C}^{2m})$ are instances of D_w^∞ $(F_\bullet \cap E^\bullet)$, by taking $F_\bullet$ to be the tautological flag on $X = Fl(\mathbb{C}^{2m})$ and $E^\bullet$ to be the standard fixed flag (both appropriately re-indexed so that the m-dimensional pieces are F_0 and E^0, respectively). To prove the second statement of the proposition, it suffices to consider this case, and in Theorem 12.3.4 we have already seen that $S_w(c; x; y)$ are characterized this way.

The general construction of $\mathbb{D}_w^\infty$ similarly reduces to that of $\mathbb{D}_{1^m \times w}$, just as in Lemma 12.3.1. Possibly after replacing X by an affine bundle, $F_\bullet$ and $E^\bullet$ determine a morphism $f \colon X \to \mathbb{E} \times^{B^-} Fl(m, \ldots, m+n; \mathbb{C}^{2m})$

such that $D_w^\infty(F_\bullet \cap E^\bullet) = f^{-1}(\mathbb{E} \times^{B^-} \Omega_{1^m \times w})$. (See §11.6 and Appendix E.) The class is then defined by $\mathbb{D}_w^\infty = f^*[\Omega_{1^m \times w}]^{B^-}$. $\qquad \square$

The characterization of $S_w(c; x; y)$ provided by Theorem 12.3.4 also shows that these polynomials are related by divided difference operators:

$$\partial_i S_w(c; x; y) = \begin{cases} S_{ws_i}(c; x; y) & \text{if } ws_i < w; \\ 0 & \text{if } ws_i > w, \end{cases}$$

and

$$\partial_i^y S_w(c; x; y) = \begin{cases} S_{s_i w}(c; x; y) & \text{if } s_i w < w; \\ 0 & \text{if } s_i w > w. \end{cases}$$

These operators are linear over $\mathbb{Z}[c]$, that is, they treat the c variables as scalars.

On the other hand, the permutations $w_\circ^{(n)} = [n, \ldots, 2, 1]$ are not stable with respect to the embedding ι'. We do not have a simple product formula for $S_{w_\circ^{(n)}}(c; x; y)$, but there are explicit determinantal formulas. Here is one such formula, based on the Kempf–Laksov formula.

Theorem 12.4.2. *For $w_\circ^{(n)} = [n, \ldots, 2, 1] \in S_\infty$, we have*

$$S_{w_\circ^{(n)}}(c; x; y) = \Delta_{(n-1, \ldots, 2, 1)}(c(1), \ldots, c(n-1))$$

$$= \det \left(c_{n-2i+j}(i) \right)_{1 \le i, j \le n-1}$$

where the entries are

$$c(p) = c \cdot c(E/E^{n-p} - F_p/F) = c \cdot \left(\prod_{i=1}^{p} \prod_{j=1}^{n-p} \frac{1 + y_j}{1 - x_i} \right).$$

Proof   Using a "diagonal trick," the problem can be reduced to a Grassmannian degeneracy locus. Let us fix a sufficiently large m, along with a vector bundle V of rank $2m$ and general flags of subbundles $E^\bullet$ and $F_\bullet$ on a variety X. Continuing the notational conventions we have used in this chapter, this means F_p has rank $m + p$ and E^q has rank $m - q$.

We consider the locus

$$D_{w_\circ^{(n)}}^\infty(F_\bullet \cap E^\bullet) = D_{1^m \times w_\circ^{(n)}}(F_\bullet \cap E^\bullet)$$

in X. The essential conditions for this locus are

$$\dim(F_p \cap E^{n-p}) \geq p, \quad p = 1, \ldots, n-1.$$

Equivalently, the locus is defined by the conditions

$$\dim(F_p \oplus E^{n-p} \cap \Delta_V) \geq p, \quad p = 1, \ldots, n-1,$$

on intersections inside the bundle $V \oplus V$, where $\Delta_V \subset V \oplus V$ is the diagonal subbundle. This is now pulled back from a Schubert locus $\Omega_{(n-1,\ldots,2,1)}(G_\bullet)$ in the Grassmann bundle $\mathbf{Gr}(2m, V \oplus V)$, using

$$G_\bullet : \cdots \subset F_1 \oplus E^{n-1} \subset \cdots \subset F_n \oplus E^0 \subset \cdots \subset V \oplus V$$

as the reference flag. The assertion then follows from the Kempf–Laksov formula (Theorem 9.2.2): the quotient bundle $\mathbb{Q}$ restricts to $(V \oplus V)/\Delta_V \cong V$, so the entries $c(p) = c(\mathbb{Q} - G_{2m+2p-n})$ become

$$c(\mathbb{Q} - G_{2m+2p-n}) = c(V - (F_p + E^{n-p}))$$
$$= c(V - (F + F_p/F + E - E/E^{n-p}))$$
$$= c(V - E - F) \cdot c(E/E^{n-p} - F_p/F),$$

as claimed. $\qquad\square$

Exercise 12.4.3. Write out the formula for $S_{321}(c; x; y)$.

The Schubert polynomials $S_w(c; x; y)$ satisfy analogues of many of the properties of $\mathfrak{S}_w(x; y)$ which we have seen in the last two chapters. Some of them are stated in the Notes below.

Notes

Various treatments of stability for Schubert polynomials appear in the literature. The geometry is often explained in terms of embeddings $Fl(\mathbb{C}^n) \subset Fl(\mathbb{C}^{n+1}) \subset \cdots$, with respect to standard inclusions $\mathbb{C}^n \subset \mathbb{C}^{n+1}$ (as the span of the first n basis vectors). The union of such a chain is an ind-variety, $Fl' = \bigcup_n Fl(\mathbb{C}^n)$, which parametrizes complete flags $F_\bullet$ which are eventually standard – that is, $F_n = \mathbb{C}^n$ for all $n \gg 0$. This is naturally a subspace of $Fl(\mathbb{C}^\infty)$, and the pullback homomorphism of cohomology rings associated to the embedding $Fl' \hookrightarrow Fl(\mathbb{C}^\infty)$ is an inclusion

$$\Lambda[x_1, x_2, \ldots] \hookrightarrow \Lambda[\![x_1, x_2, \ldots]\!]_{\mathrm{gr}}.$$

Here $H_T^* Fl' = \Lambda[\![x]\!]_{\mathrm{gr}}$ is the graded formal series ring, where expressions involving infinite sums of bounded total degree are allowed, like $x_1 + x_2 + \cdots$. This is the graded inverse limit of the rings $\Lambda[x_1, \ldots, x_n]$ (see Appendix A, §A.8). So the claim that Schubert polynomials belong to the polynomial sub-algebra $\Lambda[x] \subseteq \Lambda[\![x]\!]_{\mathrm{gr}}$ is equivalent to the assertion that the Schubert classes in $H_T^* Fl'$ appear by restriction from $H_T^* Fl(\mathbb{C}^\infty)$.

The stability property (with respect to the embeddings defining Fl') was highlighted by Fomin and Kirillov (1996b), and was taken as a definition of Schubert polynomials by Billey and Haiman (1995). We will see versions of the Billey–Haiman polynomials for symplectic flag varieties in Chapter 13.

The Sato Grassmannian appears in connection with loop groups, and parametrizes solutions to certain hierarchies of differential equations. Some general references for this point of view are the books by Pressley and Segal (1986) and by Miwa, Jimbo, and Date (2000).

Lemma 12.3.3 appears in (Buch and Rimányi, 2004, Corollary 2.5), with a more combinatorial proof.

Lam, Lee, and Shimozono (2021) launched an extensive study of the polynomials $\mathrm{S}_w(c; x; y)$, and showed that they exhibit many remarkable properties in addition to the ones mentioned here. These authors call them "back-stable" Schubert polynomials, building on ideas of Buch and Knutson. They use the characterization from Theorem 12.3.4 as a definition, and label the variables $a_1, a_2, \ldots$ and $b_1, b_2, \ldots$ by nonpositive indices, writing $x_0, x_{-1}, \ldots$ and $y_0, y_{-1}, \ldots$, respectively. The version we present here is based on (Anderson and Fulton, 2021a), a variation on (Lam et al., 2021).

Pawlowski (2021) gave a different geometric interpretation of $\mathrm{S}_w(c; x; y)$, in terms of *graph Schubert varieties*. In fact, the graph construction provided one motivation for the "back-stable Schubert polynomials"; see (Knutson et al., 2013, §7).

Using a variation on the diagonal method employed in the proof of Theorem 12.4.2, one can deduce formulas for degeneracy loci labelled by *vexillary* permutations. This idea was exploited in (Anderson and Fulton, 2012).

The polynomials $\mathrm{S}_w(c; x; y)$ may be defined more generally to allow w to be a permutation of the integers, fixing all but finitely many; this group is denoted $S_{\mathbb{Z}}$. This is the setting in (Lam et al., 2021), and in (Anderson and Fulton, 2021a). For such $w \in S_{\mathbb{Z}}$, these Schubert polynomials also depend on nonpositive variables $x_0, x_{-1}, x_{-2}, \ldots$, and they form a basis for $\Lambda[c, x]$ over Λ. They are related by divided difference operators ∂_i for $i \in \mathbb{Z}$, which act by the usual formula on the x variables, treating the c and y variables as constants – except for ∂_0, which acts by $\partial_0(c_k) = c_{k-1} + x_1 c_{k-2} + \cdots + x_1^{k-1}$.

We mention a few properties of these polynomials, without proof. The reader may try to deduce them using Theorem 12.3.4 (or consult the references).

They satisfy an extended interpolation property. For v in $S_{\mathbb{Z}}$, we specialize the x variables as usual, writing

$$x \mapsto -y^v = (-y_{v(i)})_{i \in \mathbb{Z}}.$$

The c variables specialize by

$$c \mapsto c^v := \prod_{\substack{i \leq 0 \\ v(i) > 0}} \frac{1 + y_i}{1 + y_{v(i)}}.$$

(If one writes $c = \prod_{i \leq 0} \frac{1 + y_i}{1 - x_i}$, this is compatible with the specialization of x variables.) Then $S_w(c; x; y)$ satisfy and are characterized by

$$S_w(c^v; -y^v; y) = \begin{cases} \prod_{i < j \,:\, w(i) > w(j)} (y_{w(j)} - y_{w(i)}) & \text{if } v = w; \\ 0 & \text{if } v \not\geq w \end{cases} \tag{12.1}$$

for permutations $v, w \in S_{\mathbb{Z}}$. Although it is not written in precisely this form, this property is implicit in (Lam et al., 2021).

When $v \in S_\infty$, so it permutes only the positive integers, we have $c^v = 1$, and this special case of (12.1) says

$$S_w(1; -y^v; y) = \begin{cases} \prod_{i < j \,:\, w(i) > w(j)} (y_{w(j)} - y_{w(i)}) & \text{if } v = w; \\ 0 & \text{if } v \not\geq w \end{cases} \tag{12.2}$$

for permutations $v, w \in S_\infty$. This gives another proof that $S_w(1; x; y) = \mathfrak{S}_w(x; -y)$. (However, the conditions of (12.2) for $v, w \in S_\infty$ do not suffice to determine $S_w(c; x; y)$.)

There is a duality formula for $S_w(c; x; y)$, extending the one we saw in §11.8. Let $\bar{c} = 1 + \bar{c}_1 + \bar{c}_2 + \cdots$ be defined by the relations

$$\bar{c}_k + \bar{c}_{k-1} c_1 + \cdots + c_k = 0$$

for $k > 0$, so $\bar{c} = (1 + c_1 + c_2 + c_3 + \cdots)^{-1}$. Then

$$S_w(c; x; y) = S_{w^{-1}}(\bar{c}; y; x). \tag{12.3}$$

There is also an upgraded version of the Cauchy formula from §11.8: we have

$$S_w(c; x; y) = \sum_{vu \doteq w} S_u(c'; x; t) \cdot S_v(c''; -t, y) \tag{12.4}$$

for series c, c', c'' such that $c = c' \cdot c''$, where $vu \doteq w$ means $vu = w$ and $\ell(v) + \ell(u) = \ell(w)$, as before.

As $m \to \infty$, the Schubert polynomials $\mathfrak{S}_{1^m \times w}(a_1, \ldots, a_m, 0, 0, \ldots)$ stabilize to a symmetric function F_w, known as a *Stanley symmetric function*. Theorem 12.3.4 shows that the Stanley function is obtained by specializing variables in $S_w(c; x; y)$: we have

$$F_w = S_w(c; 0; 0),$$

with the c variables evaluated as complete homogeneous symmetric functions in the a variables. Combining this with the Cauchy formula (12.4), we obtain

$$S_w(c; x; y) = \sum_{zuv \doteq w} S_v(1; x; 0) \cdot S_u(c; 0, 0) \cdot S_z(1; 0, y)$$

$$= \sum_{z^{-1}uv \doteq w} \mathfrak{S}_v(x) \cdot F_u(c) \cdot \mathfrak{S}_z(y). \tag{12.5}$$

Stanley symmetric functions are also known as *stable Schubert polynomials*, because of their characterization as stable limits of $\mathfrak{S}_{1^m \times w}$. They were introduced by Stanley (1984), who defined F_w as a certain generating function of reduced expressions for w. There are several other combinatorial formulas for them; one "Littlewood–Richardson" type formula for the Schur expansion of a Stanley function was given by Fomin and Greene (1998). Using this, (12.5) leads to an explicit expression for $S_w(c; x; y)$.

The geometry corresponding to the Schubert polynomials for $w \in S_{\mathbb{Z}}$ is that of an infinite flag variety $Fl^{\pm \infty}$ which parametrizes flags

$$\cdots \subset F_{-2} \subset F_{-1} \subset F_0 \subset F_1 \subset F_2 \subset \cdots$$

extending infinitely in both directions from $F = F_0$. The algebra and geometry are similar for the divided difference operators which compute Schubert polynomials for the symplectic flag variety, as we will see in the next chapter.

Hints for Exercises

Exercise 12.2.1. Using $c^T(\mathbb{C}_-^m - \mathbb{S}) = c^T(\mathbb{Q} - \mathbb{C}_+^m)$, the relations simply say that $c^T(\mathbb{Q} - \mathbb{C}_+^m) \cdot c^T(\mathbb{C}_+^m) = c^T(\mathbb{Q})$ vanishes in degrees greater than m. To see that the classes c_i generate, it suffices to consider the case where T is trivial, and this case was handled in Chapter 4.

Exercise 12.2.3. The element $c = 1 + c_1 + c_2 + \cdots$ is equal to $\prod_{i \geq 1} \frac{1+b_i}{1-a_i}$, so the c_k are algebraically independent and super-symmetric. So it suffices to show that the projection

$$\mathrm{Symm}(a) \otimes_R \mathrm{Symm}(b) \to \mathrm{Symm}(x), \qquad a \mapsto x, \ b \mapsto 0,$$

is injective when restricted the subring of super-symmetric functions. Using bases of complete homogeneous and elementary symmetric functions, this projection is identified with the homomorphism $R[h_1, h_2, \ldots, e_1, e_2, \ldots] \to R[h_1, h_2, \ldots]$ which sends $e \mapsto 0$. To show that no nonzero super-symmetric function lies in the ideal $(e_1, e_2, \ldots) \subseteq R[h, e]$, one can argue as follows.

Consider the homomorphism

$$\sigma \colon R[h, e] \to R[\tau, \tau', h, e]$$

defined by $h_k \mapsto h_k + \tau h_{k-1} + \tau^2 h_{k-2} + \cdots + \tau^k$ and $e_k \mapsto e_k + \tau' e_{k-1}$. Then $f \in R[h, e]$ is super-symmetric if and only if $\sigma(f)$ is independent of τ, τ'. (By writing $\tau = t - a_1$ and $\tau' = -t - b_1$, this identifies σ with the evaluation at

$a_1 = t$, $b_1 = -t$.) One checks furthermore that σ is injective. (This is not true with finite variable sets!)

Now suppose $f = \sum_{i \geq 1} f_i\, e_i$ for some $f_1, f_2, \ldots \in R[h, e]$. Then

$$\sigma(f) = \sum_{i \geq 1} \sigma(f_i)\, \sigma(e_i)$$
$$= \sigma(f_1)\, \tau' + \sum_{i \geq 1} (\sigma(f_i) + \sigma(f_{i+1})\, \tau')e_i,$$

so $\sigma(f_1) = 0$ and hence $f_1 = 0$. Continuing in this way, one sees all $f_i = 0$, so $f = 0$, as claimed.

See also (Metropolis et al., 1981; Stembridge, 1985). Many authors differ in conventions for super-symmetric functions: the more common convention is related to ours by replacing b_i with $-b_i$.

Exercise 12.4.3. The answer is

$$S_{3\,2\,1}(c; x; y) = c_2 c_1 - c_3 + c_2(x_1 + x_2 + y_1) + (c_1^2 - c_2)(x_1 + y_1 + y_2)$$
$$+ c_1(x_1 + y_1 + y_2)(x_1 + x_2 + y_1)$$
$$+ (x_1 + y_1)(x_2 + y_1)(x_1 + y_2).$$

The examples given in §9.2 may be helpful.

13

Symplectic Flag Varieties

We have seen that formulas for degeneracy loci are closely connected with flag varieties and the groups GL_n and SL_n, of Lie type A. Degeneracy loci for symmetric morphisms, or varieties of isotropic subspaces with respect to a symplectic form, are correspondingly related to the symplectic group Sp_{2n}, of Lie type C. In this chapter, we will see type C analogues of the basic facts about flag varieties. The next chapter takes up the problem of describing Schubert polynomials for such varieties.

Much of this story proceeds in parallel with the type A case, and much of it generalizes to arbitrary Lie type.

13.1 Degeneracy loci for symmetric maps

A linear map $\varphi\colon E^\vee \to E$ is *symmetric* if it is equal to its dual:

$$\varphi = \varphi^\vee \text{ as maps } E^\vee \to E^{\vee\vee} = E.$$

This is the same as saying φ lies in the subspace $\mathrm{Sym}^2 E \subseteq E \otimes E = \mathrm{Hom}(E^\vee, E)$. The group $GL(E)$ acts on $\mathrm{Sym}^2 E$ in the standard way, by $g(v \cdot w) = (gv) \cdot (gw)$. Choosing a basis, so $E = \mathbb{C}^n$, a symmetric map φ is given by a symmetric matrix A, and GL_n acts on symmetric matrices by

$$g \cdot A = gAg^\dagger.$$

Let $SM_n \subseteq M_{n,n}$ be the space of symmetric matrices, with this action of GL_n. The identification $E = \mathbb{C}^n$ induces $\mathrm{Sym}^2 E = SM_n$.

Consider the locus D_r of symmetric maps of rank at most r. This is a $GL(E)$-invariant subvariety of $\mathrm{Sym}^2 E$, and in fact, these loci are precisely the $GL(E)$-invariant subvarieties. (Any symmetric bilinear form can be diagonalized, so the $GL(E)$-orbits are precisely $D_r^\circ = \{A \mid \mathrm{rk}(A) = r\}$, and in this setting it is easy to see $D_r = \overline{D_r^\circ}$.) As before, the basic question arises: what is the polynomial corresponding to $[D_r]^{GL(E)}$ in $\Lambda_{GL(E)} = \mathbb{Z}[a_1, \ldots, a_n]$?

Exercise 13.1.1. Compute $[D_0]^{GL(E)} = c_{top}^{GL(E)}(\mathrm{Sym}^2 E)$ as a polynomial in $a_1, \ldots, a_n$.

An answer to the general question was given by Giambelli in 1906:

$$[D_r]^{GL(E)} = 2^{n-r} \Delta_{(n-r,\ldots,2,1)}(a) \quad \text{in} \quad \Lambda_{GL(E)} = \mathbb{Z}[a_1, \ldots, a_n],$$

where $a_i = c_i^{GL(E)}(E)$. Equivalently, $[D_r]^T = 2^{n-r} s_{(n-r,\ldots,2,1)}(t_1, \ldots, t_n)$ in $\Lambda_T = \mathbb{Z}[t_1, \ldots, t_n]$. As the exercise illustrates, however, this is not a special case of the degeneracy locus formula for general maps in any natural way. For example, the codimensions are wrong: $D_0 = \{0\}$ has codimension n^2 in $M_{n,n}$, but codimension $\binom{n+1}{2}$ in SM_n.

We will see variations on these formulas. As motivation for what follows, let us start by sketching an argument analogous to the one which proved the Cayley–Giambelli–Thom–Porteous formula in Chapter 11 (Proposition 11.1.1). The idea there was to replace $\varphi \colon F \to E$ by an embedding $F \hookrightarrow E \oplus F$, namely the graph subspace $\mathrm{Graph}(\varphi)$, and to interpret the conditions $\mathrm{rk}(\varphi) \le r$, or equivalently $\dim \ker(\varphi) \ge n - r$, as $\dim(\mathrm{Graph}(\varphi) \cap (0 \oplus F)) \ge n - r$, defining a Schubert variety in the Grassmannian $Gr(m, E \oplus F)$.

Here we proceed similarly: $\varphi \colon E^\vee \to E$ corresponds to a graph subspace $\mathrm{Graph}(\varphi) \subseteq E \oplus E^\vee$. There is a canonical symplectic form on $E \oplus E^\vee$, given by $\omega(v \oplus f, w \oplus g) = f(w) - g(v)$. The condition that φ be symmetric is equivalent to requiring that $\mathrm{Graph}(\varphi)$ be *isotropic* with respect to this form. The subspace $0 \oplus E^\vee$ is also isotropic, and as before there is a natural identification $\ker(\varphi) = \mathrm{Graph}(\varphi) \cap (0 \oplus E^\vee)$. We will return to this construction, and the proof of Giambelli's formula, in Chapter 14 (see §14.3).

Thus we are led to consider Schubert varieties in spaces of isotropic flags. We will begin with a quick review of the linear algebra of symplectic vector spaces.

13.2   Isotropic subspaces

We have a $2n$-dimensional vector space V, equipped with a symplectic form – that is, a skew-symmetric nondegenerate bilinear form $\omega\colon \bigwedge^2 V \to \mathbb{C}$. All such forms are equivalent, up to change of basis. We write the standard basis for V as

$$e_{\bar{n}}, \ldots, e_{\bar{1}}, e_1, \ldots, e_n,$$

regarding the barred integers as negative numbers, so that $\bar{i} = -i$ and $\bar{\bar{i}} = i$. As a convenient symplectic form with respect to this basis, we choose

$$\omega = \sum_{i=1}^{n} e_{\bar{i}}^* \wedge e_i^*$$

so for $i, j > 0$,

$$\omega(e_{\bar{i}}, e_j) = -\omega(e_j, e_{\bar{i}}) = \delta_{ij}, \quad \text{and} \quad \omega(e_i, e_j) = \omega(e_{\bar{i}}, e_{\bar{j}}) = 0.$$

This form has Gram matrix

$$\begin{pmatrix} & & & & & & 1 \\ & & & & & \iddots & \\ & & & & 1 & & \\ \hline & & -1 & & & & \\ & \iddots & & & & & \\ -1 & & & & & & \end{pmatrix}.$$

The symplectic form induces an isomorphism $V \xrightarrow{\sim} V^\vee$, by sending $v \mapsto \omega(v, \cdot)$, and these choices identify the dual basis as $e_i^* = e_{\bar{i}}$ for all i.

For a subspace $F \subseteq V$, the *orthogonal space* $F^\perp \subseteq V$ is

$$F^\perp = \Big\{ v \in V \,\Big|\, \omega(u, v) = 0 \text{ for all } u \in F \Big\}.$$

An elementary fact is that $V/F^\perp \cong F^\vee$ under the isomorphism $V \cong V^\vee$, so $\dim F^\perp = \dim V - \dim F$. A subspace is *isotropic* if $F \subset F^\perp$, or equivalently, if $\omega|_F \equiv 0$. It follows that $\dim F \leq n$ for any isotropic subspace F. For any vector $v \in V$, we have $\omega(v, v) = 0$ (since ω is skew-symmetric), so every 1-dimensional subspace of V is isotropic.

A maximal isotropic subspace $F \subset V$ is often called Lagrangian; such spaces have $\dim F = n$. A *(complete) isotropic flag* in V is a chain

$$F_\bullet : F_n \subset \cdots \subset F_1 = F \subset V$$

of isotropic subspaces, with $\dim F_p = n + 1 - p$ for all p, so $F = F_1$ is Lagrangian. An isotropic flag can also be specified by choosing a Lagrangian subspace $F \subset V$ and an ordinary complete flag in F. Our "backwards" indexing is motivated by the distinguished role of Lagrangian subspaces, and in fact is necessary for stability properties, as we will see.

A subspace $F \subseteq V$ is *co-isotropic* if $F \supseteq F^\perp$. Any complete isotropic flag $F_\bullet$ extends canonically to an ordinary complete flag in V, by appending co-isotropic spaces $F_{\bar{p}} = F_{p+1}^\perp$, to obtain

$$F_n \subset \cdots \subset F_1 \subset F_{\bar{1}} \subset \cdots \subset F_{\bar{n}} = V.$$

These co-isotropic spaces have dimension $\dim F_{\bar{p}} = n + p$.

The standard basis leads to a *standard* isotropic flag $E_\bullet$, with $E_q = \mathrm{span}\{e_n, \ldots, e_q\}$ for $q > 0$. It extends to the standard complete flag by setting

$$E_{\bar{q}} = E_{q+1}^\perp = \mathrm{span}\{e_n, \ldots, e_1, e_{\bar{1}}, \ldots, e_{\bar{q}}\}$$

for $q > 0$. The standard Lagrangian subspace is

$$E = E_1 = \mathrm{span}\{e_n, \ldots, e_1\}.$$

The *symplectic group* is the subgroup $Sp(V, \omega) \subseteq GL(V)$ preserving the form ω. When using the standard basis and standard symplectic form, we usually write this as $Sp_{2n} \subseteq GL_{2n}$. The subgroup $B \subseteq Sp_{2n}$ of upper-triangular matrices is a Borel subgroup; it is the subgroup fixing the *opposite* isotropic flag $\widetilde{E}_\bullet$, defined by $\widetilde{E}_q = \mathrm{span}\{e_{\bar{n}}, \ldots, e_{\bar{q}}\}$ for $q > 0$.

With respect to the standard basis and form ω, the symplectic group Sp_{2n} consists of block matrices

$$\left(\begin{array}{c|c} X & Y \\ \hline Z & W \end{array} \right)$$

such that

$$(XDY^\dagger)^\dagger = XDY^\dagger,$$
$$(ZDW^\dagger)^\dagger = ZDW^\dagger, \quad \text{and}$$
$$XDW^\dagger - YDZ^\dagger = D,$$

where D is the $n \times n$ matrix with 1's on the anti-diagonal and 0's else-where. These give quadratic equations defining Sp_{2n} as a subgroup of GL_{2n}.

The *Lagrangian Grassmannian* $LG(V)$ parametrizes Lagrangian sub-spaces in V, and it embeds as a closed subvariety in the ordinary Grass-mannian $LG(V) \hookrightarrow Gr(n, V)$. By Witt's theorem, the symplectic group acts transitively on Lagrangian subspaces, so there is an identification $LG(V) \cong Sp_{2n}/P$, where P is the parabolic subgroup which fixes the Lagrangian subspace $\widetilde{E}_1 = \mathrm{span}\{e_{\overline{n}}, \ldots, e_{\overline{1}}\}$.

Similarly, the *isotropic flag variety* $Fl_\omega(V)$ parametrizes complete iso-tropic flags in V. From the definition, it embeds as a closed subvariety of the partial flag variety $Fl^{(n)}(V) = Fl(1, \ldots, n; V)$. But as we saw, each isotropic flag canonically extends to a complete one, so we obtain an em-bedding $Fl_\omega(V) \hookrightarrow Fl(V)$. There is an identification $Fl_\omega(V) \cong Sp_{2n}/B$, using Witt's theorem again.

13.3   Symplectic flag bundles

The notions of symplectic form, flag variety, and Lagrangian Grassman-nian globalize immediately to the setting where V is a G-equivariant vector bundle on a variety Y with G-action. Here ω is a G-equivariant section of $\bigwedge^2 V^\vee$, locally of the form described in the previous section. We will give constructions of the isotropic flag variety parallel to the descriptions of ordinary flag varieties we saw in Chapter 4. As before, it is useful – and no more complicated – to carry this out for flag bundles.

Let G be a linear algebraic group acting on a variety Y, and let V be a G-equivariant vector bundle of rank $2n$, equipped with a symplectic form $\omega \colon \bigwedge^2 V \to \mathbb{C}_Y$ with values in the trivial line bundle. We assume ω is G-invariant; that is, $\omega(g \cdot u, g \cdot v) = \omega(u, v)$ for all $g \in G$. The isotropic flag bundle $\rho \colon \mathbf{Fl}_\omega(V) \to Y$ parametrizes complete isotropic flags of subspaces: a point in the fiber $\rho^{-1}(y)$ is an isotropic flag $F_\bullet$ in the vector space $V(y)$.

The isotropic flag bundle can be constructed as a tower of projec-tive bundles. One starts with $X_1 = \mathbb{P}(V) \to Y$, a projective bundle with fibers $\mathbb{P}^{2n-1}$; this has a tautological line bundle $\mathbb{S}_n \subseteq V$, as well as its orthogonal complement $\mathbb{S}_n^\perp$, of rank $2n - 1$. The next step is

$X_2 = \mathbb{P}(\mathbb{S}_n^\perp/\mathbb{S}_n) \to X_1$, a bundle with fibers $\mathbb{P}^{2n-3}$; this has tautological bundles $\mathbb{S}_n \subset \mathbb{S}_{n-1} \subset \mathbb{S}_{n-1}^\perp \subset \mathbb{S}_n^\perp$. Continuing in this way, one obtains

$$\mathbf{Fl}_\omega(V) = X_n = \mathbb{P}(\mathbb{S}_2^\perp/\mathbb{S}_2) \to X_{n-1} \to \cdots \to Y,$$

so the successive fibers in the tower are projective spaces of dimensions $2n-1, 2n-3, \ldots, 3, 1$, and all these maps are G-equivariant. So the isotropic flag bundle is projective, of relative dimension

$$\dim \mathbf{Fl}_\omega(V) - \dim Y = (2n-1) + (2n-3) + \cdots + 1 = n^2,$$

and $H_G^* \mathbf{Fl}_\omega(V)$ is free of rank

$$2n \cdot (2n-2) \cdots 2 = 2^n \cdot n!$$

as a module over $H_G^* Y$.

As in Chapter 4, we also obtain a presentation and a basis for the equivariant cohomology ring. Let $x_i = c_1^G(\mathbb{S}_i/\mathbb{S}_{i+1})$, and write $e_k = e_k(x_1, \ldots, x_n) = c_k^G(\mathbb{S})$, where $\mathbb{S} = \mathbb{S}_1 \subset V$ is the tautological Lagrangian subbundle.

Proposition 13.3.1. *We have*

$$H_G^* \mathbf{Fl}_\omega(V) = (H_G^* Y)[x_1, \ldots, x_n]/\big(e_k(x_1^2, \ldots, x_n^2) - c_{2k}^G(V)\big)_{1 \le k \le n},$$

where e_k is the elementary symmetric polynomial.

A basis for $H_G^* \mathbf{Fl}_\omega(V)$ as a module over $H_G^* Y$ is

$$\Big\{ x_1^{m_1} \cdots x_n^{m_n} \,\big|\, 0 \le m_i \le 2i-1 \Big\}.$$

Proof The argument is parallel to one we saw in Chapter 4 (Proposition 4.4.1). As in that situation, the basis for $H_G^* \mathbf{Fl}_\omega(V)$ comes from its description as a tower of projective bundles. To see that the relations hold, note that $V/\mathbb{S} \cong \mathbb{S}^\vee$, so the Whitney sum formula gives

$$c^G(V) = c^G(\mathbb{S}) \cdot c^G(\mathbb{S}^\vee)$$
$$= \prod_{i=1}^n (1+x_i) \prod_{i=1}^n (1-x_i)$$
$$= 1 + e_1(x_1^2, \ldots, x_n^2) + \cdots + e_n(x_1^2, \ldots, x_n^2);$$

that is, $c_{2k}^G(V) = e_k(x_1^2, \ldots, x_n^2)$. $\qquad \square$

Exercise 13.3.2. Complete the proof, by imitating the argument of Proposition 4.4.1.

Exercise 13.3.3. For a *strict partition* $\lambda = (\lambda_1 > \cdots > \lambda_s > 0)$ write $e_\lambda = e_{\lambda_1} \cdots e_{\lambda_s}$. Show that another basis for $H^*_G \mathbf{Fl}_\omega(V)$ as a module over $H^*_G Y$ is given by

$$\left\{ x_1^{a_1} \cdots x_n^{a_n} \cdot e_\lambda \,\middle|\, 0 \le a_i \le n - i, \, \lambda \text{ a strict partition}, \lambda_1 \le n \right\},$$

where $e_k = e_k(x_1, \dots, x_n) = c_k^G(\mathbb{S})$.

13.4   Lagrangian Grassmannians

In presenting the cohomology of the ordinary Grassmannian in Chapter 4, we found bases of Schur determinants. We will see an analogous basis of Schur Q-polynomials for the cohomology of the Lagrangian Grassmannian. These polynomials are defined as certain Pfaffians. Here we set up notation and quickly review the main facts, referring to Appendix C for details and references.

Given a series of formal variables $c = 1 + c_1 + c_2 + \cdots$, and integers $p, q \ge 0$, we set

$$C_{pq} = c_p c_q + 2 \sum_{i=1}^{q} (-1)^i c_{p+i} c_{q-i} \qquad (*)$$

$$= c_p c_q - 2 c_{p+1} c_{q-1} + 2 c_{p+2} c_{q-2} - \cdots + (-1)^q 2 c_{p+q}.$$

The variables generate a universal ring

$$\Gamma = \mathbb{Z}[c_1, c_2, \dots]/(C_{pp})_{p > 0},$$

where the elements c_p have degree p, and the relations are given by $C_{pp} = c_p^2 - 2 c_{p+1} c_{p-1} + 2 c_{p+2} c_{p-2} - \cdots$ as above. When the context is clear, we sometimes recycle notation and write c also for its image in Γ.

Bases for Γ are naturally indexed by *strict partitions*, that is, partitions $\lambda = (\lambda_1 > \cdots > \lambda_s > 0)$ with distinct parts. One basis is given by monomials in c:

$$\Gamma = \bigoplus_\lambda \mathbb{Z} \cdot c_{\lambda_1} \cdots c_{\lambda_s},$$

the sum ranging over all strict partitions (cf. Exercise 13.3.3 above).

A second basis is given by the Q-polynomials. Given a strict partition λ with s parts, let $m_{ij} = C_{\lambda_i, \lambda_j}$ for $i < j$, and form the corresponding $s \times s$ skew-symmetric matrix $M_\lambda(c) = (m_{ij})$. The Q-polynomial $Q_\lambda(c)$ is defined to be the Pfaffian of this matrix,

$$Q_\lambda(c) = \mathrm{Pf}(M_\lambda(c)),$$

ensuring that λ has an even number of parts by appending 0 if needed. (As before, we sometimes write $Q_\lambda(c)$ for its image in Γ.) Then

$$\Gamma = \bigoplus_\lambda \mathbb{Z} \cdot Q_\lambda(c),$$

with the sum again over strict partitions.

For $n > 0$, there is a finitely generated quotient ring

$$\begin{aligned}
\Gamma^{(n)} &= \Gamma/(c_{n+1}, c_{n+2}, \dots) \\
&= \mathbb{Z}[c_1, \dots, c_n]/(C_{pp})_{1 \le p \le n}.
\end{aligned}$$

This has bases of monomials $c_{\lambda_1} \cdots c_{\lambda_s}$ and Q-polynomials $Q_\lambda(c)$, as λ ranges over strict partitions such that $\lambda_1 \le n$. In particular, $\Gamma^{(n)}$ has rank 2^n as a $\mathbb{Z}$-module. (See Appendix C, §C.2.3.)

Now we turn to the geometry. Continuing the setting from the previous section, we have a variety Y, a G-equivariant vector bundle V of rank $2n$, and a G-invariant symplectic form $\omega \colon \bigwedge^2 V \to \mathbb{C}_Y$. The Lagrangian Grassmannian bundle $\mathbf{LG}(V) \to Y$ parametrizes maximal isotropic subspaces in the fibers of V. More precisely, it represents the functor which assigns to a morphism $Z \to Y$ the set of all Lagrangian subbundles $F \subseteq V_Z$ (where V_Z is the pullback of V).

There is a natural open cover of $\mathbf{LG}(V)$, similar to the one we saw for $\mathbf{Gr}(n, V)$ in Chapter 4. Suppose V splits so that $V = A \oplus B$ with A a Lagrangian subbundle (and then $B \cong A^\vee$), and that the symplectic form is "standard": it is given by the formula

$$\omega(a \oplus f, a' \oplus f') = f(a') - f'(a).$$

There is an open subset $U \subseteq \mathbf{LG}(V)$, defined as before by the condition that the composition $F \hookrightarrow V \twoheadrightarrow A$ be an isomorphism. The open set U is naturally identified with the vector bundle $\mathrm{Sym}^2 A^\vee \to Y$, where we regard $\mathrm{Sym}^2 A^\vee \subseteq \mathrm{Hom}(A, A^\vee)$ and the map is given by

$$(F \subset V) \mapsto (A \cong F \hookrightarrow V \twoheadrightarrow B = A^\vee).$$

Indeed, the subbundle F is isotropic if and only if the homomorphism $\varphi \colon A \to A^\vee$ defined this way is symmetric. (As before, the inverse map sends φ to its graph.)

Any vector bundle splits locally on Y, and furthermore every symplectic form is locally isomorphic to the standard one, so these open sets U give an affine covering of $\mathbf{LG}(V)$, showing that $\rho \colon \mathbf{LG}(V) \to Y$ is smooth of relative dimension $\binom{n+1}{2} = \operatorname{rk} \operatorname{Sym}^2 A^\vee$. (The closed embedding $\mathbf{LG}(V) \hookrightarrow \mathbf{Gr}(n, V)$ shows that $\mathbf{LG}(V) \to Y$ is also projective.)

There is a natural projection

$$\pi \colon \mathbf{Fl}_\omega(V) \to \mathbf{LG}(V),$$

sending an isotropic flag $F_\bullet$ to its maximal subspace $F_1 = F$. The fiber of π over a Lagrangian subspace F is the complete flag variety $Fl(F)$ in the n-dimensional space F. That is,

$$\mathbf{Fl}_\omega(V) \cong \mathbf{Fl}(\mathbb{S}) \to \mathbf{LG}(V),$$

where (as before) $\mathbb{S} = \mathbb{S}_1 \subset V$ is the tautological Lagrangian subbundle on $\mathbf{LG}(V)$. The relative dimension of $\mathbf{Fl}_\omega(V) \to Y$ is $n^2 = \binom{n}{2} + \binom{n+1}{2}$, equal to the sum of the dimensions of the ordinary flag variety and the Lagrangian Grassmannian.

Proposition 13.4.1. *The homomorphism $c_p \mapsto c_p^G(\mathbb{S}^\vee)$ defines an isomorphism*

$$H_G^* \mathbf{LG}(V) = (H_G^* Y)[c_1, \ldots, c_n]/(C_{pp} - (-1)^p c_{2p}^G(V))_{1 \le p \le n},$$

where C_{pp} is given by $(*)$ (at the beginning of this section).

As a module over $H_G^* Y$, $H_G^* \mathbf{LG}(V)$ has a basis of squarefree monomials

$$c_{\lambda_1} \cdots c_{\lambda_s},$$

for $n \ge \lambda_1 > \cdots > \lambda_s > 0$. Another basis is given by the polynomials $Q_\lambda(c)$, as λ ranges over the same set of strict partitions.

Proof As in Proposition 13.3.1, the relations come from the Whitney formula, this time writing

$$c^G(V) = c^G(\mathbb{S}) \cdot c^G(\mathbb{S}^\vee)$$
$$= (1 - c_1 + c_2 - \cdots)(1 + c_1 + c_2 + \cdots)$$
$$= 1 + (-c_1^2 + 2\,c_2) + (c_2^2 - 2\,c_3\,c_1 + 2\,c_4) + \cdots$$
$$= 1 + \sum_{p>0} (-1)^p C_{pp}.$$

We need to see that the c_p's generate, and that the claimed sets are bases. This is similar to Proposition 4.5.1. On one hand, we know $\mathbf{Fl}_\omega(V) \to \mathbf{LG}(V)$ is a flag bundle, so $H_G^* \mathbf{Fl}_\omega(V)$ is free over $H_G^* \mathbf{LG}(V)$ of rank $n!$. On the other hand, we already saw that $H_G^* \mathbf{Fl}_\omega(V)$ is free over $H_G^* Y$, of rank $2^n n!$. It follows that $H_G^* \mathbf{LG}(V)$ is a projective module over $H_G^* Y$, of rank 2^n.

Now consider the $H_G^* Y$-subalgebra $R \subseteq H_G^* \mathbf{LG}(V)$ generated by $c_1, \ldots, c_n$. By Exercise 13.3.3, we know that $H_G^* \mathbf{Fl}_\omega(V)$ has a basis over R consisting of $n!$ monomials $x_1^{a_1} \cdots x_n^{a_n}$, with $0 \le a_i \le n - i$. Since these also form a basis for $H_G^* \mathbf{Fl}_\omega(V)$ over $H_G^* \mathbf{LG}(V)$, it follows that $R = H_G^* \mathbf{LG}(V)$.

The claims about bases may be checked in the case where Y is a point and G is trivial, by graded Nakayama. Here the relations define $\Gamma^{(n)}$, and we know that the squarefree monomials and Q-polynomials form bases for this ring. $\square$

Corollary 13.4.2. *Suppose there is a G-equivariant Lagrangian sub-bundle $E \subseteq V$ on Y. Then there is a surjective homomorphism of $H_G^* Y$-algebras*

$$(H_G^* Y) \otimes \Gamma \to H_G^* \mathbf{LG}(V), \qquad c \mapsto c^G(V - \mathbb{S} - E),$$

with kernel generated by elements

$$\sum_{i=0}^n c_{k-i}\, c_i^G(E)$$

for all $k > n$.

Squarefree monomials $c_{\lambda_1} \cdots c_{\lambda_s}$ form a basis for $H_G^* \mathbf{LG}(V)$ as a module over $H_G^* Y$, as λ ranges over strict partitions with $\lambda_1 \le n$. The Pfaffians $Q_\lambda(c)$ form another basis.

Proof The second claim—about bases for $H_G^* \mathbf{LG}(V)$—can be checked in the case where Y is a point and G is trivial, and this is the same as in Proposition 13.4.1.

The relations $C_{pp} = 0$ again come from the Whitney formula. Writing $c^\vee = 1 - c_1 + c_2 - c_3 + \cdots$, we have

$$c = c^G(V - \mathbb{S} - E) = c^G(\mathbb{S}^\vee - E) = c^G(E^\vee - \mathbb{S})$$

and

$$c^\vee = c^G(V^\vee - \mathbb{S}^\vee - E^\vee) = c^G(\mathbb{S} - E^\vee) = c^G(E - \mathbb{S}^\vee),$$

so $c \cdot c^\vee = 1$. The fact that squarefree monomials in c_p form a basis implies in particular that these elements generate the algebra. □

13.5   Cohomology rings

In many applications, one has a maximal isotropic subbundle on the base, so Corollary 13.4.2 applies. For the rest of this chapter, we will focus on the case where Y is a point, and $G = T$ is a torus acting linearly on $V = \mathbb{C}^{2n}$. In this context, we will prefer the presentation given by Corollary 13.4.2.

Using the identification of $Fl_\omega(V) \to LG(V)$ with the (ordinary) flag bundle $\mathbf{Fl}(\mathbb{S}) \to LG(V)$, we obtain another presentation of $H_T^* Fl_\omega(V)$. Let us fix a standard basis $e_{\bar{n}}, \ldots, e_{\bar{1}}, e_1, \ldots, e_n$ for $V = \mathbb{C}^{2n}$. Suppose T acts by characters $-\chi_n, \ldots, -\chi_1, \chi_1, \ldots, \chi_n$, so the standard Lagrangian subspace $E \subseteq V$ has characters $\chi_1, \ldots, \chi_n$.

Corollary 13.5.1. *We have isomorphisms*

$$H_T^* LG(V) = (\Lambda_T \otimes \Gamma)/I$$

and

$$H_T^* Fl_\omega(V) = (\Lambda_T \otimes \Gamma)[x_1, \ldots, x_n]/J,$$

where $c \mapsto c^T(V - \mathbb{S} - E)$ and $x_i \mapsto c_1^T(\mathbb{S}_i/\mathbb{S}_{i+1})$, the ideal I is generated by relations

$$\sum_{i=0}^{n} c_{k-i} \cdot e_i(\chi_1, \ldots, \chi_n) = 0 \quad \text{for } k > n,$$

and J is generated by relations

$$\sum_{i=0}^{n} c_{k-i} \cdot e_i(\chi_1, \ldots, \chi_n) = e_k(-x_1, \ldots, -x_n) \quad \text{for } k > 0.$$

The squarefree monomials $c_{\lambda_1} \cdots c_{\lambda_s}$ form a basis for $H_T^* LG(V)$ over Λ_T, as λ ranges over strict partitions with $\lambda_1 \leq n$. Similarly, the monomials $x_1^{m_1} \cdots x_n^{m_n}$ form a basis for $H_T^* Fl_\omega(V)$ over Λ_T, for $0 \leq m_i \leq 2i - 1$.

The presentation for $H_T^* LG(V)$ is a special case of Corollary 13.4.2, and the one for $H_T^* Fl_\omega(V)$ follows. The claims about bases follow by the same reasoning as in Corollary 13.4.2 and Proposition 13.3.1.

There are natural embeddings $\mathbb{C}^{2m} \hookrightarrow \mathbb{C}^{2m+2} \hookrightarrow \cdots$, corresponding to the inclusions of standard bases. These induce embeddings of Lagrangian Grassmannians

$$LG(\mathbb{C}^{2m}) \hookrightarrow LG(\mathbb{C}^{2m+2}), \quad (F \subseteq \mathbb{C}^{2m}) \mapsto (\mathbb{C}e_{\overline{m+1}} \oplus F \subseteq \mathbb{C}^{2m+2}),$$

as well as similar embeddings of isotropic partial flags

$$Fl_\omega(m - n, \ldots, m; \mathbb{C}^{2m}) \hookrightarrow Fl_\omega(m + 1 - n, \ldots, m + 1; \mathbb{C}^{2m+2}).$$

In analogy with the stability discussed in Chapter 12, we can take limits as m and n tend to infinity, forming the infinite Lagrangian Grassmannian

$$LG^\infty = \bigcup_m LG(\mathbb{C}^{2m})$$

and the infinite isotropic flag variety

$$Fl_\omega^\infty = \varprojlim_n \bigcup_m Fl_\omega(m - n, \ldots, m; \mathbb{C}^{2m}).$$

Their cohomology rings are the corresponding limits of the finite-dimensional rings.

Corollary 13.5.2. *The equivariant cohomology rings of LG^∞ and Fl_ω^∞ are*

$$H_T^* LG^\infty = \Lambda_T \otimes \Gamma = \Gamma[y_1, y_2, \ldots]$$

and

$$H_T^* Fl_\omega^\infty = \Lambda_T \otimes \Gamma[x] = \Gamma[x_1, x_2, \ldots; y_1, y_2, \ldots].$$

Demanding stability with respect to these embeddings dictates many of our choices in the next chapter.

Notes

Giambelli (1906) gave a determinantal formula for $[D_r]$ which we now recognize as a Schur determinant; he refers to earlier formulas of Segre and Schubert. As usual for that era, the setting is that of a (symmetric or skew) matrix of homogeneous polynomials of specified degrees. This very naturally translates into the equivariant calculation described at the beginning of the chapter.

Partly motivated by applications to the Brill–Noether theory, the 1980s saw a renewal of interest in formulas for symmetric and skew-symmetric degeneracy loci; see (Bertram, 1987; De Concini and Pragacz, 1995). Several determinantal formulas were found for special cases (Józefiak et al., 1981; Harris and Tu, 1984; Fulton, 1996b). Pragacz seems to have been the first to realize that Schur's Q-polynomials and the corresponding Pfaffian formulas are better suited to the problem (Pragacz, 1988, 1991; Pragacz and Ratajski, 1997).

The facts about the symplectic group and isotropic subspaces reviewed in §13.2 can be found in basic algebra textbooks; see, for instance, (Jacobson, 1985, §6).

The identity

$$\mathrm{Pf}\left(\frac{x_i - x_j}{x_i + x_j}\right) = \prod_{i<j} \frac{x_i - x_j}{x_i + x_j}$$

is due to Schur (1911, §§35–36), who also defined the polynomials $Q_\lambda(x)$. Nimmo (1990) gave a formula for $Q_\lambda(x)$ as a ratio of two Pfaffians; a related formula appears in (Józefiak et al., 1981). Many variations on these polynomials can be found in Macdonald's book. In particular, the identity $Q_{(k,k-1,\dots,1)}(x) = 2^k s_{(k,k-1,\dots,1)}(x)$ follows from (Macdonald, 1995, III.8, Ex. 3(b)).

Hints for Exercises

Exercise 13.1.1. The answer is the Schur determinant $2^n \Delta_{(n,n-1,\dots,1)}(a)$. Use the diagonal torus $T \subseteq GL(E)$, acting on SM_n with weights $t_i + t_j$, to obtain

$$[D_0]^T = 2^n t_1 \cdots t_n \prod_{i<j}(t_i + t_j).$$

Then use the bialternant form of the Schur function to show $t_1 \cdots t_n \prod_{i<j}(t_i + t_j) = s_{(n,n-1,\dots,1)}(t_1,\dots,t_n)$. The a's are elementary symmetric polynomials in the t's, so the claim follows from the (dual) Jacobi–Trudi formula for the Schur polynomial. See (Macdonald, 1995, §III.8, Ex. 2).

Exercise 13.3.2. It suffices to show the monomials $x_1^{m_1} \cdots x_n^{m_n}$ span the algebra

$$A = \mathbb{Z}[x_1,\dots,x_n]/(e_1(x^2),\dots,e_n(x^2)).$$

Here we use the relation $\prod_{i=1}^{n}(1 - x_i^2 u^2) = 1$ in $A[u]$, which shows that for $1 \leq \ell \leq n$,

$$\sum_{k \geq 0} h_k(x_\ell^2, \ldots, x_n^2) u^{2k} = \prod_{i=\ell}^{n} \frac{1}{1 - x_i^2 u^2} = \prod_{i=1}^{\ell-1}(1 - x_i^2 u^2)$$

has degree $2\ell - 2$, and therefore $h_\ell(x_\ell^2, \ldots, x_n^2) = 0$ in A. This gives a relation expressing $x_\ell^{2\ell}$ in terms of other monomials, completing the proof by descending induction on ℓ.

Exercise 13.3.3. In lexicographic order, the initial monomial of e_λ is $x^\lambda = x_1^{\lambda_1} \cdots x_s^{\lambda_s}$, so the initial monomial of $x^a e_\lambda$ is $x^{a+\lambda}$. The exponents $a + \lambda$ take all values such that $0 \leq a_i + \lambda_i \leq 2n - 2i + 1$. Now use Proposition 13.3.1, with the change of variables $x_i \mapsto x_{n+1-i}$.

14

Symplectic Schubert Polynomials

In the last chapter, we computed cohomology rings of the Lagrangian Grassmannian and isotropic flag variety. Here we will study the equivariant geometry of these spaces and their Schubert varieties, by analogy with what we saw for the ordinary Grassmannians and flag varieties in Chapters 9 and 10. In particular, we find Schubert polynomials representing equivariant Schubert classes.

14.1 Schubert varieties

Continuing notation from the previous chapter, we consider a vector space $V = \mathbb{C}^{2n}$, with basis $e_{\bar{n}}, \ldots, e_{\bar{1}}, e_1, \ldots, e_n$. We have a torus $T = (\mathbb{C}^*)^n$, with its standard basis of characters $y_1, \ldots, y_n$, acting on V by characters $y_{\bar{n}}, \ldots, y_{\bar{1}}, y_1, \ldots, y_n$, where $y_{\bar{\imath}} = -y_i$. (The y_i may be specialized to arbitrary characters χ_i, as usual, so long as all $2n$ characters $\pm \chi_i$ are distinct.) The symplectic form $\omega = -\sum e_{\bar{\imath}}^* \wedge e_i^*$ is T-invariant, so T acts on the Lagrangian Grassmannian $LG(V)$ and symplectic flag variety $Fl_\omega(V)$.

14.1.1 Strict Partitions and Signed Permutations.
In §13.4, we saw that bases for $H^* LG(V)$ are indexed by strict partitions λ such that $\lambda_1 \leq n$. These are often represented by their *shifted Young diagrams*; for example, $\lambda = (5, 2, 1)$ corresponds to the diagram

.

The *length* of a strict partition is the number $s = \ell(\lambda)$ of nonzero parts, or nonempty rows in its diagram.

There is a bijection between strict partitions with $\lambda_1 \leq n$, and subsets $I = \{i_1 < \cdots < i_n\} \subseteq \{\bar{n}, \ldots, \bar{1}, 1, \ldots, n\}$ such that for each $i = \{1, \ldots, n\}$, exactly one of $\bar{i}$ or i belongs to I. To go from I to λ, one records the positive entries, written in decreasing order. For example, with $n = 5$, the subset $I = \{\bar{5}, \bar{2}, \bar{1}, 3, 4\}$ corresponds to the strict partition $\lambda = (4, 3)$.

The group of *signed permutations* W_n is the subgroup of permutations of $\{\bar{n}, \ldots, \bar{1}, 1, \ldots, n\}$ which commute with sign changes:

$$W_n = \{w \mid w(\bar{i}) = \overline{w(i)}\}.$$

By identifying S_{2n} with permutations of $\{\bar{n}, \ldots, \bar{1}, 1, \ldots, n\}$, this gives a natural embedding of W_n in S_{2n}. An element $w \in W_n$ is written in "one-line" notation by specifying its values on positive integers: $w = w(1)\, w(2) \cdots w(n)$. For example, the signed permutation $w = 5\,\bar{4}\,1\,2\,\bar{3}$ in W_5 corresponds to the permutation $3\,\bar{2}\,\bar{1}\,4\,\bar{5}\,5\,\bar{4}\,1\,2\,\bar{3}$ in S_{10}.

The group W_n is generated by simple transpositions $s_0, s_1, \ldots, s_{n-1}$, acting in one-line notation by

$$w\, s_i = w(1) \cdots w(i+1)\, w(i) \cdots w(n)$$

for $i > 0$, and

$$w\, s_0 = \overline{w(1)}\, w(2) \cdots w(n).$$

The defining relations among these generators are $s_i^2 = 1$ for all i, $s_i s_j = s_j s_i$ if $|i - j| > 1$, $s_i s_{i+1} s_i = s_{i+1} s_i s_{i+1}$ for $i > 0$, and $s_0 s_1 s_0 s_1 = s_1 s_0 s_1 s_0$.

The *length* of a signed permutation is

$$\ell(w) = \#\{i < j \mid w(i) > w(j)\} + \#\{i \leq j \mid w(i) + w(j) < 0\}.$$

For example, $w = 5\,\bar{4}\,1\,2\,\bar{3}$ has length $\ell(w) = 6+7 = 13$. The length of w is equal to the length of the shortest expression for w in terms of simple transpositions; that is, $\ell(w)$ is the smallest ℓ such that $w = s_{i_1} \cdots s_{i_\ell}$.

The *longest element* of W_n is

$$w_\circ = \bar{1}\,\bar{2}\,\cdots\bar{n}.$$

It has length $\ell(w_\circ) = n^2$, and acts by negating signs:

$$(w_\circ w)(i) = (w w_\circ)(i) = \overline{w(i)}$$

for any $w \in W_n$.

Any strict partition λ determines a *Grassmannian signed permutation* $w = w(\lambda)$, with $\ell(w(\lambda)) = |\lambda|$. If $I = \{i_1 < \cdots < i_n\}$ is the subset corresponding to λ, then $w(\lambda) = \bar{i}_n \cdots \bar{i}_1$ in one-line notation. For example, if $\lambda = (4, 3)$, so $I = \{\bar{5}, \bar{2}, \bar{1}, 3, 4\}$, then $w(\lambda) = \bar{4}\,\bar{3}\,1\,2\,5$.

Via the embedding $W_n \hookrightarrow S_{2n}$, any signed permutation w determines a permutation matrix $A_w \in Sp_{2n} \subseteq GL_{2n}$. If rows and columns are labelled by $\{\bar{n}, \ldots, \bar{1}, 1, \ldots, n\}$, this is the matrix with 1's in positions $(w(i), i)$ and 0's elsewhere. For example,

$$A_{2\bar{3}1} = \begin{bmatrix} 0 & 0 & 0 & 0 & 1 & 0 \\ 0 & 0 & 1 & 0 & 0 & 0 \\ 1 & 0 & 0 & 0 & 0 & 0 \\ 0 & 0 & 0 & 0 & 0 & 1 \\ 0 & 0 & 0 & 1 & 0 & 0 \\ 0 & 1 & 0 & 0 & 0 & 0 \end{bmatrix}.$$

Let λ be a strict partition with $\lambda_1 \leq n$, regarded as a subset of $\{1, \ldots, n\}$. The *dual* $\lambda^\vee$ is the complementary subset in $\{1, \ldots, n\}$. For example, if $n = 5$ and $\lambda = (4, 3)$ then $\lambda^\vee = (5, 2, 1)$. Using this notation, the corresponding Grassmannian signed permutation is

$$w(\lambda) = \overline{\lambda_1} \cdots \overline{\lambda_s}\, \lambda^\vee_{n-s} \cdots \lambda^\vee_1.$$

Equivalently, $\lambda^\vee$ is the complement to λ when its shifted diagram is drawn inside the "staircase shape" $(n, \ldots, 2, 1)$, as shown at right.

To summarize our running example, for $n = 5$ the following are in correspondence:

- the strict partition $\lambda = (4, 3)$;
- the n-element subset $I = \{\bar{5}, \bar{2}, \bar{1}, 3, 4\}$;
- the Grassmannian signed permutation $w = \bar{4}\,\bar{3}\,1\,2\,5$.

14.1.2 Fixed Points. First we identify the fixed points in the Lagrangian Grassmannian. The embedding $LG(V) \hookrightarrow Gr(n, V)$ is equivariant, so the fixed points of $LG(V)$ must be among the coordinate subspaces

$$E_I = \text{span}\{e_{i_1}, \ldots, e_{i_n}\} \subseteq V,$$

for $I \subset \{\overline{n}, \ldots, \overline{1}, 1, \ldots, n\}$. The subspace E_I is isotropic if and only if exactly one of $\overline{i}$ or i belongs to I, for each $i = \{1, \ldots, n\}$. So fixed points are indexed by such I. They are also in bijection with the 2^n strict partitions λ having $\lambda_1 \leq n$, as described in §14.1.1.

Writing p_λ for the fixed point corresponding to the coordinate subspace E_I under this bijection, we have $LG(V)^T = \{p_\lambda\}$. (So $\#LG(V)^T = \text{rk}\, H^*LG(V) = 2^n$, as expected.) In particular, the standard Lagrangian subspace is $E = E_{\{1,\ldots,n\}}$, corresponding to the fixed point $p_{(n,\ldots,2,1)}$.

There is a T-invariant affine open neighborhood around each fixed point p_λ in $LG(V)$. This is a subvariety of the corresponding neighborhood in $Gr(n, V)$, determined by isotropicity conditions. With $n = 3$ and $I = \{\overline{2}, \overline{1}, 3\}$, the neighborhood is represented by matrices as shown at right, where $x = b$, $y = -d$, and $z = -a$. The horizontal line is a visual aid to indicate the midpoint.

$$\begin{bmatrix} a & d & f \\ 1 & 0 & 0 \\ 0 & 1 & 0 \\ \hline b & e & y \\ c & x & z \\ 0 & 0 & 1 \end{bmatrix}$$

In general, starting with I, the matrix representatives for points in these affine open neighborhoods are written as follows. One writes a matrix just as for $Gr(n, V)$, so the submatrix on rows I is the identity matrix; the 1's in this submatrix are "pivots." Entries which are opposite and (strictly) to the right of a pivot are dependent on the other entries, because of the isotropicity requirement; other entries are free. We indicate the free entries with $*$'s, as usual, and dependent entries with $\bullet$'s. For $n = 5$ and $I = \{\overline{5}, \overline{2}, \overline{1}, 3, 4\}$, the neighborhood is shown at left.

$$\begin{bmatrix} 1 & 0 & 0 & 0 & 0 \\ * & * & * & * & * \\ * & * & * & * & \bullet \\ 0 & 1 & 0 & 0 & 0 \\ 0 & 0 & 1 & 0 & 0 \\ \hline * & * & * & \bullet & \bullet \\ * & * & \bullet & \bullet & \bullet \\ 0 & 0 & 0 & 1 & 0 \\ 0 & 0 & 0 & 0 & 1 \\ * & \bullet & \bullet & \bullet & \bullet \end{bmatrix}$$

In the notation of §13.4, this neighborhood of p_λ is the open set corresponding to the splitting $V = E_I \oplus E_I^\vee$, where $E_I^\vee$ is the span of $\{e_{\overline{i_1}}, \ldots, e_{\overline{i_n}}\}$, and it follows that this open set is equivariantly isomorphic to $\text{Sym}^2 E_I^\vee$. Since the weights of T acting on E_I are $\{y_i \mid i \in I\}$, this lets us compute the tangent weights at the fixed point p_λ: the space

$$T_{p_\lambda} LG(V) \cong \mathrm{Sym}^2 E_I^\vee$$

has weights $\{-y_i - y_j \mid i \leq j \in I\}$. Working this out explicitly for the above example, in matrix coordinates, the action is given by

$$
z \cdot
\begin{bmatrix}
1 & 0 & 0 & 0 & 0 \\
* & * & * & * & * \\
* & * & * & * & \bullet \\
0 & 1 & 0 & 0 & 0 \\
0 & 0 & 1 & 0 & 0 \\
\hline
* & * & * & \bullet & \bullet \\
* & * & \bullet & \bullet & \bullet \\
0 & 0 & 0 & 1 & 0 \\
0 & 0 & 0 & 0 & 1 \\
* & \bullet & \bullet & \bullet & \bullet
\end{bmatrix}
=
\begin{bmatrix}
z_5^{-1} & 0 & 0 & 0 & 0 \\
z_4^{-1}* & z_4^{-1}* & z_4^{-1}* & z_4^{-1}* & z_4^{-1}* \\
z_3^{-1}* & z_3^{-1}* & z_3^{-1}* & z_3^{-1}* & z_3^{-1}\bullet \\
0 & z_2^{-1} & 0 & 0 & 0 \\
0 & 0 & z_1^{-1} & 0 & 0 \\
\hline
z_1* & z_1* & z_1* & z_1\bullet & z_1\bullet \\
z_2* & z_2* & z_2\bullet & z_2\bullet & z_2\bullet \\
0 & 0 & 0 & z_3 & 0 \\
0 & 0 & 0 & 0 & z_4 \\
z_5* & z_5\bullet & z_5\bullet & z_5\bullet & z_5\bullet
\end{bmatrix}
$$

$$
=
\begin{bmatrix}
1 & 0 & 0 & 0 & 0 \\
\frac{z_5}{z_4}* & \frac{z_2}{z_4}* & \frac{z_1}{z_4}* & \frac{1}{z_3 z_4}* & \frac{1}{z_4^2}* \\
\frac{z_5}{z_3}* & \frac{z_2}{z_3}* & \frac{z_1}{z_3}* & \frac{1}{z_3^2}* & \bullet \\
0 & 1 & 0 & 0 & 0 \\
0 & 0 & 1 & 0 & 0 \\
\hline
z_1 z_5 * & z_1 z_2 * & z_1^2 * & \bullet & \bullet \\
z_2 z_5 * & z_2^2 * & \bullet & \bullet & \bullet \\
0 & 0 & 0 & 1 & 0 \\
0 & 0 & 0 & 0 & 1 \\
z_5^2 * & \bullet & \bullet & \bullet & \bullet
\end{bmatrix} .
$$

Now we turn to the isotropic flag variety. Using the description of $Fl_\omega(V)$ as a flag bundle over $LG(V)$, each T-fixed point of $Fl_\omega(V)$ must be a fixed point of $Fl(E_I)$, for some I. Such points are indexed by permutations of I; this data is the same as that of a signed permutation. In general, given $w \in W_n$, the corresponding T-fixed point p_w is the isotropic flag

$$^w E_\bullet : {}^w E_n \subset \cdots \subset {}^w E_1 \subset V,$$

where ${}^w E_q = \langle e_{\overline{w(n)}}, \ldots, e_{\overline{w(q)}} \rangle$. (Note the negative signs!) In particular, our standard flag is $E_\bullet = {}^{w_\circ} E_\bullet$.

$$\begin{bmatrix} a & f & y \\ b & g & 1 \\ 1 & 0 & 0 \\ \hline c & x & z \\ d & h & i \\ e & 1 & 0 \end{bmatrix}$$

As before, there is a T-invariant affine neighborhood around each fixed point p_w. This is represented in matrices by writing a $2n \times n$ matrix with pivot 1's in positions $(w(i), i)$ for $\bar{n} \leq i \leq \bar{1}$, and 0's to the right of these pivots. For example, for $w = 2\,\bar{3}\,1$ this neighborhood is as shown at left, where $x = -a - bh + dg + ef$, $y = gi - h$, and $z = -bi + d + egi - eh$.

Continuing the notation we used for Lagrangian Grassmannians, free entries are denoted by $*$ and dependent entries by $\bullet$. So for $w = 5\,\bar{4}\,1\,2\,\bar{3}$, the neighborhood is written as at right. In general, the free entries are in positions $(w(i), j)$, for $\bar{n} \leq j \leq \bar{1}$ and $j < i \leq \bar{j}$. The torus acts on the $(w(i), j)$ entry by the character $y_{w(i)} - y_{w(j)}$. One sees that $U_w \cong \mathbb{C}^{n^2}$, and that $T_{p_w} Fl_w(V)$ has weights $y_{w(i)} - y_{w(j)}$, for $j < 0$ and $j < i \leq \bar{j}$.

14.1.3 Schubert Cells and Schubert Varieties.

For a signed permutation $w \in W_n$, there is a dimension function similar to the one for ordinary permutations: for $1 \leq p \leq n$ and $q \in \{\bar{n}, \ldots, \bar{1}, 1, \ldots, n\}$, we define

$$\begin{bmatrix} * & * & * & * & 1 \\ * & * & * & * & \bullet \\ * & \bullet & \bullet & \bullet & \bullet \\ * & 1 & 0 & 0 & 0 \\ * & * & 1 & 0 & 0 \\ \hline * & * & * & \bullet & \bullet \\ * & * & \bullet & \bullet & \bullet \\ 1 & 0 & 0 & 0 & 0 \\ * & * & * & 1 & 0 \\ * & * & * & * & * \end{bmatrix}$$

$$k_w(p, q) = \#\{i \geq p \,|\, w(i) \leq \bar{q}\}$$
$$= \#\{i \leq \bar{p} \,|\, w(i) \geq q\}.$$

This can also be described in terms of ranks of permutation matrices. With the above conventions, so A_w is the $2n \times n$ matrix with 1's in positions $(w(i), i)$, for $\bar{n} \leq i \leq \bar{1}$, the number $k_w(p, q)$ is the rank of the southwest submatrix of A_w whose northeast corner is at position $(q, \bar{p})$. For example, with $w = 5\,\bar{4}\,1\,2\,\bar{3}$, we have $k_w(2, 3) = 2$ and $k_w(3, \bar{2}) = 3$.

Using these dimension functions and the standard flag $E_\bullet$, the Schubert cells $\Omega_w^\circ = \Omega_w^\circ(E_\bullet)$ and Schubert varieties $\Omega_w = \Omega_w(E_\bullet)$ in $Fl_w(V)$ are defined by

$$\Omega_w^\circ = \left\{ F_\bullet \,\Big|\, \dim(F_p \cap E_q) = k_w(p, q) \text{ for all } 1 \leq p \leq n, \bar{n} \leq q \leq n \right\}$$

and

$$\Omega_w = \left\{ F_\bullet \,\middle|\, \dim(F_p \cap E_q) \geq k_w(p, q) \text{ for all } 1 \leq p \leq n,\, \overline{n} \leq q \leq n \right\}.$$

(Recall that the standard flag has $E_q = \mathrm{span}\{e_q, \ldots, e_n\}$.)

Opposite Schubert cells and varieties are defined in the same way, using the opposite flag $\widetilde{E}_\bullet$ in place of the standard flag. (Recall that the opposite flag is defined by $\widetilde{E}_q = \mathrm{span}\{e_{\overline{n}}, \ldots, e_{\overline{q}}\}$.) This flag is preserved by the Borel group $B = w_\circ B^- w_\circ$. With our choice of symplectic form, B consists of upper-triangular matrices in Sp_{2n}, and B^- consists of lower-triangular matrices.

Schubert cells are B^--orbits, and opposite Schubert cells are B-orbits:

$$\Omega_w^\circ = B^- \cdot p_w \quad \text{and} \quad \widetilde{\Omega}_w^\circ = B \cdot p_{w_\circ w}.$$

They may also be written in matrix form, using the same conventions as for the affine open sets discussed above. For $w = 5\,\overline{4}\,1\,2\,\overline{3}$, so $w_\circ w = \overline{5}\,4\,\overline{1}\,\overline{2}\,3$, we have

$$\Omega_w^\circ = \begin{bmatrix}
0 & 0 & 0 & 0 & 1 \\
0 & 0 & 0 & 0 & \bullet \\
0 & 0 & 0 & 0 & \bullet \\
0 & 1 & 0 & 0 & 0 \\
0 & * & 1 & 0 & 0 \\
\hline
0 & * & * & 0 & \bullet \\
0 & * & \bullet & 0 & \bullet \\
1 & 0 & 0 & 0 & 0 \\
* & * & * & 1 & 0 \\
* & * & * & * & *
\end{bmatrix} \quad \text{and} \quad \widetilde{\Omega}_{w_\circ w}^\circ = \begin{bmatrix}
* & * & * & * & 1 \\
* & * & * & * & 0 \\
* & \bullet & \bullet & \bullet & 0 \\
* & 1 & 0 & 0 & 0 \\
* & 0 & 1 & 0 & 0 \\
* & 0 & 0 & \bullet & 0 \\
* & 0 & 0 & \bullet & 0 \\
1 & 0 & 0 & 0 & 0 \\
0 & 0 & 0 & 1 & 0 \\
0 & 0 & 0 & 0 & 0
\end{bmatrix}.$$

As this example suggests, the cells Ω_w° and $\widetilde{\Omega}_{w_\circ w}^\circ$ intersect transversally, meeting in the fixed point p_w.

Exercise 14.1.1. Show that $\Omega_w^\circ \cong \mathbb{C}^{n^2 - \ell(w)}$, so that $\Omega_w \subseteq Fl_w(V)$ has codimension $\ell(w)$. Show that the tangent and normal weights to Ω_w at p_w are as follows:

$$T_{p_w} \Omega_w^\circ \text{ has weights } \left\{ y_{w(i)} - y_{w(j)} \,\middle|\, j < i \leq \overline{j} \text{ and } w(i) > w(j) \right\};$$

$$N_{p_w} \Omega_w^\circ \text{ has weights } \left\{ y_{w(i)} - y_{w(j)} \,\middle|\, j < i \leq \overline{j} \text{ and } w(i) < w(j) \right\}.$$

Do the same for $T_{p_w}\widetilde{\Omega}^\circ_{w_\circ w}$ and $N_{p_w}\widetilde{\Omega}^\circ_{w_\circ w}$. Conclude that the intersection $\Omega^\circ_w \cap \widetilde{\Omega}^\circ_{w_\circ w} = \{p_w\}$ is indeed transverse.

As for ordinary flag varieties, each Schubert cell in $Fl_\omega(V)$ is a principal homogeneous space for a certain subgroup of B^-. For $w \in W_n$, let $U^-(w) = U^- \cap A_w B^- A_{w^{-1}}$.

Exercise 14.1.2. Show that the T-invariant affine open neighborhood of p_w is $(A_w B^- A_w^{-1}) \cdot p_w$, and that the map

$$U^-(w) \to \Omega^\circ_w$$

$$u \mapsto u \cdot p_w$$

is an isomorphism. Also show that $U^-(w) \subseteq U^-(ws_k)$ if and only if $\ell(ws_k) < \ell(w)$.

14.1.4 Bruhat Order and Poincaré Duality.

Containment among Schubert varieties is described by the *Bruhat order* on W_n. For $v, w \in W_n$, this partial order is defined by declaring $v \geq w$ if $k_v(p,q) \geq k_w(p,q)$ for all $1 \leq p \leq n$ and $\overline{n} \leq q \leq n$. This condition is equivalent to $\Omega_v \subseteq \Omega_w$.

Exercise 14.1.3. Let $W_n \hookrightarrow S_{2n}$ be the natural embedding, by regarding $w \in W_n$ as a permutation of $\{\overline{n}, \ldots, \overline{1}, 1, \ldots, n\}$. Show that $v \geq w$ in W_n if and only if $v \geq w$ in S_{2n}; that is, Bruhat order on W_n is induced as a sub-poset from Bruhat order on S_{2n}.

As for ordinary flag varieties, Schubert varieties are closures of Schubert cells: $\Omega_w = \overline{\Omega^\circ_w} \subseteq Fl_\omega(V)$. This can be proved by imitating the argument of Proposition 10.4.4. Similarly, one has

$$\Omega_w = \coprod_{v \geq w} \Omega^\circ_v.$$

In particular, the Schubert cells decompose the symplectic flag variety, so the classes $\sigma_w = [\Omega_w]^T$ form a basis for $H^*_T Fl_\omega(V)$ over Λ. The same is true for opposite Schubert cells, so the classes $\widetilde{\sigma}_w = [\widetilde{\Omega}_w]^T$ form another basis.

From this, one also sees

$$p_v \in \Omega_w \text{ iff } v \geq w \qquad \text{and} \qquad p_v \in \widetilde{\Omega}_{w_\circ w} \text{ iff } v \leq w.$$

So the T-fixed points in $\Omega_w \cap \widetilde{\Omega}_{w_\circ u}$ are those p_v such that $w \leq v \leq u$. It follows that that the bases σ_w and $\widetilde{\sigma}_{w_\circ w}$ are Poincaré dual:

Proposition 14.1.4. *Let* $\rho \colon Fl_\omega(V) \to \mathrm{pt}$. *We have*

$$\rho_*(\sigma_u \cdot \widetilde{\sigma}_{w_\circ v}) = \delta_{u,v}$$

in Λ_T.

This is proved by the same argument as for ordinary flag varieties and Grassmannians, using the computation of tangent weights in Exercise 14.1.1 to check transversality.

14.1.5 Stability. Let $Fl_\omega^{(n)}(\mathbb{C}^{2m}) = Fl_\omega(m - n, \ldots, m; \mathbb{C}^{2m})$ be the partial isotropic flag variety. At the end of §13.5, we saw embeddings

$$\iota \colon Fl_\omega(m - n, \ldots, m; \mathbb{C}^{2m}) \hookrightarrow Fl_\omega(m' - n, \ldots, m'; \mathbb{C}^{2m'}),$$

for $m \leq m'$, and projections

$$\pi \colon Fl_\omega(m - n', \ldots, m; \mathbb{C}^{2m}) \to Fl_\omega(m - n, \ldots, m; \mathbb{C}^{2m}),$$

for $n' \geq n$.

For $m \leq m'$, there is a corresponding inclusion $\iota \colon W_m \hookrightarrow W_{m'}$, defined in one-line notation by

$$\iota(w)(k) = \begin{cases} w(k) & \text{for } k \leq m; \\ k & \text{for } k > m. \end{cases}$$

We generally regard W_m as a subgroup of $W_{m'}$ by this embedding, and write $W_\infty = \bigcup_m W_m$.

For $n \leq m$, we consider the subset

$$W_m^{(n)} = \Big\{ w \in W_m \,\big|\, 0 < w(i) < w(i+1) \text{ for } i > n \Big\}.$$

Schubert varieties $\Omega_{[w]} \subseteq Fl_\omega(m - n, \ldots, m; \mathbb{C}^{2m})$ are defined just as for the complete flag variety $Fl_\omega(\mathbb{C}^{2m})$: for $w \in W_m^{(n)}$, we have

$$\Omega_{[w]} = \Big\{ F_\bullet \,\big|\, \dim(F_p \cap E_q) \geq k_w(p, q) \text{ for } 1 \leq p \leq n, \ \overline{m} \leq q \leq m \Big\}.$$

The corresponding Schubert cells decompose the partial flag variety, as w ranges over $W_m^{(n)}$, so the classes $\sigma_{[w]} = [\Omega_{[w]}]^T$ form a basis. The Schubert varieties $\Omega_{[w]}$ are stable with respect to the embeddings ι and projections π.

The situation is entirely analogous to the type A setting (see Chapter 12). The exercises below show that for each $w \in W_\infty$, there is a well-defined stable Schubert class $\sigma_w \in H_T^* Fl_\omega^\infty$, and it follows that these classes form a basis for $H_T^* Fl_\omega^\infty$ as a module over Λ_T.

Exercise 14.1.5. For $m' \geq m$, show that ι identifies $Fl_\omega^{(n)}(\mathbb{C}^{2m})$ with the opposite Schubert variety $\widetilde{\Omega}_{[w_\circ^{(m')} \cdot \iota(w_\circ^{(m)})]} \subseteq Fl_\omega^{(n)}(\mathbb{C}^{2m'})$, and for $w \in W_m^{(n)}$, show $\iota^{-1}\Omega_{[\iota(w)]} = \Omega_{[w]}$. Conclude that $\iota^*\sigma_{[u]} = 0$ unless $u = \iota(w)$ for some $w \in W_m$, and

$$\iota^*\sigma_{[\iota(w)]} = \sigma_{[w]} \quad \text{in} \quad H_T^* Fl_\omega^{(n)}(\mathbb{C}^{2m}).$$

Exercise 14.1.6. For $w \in W_m^{(n)}$ and $n' \geq n$, show $\pi^{-1}\Omega_{[w]} = \Omega_{[w]}$ in $Fl_\omega^{(n')}(\mathbb{C}^{2m})$, and therefore

$$\pi^*\sigma_{[w]} = \sigma_{[w]} \quad \text{in} \quad H_T^* Fl_\omega^{(n')}(\mathbb{C}^{2m}).$$

14.2   Double Q-polynomials and Lagrangian Schubert classes

We have seen that double Schur polynomials $s_\lambda(x|y)$ represent equivariant Schubert classes in the Grassmannian. Here we will describe the analogous polynomials representing Schubert classes in the Lagrangian Grassmannian.

As usual, we have $V = \mathbb{C}^{2n}$, with symplectic form and standard isotropic flag $E_\bullet$. The Schubert varieties in $LG(V)$ are defined by

$$\Omega_\lambda = \Omega_\lambda(E_\bullet) = \left\{ F \mid \dim(F \cap E_{\lambda_k}) \geq k \text{ for } 1 \leq k \leq s \right\}$$

for a strict partition $\lambda = (n \geq \lambda_1 > \cdots > \lambda_s > 0)$. This is the closure of a Schubert cell Ω_λ°, which is the B^--orbit of the fixed point p_λ:

$$\Omega_\lambda^\circ = B^- \cdot p_\lambda.$$

We will sometimes use the notation $\Omega_I = \Omega_\lambda$ and $p_I = p_\lambda$, where I is the n-element subset of $\{\overline{n}, \ldots, \overline{1}, 1, \ldots, n\}$ which corresponds to λ.

Opposite Schubert varieties and cells are defined analogously, with respect to the opposite flag $\widetilde{E}_\bullet$. Here $\widetilde{\Omega}_\lambda^\circ = B \cdot p_{\lambda^\vee}$, and $p_\mu \in \widetilde{\Omega}_\lambda$ if and only if $\mu \subseteq \lambda^\vee$. (Recall that $\lambda^\vee$ is the complement to λ inside the staircase $(n, \ldots, 2, 1)$.)

A partial order on strict partitions is defined just as for usual partitions, by containment of shifted Young diagrams: $\lambda \subseteq \mu$ if and only if $\lambda_k \leq \mu_k$ for all $k = 1, \ldots, s$. The corresponding order on n-element subsets I is also as before: $I \leq J$ if and only if $i_k \leq j_k$ for all $k = 1, \ldots, n$.

In Chapter 9, we saw two important cases of restriction formulas for Schubert classes. The analogous formulas for the Lagrangian Grassmannian are as follows.

Exercise 14.2.1. Show that $p_\mu \in \Omega_\lambda$ if and only if $\mu \supseteq \lambda$. Then, with I, J the sets corresponding to λ, μ, compute

$$\sigma_I|_I = \prod_{\substack{i,j \in I \\ i \geq j > \bar{\imath}}} (-y_i - y_j) \tag{$*$}$$

and

$$\sigma_I|_J = 0 \quad \text{unless} \quad J \geq I. \tag{$**$}$$

Equivalently,

$$\sigma_\lambda|_\lambda = \prod_{k=1}^{s} \left(\prod_{\substack{1 \leq \ell \leq \lambda_k \\ \ell \in \lambda}} (-y_{\lambda_k} - y_\ell) \prod_{\substack{1 \leq \ell < \lambda_k \\ \ell \notin \lambda}} (-y_{\lambda_k} + y_\ell) \right)$$

and

$$\sigma_\lambda|_\mu = 0 \quad \text{if} \quad \mu \not\supseteq \lambda.$$

Write down similar formulas for restrictions of opposite Schubert classes, $\widetilde{\sigma}_\lambda|_\mu$.

In the symplectic setting, *double Q-polynomials* play the role analogous to that of double Schur polynomials for ordinary Grassmannians. Like $s_\lambda(x|y)$, the polynomials $Q_\lambda(c|y)$ have several equivalent descriptions. We will use a "multi-Schur Pfaffian" formula which is analogous to the Jacobi–Trudi determinantal formula for Schur polynomials. More details are in Appendix C, §C.3.

Given a strict partition $\lambda = (\lambda_1 > \cdots > \lambda_s > 0)$ and series $c(1), \ldots, c(s)$, set

$$m_{ij} = c_{\lambda_i}(i) \, c_{\lambda_j}(j) + 2 \sum_{a=1}^{\lambda_j} (-1)^a c_{\lambda_i+a}(i) \, c_{\lambda_j-a}(j)$$

for $i < j$, and let $M_\lambda(\mathbf{c}) = (m_{ij})$ be the corresponding skew-symmetric matrix. The multi-Schur Pfaffian is the Pfaffian of this matrix:

$$\mathrm{Pf}_\lambda(c(1), \ldots, c(s)) := \mathrm{Pf}(M_\lambda(\mathbf{c})).$$

To define the double Q-polynomial, we fix $c = 1 + c_1 + c_2 + \cdots$ and set $c(i) = c \cdot (1 + y_1) \cdots (1 + y_{\lambda_i - 1})$. Then

$$Q_\lambda(c|y) = \mathrm{Pf}_\lambda(c(1), \ldots, c(s)).$$

When the c variables are taken to be the generators of the ring Γ, the double Q-polynomials belong to the ring $\Gamma[y]$, and they form a basis for this ring as a module over $\mathbb{Z}[y]$.

The key fact about double Q-polynomials is that they solve the same interpolation problem which characterizes Schubert classes. For any strict partition μ, we write $c \mapsto c^\mu$ for the specialization which sets c equal to

$$c^\mu = \prod_{i \in \mu} \frac{1 - y_i}{1 + y_i}.$$

Then

$$Q_\lambda(c^\lambda|y) = \prod_{k=1}^s \left(\prod_{\substack{1 \le \ell \le \lambda_k \\ \ell \in \lambda}} (-y_\ell - y_{\lambda_k}) \prod_{\substack{1 \le \ell < \lambda_k \\ \ell \notin \lambda}} (y_\ell - y_{\lambda_k}) \right) \tag{$\dagger$}$$

and

$$Q_\lambda(c^\mu|y) = 0 \qquad \text{if } \mu \not\supseteq \lambda. \tag{$\ddagger$}$$

Furthermore, the properties $(\dagger)$ and $(\ddagger)$ characterize $Q_\lambda(c|y)$ as an element of $\Gamma[y]$. (This is the main theorem of Appendix C, §C.3.)

Let y_i be the characters of T acting on $E \subseteq V$. From Corollary 13.5.1, we have a presentation

$$H_T^* LG(V) = (\Lambda_T \otimes \Gamma)/I,$$

where I is generated by $\sum_{i=0}^n c_{k-i} \cdot e_i(y)$ for $k > n$. For any μ, we have

$$c^T(V - \mathbb{S} - E)|_\mu = c^T(E_J^\vee - E) = \prod_{i \in \mu} \frac{1 - y_i}{1 + y_i},$$

where $J \subset \{\overline{n}, \ldots, \overline{1}, 1, \ldots, n\}$ is the n-element subset corresponding to μ. So the specialization $c \mapsto y^\mu$ agrees with the restriction homomorphism

$H_T^* LG(V) \to H_T^*(p_\mu) = \Lambda_T$. This leads to the main theorem of this section.

Theorem 14.2.2.   *Setting* $c = c^T(V - \mathbb{S} - E)$, *we have*

$$\sigma_\lambda = Q_\lambda(c|y)$$

in $H_T^* LG(V) = (\Lambda_T \otimes \Gamma)/I$.

Proof   Comparing the formulas (∗) and (∗∗) of Exercise 14.2.1 for restricting σ_λ, and the properties (†) and (‡) of $Q_\lambda(c|y)$, both sides solve the same interpolation problem. The claim follows, as in Chapter 9 (Lemma 9.3.1 and Corollary 9.4.3). □

The opposite Schubert variety $\widetilde{\Omega}$ is defined by replacing $E_\bullet$ with the opposite flag $\widetilde{E}_\bullet$. This has the effect of replacing y_i by $-y_i$, and $c = c^T(V - \mathbb{S} - E)$ by $\widetilde{c} = c^T(V - \mathbb{S} - E^\vee)$.

Corollary 14.2.3.   *We have* $\widetilde{\sigma}_\lambda = Q_\lambda(\widetilde{c}|-y)$.

Remark.   In Chapter 9, the x variables of the double Schur polynomials $s_\lambda(x|y)$ were evaluated as equivariant Chern roots of the dual tautological bundle. The parallel polynomial here is obtained by evaluating $Q_\lambda(c|y)$ at

$$c = c^T(V - \mathbb{S} - E) = c^T(\mathbb{S}^\vee - E) = \prod_{i=1}^n \frac{1 - x_i}{1 + y_i},$$

so the x_i are equivariant Chern roots of $\mathbb{S}$. With this notation, restriction to p_μ is given by further specializing the x variables by

$$x_i \mapsto \begin{cases} y_i & \text{if } i \in \mu; \\ -y_i & \text{if } i \notin \mu. \end{cases}$$

(These are the torus weights on the space E_J corresponding to p_μ.) See Appendix C for more discussion on these and related polynomials in x variables.

14.3   Symplectic degeneracy loci

As in Chapters 11 and 12, the formula for equivariant Schubert classes implies a formula for general degeneracy loci. Suppose V is a vector bundle on a variety X, with a complete isotropic flag $E_\bullet$ and a maximal isotropic subbundle $F \subseteq V$. For a strict partition λ, there is a degeneracy locus

$$D_\lambda(F \cap E_\bullet) = \left\{ x \in X \mid \dim(F \cap E_{\lambda_k}) \geq k \text{ for } 1 \leq k \leq s \right\},$$

and a class $\mathbb{D}_\lambda \in H^{2|\lambda|}X$, stable under pullbacks and equal to the class of $D_\lambda(F \cap E_\bullet)$ when the latter has codimension $|\lambda|$ in X.

Corollary 14.3.1. *In $H^{2|\lambda|}X$, we have*

$$\mathbb{D}_\lambda = \mathrm{Pf}_\lambda(c(1), \dots, c(s)),$$

where $c(k) = c(V - F - E_{\lambda_k})$ for each $1 \leq k \leq s$.

Example 14.3.2. The Corollary gives a formula for the class of the locus $D_r \subseteq \mathrm{Sym}^2 E = SM_n$ of symmetric maps with rank at most r, considered in §13.1.

Consider $V = E \oplus E^\vee$ as a vector bundle on $\mathrm{Sym}^2 E$, with its canonical symplectic form ω defined by

$$\omega(v \oplus f, w \oplus g) = f(w) - g(v).$$

A morphism $\varphi \colon E^\vee \to E$ is symmetric if and only if $\mathrm{Graph}(\varphi) \subseteq E \oplus E^\vee$ is isotropic with respect to ω. Over $\mathrm{Sym}^2 E$, there is the universal homomorphism $\Phi \colon E^\vee \to E$, whose graph is a maximal isotropic subbundle, $\mathrm{Graph}(\Phi) \subseteq V$. The subbundle $0 \oplus E^\vee$ is also isotropic, and the locus defined by

$$\dim(\mathrm{Graph}(\Phi) \cap (0 \oplus E^\vee)) \geq n - r$$

is precisely the locus $D_r \subseteq \mathrm{Sym}^2 E$.

Let $G_i = 0 \oplus E_i^\vee$, where $E_i^\vee \subseteq E^\vee$ is any T-invariant subspace of codimension $i - 1$. In particular, $G_1 = 0 \oplus E^\vee$, so $D_r = D_\lambda(E_\Phi^\vee \cap G_\bullet)$, where $\lambda = (n - r, \dots, 2, 1)$. The formula from Corollary 14.3.1 is

$$[D_r]^T = \mathrm{Pf}_{(n-r,\dots,2,1)}(c(1), \dots, c(n-r)),$$

where $c(k) = c^T(V - E_\Phi^\vee - G_{n-r-k+1}) = c^T(E - E_{n-r-k+1}^\vee)$. Writing $y_i = -c_1^T(E_i^\vee/E_{i+1}^\vee)$ and $c = c^T(E - E^\vee)$, this is

$$c(k) = c \cdot \prod_{i=1}^{n-r-k} (1 - y_i).$$

Using

$$c(k)_{n-r-k+1} = \sum_{i=0}^{n-r-k} (-1)^i c_{n-r-k+1-i} \cdot e_i(y_1, \ldots, y_{n-r-k})$$

the Pfaffian of $M_\lambda(c)$ is equal to that of $M_\lambda(c)$, by multilinearity. So

$$[D_r]^T = Q_\lambda(c).$$

Since $y_1, \ldots, y_n$ are the equivariant Chern roots of E, we have

$$c = \prod_{i=1}^{n} \frac{1 + y_i}{1 - y_i},$$

so $Q_\lambda(c) = Q_\lambda(y_1, \ldots, y_n)$ is the classical Schur Q-polynomial. Giambelli's formula for $[D_r]^T$ follows from the identity $Q_{(n-r,\ldots,2,1)}(y_1, \ldots, y_n) = 2^{n-r} s_{(n-r,\ldots,2,1)}(y_1, \ldots, y_n)$. (See Appendix C, §C.2.4.)

Example 14.3.3. The same formula arises by restricting a Lagrangian Schubert class to a fixed point. Consider the $2n$-dimensional vector space $V = E \oplus E^\vee$ as in the previous example, and let $U \subseteq LG(V)$ be the standard affine open neighborhood of $0 \oplus E^\vee \subseteq V$, as described in §13.4, so $U \cong \mathrm{Sym}^2 E = SM_n$.

The degeneracy locus $D_r \subseteq SM_n = U$ is the intersection of U with a Schubert variety,

$$D_r = U \cap \Omega_{(n-r,\ldots,2,1)}.$$

Since U is an affine space – in fact, an opposite Schubert cell – restricting a Schubert class to U is the same as restricting to the fixed point corresponding to $E^\vee$, which is $p_{(n,\ldots,2,1)}$. (Compared with the notation of §14.1.2, the roles of E and $E^\vee$ are swapped.) So

$$[D_r]^T = \sigma_{(n-r,\ldots,2,1)}|_{(n,\ldots,2,1)}.$$

Using $\sigma_\lambda = Q_\lambda(c|y)$, with $c = c^T(V - \mathbb{S} - E^\vee)$, this recovers Giambelli's formula as before, since $c|_{(n,\ldots,2,1)} = c^T(E - E^\vee) = \prod \frac{1+y_i}{1-y_i}$.

14.4 Type C Schubert polynomials

Finally we turn to the problem of finding Schubert polynomials for the isotropic flag variety. The goal is the same as before: we wish to express

$\sigma_w = [\Omega_w]^T$ as a polynomial in Chern classes. More precisely, using the presentation $H_T^* Fl_\omega(V) = (\Lambda_T \otimes \Gamma)[x_1, \ldots, x_n]/J$, we seek Schubert polynomials $S_w^C(c; x, y) \in \Gamma[x, y]$ mapping to σ_w under appropriate evaluations of the variables. As noted in §14.1.5, there are stable Schubert classes in $H_T^* Fl_\omega^\infty = \Gamma[x, y]$, so such polynomials exist. The problem is to find formulas for them.

The necessity of working in $\Gamma[x; y]$, rather than $\mathbb{Z}[x; y]$, comes from our stability requirement. For instance, the formula for

$$\Omega_{\bar{1}} = \left\{ F_\bullet \mid \dim(F_1 \cap E_1) \geq 1 \right\}$$

in $H_T^* Fl_\omega(\mathbb{C}^{2n})$ is

$$\sigma_{\bar{1}} = -x_1 - \cdots - x_n - y_1 - \cdots - y_n.$$

In terms of x and y, this is not stable as $n \to \infty$, but it can be written as

$$\sigma_{\bar{1}} = c_1,$$

where $c = c^T(V - \mathbb{S} - E)$, and this is the version we prefer.

The general approach is very similar to what we have already seen in type A. We will define the Schubert polynomials inductively for each $w \in W_n$, starting with $w_\circ^{(n)} = \bar{1} \cdots \bar{n}$.

As before, the basic tools are divided difference operators ∂_k, for $k \geq 0$, acting on the ring $\Gamma[x, y]$, with the y variables appearing as scalars. For $k > 0$, the operator ∂_k is defined exactly as in type A, acting only on the x variables. To define the operator ∂_0, we first say how the simple transposition s_0 acts. This is the algebra homomorphism determined by

$$s_0 \cdot x_1 = -x_1 \qquad \text{and} \qquad s_0 \cdot c_p = c_p + 2x_1 c_{p-1} + \cdots + 2x_1^p.$$

(The second formula is equivalent to $s_0 \cdot c = \frac{1+x_1}{1-x_1} \cdot c$, which is compatible with the first formula and the specialization $c = \prod \frac{1-x_i}{1+y_i}$. This also satisfies the relations defining Γ, so s_0 is indeed a well-defined algebra homomorphism.) Then we define

$$\partial_0 f = \frac{f - s_0 f}{-2x_1}$$

for any $f \in \Gamma[x, y]$. In particular, $\partial_0 c_p = c_{p-1} + x_1 c_{p-2} + \cdots + x_1^{p-1}$.

The operator ∂_0 obeys the same Leibniz-type rule as the other difference operators. That is,

$$\partial_k(f \cdot g) = (\partial_k f) \cdot g + (s_k f) \cdot (\partial_k g)$$

for all $f, g \in \Gamma[x, y]$ and all $k \geq 0$.

The *type C double Schubert polynomials* are defined as follows, for $w \in W_n$. First, the polynomial for $w_\circ^{(n)}$ has an explicit formula as a multi-Schur Pfaffian:

$$S^C_{w_\circ^{(n)}}(c; x; y) = \mathrm{Pf}_{(2n-1, 2n-3, \ldots, 3, 1)}(c(1), \ldots, c(n)), \qquad (14.1)$$

where

$$c(k) = c \cdot \prod_{i=1}^{n-k}(1 + x_i) \prod_{i=1}^{n-k}(1 + y_i).$$

The other polynomials are determined inductively by

$$S^C_{ws_k}(c; x; y) = \partial_k S^C_w(c; x; y) \qquad (14.2)$$

whenever $\ell(ws_k) < \ell(w)$. Setting $y = 0$, one has *single Schubert polynomials*

$$S^C_w(c; x) = S^C_w(c; x; 0)$$

in $\Gamma[x]$.

The double Schubert polynomials for W_2 are shown in Figure 14.1. Here we are using the additive basis $Q_\lambda = Q_\lambda(c)$ for Γ, so in particular $Q_{(p)} = c_p$.

Theorem 14.4.1. *Evaluating the variables as*

$$c = c^T(V - \mathbb{S} - E), \quad x_i = c_1^T(\mathbb{S}_i/\mathbb{S}_{i+1}), \quad and \quad y_i = c_1^T(E_i/E_{i+1}),$$

we have

$$\sigma_w = S^C_w(c; x; y)$$

in $H_T^* Fl_\omega(V)$.

The proof of the theorem takes up the remainder of this section. As in Chapter 10, we realize difference operators geometrically by pullback and pushforward along $\mathbb{P}^1$-bundles. For $V = \mathbb{C}^{2n}$ as usual, we write $Fl_\omega = Fl_\omega(V)$ for the symplectic flag variety, and for $k \geq 0$,

$$Fl_\omega(\widehat{k}) = Fl_\omega(\ldots, n - k - 1, n - k + 1, \ldots, n; \mathbb{C}^{2n})$$

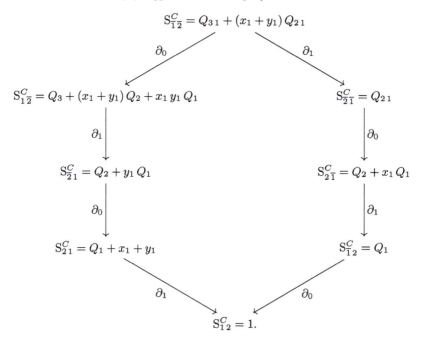

Figure 14.1 Schubert polynomials for W_2.

denotes the partial flag variety omitting the $(n-k)$-dimensional isotropic subspace. So $Fl(\widehat{0})$ parametrizes flags in which all but the maximal isotropic space appear. In the diagram

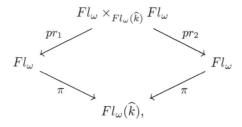

the fiber product parametrizes pairs of isotropic flags $(F_\bullet^{(1)}, F_\bullet^{(2)})$ such that $F_i^{(1)} = F_i^{(2)}$ for all $i \neq k+1$. When $k > 0$, this identifies pr_2 as the $\mathbb{P}^1$-bundle $\mathbb{P}(\mathbb{S}_k/\mathbb{S}_{k+2}) \to Fl$. When $k = 0$, the projection pr_2 is identified with the $\mathbb{P}^1$-bundle

$$\mathbb{P}(\mathbb{S}_2^\perp/\mathbb{S}_2) \to Fl_\omega,$$

whose tautological line bundle is

$$\mathbb{L} = pr_1^*(\mathbb{S}_1/\mathbb{S}_2) \subseteq \mathbb{S}_2^\perp/\mathbb{S}_2.$$

As for ordinary flag varieties, we will consider the homomorphisms $pr_{2*}pr_1^*: H_T^*Fl_\omega(V) \to H_T^*Fl_\omega(V)$. In general, the $k > 0$ case is similar to what we saw in Chapter 10, so we will focus on $k = 0$.

Lemma 14.4.2.   *Under the isomorphism* $H_T^*Fl_\omega(V) \cong (\Lambda \otimes \Gamma)[x]/J$, *we have* $pr_{2*}pr_1^* = \partial_k$.

Proof For $k > 0$, this is the same as for ordinary flag varieties (Lemma 10.6.5). For $k = 0$, we compute pr_{2*}. Using Corollary 13.5.1, any class $\alpha \in H_T^*Fl_\omega$ can be written as $\alpha = a + bx_1$, where a and b do not involve c or x_1. Now $pr_1^*\alpha = a + b\, c_1^T(\mathbb{L})$, so by basic properties of Chern classes, we have $pr_{2*}pr_1^*\alpha = -b$. Since $\partial_0(x_1) = -1$, this agrees with $\partial_0(a + bx_1)$. $\qquad\square$

Lemma 14.4.3.   *For* $w \in W_n$,

$$pr_{2*}pr_1^*\sigma_w = \begin{cases} \sigma_{ws_k} & \text{if } \ell(ws_k) < \ell(w); \\ 0 & \text{otherwise.} \end{cases}$$

The proof of this lemma goes exactly as before (Lemma 10.6.6), and we leave it as an exercise.

To complete the proof of Theorem 14.4.1, we need a formula for $\sigma_{w_\circ^{(n)}}$.

Proposition 14.4.4.   *For* $w_\circ^{(n)} = \overline{1}\,\overline{2}\cdots\overline{n} \in W_\infty$, *and* $Fl_\omega(V) = Fl_\omega$ $(\mathbb{C}^{2m})$ *with* $m \geq n$, *we have*

$$\sigma_{w_\circ^{(n)}} = \mathrm{Pf}_{(2n-1,\dots,3,1)}(c(1),\dots,c(n))$$

in $H_T^*Fl_\omega(V)$, *where* $c(k) = c(V - E_{n+1-k} - \mathbb{S}_{n+1-k})$.

Proof   The argument is the same as the one we used in Theorem 12.4.2: reduce to a Lagrangian Grassmannian degeneracy locus, and apply Corollary 14.3.1.

Consider the vector bundle $V \oplus V$ on $Fl_\omega(V)$, with the symplectic form ω' defined by

$$\omega'(v_1 \oplus w_1, v_2 \oplus w_2) = \omega(v_1, v_2) - \omega(w_1, w_2).$$

Then the flag

$$G_\bullet : E_n \oplus \mathbb{S}_n \subset E_{n-1} \oplus \mathbb{S}_{n-1} \subset \cdots \subset E_1 \oplus \mathbb{S}_1 \subset V \oplus V$$

is isotropic with respect to the form ω', as is the diagonal subbundle $\Delta_V \subseteq V \oplus V$. Using notation from §14.3, the Lagrangian degeneracy locus $D_{(2n-1,\ldots,3,1)}(\Delta_V, G_\bullet) \subseteq Fl_\omega(V)$ is defined by the conditions

$$\dim(\Delta_V \cap (E_{n+1-k} \oplus \mathbb{S}_{n+1-k})) \geq k$$

for $1 \leq k \leq n$. The equivalent conditions

$$\dim(E_{n+1-k} \cap F_{n+1-k}) \geq k$$

define $\Omega_{w_\circ^{(n)}}$, so $D_{(2n-1,\ldots,3,1)}(\Delta_V, G_\bullet) = \Omega_{w_\circ^{(n)}}$. It follows that this degeneracy locus is irreducible of codimension $\ell(w_\circ^{(n)}) = (2n-1) + \cdots + 3 + 1 = n^2$, and therefore

$$\sigma_{w_\circ^{(n)}} = \mathbb{D}_{(2n-1,\ldots,3,1)}.$$

Applying Corollary 14.3.1 proves the proposition, since

$$c(V \oplus V - \Delta_V - E_{n+1-k} - \mathbb{S}_{n+1-k}) = c(V - E_{n+1-k} - \mathbb{S}_{n+1-k}),$$

as claimed. $\square$

14.5 Properties of type C Schubert polynomials

The type C Schubert polynomials $\mathrm{S}_w^C(c; x; y)$ satisfy properties analogous to those of the type A polynomials $\mathfrak{S}_w$ and S_w (Chapters 10 and 12). To conclude this chapter, we mention some of these properties. Our treatment is brief; much more can be found in the references.

14.5.1 Basis. The classes of Schubert varieties in $Fl_\omega(V)$ form a basis for the cohomology, and as noted in §14.1.5, it follows that the classes form a basis in the infinite-dimensional limit.

As w ranges over W_∞, the Schubert polynomials $\mathrm{S}_w^C(c; x; y)$ form a basis for $\Gamma[x, y]$ over $\mathbb{Z}[y]$.

Let V be a $2n$-dimensional vector space with symplectic form ω. The Schubert polynomials also form a basis for the ideal J in the presentation $\Gamma[x, y]/J \xrightarrow{\sim} H_T^* Fl_\omega(V)$ from Corollary 13.5.1.

> *The ideal J has basis $\{S_w^C(c; x; y) \mid w \in W_\infty, w \notin W_n\}$ as a module over $\mathbb{Z}[y]$.*

The proof is the same as in type A, using Exercise 14.1.5.

14.5.2 Multiplication.   The polynomials $S_w^C(c; x; y)$ multiply in the same way as Schubert classes. Since they form a basis, we can write

$$S_u^C \cdot S_v^C = \sum_{w \in W_\infty} c_{uv}^w \, S_w^C$$

for polynomials $c_{uv}^w \in \mathbb{Z}[y]$.

> *We have*
>
> $$\sigma_u \cdot \sigma_v = \sum_{w \in W_n} c_{uv}^w \, \sigma_w$$
>
> *in $H_T^* Fl_\omega(V)$.*

14.5.3 Localization and Interpolation.   The polynomials $S_w^C(c; x; y)$ satisfy and are characterized by an interpolation property, as in type A. For any $w \in W_\infty$, the c variables specialize as

$$c^w = \prod_{i>0} \frac{1 + y_{w(i)}}{1 + y_i} = \prod_{\substack{i>0 \\ w(i)<0}} \frac{1 + y_{w(i)}}{1 - y_{w(i)}},$$

recalling that $y_{\bar{\imath}} = -y_i$. For instance, $c^{5\,\bar{4}\,1\,2\,\bar{3}} = \frac{(1-y_3)(1-y_4)}{(1+y_3)(1+y_4)}$. The x variables specialize as $x \mapsto -y^w$, by $x_i \mapsto -y_{w(i)}$ for all i.

> *For $w \in W_\infty$, the type C Schubert polynomial S_w^C specializes as*

$$S_w^C(c^w; -y^w; y) = \prod_{\substack{j<i\leq \bar{\jmath} \\ w(i)<w(j)}} (y_{w(i)} - y_{w(j)}) \qquad (14.3)$$

> *and*

$$S_w^C(c^v; -y^v; y) = 0 \quad \text{if} \quad v \not\geq w, \qquad (14.4)$$

and it is the unique polynomial in $\Gamma[x, y]$ of degree $\ell(w)$ with these properties.

This can be proved by comparing with the corresponding restrictions of Schubert classes. Just as in type A (see Lemma 9.3.1), the classes σ_w are characterized by their restrictions to fixed points:

$$\sigma_w|_w = \prod_{\substack{j < i \leq \bar{\jmath} \\ w(i) < w(j)}} (y_{w(i)} - y_{w(j)}) \tag{14.5}$$

and

$$\sigma_w|_v = 0 \quad \text{if} \quad v \not\geq w. \tag{14.6}$$

These formulas follow from the computations of tangent weights in Exercise 14.1.1.

Now setting $c = c^T(V - \mathbb{S} - E)$ and $x_i = c_1^T(\mathbb{S}_i/\mathbb{S}_{i+1})$, we have $\mathrm{S}_w^C(c; x; y) = \sigma_w$ in $\Gamma[x, y] = H_T^* Fl_\omega(V)$. Restricting to the fixed point p_v sends x_i to $c_1^T({}^v E_i/{}^v E_{i+1}) = -y_{v(i)}$, and from $c = \prod_{i>0} \frac{1-x_i}{1+y_i}$, we have $c \mapsto c^v$. (See §14.1.2.) So $\mathrm{S}_w^C(c; x; y)|_{p_v} = \mathrm{S}_w^C(c^v; -y^v; y)$, and the restriction formulas follow from Equations (14.5) and (14.6). $\qquad\square$

14.5.4 Comparison with Type A. Ordinary permutations form a subgroup of signed permutations, and for $w \in S_\infty \subseteq W_\infty$, the type A (Lascoux–Schützenberger) Schubert polynomial is a specialization of $\mathrm{S}_w^C(c; x; y)$.

For $w \in W_\infty$, we have

$$\mathrm{S}_w^C(1; x; y) = \begin{cases} \mathfrak{S}_w(x, -y) & \text{if } w \in S_\infty; \\ 0 & \text{if } w \notin S_\infty. \end{cases}$$

Using the interpolation characterization of $\mathfrak{S}_w$ and comparing the restriction formulas above with those in §10.10.5, one sees $\mathrm{S}_w^C(1; x; y) = \mathfrak{S}_w(x, -y)$ when $w \in S_\infty$. (Note that $c^w = 1$ for any $w \in S_\infty$.)

Next consider the embedding $f: Fl(\mathbb{C}^n) \hookrightarrow Fl_\omega(\mathbb{C}^{2n})$ by regarding $\mathbb{C}^n = \mathrm{span}\{e_{\bar{n}}, \ldots, e_{\bar{1}}\}$. This identifies $Fl(\mathbb{C}^n)$ with the (opposite) Schubert variety $\widetilde{\Omega}_{\bar{n} \cdots \bar{2} \bar{1}} \subseteq Fl_\omega(\mathbb{C}^{2n})$, so $f^{-1}\Omega_w = \emptyset$ and $f^*\sigma_w = 0$ if $w \notin S_n$

(see §14.1.4). On the other hand, the series c pulls back to $f^*c = 1$, and by taking n large enough, one can arrange that there are no relations among the x and y variables, so $f^* \mathrm{S}_w^C(c; x; y) = \mathrm{S}_w^C(1; x; y)$.   □

14.5.5 Grassmannian Signed Permutations.

For each strict partition λ, there is a Grassmannian signed permutation $w(\lambda)$ (§14.1.1).

The Schubert polynomial for a Grassmannian signed permutation is equal to the corresponding double Q-polynomial:

$$\mathrm{S}_{w(\lambda)}^C(c; x; y) = Q_\lambda(c|y).$$

Writing $\pi \colon Fl_\omega(V) \to LG(V)$ for the projection, this follows by observing $\pi^{-1}\Omega_\lambda = \Omega_{w(\lambda)}$, for example by comparing fixed points or the defining rank conditions.   □

Notes

Ivanov (2005) extended Nimmo's ratio formula for $Q_\lambda(x)$ to a similar formula for the multi-parameter versions $Q_\lambda(x|y)$, and proved identities relating the several different versions of these polynomials described in §14.2. Ikeda (2007) established the equivariant Giambelli formula for Lagrangian Grassmannians, using Ivanov's double Q-polynomials. Our proof of Theorem 14.2.2 uses a different interpretation of the variables and appears to be new.

As in type A, the rank conditions defining Ω_w in the isotropic flag variety are often redundant. An essential set of conditions can be determined using a variation on the Rothe diagram for signed permutations (Anderson, 2018). Alternative descriptions of Bruhat order for W_n and related groups can be found in (Björner and Brenti, 2005, §8).

There is an analogous story for Schubert varieties and degeneracy loci in the presence of a nondegenerate quadratic form on V. These correspond to Lie types B (when $\dim V = 2n + 1$) and D (when $\dim V = 2n$). As in other types, Schubert cells Ω_w° are Borel orbits, but in general one should define the Schubert varieties as the closures of these cells, $\Omega_w := \overline{\Omega_w^\circ}$, rather than directly by rank inequalities. In type D, the locus naively defined by conditions $\dim(E_p \cap F_q) \geq k$ may be larger. This is related to the fact that the corresponding Bruhat order is not induced as a sub-poset of the symmetric group. See (Fulton and Pragacz, 1998; Buch et al., 2015) for examples.

Starting in the 1990s, many authors sought representatives of Schubert classes in other classical types which are appropriately analogous to the Lascoux–Schützenberger Schubert polynomials. Fomin and Kirillov (1996b) showed that stability cannot be achieved in $\mathbb{Z}[x]$. Billey and Haiman (1995), meanwhile, gave formulas in $\Gamma[x]$. Different possibilities were investigated by several others, e.g., (Fulton, 1996a,b; Lascoux and Pragacz, 1998; Kresch and Tamvakis, 2002). The Billey–Haiman polynomials turn out to satisfy the most straightforward extensions of the properties of type A Schubert polynomials; in our notation, they are equal to $\mathrm{S}_w^C(c; x; 0)$.

The double Schubert polynomials $\mathrm{S}_w^C(c; x; y)$ were introduced by Ikeda, Mihalcea, and Naruse (2011) as representatives for Schubert classes in $H_T^* Fl_\omega(V)$ (as well as versions in types B and D). These authors established basic properties of these polynomials, including those listed in §14.5 above. Following Kazarian (2000), they also interpreted the c variables to obtain Chern class formulas for symplectic degeneracy loci. More refined formulas and different arguments were given in (Anderson and Fulton, 2012, 2018).

While the formulas given in §14.4 for the operators ∂_k suffice to define them on $\Gamma[x, y]$, when computing Schubert polynomials it is useful to know a direct formula for $\partial_0 Q_\lambda(c)$:

$$\partial_0 Q_\lambda(c) = \sum_\mu x_1^{b(\lambda/\mu)} (2x_1)^{k(\lambda/\mu)} Q_\mu(c),$$

where the sum is over all strict partitions μ obtained from λ by removing a border strip; $b(\lambda/\mu)$ is the number of lines separating boxes in such a strip; and $k(\lambda/\mu)$ is the number of connected components of the strip. This formula follows from (Billey and Haiman, 1995, Corollary 4.5), and a refined version is proved in (Anderson and Fulton, 2021b).

Grassmannians of non-maximal isotropic subspaces are somewhat more complicated than Lagrangian Grassmannians. Formulas for Schubert classes in these spaces were given by Buch, Kresch, and Tamvakis (2017, 2015). Equivariant formulas, applications to degeneracy loci, and combinatorial refinements appear in (Ikeda and Matsumura, 2015; Tamvakis, 2016a,b; Tamvakis and Wilson, 2016; Anderson and Fulton, 2018).

As we have emphasized in our notation, the type C Schubert polynomials $\mathrm{S}_w^C(c; x; y)$ are closely related to the "infinite" type A Schubert polynomials $\mathrm{S}_w^A(c; x; y)$ of the previous chapter. In fact, using the interpolation characterization of the type A polynomials (Chapter 12, Notes), one can prove that for $w \in S_\infty$, the natural surjection $\mathbb{Z}[c, x, y] \twoheadrightarrow \Gamma[x, y]$ sends $\mathrm{S}_w^A(c; x; y)$ to $\mathrm{S}_w^C(c; x; y)$. This refines the assertion of §14.5.4 that $\mathrm{S}_w^C(1; x; y) = \mathfrak{S}_w(x; -y)$. A geometric proof is given in (Anderson and Fulton, 2021a).

Other properties of the type A polynomials also hold in type C. There is a duality formula $\mathrm{S}_w^C(c; x; y) = \mathrm{S}_{w^{-1}}^C(c; y; x)$, and a Cauchy formula

$$\mathrm{S}_w^C(c; x; y) = \sum_{vu \doteq w} \mathrm{S}_u^C(c'; x; t) \cdot \mathrm{S}_v^C(c''; -t, y),$$

where $c = c' \cdot c''$ and all three series satisfy the relations defining Γ. The analogous Stanley functions $F_w = \mathrm{S}_w^C(c; x; y) \in \Gamma$ were introduced in (Billey and Haiman, 1995). Using the Cauchy formula one can express $S_w^C(c; x; y)$ in terms of these type C Stanley functions and the type A (single) Schubert polynomials $\mathfrak{S}_u$.

Hints for Exercises

Exercise 14.1.2. Work out the type A case (Exercise 10.2.1), and keep track of the conditions imposed by the symplectic group.

15

Homogeneous Varieties

A homogeneous variety is one on which an algebraic group acts transitively. We have studied the Grassmannian and flag variety, which are homogeneous for SL_n (or GL_n), as well as the Lagrangian Grassmannian and isotropic flag variety, which are homogeneous for the symplectic group Sp_{2n}. Naturally generalizing and unifying these examples are the projective homogeneous varieties for a complex semisimple (or reductive) Lie group G.

The remaining chapters deal with the equivariant cohomology of these spaces; as we will see, they play a central role in the general story of G-spaces. In this chapter, we review the basic structure of semisimple groups and their flag varieties. We will state most facts without proofs.

Some prior familiarity with Lie theory may be helpful from this point forward, but we hope the present chapter, along with its references, may provide a useful first exposure to a reader who is willing to accept statements and examples in lieu of proofs. For the reader encountering this language for the first time, a useful exercise is to work out every statement explicitly for the cases SL_n or Sp_{2n}, which we have seen in the previous chapters.

15.1 Linear algebraic groups

Let G be a *complex semisimple linear algebraic group*, with a *Borel subgroup* (i.e., maximal connected solvable subgroup) and *maximal torus*,

$G \supset B \supset T$. Let $\mathfrak{g} \supset \mathfrak{b} \supset \mathfrak{t}$ be the corresponding Lie algebras. There is a unique *maximal unipotent group* $U \subseteq B$, so that $B = U \cdot T$ (and in fact this is a semidirect product). The basic example is that of $G = SL_n$, with B the subgroup of upper-triangular matrices, T the diagonal torus, and $U = U_n$ the group of upper-triangular matrices with 1's on the diagonal.

The class of *reductive* groups is slightly larger, and with a few technical changes, most of our discussion of semisimple groups applies equally well to reductive groups. The main example of a reductive group is GL_n, along with its Borel subgroup B_n of upper-triangular matrices and diagonal torus T_n. Any semisimple or reductive group G embeds as a closed subgroup $G \subseteq GL_n$, so that $B = G \cap B_n$ and $T = G \cap T_n$.

Every linear algebraic group G has a maximal normal unipotent group, its unipotent radical $R_u(G)$, so that $G/R_u(G)$ is reductive. The main examples are $R_u(B) = U$ for a Borel subgroup, but we will see others when dealing with parabolic subgroups.

To streamline the presentation, we focus on the semisimple case from now on.

15.1.1 Roots and Weights.

The *weight lattice* of T is the character group $M = \mathrm{Hom}(T, \mathbb{C}^*)$. The set of *roots* $R \subset M$ consists of the nonzero weights of T acting on $\mathfrak{g}$ by the adjoint action. This gives a decomposition

$$\mathfrak{g} = \mathfrak{t} \oplus \bigoplus_{\beta \in R} \mathfrak{g}_\beta,$$

where $\mathfrak{g}_\beta$ is the weight space for β, and $\mathfrak{t}$ is the 0-weight space. It is a basic fact that each $\mathfrak{g}_\beta$ is one-dimensional.

The *positive roots* R^+ are the nonzero weights for the action of T on $\mathfrak{b}$, so

$$\mathfrak{b} = \mathfrak{t} \oplus \bigoplus_{\beta \in R^+} \mathfrak{g}_\beta.$$

This partitions the set of roots as $R = R^+ \coprod R^-$, where the *negative roots* are $R^- = -R^+ = \{\alpha \in R \mid -\alpha \in R^+\}$. There is an *opposite Borel subgroup* B^- whose Lie algebra $\mathfrak{b}^-$ has weights in R^-.

The roots span the real vector space $M_{\mathbb{R}} = M \otimes_{\mathbb{Z}} \mathbb{R}$. There is a unique basis of *simple roots* $\Delta \subseteq R^+$ such that every positive root is a nonnegative combination of simple roots. The integral span of the roots is the *root lattice* $M_{\mathrm{rt}} \subseteq M$.

Letting $M^\vee = \mathrm{Hom}(M, \mathbb{Z})$ be the dual lattice, the natural pairing is denoted $\langle\ ,\ \rangle \colon M \times M^\vee \to \mathbb{Z}$. For each root β, there is a *coroot* $\beta^\vee$ in $M^\vee$, such that $\langle \alpha, \beta^\vee \rangle$ is an integer for all roots $\alpha \in R$, $\langle \beta, \beta^\vee \rangle = 2$, and the *reflection* s_β defined on M by

$$s_\beta(\lambda) = \lambda - \langle \lambda, \beta^\vee \rangle \beta$$

preserves R. Note that $s_\beta(\beta) = -\beta$. (In fact, the coroots are uniquely determined by these properties.)

The coroots $\alpha^\vee$ associated to simple roots α form a basis for $M_{\mathbb{R}}^\vee$. The *fundamental weights* ϖ_α are the dual basis for $M_{\mathbb{R}}$, so if α and β are simple roots we have $\langle \varpi_\alpha, \beta^\vee \rangle = \delta_{\alpha\beta}$. These span the *fundamental weight lattice* $M_{\mathrm{wt}} \subseteq M_{\mathbb{R}}$. In general, one has $M_{\mathrm{rt}} \subseteq M \subseteq M_{\mathrm{wt}}$.

Each root $\beta \in R$ corresponds to a one-dimensional *root subgroup* $U_\beta \subseteq G$, isomorphic to the additive group $\mathbb{C}$, with Lie algebra $\mathfrak{g}_\beta \subseteq \mathfrak{g}$. There is a T-equivariant isomorphism

$$\mathbb{C} = \mathfrak{g}_\beta \xrightarrow{\sim} U_\beta,$$

with T acting on $\mathfrak{g}_\beta$ by the character β, and on U_β by conjugation. There is also a T-equivariant isomorphism

$$\prod_{\beta \in R^+} U_\beta \xrightarrow{\sim} U \subseteq B,$$

so $U \cong \mathbb{A}^N$ is isomorphic to affine space (as varieties, not groups), with $N = \#R^+$. There is a similar isomorphism for the opposite unipotent group U^-, using the negative roots R^-.

Example 15.1.1. Let $G = SL_n$, with the standard Borel B and torus T. Let $t_1, \ldots, t_n$ be the characters which pick out the diagonal entries, so the weight lattice M is generated by the t_i's, modulo the relation $t_1 + \cdots + t_n = 0$. The roots are $R = \{t_i - t_j \mid i \neq j\}$, the positive roots are $R^+ = \{t_i - t_j \mid i < j\}$, and the simple roots are $\Delta = \{t_1 - t_2, \ldots, t_{n-1} - t_n\}$. For $\beta = t_i - t_j$, the corresponding root subgroup U_β is the set of matrices with 1's along the diagonal and a free entry in position (i, j).

Exercise 15.1.2. For $G = SL_n$ as above, the root lattice $M_{rt} \subset M$ has index n, and $M/M_{rt} \cong \mathbb{Z}/n\mathbb{Z}$. Write down the fundamental weights in terms of the characters t_i.

Exercise 15.1.3. For $G = Sp_{2n}$ and the diagonal torus as in Chapter 13, work out the roots. What is M/M_{rt} in this case?

Different semisimple groups may have isomorphic Lie algebras. For any given semisimple Lie algebra, there are always two distinguished semisimple groups: the *simply connected* group, whose maximal torus has weight lattice $M = M_{wt}$, and the *adjoint* group, which has $M = M_{rt}$. As noted above, the weight lattice of the maximal torus is always an intermediate lattice, so the simply connected and adjoint groups are the extreme cases.

Example 15.1.4. The groups SL_n and PGL_n have isomorphic Lie algebras; SL_n is the simply connected group, and PGL_n is the adjoint group.

The simply connected and adjoint groups coincide if $M_{rt} = M_{wt}$. This happens, for example, in the root system of type G_2.

15.1.2 The Weyl Group.

The *Weyl group* is $W = N(T)/T$, where $N(T)$ is the normalizer of T in G. This is a finite group, and it acts on M by

$$(w \cdot \lambda)(t) = \lambda(\dot{w}^{-1}t\dot{w}),$$

for any lift $\dot{w} \in N(T)$. (An easy calculation shows this is independent of the choice of lift.) This action preserves the set of roots.

For each root β, the reflection s_β belongs to the Weyl group. The *simple reflections* s_α (associated to simple roots) generate W. A *reduced expression* for $w \in W$ is an expression $w = s_{\alpha_1} \cdots s_{\alpha_\ell}$, where $\alpha_1, \ldots, \alpha_\ell$ are simple roots and $\ell = \ell(w)$ is minimal among all such expressions. The sequence $\alpha_1, \ldots, \alpha_\ell$ is sometimes called a *reduced word*.

The number $\ell(w)$ is called the *length* of w. It may be characterized as

$$\ell(w) = \#(w(R^-) \cap R^+).$$

There is a unique *longest element* $w_\circ \in W$, with $w_\circ(R^-) = R^+$, so $\ell(w_\circ) = \#R^+$.

A root β is an *inversion* of w if $\beta \in R^+$ and $w(\beta) \in R^-$. That is, the inversions of w are the roots in $R^+ \cap w^{-1}(R^-)$. Since

$$w(R^-) \cap R^+ = -w(R^+ \cap w^{-1}(R^-)),$$

$\ell(w)$ is equal to the number of inversions of w. (One also sees that $\ell(w) = \ell(w^{-1})$, but this is evident from the definition of length.)

Example 15.1.5. For SL_n, the Weyl group is the symmetric group S_n. The simple reflection associated to the root $t_i - t_{i+1}$ is the simple transposition s_i. A positive root $\beta = t_i - t_j$ has $w(\beta) \in R^-$ if and only if $i < j$ and $w(i) > w(j)$. (So the general notion of inversion recovers the one for S_n defined in Chapter 10.) The longest element is the permutation $w_o = n \cdots 2\,1$.

A choice of reduced word induces a useful ordering of the roots:

Lemma 15.1.6. *Given a reduced word* $\alpha_1, \dots, \alpha_\ell$ *for* w, *we have*

$$w(R^-) \cap R^+ = \{\beta_1, \dots, \beta_\ell\},$$

with

$$\beta_i = s_{\alpha_1} \cdots s_{\alpha_{i-1}}(\alpha_i),$$

so $\beta_1 = \alpha_1$, $\beta_2 = s_{\alpha_1}(\alpha_2)$, *etc.*

The action of W on M induces an action on the dual lattice $M^\vee$, preserving the coroots. The natural pairing is W-invariant:

$$\langle w(\alpha), w(\beta^\vee) \rangle = \langle \alpha, \beta^\vee \rangle$$

for any root α and coroot $\beta^\vee$.

Remark 15.1.7. Since W is finite, one can choose an invariant inner product $(\ ,\ )$ on the vector space $M_\mathbb{R}$ and use this to identify $M_\mathbb{R}$ with $M_\mathbb{R}^\vee$. With respect to such an identification, one has

$$\beta^\vee = \frac{2\beta}{(\beta, \beta)}$$

for any root β. In particular, writing $\beta = \sum_{\alpha \in \Delta} n_\beta^\alpha \cdot \alpha$ in terms of simple roots, we obtain

$$\langle \varpi_\alpha, \beta^\vee \rangle = n_\beta^\alpha \cdot \frac{(\alpha, \alpha)}{(\beta, \beta)}.$$

Basic facts about the action of W on root systems show these formulas are independent of the choice of invariant inner product. The W-invariant inner product can be helpful for intuition, but we will not need to reason with it.

15.1.3 Bruhat Decomposition.   For any ordering of the set $w(R^-) \cap R^+ = \{\beta_1, \ldots, \beta_\ell\}$, the multiplication map

$$U_{\beta_1} \times \cdots \times U_{\beta_\ell} \to G$$

defines an isomorphism (of varieties) onto the subgroup

$$U(w) := U \cap \dot{w} U^- \dot{w}^{-1},$$

where $U \subseteq B$ and $U^- \subseteq B^-$ are maximal unipotent groups. This isomorphism is T-equivariant, where T acts by conjugation on each factor.

Lemma 15.1.8.   *Writing $C(w) = B\dot{w}B \subseteq G$, the map $(u, b) \mapsto u\dot{w}b$ is an isomorphism*

$$U(w) \times B \xrightarrow{\sim} C(w).$$

In particular, $C(w) \cong \prod U_{\beta_i} \times B \cong \mathbb{A}^\ell \times B$.

The above lemma is the key ingredient in Bruhat decomposition.

Proposition 15.1.9 (Bruhat decomposition).   *We have*

$$G = \coprod_{w \in W} C(w),$$

with each $C(w)$ a locally closed subvariety of G.

For $u, v, w \in W$, the notation $uv \doteq w$ denotes a length-additive product, so $uv = w$ and $\ell(u) + \ell(v) = \ell(w)$.

Lemma 15.1.10.   *If $uv \doteq w$, then $C(u) \cdot C(v) = C(w)$. For a simple root α, if $\ell(s_\alpha v) < \ell(v)$, then $C(s_\alpha) \cdot C(v) = C(s_\alpha v) \cup C(v)$. Similarly, if $\ell(v s_\alpha) < \ell(v)$, then $C(v) \cdot C(s_\alpha) = C(v) \cup C(v s_\alpha)$.*

15.2   Flag varieties

The quotient space G/B is a smooth projective variety of dimension

$$\dim G/B = N = \#R^+ = \ell(w_\circ).$$

It is called the *(generalized) flag variety* of G. For each $w \in W$, there is a point $p_w = \dot{w}B \in G/B$. The torus T acts on G/B by left multiplication, and the points p_w are precisely the T-fixed points.

We quickly review the geometry of G/B, which naturally generalizes that of $SL_n/B = Fl(\mathbb{C}^n)$ and $Sp_{2n}/B = Fl_\omega(\mathbb{C}^{2n})$. As usual, we invite the reader to unpack each statement for these cases.

15.2.1 Schubert Varieties.
The Borel subgroups B and B^- also act on G/B by left multiplication. The orbit

$$X(w)^\circ = B \cdot p_w = B\dot{w}B/B$$

is called a *Schubert cell*, and the orbit

$$Y(w)^\circ = B^- \cdot p_w = B^-\dot{w}B/B$$

is an *opposite Schubert cell*. Since $B^- = \dot{w}_\circ B \dot{w}_\circ^{-1}$, these are related by $Y(w)^\circ = w_\circ \cdot X(w_\circ w)^\circ$.

The Bruhat decomposition gives T-equivariant isomorphisms

$$\mathbb{A}^{\ell(w)} \cong \prod_{\beta \in w(R^-) \cap R^+} U_\beta \xrightarrow{\sim} X(w)^\circ \tag{15.1}$$

$$(u_1, \ldots, u_\ell) \mapsto u_1 \cdots u_\ell \cdot p_w,$$

and similarly

$$\mathbb{A}^{N-\ell(w)} \cong \prod_{\beta \in w(R^-) \cap R^-} U_\beta \xrightarrow{\sim} Y(w)^\circ \tag{15.2}$$

$$(u_1, \ldots, u_{N-\ell}) \mapsto u_1 \cdots u_{N-\ell} \cdot p_w.$$

In each case, the origin in the affine space maps to p_w in the cell.

The orbit closures

$$X(w) = \overline{X(w)^\circ} \quad \text{and} \quad Y(w) = \overline{Y(w)^\circ}$$

are called *Schubert varieties* and *opposite Schubert varieties*, respectively. From the definitions, $X(w)$ is B-invariant, of dimension $\ell(w)$, and $Y(w)$ is B^--invariant, of codimension $\ell(w)$.

Remark. Our terminology for Schubert varieties versus opposite Schubert varieties in G/B conforms to standard usage, but conflicts with what we used in previous chapters. In our earlier notation, for $G = SL_n$ or $G = Sp_{2n}$, the (opposite) Schubert cells are

$$X(w)^\circ = \widetilde{\Omega}^\circ_{w_\circ w} \quad \text{and} \quad Y(w)^\circ = \Omega^\circ_w,$$

and similarly for the closures of these cells.

15.2.2 Partial Order.

15.2.2 Partial Order. The *Chevalley–Bruhat order* on W describes containment of Schubert varieties. It is defined by

$$v \leq w \quad \text{iff} \quad X(v) \subseteq X(w).$$

Multiplication by $w_\circ$ reverses this partial order, and it follows that $v \leq w$ if and only if $Y(v) \supseteq Y(w)$. It also follows that

$$X(w) = \coprod_{v \leq w} X(v)^\circ \quad \text{and} \quad Y(w) = \coprod_{v \geq w} Y(v)^\circ.$$

In the cases $W = S_n$ or $W = W_n$, this agrees with the Bruhat order we saw in Chapters 10 and 13. As before, there are many equivalent characterizations. Here are some.

Lemma 15.2.1. *For $v, w \in W$, the following are equivalent.*

(1) $v \leq w$;

(2) $X(v) \subseteq X(w)$;

(3) $p_v \in X(w)$;

(4) $Y(w) \subseteq Y(v)$;

(5) $p(w) \in Y(v)$;

(6) $w_\circ w \leq w_\circ v$;

(7) $v^{-1} \leq w^{-1}$;

(8) *there is a reduced word $(\alpha_1, \ldots, \alpha_\ell)$ for w and a subword $(\alpha_{i_1}, \ldots, \alpha_{i_k})$ such that $s_{\alpha_{i_1}} \cdots s_{\alpha_{i_k}} = v$;*

(9) *there is a reduced word $(\alpha_1, \ldots, \alpha_\ell)$ for w and a subword $(\alpha_{i_1}, \ldots, \alpha_{i_k})$ which is a reduced word for v.*

The nontrivial equivalence is (1) $\Leftrightarrow$ (9); the rest are straightforward from the others and what we have seen.

15.2.3 Bases for Cohomology.

15.2.3 Bases for Cohomology. The Schubert varieties and opposite Schubert varieties determine bases for the equivariant cohomology of G/B, for the same reasons we have seen for $Fl(V)$ and $Fl_\omega(V)$.

Setting $F_p = \bigcup_{\ell(w) \le p} X(w)$ defines a filtration

$$\emptyset \subset F_0 \subset \cdots \subset F_N = G/B$$

by closed B-invariant subsets, and the Bruhat decomposition shows that each $F_p \smallsetminus F_{p-1}$ is a disjoint union of affine spaces of dimension p. Using the cell-decomposition lemma (Proposition 4.7.1), it follows that the classes $[X(w)]^B$ form a basis for $H_B^*(G/B)$ over $\Lambda_B = \Lambda_T$. For the same reason, the classes $[Y(w)]^{B^-}$ form a (different) basis for $H_{B^-}^*(G/B)$.

The subvarieties $X(w)$ and $Y(w)$ are T-invariant, so the classes

$$x(w) = [X(w)]^T \quad \text{and} \quad y(w) = [Y(w)]^T$$

form two distinct bases for $H_T^*(G/B)$ over Λ_T. From our dimension conventions, $x(w) \in H_T^{2N-2\ell(w)}(G/B)$ and $y(w) \in H_T^{2\ell(w)}(G/B)$.

15.2.4 Tangent Spaces and Poincaré Duality.

The tangent space to G/B at the base point $p_e = \dot{e}B$ is

$$T_{p_e} G/B = \mathfrak{g}/\mathfrak{b} = \bigoplus_{\beta \in R^-} \mathfrak{g}_\beta.$$

Left multiplication by $\dot{w}$ defines an automorphism of G/B taking p_e to p_w, and twists the T-action so that

$$T_{p_w} G/B = \bigoplus_{\beta \in w(R^-)} \mathfrak{g}_\beta.$$

The isomorphisms of (15.1) and (15.2) are T-equivariant, which implies the following:

Lemma 15.2.2. *The T-fixed point p_w is a nonsingular point of $X(w)$ and of $Y(w)$, and there are canonical T-equivariant isomorphisms*

$$T_{p_w} X(w) = T_{p_w} X(w)^\circ = \bigoplus_{\beta \in w(R^-) \cap R^+} \mathfrak{g}_\beta$$

and

$$T_{p_w} Y(w) = T_{p_w} Y(w)^\circ = \bigoplus_{\beta \in w(R^-) \cap R^-} \mathfrak{g}_\beta.$$

In particular, since $w(R^-) = (w(R^-) \cap R^+) \amalg (w(R^-) \cap R^-)$, the varieties $X(w)$ and $Y(w)$ intersect transversally at the point p_w.

It follows that the bases $\{x(w)\}$ and $\{y(w)\}$ are Poincaré dual.

Proposition 15.2.3. *Writing $\rho\colon G/B \to$ pt as before, for $v, w \in W$, we have*

$$\rho_*(x(v) \cdot y(w)) = \delta_{v,w}$$

in $H_T^*(\mathrm{pt}) = \Lambda_T$.

The proof follows the same pattern we have seen before.

15.3   Parabolic subgroups and partial flag varieties

A closed subgroup P with $G \supseteq P \supseteq B$ is called a *(standard) parabolic subgroup*. These are in bijection with subsets of the simple roots Δ. A standard parabolic P corresponds to the subset

$$\Delta_P = \{\alpha \in \Delta \mid \mathfrak{g}_{-\alpha} \subseteq \mathfrak{p}\}$$

where $\mathfrak{p}$ is the Lie algebra of P. Conversely, given a subset $J \subset \Delta$, the parabolic P_J is generated by B together with the root subgroups $U_{-\beta}$ for all $\beta \in R^+$ which are in the linear span of J. The set of such β is written $R_P^+ \subseteq R^+$.

The extreme cases are $P_\emptyset = B$ and $P_\Delta = G$. Excluding these, a *minimal parabolic* corresponds to a set consisting of a single simple root, $P_\alpha = P_{\{\alpha\}}$; and a *maximal parabolic* corresponds to a set which omits a single simple root, $P_{\widehat{\alpha}} = P_{\Delta \setminus \{\alpha\}}$.

The Lie algebra of P decomposes as

$$\mathfrak{p} = \mathfrak{b} \oplus \bigoplus_{\beta \in R_P^+} \mathfrak{g}_{-\beta}$$

$$= \mathfrak{t} \oplus \bigoplus_{\beta \in R_P^+} (\mathfrak{g}_{-\beta} \oplus \mathfrak{g}_\beta) \oplus \bigoplus_{\beta \in R^+ \setminus R_P^+} \mathfrak{g}_\beta.$$

The unipotent radical $U_P = R_u(P)$ has Lie algebra $\mathfrak{u}_P = \bigoplus_{\beta \in R^+ \setminus R_P^+} \mathfrak{g}_\beta$, and there is an equivariant isomorphism

$$U_P = \prod_{\beta \in R^+ \setminus R_P^+} U_\beta,$$

for any ordering of the roots in $R^+ \smallsetminus R_P^+$. Writing $R_P^- = -R_P^+$, there is an opposite unipotent group U_P^-, with

$$U_P^- = \prod_{\beta \in R^- \smallsetminus R_P^-} U_\beta,$$

with Lie algebra $\mathfrak{u}_P^-$.

There is an associated Weyl group $W_P = N_P(T)/T$, equal to the subgroup of W generated by the simple reflections s_α for $\alpha \in \Delta_P$. Every coset $[w]$ in W/W_P has a unique representative $w^{\min} \in W$ of minimal length, characterized by the property that $w^{\min}(\alpha) \in R^+$ for each $\alpha \in \Delta_P$. Similarly, there is a unique representative $w^{\max} \in W$ of maximal length, characterized by $w^{\max}(\alpha) \in R^-$ for each $\alpha \in \Delta_P$. The element $w^{\min}$ is the *minimal representative* of $[w]$, and $w^{\max}$ is the *maximal representative*. Writing $w_\circ^P$ for the longest element of W_P, we have $w^{\max} \doteq w^{\min} w_\circ^P$, and

$$\ell(w^{\max}) - \ell(w^{\min}) = \ell(w_\circ^P) = \#R_P^+.$$

For a parabolic subgroup $P \subseteq G$, the homogeneous variety G/P is called a *partial flag variety* for G. It is projective, and this property characterizes parabolic subgroups. Just as for G/B, the groups T, B, and B^- act on G/P by left multiplication. For each $[w] \in W/W_P$, there is a point $p_{[w]} = \dot{w}P \in G/P$, and these are precisely the T-fixed points of G/P.

There is a canonical isomorphism

$$T_{[e]}G/P = \mathfrak{g}/\mathfrak{p} = \mathfrak{u}_P^- = \bigoplus_{\beta \in R^- \smallsetminus R_P^-} \mathfrak{g}_\beta.$$

In particular, $\dim G/P = N_P = \#(R^- \smallsetminus R_P^-) = \#(R^+ \smallsetminus R_P^+)$. Multiplying by any coset representative for $[w] \in W/W_P$, we have

$$T_{[w]}G/P = \bigoplus_{\beta \in w(R^- \smallsetminus R_P^-)} \mathfrak{g}_\beta.$$

(The subgroup W_P preserves the roots R_P^-, so the set $w(R^- \smallsetminus R_P^-)$ depends only on the coset $[w] \in W/W_P$.) A refinement of Bruhat decomposition shows that there is a T-equivariant isomorphism

$$\mathbb{A}^{N_P} \cong \prod_{\beta \in w(R^- \smallsetminus R_P^-)} U_\beta \xrightarrow{\sim} V_{[w]} \tag{15.3}$$

$$(u_1, \ldots, u_{N_P}) \mapsto u_1 \cdots u_{N_P} \cdot p_{[w]},$$

where $V_{[w]} \subseteq G/P$ is an open neighborhood of $p_{[w]}$. In fact, this establishes an equivariant isomorphism $\dot{w} U_P^- \dot{w}^{-1} \xrightarrow{\sim} V_{[w]}$.

Example 15.3.1. For $G = SL_n$ and the simple root $\alpha_d = t_d - t_{d+1}$, consider the maximal parabolic $P = P_{\widehat{\alpha_d}}$. In block matrices, one has

$$U_P = \left[\begin{array}{c|c} I_d & * \\ \hline 0 & I_{n-d} \end{array} \right] \quad \text{and} \quad U_P^- = \left[\begin{array}{c|c} I_d & 0 \\ \hline * & I_{n-d} \end{array} \right],$$

where I_k is the $k \times k$ identity matrix. Both U_P and U_P^- are isomorphic to $\mathbb{A}^{d(n-d)}$. The flag variety is $G/P = Gr(d, \mathbb{C}^n)$. (Compare Example 3.3.7.)

Schubert cells and varieties are defined as before. For each coset $[w] \in W/W_P$, the orbit

$$X[w]^\circ = B \cdot p_{[w]}$$

is a Schubert cell, and

$$Y[w]^\circ = B^- \cdot p_{[w]}$$

is an opposite Schubert cell. The Schubert varieties and opposite Schubert varieties are the respective orbit closures:

$$X[w] = \overline{X[w]^\circ} \quad \text{and} \quad Y[w] = \overline{Y[w]^\circ}.$$

From the characterizations of $w^{\min}$ and $w^{\max}$, one sees

$$w^{\min}(R^- \smallsetminus R_P^-) \cap R^+ = w^{\min}(R^-) \cap R^+$$

and

$$w^{\max}(R^- \smallsetminus R_P^-) \cap R^- = w^{\max}(R^-) \cap R^-.$$

So choosing any representative w of the coset $[w]$, we have

$$\#\big(w(R^- \smallsetminus R_P^-) \cap R^+\big) = \ell(w^{\min})$$

and

$$\#\big(w(R^- \smallsetminus R_P^-) \cap R^-\big) = N - \ell(w^{\max}) = N_P - \ell(w^{\min}),$$

using $R^+ = (R^+ \smallsetminus R_P^+) \amalg R_P^+$ to see the last equality.

There are equivariant isomorphisms

$$\mathbb{A}^{\ell(w^{\min})} \cong \prod_{\beta \in w(R^- \smallsetminus R_P^-) \cap R^+} U_\beta \xrightarrow{\sim} X[w]^\circ \qquad (15.4)$$

and

$$\mathbb{A}^{N_P - \ell(w^{\min})} \cong \prod_{\beta \in w(R^- \smallsetminus R_P^-) \cap R^-} U_\beta \xrightarrow{\sim} Y[w]^\circ, \qquad (15.5)$$

so

$$\dim X[w] = \operatorname{codim} Y[w] = \ell(w^{\min}).$$

As for G/B, the Schubert cells decompose the flag variety G/P, and their classes

$$x[w] = [X[w]]^T \quad \text{and} \quad y[w] = [Y[w]]^T$$

form two bases for $H_T^*(G/P)$, as $[w]$ ranges over W/W_P. We have $x[w] \in H_T^{2N_P - 2\ell(w^{\min})}(G/P)$ and $y[w] \in H_T^{2\ell(w^{\min})}(G/P)$.

The isomorphisms (15.4) and (15.5) lead to a description of tangent spaces, as before.

Lemma 15.3.2. *The T-fixed point $p_{[w]}$ is a nonsingular point of $X[w]$ and of $Y[w]$, and there are canonical T-equivariant isomorphisms*

$$T_{p_{[w]}} X[w] = T_{p_{[w]}} X[w]^\circ = \bigoplus_{\beta \in w(R^- \smallsetminus R_P^-) \cap R^+} \mathfrak{g}_\beta$$

and

$$T_{p_{[w]}} Y[w] = T_{p_{[w]}} Y[w]^\circ = \bigoplus_{\beta \in w(R^- \smallsetminus R_P^-) \cap R^-} \mathfrak{g}_\beta.$$

In particular, the varieties $X[w]$ and $Y[w]$ intersect transversally at the point $p_{[w]}$.

Let $\pi = \pi_P$ be the projection $G/B \to G/P$; this is proper, smooth, and G-equivariant. It determines an isomorphism of Schubert cells $X(w^{\min})^\circ \to X[w]^\circ$, so by basic properties of Gysin homomorphisms, we have

$$\pi_*(x(w^{\min})) = x[w] \qquad (15.6)$$

in $H_T^* G/P$. Similarly, we have an isomorphism $Y(w^{\max})^\circ \to Y[w]^\circ$, so

$$\pi_*(y(w^{\max})) = y[w]. \qquad (15.7)$$

On the other hand, we have $\pi^{-1}X[w] = X(w^{\max})$ (as one sees by tracking T-fixed points), so

$$\pi^*x[w] = x(w^{\max}). \qquad (15.8)$$

Similarly, $\pi^{-1}Y[w] = Y(w^{\min})$, and

$$\pi^*y[w] = y(w^{\min}). \qquad (15.9)$$

From these considerations, one can establish Poincaré duality for G/P. It can be deduced from G/B using functoriality: writing $\rho^P : G/P \to$ pt and $\rho : G/B \to$ pt, we have

$$
\begin{aligned}
\rho^P_*\left(x[v] \cdot y[w]\right) &= \rho^P_*\left(x[v] \cdot \pi_* y(w^{\max})\right) \\
&= \rho^P_* \pi_*\left(\pi^*x[v] \cdot y(w^{\max})\right) \\
&= \rho_*\left(x(v^{\max}) \cdot y(w^{\max})\right) \\
&= \delta_{[v],[w]}.
\end{aligned}
$$

Another proof of Poincaré duality can be given by analyzing the induced partial order on W/W_P.

Exercise 15.3.3. Show that $p_{[v]} \in X[w]$ if and only if $v^{\min} \leq w^{\min}$ in W, and similarly $p_{[v]} \in Y[w]$ if and only if $v^{\max} \geq w^{\max}$. Using this, imitate earlier arguments to show that the bases $\{x[w]\}$ and $\{y[w]\}$ are Poincaré dual.

Exercise 15.3.4. For any $[v], [w] \in W/W_P$, show that $v^{\min} \leq w^{\min}$ if and only if $v^{\max} \leq w^{\max}$.

Example 15.3.5. Consider $G = SL_n$ and the maximal parabolic omitting the dth simple root, so $P = P_{\widehat{\alpha_d}}$ and $G/P = Gr(d, \mathbb{C}^n)$. The Weyl groups are $W = S_n$ and $W_P = S_d \times S_{n-d}$. For $w = w_1 w_2 \cdots w_n$, the minimal coset representative of $[w]$ is obtained by sorting $\{w_1, \ldots, w_d\}$ and $\{w_{d+1}, \ldots, w_n\}$ into increasing order. Similarly, the maximal coset representative is obtained by sorting these sets into decreasing order.

For instance, with $n = 9$, $d = 4$ and $w = 3\,2\,9\,6\,1\,8\,5\,7\,4$, we have

$$w^{\min} = 2\,3\,6\,9\,1\,4\,5\,7\,8$$

and

$$w^{\max} = 9\,6\,3\,2\,8\,7\,5\,4\,1.$$

Comparing with notation from Chapters 9 and 10, the partition $\lambda = (5, 3, 1, 1)$ corresponds to the subset $I = \{2, 3, 6, 9\}$, and $w^{\min} = w(\lambda)$ is the corresponding Grassmannian permutation. The complementary partition is $\lambda^\vee = (4, 4, 2, 0)$, and its Grassmannian permutation is $w(\lambda^\vee) = 1\,4\,7\,8\,2\,3\,5\,6\,9 = w_\circ w^{\max}$. We have

$$X[w] = \widetilde{\Omega}_{\lambda^\vee} \quad \text{and} \quad Y[w] = \Omega_\lambda.$$

Example 15.3.6. For $G = SL_n$ again, and the minimal parabolic $P = P_{\alpha_d}$, we have $G/P = Fl(1, \ldots, \widehat{d}, \ldots, n - 1; \mathbb{C}^n)$, the partial flag variety which omits the d-dimensional space. The Weyl group W_P is generated by the single reflection s_d, so for $w = w_1 \cdots w_n \in S_n$, we have

$$w^{\min} = \begin{cases} w & \text{if } w_d < w_{d+1}; \\ ws_d & \text{if } w_d > w_{d+1}, \end{cases}$$

and vice versa for $w^{\max}$.

15.4 Invariant curves

The T-invariant curves in G/P can be described explicitly, using a small application of Bruhat decomposition. We will need a general lemma about reflections.

Lemma 15.4.1. *Let γ be a root, and $\beta = w(\gamma)$ for some $w \in W$. Then*

$$s_\beta = w\, s_\gamma\, w^{-1} \quad \text{and} \quad U_\beta = \dot{w} U_\gamma \dot{w}^{-1}.$$

We have seen that the tangent weights at a T-fixed point $p_{[w]}$ are $w(R^- \setminus R_P^-)$. Since the weights are distinct, it follows from general equivariant geometry that there are finitely many invariant curves through $p_{[w]}$, one for each $\beta \in w(R^- \setminus R_P^-)$ (see §7.2). For each such root β, there is a reductive group

$$G_\beta = C_G(\ker(\beta)^\circ),$$

the centralizer of the identity component of $\ker(\beta \colon T \to \mathbb{C}^*)$. (So $G_\beta = G_{-\beta}$.) It has Lie algebra

$$\mathrm{Lie}(G_\beta) = \mathfrak{t} \oplus \mathfrak{g}_\beta \oplus \mathfrak{g}_{-\beta},$$

and its semisimple part is isomorphic to SL_2 or PGL_2.

Proposition 15.4.2. *The T-invariant curves in G/P through $p_{[w]}$ are the orbits*

$$G_\beta \cdot p_{[w]} \cong \mathbb{P}^1$$

for $\beta \in w(R^- \smallsetminus R_P^-)$. Such a curve contains two T-fixed points, $p_{[w]}$ and $p_{[s_\beta w]}$. Its tangent weights at these fixed points are β and $-\beta$, respectively.

Proof Let $E \subseteq G/P$ be an invariant curve containing $p_{[w]}$, and let $V_{[w]} \subseteq G/P$ be the T-invariant affine open neighborhood of $p_{[w]}$, described above in §15.3, Equation (15.3). Then $E \cap V_{[w]} = U_\beta \cdot p_{[w]}$, for some $\beta \in w(R^- \smallsetminus R_P^-)$. Since $TU_{-\beta} \subset G_\beta$ is a Borel subgroup, Bruhat decomposition for G_β gives

$$G_\beta = U_\beta T U_{-\beta} \amalg U_\beta \dot{s}_\beta T U_{-\beta}.$$

Let $\gamma = w^{-1}(\beta) \in R^- \smallsetminus R_P^-$, so $\beta = w(\gamma)$. By Lemma 15.4.1, we have $U_{-\beta} = \dot{w} U_{-\gamma} \dot{w}^{-1}$, so $U_{-\beta}$ fixes $p_{[w]}$. Similarly, $U_\beta = U_{-s_\beta(\beta)} = U_{-s_\beta w(\gamma)}$ fixes $p_{[s_\beta w]} = \dot{s}_\beta \cdot p_{[w]}$. From the decomposition of G_β, we see

$$G_\beta \cdot p_{[w]} = U_\beta \cdot p_{[w]} \amalg \{p_{[s_\beta w]}\},$$

so $E \cong G_\beta \cdot p_{[w]} \cong G_\beta/(TU_{-\beta})$ and the proposition follows. $\qquad\square$

A refinement of the proposition describes invariant curves in Schubert varieties. Let $X[v] \subseteq G/P$ be a Schubert variety, and let $[w] \in W/W_P$ be such that $w^{\min} \le v^{\min}$, so $p_{[w]} \in X[v]$.

Corollary 15.4.3. *The T-invariant curves in $X[v]$ through $p_{[w]}$ are the orbits $G_\beta \cdot p_{[w]}$ such that $\beta \in w(R^- \smallsetminus R_P^-)$ and $(s_\beta w)^{\min} \le v^{\min}$.*

Proof Since $p_{[s_\beta w]} \in X[v]$ if and only if $(s_\beta w)^{\min} \le v^{\min}$, the condition is clearly necessary for $G_\beta \cdot p_{[w]} \subseteq X[v]$. If $(s_\beta w)^{\min} \le w^{\min}$, then this curve lies in $X[w] \subseteq X[v]$, since β is a tangent weight for $T_{p_{[w]}} X[w]$. Similarly, if $w^{\min} \le (s_\beta w)^{\min} \le v^{\min}$, then the curve lies in $X[s_\beta w] \subseteq X[v]$, since $-\beta$ is a tangent weight for $T_{p_{[s_\beta w]}} X[s_\beta w]$. $\qquad\square$

Exercise 15.4.4. Let $\alpha \in \Delta$ be a simple root. For the corresponding minimal parabolic subgroup P_α and projection $\pi \colon G/B \to G/P_\alpha$, show that the fiber $\pi^{-1}(p_{[w]})$ is the T-invariant curve $G_\beta \cdot p_w \subseteq G/B$, where $\beta = w(-\alpha)$. In particular, the fixed points are $\{p_w, p_{ws_\alpha}\}$, since $s_\beta w =$

ws_α by Lemma 15.4.1. The tangent weights to this curve at p_w and p_{ws_α} are $\beta = -w(\alpha)$ and $-\beta = w(\alpha)$, respectively.

15.5 Compact groups

Every complex semisimple Lie group G contains a *maximal compact subgroup* $K \subset G$, which is unique up to conjugacy. Conversely, G is the *complexification* of K, sometimes written $G = K_{\mathbb{C}}$. For example, we have the special unitary group $SU(n) \subset SL_n$, the compact symplectic group $Sp(n) \subset Sp_{2n}$, and the real orthogonal groups $SO(n) \subset SO_n$. This also holds for reductive groups. For example, one has the unitary group $U(n) \subset GL_n$, and the compact torus $(S^1)^n \subset (\mathbb{C}^*)^n$.

More generally, any linear algebraic group G has a maximal compact subgroup K (unique up to conjugacy). But when G is not reductive, the complexification of K need not be G. For example, consider the Borel subgroup of upper-triangular matrices $B \subset GL_n$: a compact subgroup is a torus $(S^1)^n$, whose complexification is complex torus in B.

A basic fact is that K is a deformation retract of G, and consequently

$$H_K^* X = H_G^* X$$

for any space X on which G acts. In Chapter 1, we saw a simple instance of this, for $S^1 \subset \mathbb{C}^*$.

For a semisimple (or reductive) G with maximal compact subgroup K, one obtains a maximal compact torus $S \subset K$ from a choice of Borel subgroup $B \subset G$, by

$$S = B \cap K.$$

The complexification of S is a maximal torus $T \subset B$, and in fact $S = T \cap K$.

The inclusions determine a canonical homeomorphism

$$K/S \xrightarrow{\sim} G/B,$$

and we use this to endow K/S with a complex algebraic structure.

We have $N_K(S) = N_G(T) \cap K$, so the Weyl group is isomorphic to $N_K(S)/S$. This identification yields a natural right action of W on $K/S = G/B$, defined by

$$(kS) \cdot w = k\dot{w}S$$

for any representative $\dot{w} \in N_K(S)$. This action is not algebraic or holomorphic: if $\ell(w)$ is odd, the automorphism reverses orientation.

On the other hand, using the identification $K/S = G/B$, the inclusion $K/S \hookrightarrow G/T$ determines a section of the projection $G/T \to G/B$. This section is not algebraic, or even holomorphic. But there is a natural (algebraic) right action of $W = N_G(T)/T$ on G/T, and restricting to the non-holomorphically embedded $K/S \subset G/T$ recovers the above non-holomorphic action of W on G/B.

Example 15.5.1. For $G = GL_n$, reducing to the compact subgroup $K = U(n)$ corresponds to choosing a Hermitian metric on $\mathbb{C}^n$. The identification $Fl(\mathbb{C}^n) = GL_n/B = U(n)/(S^1)^n$ realizes the flag variety as the space of decompositions of $\mathbb{C}^n$ into n perpendicular lines, spanned by unit vectors. The right action of $W = S_n$ permutes these lines.

From this description, one sees that $Fl(\mathbb{C}^n) = U(n)/(S^1)^n$ embeds (non-holomorphically!) in $(\mathbb{P}^{n-1})^n$, sending each column of a matrix in $U(n)$ to the line it spans in $\mathbb{C}^n$. The action of S_n on $Fl(n)$ is then the restriction of the natural action on $(\mathbb{P}^{n-1})^n$, permuting the n factors.

Any reductive G embeds in GL_n, inducing compatible embeddings $G/B = K/S \hookrightarrow Fl(\mathbb{C}^n)$ and $W \hookrightarrow S_n$. The above construction then gives another realization of the right action of W on G/B as the restriction of an algebraic action via a non-holomorphic embedding, this time in $(\mathbb{P}^{n-1})^n$ rather than G/T.

Example 15.5.2. Using the compact symplectic group $Sp(n) \subset Sp_{2n}$, we can complete the computation of $\Lambda_{Sp(n)} = \Lambda_{Sp_{2n}}$ begun in §2.5. To do this, we give another construction of the approximation spaces.

The symplectic group can be described as a quaternionic unitary group, that is,

$$Sp(n) = \left\{ A \in M_{n,n}(\mathbb{H}) \,\middle|\, A \cdot \overline{A}^\dagger = I_n \right\},$$

where $\mathbb{H}$ is the algebra of quaternions (over $\mathbb{R}$), and $\overline{A}^\dagger$ is the quaternionic conjugate transpose. For $m \geq n$, let

$$\mathbb{E}_m Sp(n) = \mathrm{Emb}(\mathbb{H}^n, \mathbb{H}^m)$$

be the space of $\mathbb{H}$-linear embeddings, with $\mathbb{H}$ acting by right multiplication on $\mathbb{H}^n$ and $\mathbb{H}^m$, and $Sp(n)$ acting by left multiplication on these spaces. The right action of $Sp(n)$ on $\mathbb{E}_m Sp(n)$ is defined in the usual way, by $(g \cdot \varphi)(v) = \varphi(g^{-1}v)$, and the quotient space is

$$\mathbb{B}_m Sp(n) = Gr_{\mathbb{H}}(n, \mathbb{H}^m),$$

the *quaternionic Grassmannian*. The cellular structure of this space is the same as that of the complex Grassmannian $Gr(n, \mathbb{C}^m)$, but with the dimension of each cell doubled, since cells $\mathbb{C}^{|\lambda|}$ are replaced by $\mathbb{H}^{|\lambda|}$. In the limit as $m \to \infty$, we have

$$\Lambda_{Sp(n)} = H^* \mathbb{B} Sp(n) = \mathbb{Z}[c_2, c_4, \ldots, c_{2n}],$$

a polynomial ring on n generators c_{2k} in degree $4k$.

The embeddings $Sp(n) \subset U(2n)$ give compatible embeddings $Gr_{\mathbb{H}}$ $(n, \mathbb{H}^m) \hookrightarrow Gr(2n, \mathbb{C}^{2m})$, and the corresponding homomorphism on cohomology

$$\Lambda_{U(2n)} = \mathbb{Z}[c_1, c_2, \ldots, c_{2n}] \to \Lambda_{Sp(n)} = \mathbb{Z}[c_2, \ldots, c_{2n}]$$

is given by $c_i \mapsto c_i$ for i even, and $c_i \mapsto 0$ for i odd.

15.6 Borel presentation and equivariant line bundles

Using cohomology with $\mathbb{Q}$ coefficients, there are uniform descriptions of $H^*(G/B)$ and $H_T^*(G/B)$, due to Borel. *In this section all cohomology is taken with $\mathbb{Q}$ coefficients.* For instance, we will write $\Lambda_G = H^*(\mathbb{B}G; \mathbb{Q})$.

Let G be a semisimple or reductive group, with Borel subgroup B and maximal torus T; let $K \supset S$ be corresponding compact groups, as in the previous section. Suppose X is a space with left G-action. There is a canonical isomorphism

$$H_T^* X = H_G^*(G/B \times X),$$

with G acting diagonally on $G/B \times X$ (Example 3.4.4). Using compact groups, this is $H_S^* X = H_K^*(K/S \times X)$. The Weyl group acts on $K/S \times X$ via its right action on K/S (and a trivial action on X), and this induces a left action of W on $H_T^* X = H_S^* X$. The main goal of this section is to investigate the W-invariant subring $(H_T^* X)^W$.

The projection $G/B \times X \to X$ is G-equivariant, the pullback by this projection coincides with the change-of-groups homomorphism $H_G^* X \to H_T^* X$. Since W acts trivially on X, we see that the change-of-groups map factors through W-invariants:

$$H_G^* X \to (H_T^* X)^W \subseteq H_T^* X.$$

The main theorem is that the first map is an isomorphism.

Theorem 15.6.1 (Borel). *For any G-space X, there is a canonical isomorphism*

$$H_G^* X = (H_T^* X)^W.$$

Similarly, $H_K^* X = (H_S^* X)^W$ for any K-space.

To prove the theorem, we will use a basic fact from topology.

Proposition 15.6.2. *Let W be a finite group acting freely (on the right) on a topological space Y. There is a canonical isomorphism*

$$H^*(Y/W) = (H^* Y)^W,$$

using $\mathbb{Q}$ coefficients. $\square$

Proof of Theorem 15.6.1 First we consider the right action of W on K/S. This is free, with quotient $K/N_K(S)$, so by Proposition 15.6.2, we have

$$H^*(K/N_K(S)) = H^*(K/S)^W.$$

Furthermore, the map $K/S \to K/N_K(S)$ is a covering space, with covering group W, so the Euler characteristics are related by

$$\chi(K/N_K(S)) = \frac{1}{\#W}\chi(K/S).$$

Since $K/S = G/B$ has a decomposition into Schubert cells, its Euler characteristic is equal to $\#W$. It follows that $\chi(K/N_K(S)) = 1$, and that

$$H^*(K/N_K(S)) = \mathbb{Q}.$$

That is, $K/N_K(S)$ is $\mathbb{Q}$-acyclic – it has the rational cohomology of a point.

Next consider the quotient map

$$\mathbb{E}K \times^{N_K(S)} X \to \mathbb{E}K \times^K X,$$

a fiber bundle with fiber $K/N_K(S)$. Since $K/N_K(S)$ is acyclic, the Leray–Hirsch theorem says the corresponding pullback map on cohomology is an isomorphism, and we have

$$H_K^* X = H^*(\mathbb{E}K \times^K X) \xrightarrow{\sim} H^*(\mathbb{E}K \times^{N_K(S)} X).$$

Finally, the map

$$\mathbb{E}K \times^S X \to \mathbb{E}K \times^{N_K(S)} X$$

is a covering space, with group W, so applying Proposition 15.6.2 again, we have

$$H^*(\mathbb{E}K \times^{N_K(S)} X) = H^*(\mathbb{E}K \times^S X)^W = (H_S^* X)^W.$$

Combining these isomorphisms establishes the theorem. $\square$

Taking $X = \mathrm{pt}$, we obtain a description of Λ_G.

Corollary 15.6.3. *For any reductive group G with Weyl group W and maximal torus T, we have $\Lambda_G = (\Lambda_T)^W$.*

For example, taking $G = GL_n$, this says

$$\Lambda_{GL_n} = \mathbb{Q}[c_1, \ldots, c_n] = \mathbb{Q}[t_1, \ldots, t_n]^{S_n} = (\Lambda_T)^{S_n},$$

as we have seen before.

Next we will deduce presentations for the cohomology rings of G/B. The first will be given in terms of equivariant line bundles, which will play a significant role in what follows.

Any character $\lambda \in M = \mathrm{Hom}(T, \mathbb{C}^*)$ extends to a character of $B = T \cdot U$, by $\lambda(t \cdot u) = \lambda(t)$. Let $\mathbb{C}_\lambda$ be the corresponding one-dimensional representation of B. Then

$$\mathcal{L}_\lambda := G \times^B \mathbb{C}_\lambda$$

is a G-equivariant line bundle on G/B. There is a ring homomorphism

$$b\colon \Lambda_T \to H_T^*(G/B),$$

defined by $b(\lambda) = c_1^T(\mathcal{L}_\lambda)$ for $\lambda \in M$.

In fact, the homomorphism b is simply a change-of-groups homomorphism. Using the canonical identifications $\Lambda_T = \Lambda_B = H_G^*(G/B)$ (see Example 3.4.3) and $H_T^*(G/B) = H_B^*(G/B)$, we can write

$$b\colon H_G^*(G/B) \to H_B^*(G/B)$$

as $c_1^G(\mathcal{L}_\lambda) \mapsto c_1^B(\mathcal{L}_\lambda)$.

Corollary 15.6.4. *Let $\Lambda = \Lambda_T$. The homomorphism $b\colon \Lambda \to H_T^*(G/B)$ descends to a canonical isomorphism*

$$\Lambda/\Lambda_+^W \xrightarrow{\sim} H^*(G/B),$$

where Λ_+^W is the positive-degree part of Λ^W.

Proof Consider the fiber bundle

$$\mathbb{E}G \times^G G/B \to \mathbb{B}G,$$

with fiber G/B. The base and fiber have no odd cohomology (over $\mathbb{Q}$), so the associated spectral sequence degenerates. The homomorphism $H_G^*(G/B) \to H^*(G/B)$ comes from restriction to a fiber, and degeneration of the spectral sequence implies

$$H_G^*(G/B)/H^*(\mathbb{B}G)_+ = H^*(G/B).$$

We have $H_G^*(G/B) = \Lambda_B = \Lambda$, and $H^*\mathbb{B}G = \Lambda^W$ (by Corollary 15.6.3 above), so the left-hand side is Λ/Λ_+^W.

It remains to observe that restriction to a fiber is the same as composing b with the forgetful homomorphism $H_T^*(G/B) \to H^*(G/B)$. Indeed, considering the diagram

$$
\begin{array}{ccccc}
G/B & \hookrightarrow & \mathbb{E}G \times^B G/B & \longrightarrow & \mathbb{E}G \times^G G/B \\
\downarrow & & \downarrow & & \downarrow \\
\mathrm{pt} & \hookrightarrow & \mathbb{B}B & \longrightarrow & \mathbb{B}G,
\end{array}
$$

the line bundle $\mathbb{E} \times^G \mathcal{L}_\lambda$ on $\mathbb{E}G \times^G G/B$ pulls back to $\mathbb{E}G \times^B \mathcal{L}_\lambda$ on $\mathbb{E}G \times^B G/B$, and then restricts to $\mathcal{L}_\lambda$ on the fiber G/B. $\square$

Proposition 15.6.5.   *For any G-space X, there is a canonical isomorphism*

$$H_T^* X = \Lambda \otimes_{\Lambda^W} H_G^* X,$$

where $\Lambda = \Lambda_T$.

Proof   Consider the fiber bundle $\mathbb{E}G \times^B X \to \mathbb{E}G \times^G X$, with fiber G/B, and let

$$H_B^* X = H^*(\mathbb{E}G \times^B X) \to H^*(G/B)$$

be restriction to a fiber. This is surjective, because the composition $\Lambda = \Lambda_B \to H_B^* X \to H^*(G/B)$ is surjective, as we have just seen.

Take elements $x_w \in \Lambda$ which lift a basis of $H^*(G/B) = \Lambda/\Lambda_+^W$, and let $\overline{x}_w \in H_B^* X$ be their images under $\Lambda \to H_B^* X$. By the Leray–Hirsch theorem, $H_T^* X = H_B^* X$ is free over $H_G^* X$ with basis given by the elements $\overline{x}_w$. $\square$

Taking $X = G/B$ and using $H_G^*(G/B) = \Lambda$, we obtain a description of the equivariant cohomology of G/B.

Corollary 15.6.6.   *We have $H_T^*(G/B) = \Lambda \otimes_{\Lambda^W} \Lambda$.*

In the next chapter, we will see some consequences of the symmetry manifested in this description of $H_T^*(G/B)$.

Exercise 15.6.7.   Show that the action of W on $\Lambda_T = H_G^*(G/B)$, as described at the beginning of this section, is the same as the one induced by the natural action of W on the weight lattice of T.

Example 15.6.8.   For $G = GL_n$, so $G/B = Fl(\mathbb{C}^n)$, we have seen

$$H^* Fl(\mathbb{C}^n) = \mathbb{Q}[x]/(e_1(x), \dots, e_n(x))$$

and

$$H_T^* Fl(\mathbb{C}^n) = \mathbb{Q}[x; y]/(e_1(x) - e_1(y), \dots, e_n(x) - e_n(y)),$$

where the variables are $x = (x_1, \dots, x_n)$ and $y = (y_1, \dots, y_n)$ (Proposition 4.4.1).

Exercise 15.6.9.   For $G = Sp_{2n}$, consider the group of signed permutations W_n acting on $\Lambda = \mathbb{Q}[t_1, \dots, t_n]$. Show that the invariant ring is generated by the elementary symmetric functions in the squares of the t variables: $e_1(t_1^2, \ldots, t_n^2), \ldots, e_n(t_1^2, \ldots, t_n^2)$. This recovers the presentation for $H^* Fl_\omega(\mathbb{C}^{2n})$ given in Proposition 13.3.1.

Notes

General references for linear algebraic groups are the books by Borel (1991), Humphreys (1981), and Springer (1998). For Weyl groups and Coxeter groups, standard references are Bourbaki (1981) and Humphreys (1990). More detail about Schubert varieties appears in (Jantzen, 2003, §13). The Bruhat decomposition is described in (Borel, 1991, §14.12), (Humphreys, 1981, §28.3), (Jantzen, 2003, §13.2), or (Springer, 1998, §8.3). Lemma 15.1.10 is proved in (Springer, 1998, Lemma 8.3.7).

A proof of the nontrivial equivalence in the characterization of Bruhat order (Lemma 15.2.1) is given in (Jantzen, 2003, §13.7). Other equivalent conditions are in (Humphreys, 1990, §§5.9–5.10).

Lemma 15.4.1 appears in (Humphreys, 1981, §26.3).

A general reference for compact and complex Lie groups is (Helgason, 2001, Chapter VI). In particular, the fact that $K \subset G$ is a deformation retract may be found in (Helgason, 2001, §VI.2).

Although Borel did not use the language of equivariant cohomology at the time, Theorem 15.6.1 comes from (Borel, 1953, Proposition 27.3).

Proposition 15.6.2 can be found in (Hatcher, 2002, Proposition 3G.1); see also (Grothendieck, 1957, 5.2.3, Corollaire), which draws the same conclusion without requiring that G act freely. One consequence is that $H_W^*(\mathrm{pt}, \mathbb{Q}) = \mathbb{Q}$ for any finite group W. (Take $Y = \mathbb{E}W$ in the proposition.)

The homomorphism $b \colon \Lambda \to H^*(G/B)$ was studied extensively in the 1958 Chevalley seminar, using $\mathbb{Z}$ coefficients. Serre calls a group G *special* if every principal G-bundle which is locally trivial in the étale topology is also locally trivial in the Zariski topology. Grothendieck shows that special groups are precisely the connected linear algebraic groups for which b is surjective (over $\mathbb{Z}$); furthermore, among semisimple groups, the only special groups are direct products of copies of SL_n and Sp_{2n} (Grothendieck, 1958, Théorème 3).

The presentations of $H^* Fl(\mathbb{C}^n)$ and $H^* Fl_\omega(\mathbb{C}^{2n})$ (from Example 15.6.8 and Exercise 15.6.9) are valid with $\mathbb{Z}$ coefficients; since SL_n and Sp_{2n} are special groups, this is not a coincidence, and these are essentially the only examples.

Hints for Exercises

Exercise 15.1.3. Writing the characters of the diagonal torus as $-t_n$, $\ldots, -t_1, t_1, \ldots, t_n$, the roots are

$$R = \{\pm t_i \pm t_j \mid i \neq j\} \ \amalg \ \{\pm 2t_1, \ldots, \pm 2t_n\}.$$

These generate an index 2 sublattice of M (which has basis $t_1, \ldots, t_n$). So $M/M_{\mathrm{rt}} \cong \mathbb{Z}/2\mathbb{Z}$. The positive roots are

$$R^+ = \{t_i - t_j \mid i < j\} \amalg \{-t_i - t_j \mid i \leq j\}.$$

Exercise 15.3.4. Since $w^{\max} \doteq w^{\min} w_\circ^P$, this follows from the subword characterization of Bruhat order in Lemma 15.2.1.

Exercise 15.6.7. Under the automorphism $kS \mapsto k\dot{w}S$ of $G/B = K/S$, the line bundle $\mathcal{L}_\lambda$ pulls back to $\mathcal{L}_{w \cdot \lambda}$, as one can check by examining the fiber over the identity point.

16

The Algebra of Divided Difference Operators

When studying Schubert polynomials for the ordinary and symplectic flag varieties, we saw algebras of divided difference operators acting on equivariant cohomology. Similar operators act on the cohomology of any G/B, compatibly with localization as well as with the presentations we saw in the last chapter. These operators give an explicit, algorithmic means of computing in $H_T^*(G/B)$, although in practice one often requires a more efficient method.

The general divided difference operators – and variants we will see in this chapter – have been used by many authors as an algebraic model for $H_T^*(G/B)$, often including the infinite-dimensional Kac–Moody flag varieties. Here, only considering finite-dimensional flag varieties, we will see geometric constructions which simplify the algebraic arguments.

We work with the usual data from the last chapter: a semisimple (or reductive) group G, with maximal torus T and Borel subgroup B, with simple roots Δ and Weyl group W. We write $\Lambda = \Lambda_T$.

16.1   Push-Pull operators

For a subset of simple roots $J \subseteq \Delta$, with corresponding standard parabolic subgroup $P = P_J$, let π_J be the projection $G/B \to G/P$. We will consider the Λ-linear map $D_J \colon H_T^*(G/B) \to H_T^*(G/B)$ defined by

$$D_J = \pi_J^* \circ \pi_{J*}.$$

This lowers degree by twice the relative dimension of the morphism $\pi_J \colon G/B \to G/P$, that is, by $2 \cdot \#R_P^+$.

For a coset $[w] \in W/W_P$, recall that $w^{\min}$ (respectively, $w^{\max}$) is the unique representative in W of minimal (resp., maximal) length.

Lemma 16.1.1. *Let $w \in W$ be a representative for the coset $[w] \in W/W_P$. We have*

$$\mathrm{D}_J(x(w)) = \begin{cases} x(w^{\max}) & \text{if } w = w^{\min}; \\ 0 & \text{if } w \neq w^{\min}, \end{cases}$$

and

$$\mathrm{D}_J(y(w)) = \begin{cases} y(w^{\min}) & \text{if } w = w^{\max}; \\ 0 & \text{if } w \neq w^{\max}. \end{cases}$$

Proof For any w, we have

$$\pi_J(X(w)) = X[w] \quad \text{and} \quad \pi_J(Y(w)) = Y[w]$$

in G/P. If $w \neq w^{\min}$, then $\dim X(w) > \dim X[w]$ so the pushforward is zero; and similarly $\dim Y(w) > \dim Y[w]$ if $w \neq w^{\max}$. The formulas $\mathrm{D}_J(x(w^{\min})) = x(w^{\max})$ and $\mathrm{D}_J(y(w^{\max})) = y(w^{\min})$ follow from Chapter 15, Equations (15.6)–(15.9). □

Our main interest is in the operators associated to one simple root. We will write $\pi_\alpha = \pi_{\{\alpha\}}$ and $\mathrm{D}_\alpha = \mathrm{D}_{\{\alpha\}}$ for this case. Here $P = P_\alpha$ is a minimal parabolic, and $W_P = \{e, s_\alpha\}$. So each coset in W/W_P has two representatives, w and ws_α, whose lengths differ by 1. In this case, the lemma takes the following form:

$$\mathrm{D}_\alpha(x(w)) = \begin{cases} x(ws_\alpha) & \text{if } \ell(ws_\alpha) > \ell(w); \\ 0 & \text{otherwise,} \end{cases} \tag{16.1}$$

and

$$\mathrm{D}_\alpha(y(w)) = \begin{cases} y(ws_\alpha) & \text{if } \ell(ws_\alpha) < \ell(w); \\ 0 & \text{otherwise.} \end{cases} \tag{16.2}$$

Equation (16.2) generalizes the analogous formulas we have seen for Schubert classes in the ordinary and symplectic flag varieties (Lemma 10.6.6 and Lemma 14.4.3).

For any sequence of simple roots $\alpha_1, \ldots, \alpha_\ell$, the composition

$$D_{\alpha_1} \circ \cdots \circ D_{\alpha_\ell} : H^*_T(G/B) \to H^{*-2\ell}_T(G/B)$$

takes $x(w)$ to $x(ws_{\alpha_\ell} \cdots s_{\alpha_1})$ if $\ell(ws_{\alpha_\ell} \cdots s_{\alpha_1}) = \ell(w) + \ell$, and sends $x(w)$ to 0 otherwise. The effect on $y(w)$'s is similar: the operator takes $y(w)$ to $y(ws_{\alpha_\ell} \cdots s_{\alpha_1})$ if $\ell(ws_{\alpha_\ell} \cdots s_{\alpha_1}) = \ell(w) - \ell$, and sends $y(w)$ to 0 otherwise.

For any $v \in W$, we may choose any reduced word $(\alpha_1, \ldots, \alpha_\ell)$ and define

$$D_v := D_{\alpha_1} \circ \cdots \circ D_{\alpha_\ell},$$

so from the considerations of the previous paragraph it follows that

$$D_v(x(w)) = \begin{cases} x(wv^{-1}) & \text{if } \ell(wv^{-1}) = \ell(w) + \ell(v^{-1}); \\ 0 & \text{otherwise,} \end{cases} \tag{16.3}$$

and similarly for $D_v(y(w))$. Since the elements $x(w)$ form a basis for $H^*_T(G/B)$ over Λ, this shows that D_v is well defined and independent of the choice of reduced word for v. (If $v = s_{\alpha_1} \cdots s_{\alpha_\ell}$ is not a reduced expression, i.e., $\ell(v) < \ell$, then the operator $D_{\alpha_1} \circ \cdots \circ D_{\alpha_\ell}$ is identically zero.) This also shows

$$D_u \circ D_v = \begin{cases} D_{uv} & \text{if } \ell(uv) = \ell(u) + \ell(v); \\ 0 & \text{otherwise.} \end{cases} \tag{16.4}$$

The algebra of Λ-linear endorphisms of $H^*_T(G/B)$ generated by these operators is called the *nil-Hecke algebra*.

The operators D_v may also be constructed via a correspondence. Let $Z(v^{-1}) \subseteq G/B \times G/B$ be the G-orbit closure

$$Z(v^{-1}) = \overline{G \cdot (p_e, p_{v^{-1}})} = \overline{G \cdot (p_v, p_e)}$$

where G acts diagonally. (For $G = SL_n$, this is the "double Schubert variety" $\Omega_{v^{-1}}$ considered in §10.5.) Let pr_1 and pr_2 be the projections $Z(v^{-1}) \to G/B$.

Proposition 16.1.2. *We have* $D_v = pr_{2*} \circ pr_1^*$.

Proof We temporarily denote the operators $pr_{2*} \circ pr_1^*$ by D'_v.

First consider a simple root α. Then we have $Z(\alpha) = Z(s_\alpha)$ is the fiber product $G/B \times_{G/P_\alpha} G/B$, and the claim follows from the diagram of $\mathbb{P}^1$-bundles

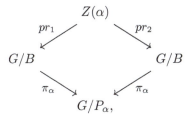

since $\pi_\alpha^* \pi_{\alpha*} = pr_{2*}pr_1^*$. So $D'_\alpha = D_\alpha$.

For the general case, we use induction on length. Supposing $\ell(s_\alpha v) = \ell(v) + 1$, we consider the fiber product $\widetilde{Z}(v^{-1}s_\alpha)$ sitting in the diagram

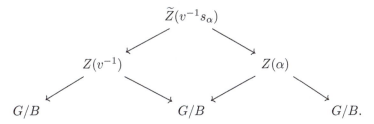

The projection $pr_{13} \colon \widetilde{Z}(v^{-1}s_\alpha) \to G/B \times G/B$ (onto the first and third factors) is birational onto its image, which is $Z(v^{-1}s_\alpha)$, and it follows that $D'_{s_\alpha v} = D'_\alpha \circ D'_v$. On the other hand, if $\ell(s_\alpha v) = \ell(v) - 1$, then $pr_{13}(\widetilde{Z}(v^{-1}s_\alpha)) = Z(v^{-1})$, so the dimension drops and $D'_\alpha \circ D'_v = 0$ in this case. So these operators satisfy the same relations as the D_v, and therefore $D'_v = D_v$ for all v. $\qquad\square$

Example 16.1.3. For $G = SL_n$, so $G/B = Fl(\mathbb{C}^n)$, and a simple root $\alpha = t_k - t_{k+1}$, the action of D_α on $H^*_T Fl(\mathbb{C}^n)$ agrees with that of ∂_k (§10.6).

Exercise 16.1.4. Suppose α is a simple root and $u, v \in W$ are such that $\ell(us_\alpha) > \ell(u)$ and $\ell(vs_\alpha) < \ell(v)$. Show that

$$D_\alpha(y(u) \cdot y(v)) = y(u) \cdot y(vs_\alpha).$$

In particular, $D_\alpha(y(u) \cdot y(s_\alpha)) = y(u)$ if $\ell(us_\alpha) > \ell(u)$.

Remark. For a minimal parabolic P_α, there is a distinguished homomorphism $SL_2 \to P_\alpha$, whose image has Lie algebra

$$\mathfrak{g}_{-\alpha} \oplus [\mathfrak{g}_{-\alpha}, \mathfrak{g}_\alpha] \oplus \mathfrak{g}_\alpha \subseteq \mathfrak{p}.$$

The standard 2-dimensional representation $\mathbb{C}^2$ of SL_2 determines a rank 2 vector bundle V_α on G/P_α. (Without changing the flag variety, one may assume G is simply connected, which implies $SL_2 \hookrightarrow P_\alpha$, and the SL_2-representation $\mathbb{C}^2$ extends to P_α. Then $V_\alpha = G \times^{P_\alpha} \mathbb{C}^2$.) The map $\pi_\alpha \colon G/B \to G/P_\alpha$ is the $\mathbb{P}^1$-bundle $\mathbb{P}(V_\alpha) \to G/P_\alpha$.

16.2   Restriction to fixed points

The localization homomorphism

$$H_T^*(G/B) \to H_T^*(G/B)^T = \bigoplus_{w \in W} H_T^*(p_w)$$

is injective, since $H_T^*(G/B)$ is free over Λ (Theorem 5.1.8). We wish to extend the operators D_α to the ring $H_T^* X^T$.

It will be useful to identify $H_T^* X^T$ with the ring $F(W, \Lambda)$ of Λ-valued functions on the Weyl group. In this notation, the localization homomorphism is

$$x \mapsto \psi_x,$$

where ψ_x is the function $\psi_x(w) = x|_w$. It will also be useful to further extend to the fraction field Q of Λ, and consider the Q-algebra $F(W, Q)$ of Q-valued functions on W. Since Λ is a domain, we have $F(W, \Lambda) \subset F(W, Q)$.

One reason to prefer the function notation is that it immediately suggests natural W-actions. Here we will use the action induced by right multiplication on itself: for $v \in W$ and $\psi \in F(W, Q)$, we have

$$(v \cdot \psi)(w) = \psi(wv).$$

(In §16.5 we will see another natural action.)

There is a Q-linear map $\mathrm{A}_\alpha \colon F(W, Q) \to F(W, Q)$ for each simple root α, defined by the formula

$$(\mathrm{A}_\alpha \psi)(w) = \frac{\psi(w s_\alpha) - \psi(w)}{w(\alpha)}. \tag{16.5}$$

In terms of the W-action on $F(W, Q)$, this is $\mathrm{A}_\alpha \psi = \frac{1}{w(\alpha)} (s_\alpha \cdot \psi - \psi)$.

Exercise 16.2.1. Show that the operators A_α satisfy a Leibniz-type formula:

$$A_\alpha(\psi \cdot \xi) = A_\alpha(\psi) \cdot \xi + (s_\alpha \cdot \psi) \cdot A_\alpha(\xi)$$

for any $\psi, \xi \in F(W, Q)$.

Proposition 16.2.2. *The diagram*

$$
\begin{array}{ccc}
H_T^* X & \lhook\joinrel\longrightarrow & F(W, Q) \\
{\scriptstyle D_\alpha} \downarrow & & \downarrow {\scriptstyle A_\alpha} \\
H_T^* X & \lhook\joinrel\longrightarrow & F(W, Q)
\end{array}
$$

commutes.

Proof For $\pi = \pi_\alpha \colon G/B \to G/P_\alpha$, we compute π_* by localization, using the integration formula (Example 5.2.3). The fiber over $p_{[w]} \in G/P_\alpha$ is the T-invariant curve $E = G_{w(\alpha)} \cdot p(w)$, with fixed points p_w and p_{ws_α} (Exercise 15.4.4). The tangent weights are $-w(\alpha)$ and $w(\alpha)$, respectively. For any $x \in H_T^*(G/B)$, then, we have

$$
\begin{aligned}
\pi_*(x)|_{[w]} &= \frac{x|_w}{-w(\alpha)} + \frac{x|_{ws_\alpha}}{w(\alpha)} \\
&= \frac{x|_{ws_\alpha} - x|_w}{w(\alpha)}.
\end{aligned}
$$

The pullback map $\pi^* \colon H_T^*(G/P)^T \to H_T^*(G/B)^T$ is determined by

$$H_T^*(p_{[w]}) \to H_T^*(p_w) \oplus H_T^*(p_{ws_\alpha})$$

$$y \mapsto (y, y).$$

It follows that

$$D_\alpha(x)|_w = \pi^*\pi_*(x)|_w = \frac{x|_{ws_\alpha} - x|_w}{w(\alpha)} = (A_\alpha \psi_x)(w),$$

as claimed. $\qquad\square$

The inclusion $H_T^* X \hookrightarrow F(W, Q)$ becomes an isomorphism after tensoring with Q, and the proposition says the operators A_α are the ones induced by D_α on $H_T^*(G/B) \otimes_\Lambda Q$. It follows that they satisfy similar properties. For a sequence of simple roots $(\alpha_1, \dots, \alpha_\ell)$, the composition $A_{\alpha_1} \circ \cdots \circ A_{\alpha_\ell}$ depends only on $v = s_{\alpha_1} \cdots s_{\alpha_\ell}$ if $\ell(v) = \ell$ (i.e., if the word is reduced), and vanishes otherwise. Writing

$$A_v = A_{\alpha_1} \circ \cdots \circ A_{\alpha_\ell}$$

for any reduced word $(\alpha_1, \ldots, \alpha_\ell)$, we have

$$A_u \circ A_v = \begin{cases} A_{uv} & \text{if } \ell(uv) = \ell(u) + \ell(v); \\ 0 & \text{otherwise.} \end{cases}$$

The W-action on $F(W, Q)$ preserves the subalgebra $H_T^*(G/B)$. This is the content of the following proposition.

Proposition 16.2.3. *There is a unique left Λ-linear action of W on $H_T^*(G/B)$, preserving grading and satisfying*

$$s_\alpha \cdot x = D_\alpha(x \cdot y(s_\alpha)) - D_\alpha(x) \cdot y(s_\alpha)$$

for all $x \in H_T^*(G/B)$ and all simple roots α. This action also satisfies and is determined by the formula

$$\psi_{v \cdot x} = v \cdot \psi_x$$

for all $v \in W$ and $x \in H_T^*(G/B)$, where $x \mapsto \psi_x$ is the inclusion $H_T^*(G/B) \hookrightarrow F(W, Q)$.

Proof By the Leibniz rule for A_α, for any $x \in H_T^*(G/B)$ we have

$$A_\alpha(\psi_x \cdot \psi_{y(s_\alpha)}) = A_\alpha(\psi_x) \cdot \psi_{y(s_\alpha)} + (s_\alpha \cdot \psi_x) \cdot A_\alpha(\psi_{y(s_\alpha)}).$$

By the previous proposition, we have $A_\alpha(\psi_x) = \psi_{D_\alpha(x)}$ for any x in $H_T^*(G/B)$. In particular, $A_\alpha(\psi_{y(s_\alpha)}) = 1$, so the above equation becomes

$$s_\alpha \cdot \psi_x = A_\alpha(\psi_x \psi_{y(s_\alpha)}) - A_\alpha(\psi_x)\psi_{y(s_\alpha)}.$$

The functions ψ_x and $\psi_{y(s_\alpha)}$ lie in the subalgebra $H_T^*(G/B) \subseteq F(W, Q)$, so this implies the claimed relation, and proves that W preserves this subalgebra. The same calculations also show $\psi_{s_\alpha \cdot x} = s_\alpha \cdot \psi_x$, which suffices to prove the second statement. □

It follows that the operators D_α satisfy the same Leibniz-type formula as the A_α:

$$D_\alpha(x \cdot y) = D_\alpha(x) \cdot y + (s_\alpha \cdot x) D_\alpha(y)$$

for any $x, y \in H_T^*(G/B)$.

We will use the operators A_α to recursively compute localizations of Schubert classes. There is a simple formula for certain restrictions, generalizing what we have seen for Schubert classes in ordinary and symplectic flag varieties. It follows immediately from the descriptions of tangent weights (Lemma 15.3.2).

Lemma 16.2.4. *For $[v] \in W/W_P$, we have*

$$x[v]|_{[v]} = \prod_{\beta \in v(R^- \smallsetminus R_P^-) \cap R^-} \beta$$

and

$$y[v]|_{[v]} = \prod_{\beta \in v(R^- \smallsetminus R_P^-) \cap R^+} \beta.$$

Focusing on G/B, we will write $\psi_v \in F(W, \Lambda)$ for the function given by

$$\psi_v(w) = y(v)|_w,$$

so ψ_v is the image of $y(v)$ under the inclusion $H_T^*(G/B) \hookrightarrow F(Q, \Lambda)$. We use the same notation for the corresponding function in $F(W, Q)$.

Proposition 16.2.5. *The functions ψ_v satisfy and are uniquely characterized by the following properties:*

(i) $\psi_v(w) = 0$ *unless* $v \leq w$.

(ii) $\displaystyle \psi_v(v) = \prod_{\beta \in v(R^-) \cap R^+} \beta.$

(iii) $A_\alpha \psi_v = \begin{cases} \psi_{vs_\alpha} & \text{if } \ell(vs_\alpha) < \ell(v); \\ 0 & \text{if } \ell(vs_\alpha) > \ell(v). \end{cases}$

The proof is straightforward from what we already know.

Proof Property (i) holds since $y(v)|_w = 0$ if $p_w \notin Y(v)$. Property (ii) comes from Lemma 16.2.4. Property (iii) comes from the corresponding formula for $D_\alpha y(v)$ (Equation (16.2) above).

To see that these properties characterize ψ_v, first note that Property (i) says $\psi_{w_\circ}$ has only one nonzero value, which is specified by Property (ii). Any $v \in W$ may be written as $v = w_\circ s_{\alpha_\ell} \cdots s_{\alpha_1}$ with $\ell(v) = \ell(w_\circ) - \ell$, so ψ_v may be computed from $\psi_{w_\circ}$ using Property (iii). $\qquad\square$

The proposition makes it possible to calculate any restriction of a Schubert class, by starting with $w_\circ$ and working inductively in Bruhat order. In practice, however, one often wants a more direct way of computing. In Chapter 18 we will see an explicit formula for every value $\psi_v(w)$. Here we give a different formula, for the simple case of a divisor.

Recall that for each simple root α there is a fundamental weight ϖ_α, characterized by $\langle \varpi_\alpha, \beta^\vee \rangle = \delta_{\alpha\beta}$.

Lemma 16.2.6.   *The Schubert divisors restrict as follows:*

$$y(s_\alpha)|_w = \varpi_\alpha - w(\varpi_\alpha)$$

and

$$x(w_\circ s_\alpha)|_w = w_\circ(\varpi_\alpha) - w(\varpi_\alpha).$$

(If G is not simply connected, so $M \subsetneq M_{\mathrm{wt}}$, then ϖ_α may not lie in M, but the differences on the right-hand sides always lie in the root lattice $M_{\mathrm{rt}} \subseteq M$.)

Proof   We will prove the first formula; the second is similar. Let us temporarily write ψ_α for ψ_{s_α}. We know

$$\psi_\alpha(e) = 0 \qquad\qquad (*)$$

by Property (i) (that is, $p_e \notin Y(s_\alpha)$). For any simple root $\beta \neq \alpha$, we have $A_\beta \psi_\alpha = 0$ by Property (iii), so

$$\psi_\alpha(w s_\beta) = \psi_\alpha(w). \qquad\qquad (**)$$

Similarly, $A_\alpha \psi_\alpha = \psi_e = 1$, so

$$\psi_\alpha(w s_\alpha) = \psi_\alpha(w) + w(\alpha). \qquad\qquad (***)$$

The function $\psi_\alpha \colon W \to \Lambda$ is uniquely determined by $(*)$, $(**)$, and $(***)$. (By $(**)$ and $(***)$, the difference of two such functions would take the same values at w and $w s_\beta$ for every simple root β, so it must be constant; by $(*)$, the constant is zero.)

On the other hand, the function $f_\alpha(w) = \varpi_\alpha - w(\varpi_\alpha)$ clearly satisfies $(*)$, and it satisfies $(**)$ and $(***)$ since

$$f_\alpha(ws_\beta) = \varpi_\alpha - ws_\beta(\varpi_\alpha)$$
$$= \varpi_\alpha - w(\varpi_\alpha - \delta_{\alpha\beta}\beta)$$
$$= f_\alpha(w) + \delta_{\alpha\beta}w(\beta).$$

It follows that $f_\alpha = \psi_\alpha$, as claimed. $\qquad\qquad\qquad\qquad\square$

16.3   Difference operators and line bundles

In §15.6, we saw G-equivariant line bundles

$$\mathcal{L}_\lambda = G \times^B \mathbb{C}_\lambda$$

for each character $\lambda \in M$, inducing a homomorphism

$$b \colon \Lambda = \mathrm{Sym}^* M \to H_T^*(G/B)$$

of graded rings. (There we used $\mathbb{Q}$ coefficients, but the definition of b is the same for $\mathbb{Z}$ coefficients.) And W acts on Λ, induced by its action on the character group M.

Exercise 16.3.1. Show that the fiber of $\mathcal{L}_\lambda$ at the T-fixed point p_w is the representation $\mathbb{C}_{w(\lambda)}$, so $c_1^T(\mathcal{L}_\lambda)|_w = w(\lambda)$. Conclude that the homomorphism $b \colon \Lambda \to H_T^*(G/B)$ is injective.

Exercise 16.3.2. Show that

$$c_1^T(\mathcal{L}_\lambda) = \sum_{\alpha \in \Delta} -\langle \lambda, \alpha^\vee \rangle y(s_\alpha) + \lambda$$

for any $\lambda \in M$.

We have seen operators D_α on $H_T^*(G/B)$ and A_α on $F(W, Q)$. The "classical" divided difference operators on Λ are defined by

$$\partial_\alpha(f) = \frac{s_\alpha(f) - f}{\alpha} \tag{16.6}$$

for each simple root α.

Exercise 16.3.3. The operators ∂_α also satisfy a Leibniz rule:

$$\partial_\alpha(f \cdot g) = \partial_\alpha(f) \cdot g + (s_\alpha f) \cdot \partial_\alpha(g)$$

for any $f, g \in \Lambda$.

Proposition 16.3.4.   *For any $f \in \Lambda$, we have $D_\alpha b(f) = b(\partial_\alpha f)$. That is, the diagram*

$$
\begin{array}{ccccc}
\Lambda & \overset{b}{\hookrightarrow} & H_T^*(G/B) & \longrightarrow & F(W,Q) \\
\Big\downarrow{\partial_\alpha} & & \Big\downarrow{D_\alpha} & & \Big\downarrow{A_\alpha} \\
\Lambda & \overset{b}{\hookrightarrow} & H_T^*(G/B) & \longrightarrow & F(W,Q)
\end{array}
$$

commutes.

Proof It suffices to show that the outer rectangle commutes, since we already know the right square does (Proposition 16.2.2). That is, writing $b': \Lambda \to F(W,Q)$ for the composition, we must show $A_\alpha b'(f) = b'(\partial_\alpha f)$ for any $f \in \Lambda$.

First consider the case $f = \lambda \in M$. By Exercise 16.3.1, the corresponding function $b'(\lambda) \in F(W,Q)$ is $w \mapsto w(\lambda)$. The operator A_α takes this to the constant function sending w to

$$
\frac{ws_\alpha(\lambda) - w(\lambda)}{w(\alpha)} = -\langle \lambda, \alpha^\vee \rangle.
$$

On the other hand,

$$
\partial_\alpha(\lambda) = \frac{s_\alpha(\lambda) - \lambda}{\alpha} = -\langle \lambda, \alpha^\vee \rangle,
$$

so the claim is true in this case.

For the general case, we use induction on degree: since all maps are additive, it suffices to prove that knowing the asserted formula for λ and f implies it for λf. For this, we apply the Leibniz formulas

$$
\partial_\alpha(\lambda f) = \partial_\alpha(\lambda)\, f + s_\alpha(\lambda)\, \partial_\alpha(f)
$$

and

$$
A_\alpha(\lambda \psi) = A_\alpha(\lambda)\, \psi + s_\alpha(\lambda)\, A_\alpha(\psi),
$$

together with the fact that b' respects the action of W: for $v \in W$ and $f \in \Lambda$, one computes $v \cdot b'(f) = b'(v \cdot f)$. $\square$

It follows that the operators ∂_α satisfy the same properties as D_α and A_α. For a sequence of simple roots $(\alpha_1, \ldots, \alpha_\ell)$, the composition $\partial_{\alpha_1} \circ \cdots \circ \partial_{\alpha_\ell}$ depends only on $v = s_{\alpha_1} \cdots s_{\alpha_\ell}$ if $\ell(v) = \ell$ (i.e., if the word is reduced), and vanishes otherwise. Writing

$$\partial_v = \partial_{\alpha_1} \circ \cdots \circ \partial_{\alpha_\ell}$$

for any reduced word $(\alpha_1, \ldots, \alpha_\ell)$, we have

$$\partial_u \circ \partial_v = \begin{cases} \partial_{uv} & \text{if } \ell(uv) = \ell(u) + \ell(v); \\ 0 & \text{otherwise.} \end{cases}$$

Example 16.3.5. In terms of the W-action on $H_T^*(G/B)$ described in the last section, we have

$$D_\alpha(x) = \frac{s_\alpha \cdot x - x}{b(\alpha)}.$$

In particular, taking $x = x(s_\alpha)$ and using $D_\alpha(x(e)) = x(s_\alpha)$, we see

$$s_\alpha \cdot x(e) = -x(e).$$

Since $x(e)$ is the class of a point, this means that s_α cannot come from any holomorphic action on G/B.

16.4   The right W-action

In Proposition 16.2.3, we have seen one way in which W acts on $H_T^*(G/B)$, equivariantly for the inclusion $H_T^*(G/B) \hookrightarrow F(W, Q)$ and the homomorphism $b \colon \Lambda \to H_T^*(G/B)$. This action is by Λ-algebra automorphisms of $H_T^*(G/B)$, but as noted in Example 16.3.5, it does not come from an algebraic action of W on G/B.

In fact, this W-action is the same as the one induced by the action on $G/B = K/S$ discussed in §15.6. We can see this algebraically, using the projection $G/T \to G/B$. This is a G-equivariant fiber bundle with fibers isomorphic to $B/T \cong U \cong \mathbb{A}^N$, so the pullback maps

$$H_G^*(G/B) \to H_G^*(G/T) \quad \text{and} \quad H_T^*(G/B) \to H_T^*(G/T)$$

are isomorphisms. The Weyl group $W = N(T)/T$ acts naturally on G/T by right multiplication, so $gT \cdot \dot{w} = g\dot{w}T$. (Choosing a lift $\dot{w} \in N_K(S) \subset N_G(T)$, this evidently restricts to the right action of W on K/S from Chapter 15.) The right action of W commutes with the left action of G, so it induces left actions on $\Lambda = H_G^*(G/B) = H_G^*(G/T)$ and $H_T^*(G/B) = H_T^*(G/T)$.

Another way of realizing the same action will be useful. For $w \in W$, let $B^w = \dot{w}B\dot{w}^{-1} \subseteq G$. This is the stabilizer of the point $p_w \in G/B$. There is a canonical G-equivariant isomorphism Φ^w fitting into a commuting diagram

$$
\begin{array}{ccc}
G/T & \xrightarrow{\cdot w^{-1}} & G/T \\
\downarrow & & \downarrow \\
G/B & \xrightarrow[\Phi^w]{\sim} & G/B^w,
\end{array}
\qquad (16.7)
$$

defined by $\Phi^w(gB) = g\dot{w}^{-1}B^w$.

Lemma 16.4.1.   *The diagram*

$$
\begin{array}{ccc}
H_T^*(G/B^w) & \hookrightarrow & F(W,Q) \\
(\Phi^w)^* \downarrow & & \downarrow w\cdot \\
H_T^*(G/B) & \hookrightarrow & F(W,Q)
\end{array}
$$

commutes, where the right vertical arrow is $\psi \mapsto w\cdot\psi$, with the W-action as in §16.2.

Proof For $v \in W$, let $p_v^w = vB^w \in G/B^w$ be the corresponding T-fixed point in G/B^w. It suffices to consider the fixed-point classes $[p_v^w]^T \in H_T^*(G/B^w)$, since their images in $F(W,Q)$ form a Q-linear basis. On one hand, we have

$$
(\Phi^w)^*[p_v^w]^T = [(\Phi^w)^{-1}(p_v^w)]^T = [p_{vw}]^T.
$$

On the other hand, letting $\psi_{p_v^w} \in F(W,Q)$ be the function given by $\psi_{p_v^w}(u) = [p_v^w]^T|_u$, one computes $w \cdot \psi_{p_v^w} = \psi_{p_{vw}}$. □

Proposition 16.4.2.   *The W-action on $H_T^*(G/B)$ induced by the right action on G/T is the same as the action defined in Proposition 16.2.3. The W-action on $\Lambda = H_G^*(G/B)$ is the usual one induced by the action on M.*

Proof The first statement follows from Lemma 16.4.1, using the diagram (16.7). The second statement is Exercise 15.6.7. □

16.5 Left-Handed actions and operators

We will see another W-action on $H_T^*(G/B)$, along with corresponding operators $\widetilde{D}_\alpha$ and $\widetilde{A}_\alpha$. In the next section, we use a geometric construction to realize the symmetry between the two W-actions, and these two sets of operators.

This W-action comes from left multiplication of $N(T) \subset G$ on G/B. For $w \in W$, let

$$\tau_w \colon H_T^*(G/B) \to H_T^*(G/B)$$

be the pullback homomorphism corresponding to the automorphism $gB \mapsto \dot{w}^{-1} gB$ of G/B. (As usual, this is independent of the choice of representative $\dot{w} \in N(T)$.) This automorphism of G/B is equivariant with respect to the automorphism $t \mapsto \dot{w}^{-1} t \dot{w}$ of T. So τ_w is not a Λ-algebra automorphism; it intertwines the natural action of W on Λ. That is, for $f \in \Lambda$ and $x \in H_T^*(G/B)$,

$$\tau_w(f \cdot x) = w(f) \cdot \tau_w(x), \tag{16.8}$$

where W acts on $\Lambda = \mathrm{Sym}^* M$ via its usual action on M.

We will call this the *left-handed action* of W on $H_T^*(G/B)$, to distinguish it from the one defined by Proposition 16.2.3, which we call the *right-handed action*. The terminology will be justified in the next section, by Theorem 16.6.3.

The left-handed action on $H_T^*(G/B)$ extends to an action of W on $F(W, Q)$, as follows. For $\psi \in F(W, Q)$ and $w \in W$, define $\tau_w \psi$ by

$$(\tau_w \psi)(v) = w(\psi(w^{-1}v))$$

for each $v \in W$. So τ_w intertwines the Q-algebra structure of $F(W, Q)$ by the same twist (16.8), and preserves $F(W, \Lambda) \subseteq F(W, Q)$.

Exercise 16.5.1. Verify that the diagram

$$
\begin{array}{ccc}
H_T^*(G/B) & \hookrightarrow & F(W, Q) \\
{\scriptstyle \tau_w} \downarrow & & \downarrow {\scriptstyle \tau_w} \\
H_T^*(G/B) & \hookrightarrow & F(W, Q)
\end{array}
$$

commutes, so the action of τ_w on $F(W, Q)$ extends the one on $H_T^*(G/B)$.

For a simple root α, we saw formulas for the operators D_α and A_α in terms of the right-handed W-action. Operators $\widetilde{D}_\alpha$ and $\widetilde{A}_\alpha$ are defined analogously, using the left-handed action. Given elements $x \in H_T^*(G/B)$ and $\psi \in F(W, Q)$, let

$$\widetilde{D}_\alpha(x) = \frac{\tau_\alpha(x) - x}{\alpha}$$

and

$$\widetilde{A}_\alpha(\psi) = \frac{\tau_\alpha(\psi) - \psi}{\alpha},$$

where we write τ_α for τ_{s_α}.

The analogue of Proposition 16.3.4 holds, with $b \colon \Lambda \to H_T^*(G/B)$ replaced by $\rho^* \colon \Lambda \to H_T^*(G/B)$, where $\rho \colon G/B \to \mathrm{pt}$ (so ρ^* gives the usual Λ-algebra structure).

Proposition 16.5.2. *The diagram*

$$
\begin{array}{ccccc}
\Lambda & \xhookrightarrow{\rho^*} & H_T^*(G/B) & \longhookrightarrow & F(W, Q) \\
\Big\downarrow{\scriptstyle\partial_\alpha} & & \Big\downarrow{\scriptstyle\widetilde{D}_\alpha} & & \Big\downarrow{\scriptstyle\widetilde{A}_\alpha} \\
\Lambda & \xhookrightarrow{\rho^*} & H_T^*(G/B) & \longhookrightarrow & F(W, Q)
\end{array}
$$

commutes.

Exercise 16.5.3. Prove the above proposition, following the proof of Proposition 16.3.4.

The left-handed operators behave similarly to the right-handed operators. The action on Schubert bases is given by

$$\widetilde{D}_\alpha(x(w)) = \begin{cases} x(s_\alpha w) & \text{if } \ell(s_\alpha w) > \ell(w); \\ 0 & \text{otherwise,} \end{cases} \tag{16.9}$$

and

$$\widetilde{D}_\alpha(y(w)) = \begin{cases} y(s_\alpha w) & \text{if } \ell(s_\alpha w) < \ell(w); \\ 0 & \text{otherwise.} \end{cases} \tag{16.10}$$

It follows that the compositions $\widetilde{D}_{\alpha_1} \circ \cdots \circ \widetilde{D}_{\alpha_\ell}$ and $\widetilde{A}_{\alpha_1} \circ \cdots \circ \widetilde{A}_{\alpha_\ell}$ are zero when the word $(\alpha_1, \ldots, \alpha_\ell)$ is not reduced, and they depend only on $v = s_{\alpha_1} \cdots s_{\alpha_\ell}$ when this is a reduced expression. These facts may be proved directly, but we will deduce them from a geometric construction which illustrates the symmetry between left- and right-handed operators.

Example 16.5.4. Consider $G = SL_n$, so $G/B = Fl(\mathbb{C}^n)$, and a simple root $\alpha = t_k - t_{k+1}$. Using the identification $y(w) = [\Omega_w]^T$ and comparing (16.10) with Corollary 11.4.4, it follows that $-\widetilde{D}_\alpha$ corresponds to the "y-variable" divided difference operator ∂_k^y.

16.6 The convolution algebra

We have canonical isomorphisms

$$H_G^*(G/B) = H_B^*(\text{pt}) = \Lambda$$

and

$$H_G^*(G/B \times G/B) = H_B^*(G/B) = H_T^*(G/B),$$

where G acts diagonally on G/B. In particular, the identification with $H_G^*(G/B \times G/B)$ offers a more symmetric picture of $H_T^*(G/B)$, and also lets us exploit a general algebraic structure on the cohomology of a product.

Definition 16.6.1. Let X be a nonsingular compact variety with an action of a linear algebraic group G, so G acts diagonally on $X \times X$. The *convolution product* on $H_G^*(X \times X)$ is defined by

$$x \star y = pr_{13*}(pr_{12}^*x \cdot pr_{23}^*y)$$

for $x, y \in H_G^*(X \times X)$, where the maps $pr_{ij} \colon X \times X \times X \to X \times X$ are the projections.

The convolution product is not commutative, but it is associative and compatible with the Λ_G-algebra structure. (For $c \in \Lambda_G^p$ and $x \in H_G^q(X \times X)$, we have $c \cdot (x \star y) = (c \cdot x) \star y = (-1)^{pq} x \star (c \cdot y)$.)

Exercise 16.6.2. Let ω be the involution which exchanges factors of $X \times X$, that is, $\omega(a, b) = (b, a)$. Show that

$$\omega^*(\omega^*(x) \star y) = (-1)^{pq}\omega^*(y) \star x$$

for $x \in H_G^p(X \times X)$ and $y \in H_G^q(X \times X)$.

For $X = G/B$, the convolution algebra gives a symmetric description of the left- and right-handed actions and operations.

Theorem 16.6.3.

(a) *Let* $pr_i \colon G/B \times G/B \to G/B$ *and* $\rho \colon G/B \to$ pt *be the projections, and let* $b \colon \Lambda \to H_T^*(G/B)$ *be the homomorphism defined by* $b(\lambda) = c_1^T(\mathcal{L}_\lambda)$. *Under the usual identifications* $\Lambda = H_G^*(G/B)$ *and* $H_T^*(G/B) = H_G^*(G/B \times G/B)$, *we have*

$$\rho^* = pr_1^* \quad and \quad b = pr_2^*$$

as homomorphisms $\Lambda \to H_T^*(G/B)$.

(b) *For* $w \in W$ *and* $x \in H_T^*(G/B)$, *write* $\tau_w(x)$ *and* $w \cdot x$ *for the left- and right-handed actions, respectively. Let* $W \times W$ *act on* $G/T \times G/T$ *by right multiplication, and for* $w, v \in W$ *write* $(w \times v)(x)$ *for the induced action on* $H_G^*(G/T \times G/T)$. *Identifying* $H_T^*(G/B) = H_G^*(G/T \times G/T)$, *we have*

$$\tau_w(x) = (w \times 1)(x) \quad and \quad w \cdot x = (1 \times w)(x).$$

(c) *Let* $Z(\alpha) = \overline{G \cdot (p_e, p_{s_\alpha})} \subseteq G/B \times G/B$, *and let* $z(\alpha) = [Z(\alpha)]^G$ *in* $H_G^*(G/B \times G/B)$. *We have*

$$\widetilde{D}_\alpha(x) = z(\alpha) \star x \quad and \quad D_\alpha(x) = x \star z(\alpha)$$

for $x \in H_T^*(G/B)$.

Part (b) says the left-handed action of W on $H_T^*(G/B)$ comes from the first factor of $W \times W$, and the right-handed action comes from the second factor. Part (c) says the actions of $\widetilde{D}_\alpha$ and D_α on $H_T^*(G/B)$ are given by left and right convolution with $z(\alpha)$, respectively.

Proof The G-equivariant isomorphism $G \times^B G/B \cong G/B \times G/B$ establishes part (a).

For (b), one uses the commutative diagram

$$
\begin{array}{ccc}
\mathbb{E} \times^G (G/T \times G/T) & \xrightarrow{\ \sim\ } & \mathbb{E} \times^T (G/T) \\
\downarrow & & \downarrow \\
\mathbb{E} \times^G (G/T \times G/T) & \xrightarrow{\ \sim\ } & \mathbb{E} \times^T (G/T)
\end{array}
$$

given by

$$[e, g_1, g_2] \longmapsto [eg_1, g_1^{-1}g_2]$$

$$\downarrow \qquad\qquad \downarrow$$

$$[e, g_1\dot{w}, g_2] \longmapsto [eg_1\dot{w}, \dot{w}^{-1}g_1^{-1}g_2]$$

to see $\tau_w(x) = (1 \times w)(x)$. A similar diagram describes the right-handed W-action.

In (c), it follows from Proposition 16.1.2 that $D_\alpha(x) = x \star z(\alpha)$. For the left-handed operator $\widetilde{D}_\alpha$, we use the involution ω to exchange factors of $G/B \times G/B$. From (a), we have

$$\omega^*(b(f)) = \rho^*(f)$$

for any $f \in \Lambda$. From (b) we have

$$\omega(w \cdot \omega^*(x)) = \tau_w(x)$$

for any $x \in H_T^*(G/B)$ and $w \in W$. We compute:

$$(\omega^* \circ D_\alpha \circ \omega^*)(x) = \omega^* \left(\frac{s_\alpha \cdot (\omega^* x) - \omega^* x}{b(\alpha)} \right)$$

$$= \frac{\omega^*(s_\alpha \cdot (\omega^* x)) - \omega^*(\omega^* x)}{\omega^*(b(\alpha))}$$

$$= \frac{\tau_\alpha(x) - x}{\alpha}$$

$$= \widetilde{D}_\alpha(x).$$

So it suffices to observe

$$\omega^*(\omega^*(x) \star z(\alpha)) = z(\alpha) \star x,$$

which follows from Exercise 16.6.2 together with $\omega^* z(\alpha) = z(\alpha)$. $\qquad \square$

Exercise 16.6.4. For $w \in W$, let $z(w) = [Z(w)]^G$ in $H_G^*(G/B \times G/B)$. Show that $z(w) = x(w)$ under the identification of $H_G^*(G/B \times G/B)$ with $H_T^*(G/B)$. Conclude that $\omega^* x(w) = x(w^{-1})$.

It follows that the convolution algebra has the same structure as the nil-Hecke algebra of difference operators.

Corollary 16.6.5.   *Using $H_T^*(G/B) = H_G^*(G/B \times G/B)$, the convolution algebra structure is determined by*

$$x(u) \star x(v) = \begin{cases} x(uv) & \text{if } \ell(uv) = \ell(u) + \ell(v); \\ 0 & \text{otherwise,} \end{cases} \tag{$*$}$$

and

$$(f \cdot x) \star (f' \cdot x') = (f \cdot b(f') \cdot x) \star x' \tag{$**$}$$

for $f, f' \in \Lambda$ and $x, x' \in H_T^*(G/B)$.

Proof   The classes $x(w)$ form a basis for $H_T^*(G/B)$ over Λ, so the formulas $(*)$ and $(**)$ characterize the product. To see that $(*)$ holds, use Theorem 16.6.3(c) together with Exercise 16.6.4 and induction on $\ell(v)$. To see $(**)$ holds, write

$$(f \cdot x) \star (f' \cdot x') = (f \cdot x) \star (\rho^*(f')x')$$

and apply Theorem 16.6.3(a). $\square$

The formulas for $\widetilde{D}_\alpha$ in (16.9) and (16.10) are immediate consequences, as is the fact that $\widetilde{D}_v = \widetilde{D}_{\alpha_1} \circ \cdots \circ \widetilde{D}_{\alpha_\ell}$ is independent of the choice of reduced word for v: we have

$$\widetilde{D}_v(x) = x(v) \star x \quad \text{and} \quad D_v(x) = x \star x(v^{-1})$$

for any $v \in W$ and $x \in H_T^*(G/B)$.

Remark.   The nil-Hecke algebra acts as operators on $H_T^* X$, for any left G-space X. One can see this by identifying

$$H_T^* X = H_G^*(G/B \times X).$$

Then for any $z \in H_G^*(G/B \times G/B)$ and $x \in H_T^* X$, one has a convolution action defined by

$$z \star x = pr_{13*}(pr_{12}^*(z) \cdot pr_{23}^*(x)),$$

where pr_{ij} are the projections from $G/B \times G/B \times X$.

In particular, for each simple root α there are left-handed difference operators $\widetilde{D}_\alpha$ acting on $H_T^* X$ by $\widetilde{D}_\alpha(x) = z(\alpha) \star x$.

Notes

The operators D_α were introduced in the 1970s by Bernstein, Gelfand, and Gelfand (1973) and Demazure (1974) to study the ordinary cohomology, Chow groups, and K-theory of G/B. Variations for the equivariant setting were developed in the 1980s by Kostant and Kumar (1986), who made extensive use of localization; they were also studied by Arabia (1986; 1989). Many of the formulas appearing in this chapter may be found in their work. The term "nil-Hecke ring" was coined by Kostant and Kumar. It is also sometimes called the *nil-Coxeter algebra*; see, e.g., (Fomin and Stanley, 1994).

One can define operators $\mathscr{L}_w \colon H_T^*(G/B) \to \Lambda$ by $\mathscr{L}_w(x) = \rho_*(x \cdot x(w))$. Equivalently (via Poincaré duality), this operator picks out the coefficient of $y(w)$ in the expansion of x, that is,

$$x = \sum_{w \in W} \mathscr{L}_w(x) \cdot y(w).$$

Arabia considers these operators when x lies in the subalgebra $H_G^*(G/B)$, which is included in $H_T^*(G/B)$ via the homomorphism b. Using the identifications $H_T^*(G/B) = H_G^*(G/B \times G/B)$, $x(w) = z(w)$, $b = pr_2^*$, and $\rho_* = pr_{1*}$, we have

$$\mathscr{L}_w(x) = pr_{1*}(pr_2^* x \cdot z(w)).$$

By applying the involution ω, the same formula defines $D_w(x)$, for any x in $H_T^*(G/B)$ (Proposition 16.1.2). For $x \in H_G^*(G/B)$, it follows from Proposition 16.3.4 that $\mathscr{L}_w = \partial_w$ as endomorphisms of $H_G^*(G/B) = \Lambda$. A different proof of this formula is given in (Arabia, 1986).

The functions ψ_v are often denoted ξ^v, especially in the literature stemming from (Kostant and Kumar, 1986). They were studied in detail by Billey (1999).

The left- and right-handed actions of W on $H_T^*(G/B)$ are sometimes simply called left and right actions; as group actions they are both left actions. They were used by Knutson (2003) to obtain recurrence formulas for multiplying Schubert classes in $H_T^*(G/B)$. They are also referred to as the "dot" and "star" actions by Tymoczko (2008a), and by others following her, e.g. (Brosnan and Chow, 2018).

We learned much about computing Schubert classes via convolution from (unpublished) work of William Graham. Subtler versions of convolution algebras have many applications in representation theory. For instance, they are used extensively in the book by Chriss and Ginzburg (1997).

Hints for Exercises

Exercise 16.1.4. Apply the projection formula, using $y(u) = \pi_\alpha^* y[u]$, $y[v] = \pi_{\alpha*} y(v)$, and $\pi_\alpha^* y[v] = y(vs_\alpha)$.

Exercise 16.3.1. The first statement is an easy calculation: for $t \in T$ and $z \in \mathbb{C}_\lambda$,

$$t \cdot [\dot{w}, z] = [t\dot{w}, z] = [\dot{w}(\dot{w}^{-1}t\dot{w}), z] = [\dot{w}, \lambda(\dot{w}^{-1}t\dot{w})z] = [\dot{w}, w(\lambda)(t)z].$$

The composition $\Lambda \to H_T^*(G/B) \hookrightarrow F(W, \Lambda)$ is injective, since it sends any $f \in \Lambda$ to the function ψ_f defined by $\psi_f(w) = w(f)$.

Exercise 16.3.2. Since the fundamental weights are a basis, it suffices to do this for $\lambda = \varpi_\alpha$. Then one sees $c_1^T(\mathcal{L}_{\varpi_\alpha}) = \varpi_\alpha - y(s_\alpha)$ by restricting both sides to p_w, using the formula for $y(s_\alpha)|_w$ from Lemma 16.2.6.

Exercise 16.5.1. The fixed-point classes $[p_v]^T \in H_T^*(G/B)$ restrict to a Q-basis for $F(W, Q)$, so it suffices to work with these classes. The corresponding function ψ_{p_v} is given by $\psi_{p_v}(v) = \prod_{v(R^-)} \beta$ and $\psi_{p_v}(u) = 0$ for $u \neq v$. One checks $\tau_w([p_v]^T) = [p_{wv}]^T$, so the claim is that

$$\tau_w \cdot \psi_{p_v} = \psi_{p_{wv}},$$

which is a straightforward calculation.

Exercise 16.5.3. Using Exercise 16.5.1, the right square commutes, so it suffices to show the outer square commutes. For $f \in \Lambda$, let $\psi_f \in F(W, Q)$ be the constant function. Then one computes

$$
\begin{aligned}
(\widetilde{A}_\alpha \psi_f)(v) &= \frac{(\tau_\alpha \psi_f)(v) - \psi_f(v)}{\alpha} \\
&= \frac{s_\alpha(\psi_f(s_\alpha^{-1}v)) - \psi_f(v)}{\alpha} \\
&= \frac{s_\alpha(f) - f}{\alpha}
\end{aligned}
$$

which is $\psi_{\partial_\alpha(f)}$, as claimed.

Exercise 16.6.2. Use $pr_{12} = \omega \circ pr_{13} \circ \omega_{13}$, where ω_{13} exchanges the first and third factors of $X \times X \times X$. Since ω is an involution, we have $\omega^* = \omega_*$ on cohomology, and one can compute

$$
\begin{aligned}
\omega^*(\omega^*(x) \star y) &= \omega^* pr_{13*}(pr_{12}^*\omega^*(x) \cdot pr_{23}^*(y)) \\
&= (\omega \circ pr_{13})_*(\omega_{13}^* pr_{23}^*(x) \cdot \omega_{13}^* pr_{12}^*\omega^*(y)) \\
&= (\omega \circ pr_{13} \circ \omega_{13})_*(pr_{23}^*(x) \cdot pr_{12}^*\omega^*(y)) \\
&= pr_{13*}(pr_{23}^*(x) \cdot pr_{12}^*\omega^*(y)) \\
&= (-1)^{pq} pr_{13*}(pr_{12}^*\omega^*(y) \cdot pr_{23}^*(x)) \\
&= (-1)^{pq} \omega^*(y) \star x.
\end{aligned}
$$

17

Equivariant Homology

We have focused on cohomology, but in fact the perspective we emphasize – using finite-dimensional approximation spaces – is well suited to homology, and in particular, equivariant Borel–Moore homology and equivariant Chow groups. Here we introduce these constructions, as well as the related Segre classes and equivariant multiplicities. In the next chapter, we will apply properties of equivariant multiplicities to study singularities of Schubert varieties.

In this chapter, G is a complex linear algebraic group. When discussing localization and equivariant multiplicities, we will restrict attention to tori.

17.1 Equivariant Borel–Moore homology and Chow groups

Let $\mathbb{E} \to \mathbb{B}$ be a map of nonsingular algebraic varieties which is the principal G-bundle, with $\widetilde{H}^i \mathbb{E} = 0$ for $i < N$, so these serve as finite-dimensional approximation spaces for defining equivariant cohomology. In Chapter 3, we also saw relative groups for a G-invariant subspace $Y \subseteq X$, defined by $H_G^i(X, Y) = H^i(\mathbb{E} \times^G X, \mathbb{E} \times^G Y)$ for $i < N$. These have long exact sequences

$$\cdots \to H_G^i(X, Y) \to H_G^i(X) \to H_G^i(Y) \to H_G^{i+1}(X, Y) \to \cdots$$

coming from the corresponding sequence of the pair $(\mathbb{E} \times^G X, \mathbb{E} \times^G Y)$.

The homology groups most useful for algebraic varieties are the Borel–Moore homology groups, which we denote by $\overline{H}_i X$. When X is embedded as a closed subspace of an oriented manifold M of (real) dimension n, there are canonical isomorphisms

$$\overline{H}_i X = H^{n-i}(M, M \smallsetminus X).$$

Borel–Moore groups are covariant for proper maps $X \to Y$. For certain maps $f \colon X \to Y$, there are Gysin pullback homomorphisms $f^* \colon \overline{H}_i Y \to \overline{H}_{i+d} X$, where $d = \dim_{\mathbb{R}} X - \dim_{\mathbb{R}} Y$. For example, these exist if f is an open embedding, or more generally if it is a smooth morphism. When X is a compact algebraic variety, Borel–Moore homology agrees with singular homology. More about Borel–Moore homology may be found in Appendix A, §A.2.

Now suppose a G-space X can be embedded equivariantly as a closed invariant subspace of an oriented n-dimensional G-manifold. One expects the equivariant Borel–Moore homology groups of X to satisfy

$$\overline{H}_i^G(X) = H_G^{n-i}(M, M \smallsetminus X). \tag{17.1}$$

This means

$$\overline{H}_i^G(X) = H^{n-i}\big(\mathbb{E} \times^G M, (\mathbb{E} \times^G M) \smallsetminus (\mathbb{E} \times^G X)\big)$$

when $n - i < N$, so $i > n - N$. The space $\mathbb{E} \times^G M$ is an oriented manifold of dimension $n + \dim_{\mathbb{R}} \mathbb{B} = n + \dim_{\mathbb{R}} \mathbb{E} - \dim_{\mathbb{R}} G$, and it contains $\mathbb{E} \times^G X$ as a closed subspace. So the group on the right-hand side of the above expression is $\overline{H}_{n+\dim_{\mathbb{R}} \mathbb{B} - (n-i)}(\mathbb{E} \times^G X)$. This suggests the general definition, where we do not require X to be equivariantly embedded in a manifold.

Definition 17.1.1. Let X be an algebraic variety with (left) G-action, and let $\mathbb{E} \to \mathbb{B}$ be an approximation space as above. The *equivariant Borel–Moore homology groups* are defined as

$$\overline{H}_i^G(X) = \overline{H}_{i+\dim_{\mathbb{R}} \mathbb{B}}(\mathbb{E} \times^G X), \tag{17.2}$$

whenever $i > -N$.

These groups are typically nonzero for $i < 0$. For instance, we have

$$\overline{H}_i^G(\mathrm{pt}) = H_G^{-i}(\mathrm{pt}).$$

More generally, when X is a nonsingular variety of (complex) dimension n, there are canonical isomorphisms

$$\overline{H}_i^G(X) = H_G^{2n-i}(X)$$

for all i.

To compare groups for approximations $\mathbb{E} \to \mathbb{B}$ and $\mathbb{E}' \to \mathbb{B}'$, one uses Gysin pullback maps for smooth projections. Assume that $\widetilde{H}^i\mathbb{E} = \widetilde{H}^i\mathbb{E}' = 0$ for $i < N$. Writing $b = \dim_{\mathbb{R}} \mathbb{B}$ and $b' = \dim_{\mathbb{R}} \mathbb{B}'$, the maps

$$\overline{H}_{i+b}(\mathbb{E} \times^G X) \to \overline{H}_{i+b+b'+\dim_{\mathbb{R}} G}((\mathbb{E} \times \mathbb{E}') \times^G X) \leftarrow \overline{H}_{i+b'}(\mathbb{E}' \times^G X)$$

are isomorphisms for $i > -N$. To determine an element of $\overline{H}_i^G(X)$, one must give elements of each of these groups, which map to one another via these Gysin homomorphisms. For example, a k-dimensional G-invariant closed subvariety Z of a complex variety X determines a *fundamental class*

$$[Z]^G = [\mathbb{E} \times^G Z]^G \quad \text{in} \quad \overline{H}_{2k}^G(X) = \overline{H}_{2k+\dim_{\mathbb{R}} \mathbb{B}}(\mathbb{E} \times^G X).$$

The functorial properties of equivariant Borel–Moore groups are similar to those of the non-equivariant ones. A G-equivariant proper morphism $f \colon X \to Y$ determines pushforward homomorphisms $f_* \colon \overline{H}_i^G(X) \to \overline{H}_i^G(Y)$. There are cap product actions

$$H_G^i(X) \otimes \overline{H}_j^G(X) \to \overline{H}_{j-i}^G(X),$$

with the usual projection formula.

An advantage of working with Borel–Moore homology groups is that they fit into long exact sequences. The same is true in the equivariant setting. If $Z \subseteq X$ is a closed G-invariant subspace, with complement $U = X \smallsetminus Z$, there is a long exact sequence

$$\cdots \to \overline{H}_i^G(Z) \to \overline{H}_i^G(X) \to \overline{H}_i^G(U) \to \overline{H}_{i-1}^G(Z) \to \cdots.$$

(Using approximation spaces, this comes from the corresponding sequence for the closed subspace $\mathbb{E} \times^G Z \subseteq \mathbb{E} \times^G X$.)

Borel–Moore groups are not homotopy invariant. For example, with G acting on an n-dimensional vector space V via a representation, one has

$$\overline{H}_i^G(V) = \overline{H}_{i-2n}^G(\text{pt}) = H_G^{2n-i}(\text{pt})$$

for all i. (Using approximation spaces, this corresponds to the Thom isomorphism for the vector bundle $\mathbb{E} \times^G V \to \mathbb{B}$.) Via the cap product, this makes $\overline{H}_i^G(V)$ into a free module over Λ_G, generated by the class $[V]^G$.

The same ideas are used to define *equivariant Chow groups*, in the context of algebraic geometry over any ground field. One sets

$$A_i^G(X) = A_{i+b}(\mathbb{E} \times^G X), \qquad (17.3)$$

where $b = \dim \mathbb{B} = \dim \mathbb{E} - \dim G$, and the groups on the right are ordinary Chow groups. Here one takes a representation V of G, with $\mathbb{E} \subset V$ a Zariski-open subset on which G acts freely. The definition (17.3) is independent of $\mathbb{E}$ when $\text{codim}(V \smallsetminus \mathbb{E}, V)$ is sufficiently large. Indeed, for fixed i, the Gysin maps (smooth pullback)

$$A_{i+b}(\mathbb{E} \times^G X) \to A_{i+b+b'+\dim G}((\mathbb{E} \times \mathbb{E}') \times^G X) \leftarrow A_{i+b'}(\mathbb{E}' \times^G X)$$

are isomorphisms when $\text{codim}(V \smallsetminus \mathbb{E}, V)$ and $\text{codim}(V' \smallsetminus \mathbb{E}', V')$ are large enough.

As before, a G-invariant k-dimensional closed subvariety $Z \subseteq X$ determines a class $[Z]^G \in A_k^G(X)$. Equivariant Chow groups are functorial for proper equivariant maps $f \colon X \to Y$.

We have $A_i^G(X) = 0$ for $i > \dim(X)$, but as with the equivariant homology groups, the equivariant Chow groups can be nonzero for $i < 0$. For complex varieties, $\overline{H}_i^G(X) = 0$ for $i > 2\dim(X)$, and there are cycle class homomorphisms

$$A_i^G(X) \to \overline{H}_{2i}^G(X),$$

with the usual properties of the non-equivariant cycle class maps.

There is a cell decomposition lemma:

Proposition 17.1.2. *Suppose there is a filtration by G-invariant closed subsets*

$$\emptyset \subset X_0 \subset X_1 \subset \cdots \subset X_m = X$$

such that each $X_p \smallsetminus X_{p-1} = \coprod_j U_{p,j}$, with $U_{p,j} \cong \mathbb{C}^{n(p)}$. Then the classes $[\overline{U_{p,j}}]^G$ form a basis for $\overline{H}_*^G(X)$ as a Λ_G-module.

This generalizes what we have seen for equivariant cohomology (Proposition 4.7.1). The proof is similar, and also establishes the analogous fact for equivariant Chow groups $A_*^G(X)$.

Example 17.1.3. For $X = G/P$, there is a filtration by the B-invariant Schubert varieties, with X_p being the union of all $X[w]$ of dimension p. It follows that

$$A_*^T X = \bigoplus_{[w]} \Lambda_T \cdot [X[w]]^T \quad \text{and} \quad \overline{H}_*^T X = \bigoplus_{[w]} \Lambda_T \cdot [X[w]]^T,$$

and the cycle class map is an isomorphism. (In identifying $\Lambda_T = \operatorname{Sym}^* M$ with $A_T^*(\mathrm{pt})$, the characters have degree 1. In other words, the cycle class map $A_T^*(\mathrm{pt}) \xrightarrow{\sim} H_T^*(\mathrm{pt})$ simply doubles degrees in Λ_T.)

17.2 Segre classes

The usual construction of Segre classes extends directly to the equivariant setting. This class naturally lies in the product $\prod \overline{H}_i^G(X)$ (or $\prod A_i^G(X)$), which may be viewed as a completion of the direct sum $\overline{H}_*^G(X)$ (respectively, $A_*^G(X)$). We will use notation for Borel–Moore groups, but everything in this section applies to Chow groups, as well.

First we review some basic notions about cones. Let X be a scheme, and $\mathcal{S}_\bullet = \bigoplus \mathcal{S}_k$ a graded sheaf of $\mathcal{O}_X$-algebras. We will always assume $\mathcal{S}_0 \cong \mathcal{O}_X$, and that $\mathcal{S}_1$ is coherent and generates $\mathcal{S}_\bullet$ as an $\mathcal{O}_X$-algebra. The scheme

$$C = \operatorname{Spec} \mathcal{S}_\bullet \to X$$

is a *cone* over X. There is a natural "zero" section $X \to C$, defined by the quotient $\mathcal{S}_\bullet \to \mathcal{S}_0 = \mathcal{O}_X$.

For any cone $C = \operatorname{Spec} \mathcal{S}_\bullet$, we define a cone $C \oplus 1 = \operatorname{Spec} \mathcal{S}_\bullet[z]$, where $\mathcal{S}_\bullet[z]$ is graded so that the degree k piece is

$$(\mathcal{S}_\bullet[z])_k = \mathcal{S}_k \oplus \mathcal{S}_{k-1} \cdot z \oplus \mathcal{S}_{k-2} \cdot z^2 \oplus \cdots \oplus \mathcal{S}_0 \cdot z^k.$$

The *projective cone* associated to C is $\mathbb{P}(C) = \operatorname{Proj} \mathcal{S}_\bullet$, and the *projective completion* of C is $\mathbb{P}(C \oplus 1)$. Thus C is an open subset of $\mathbb{P}(C \oplus 1)$, and the section $X \to C \subseteq \mathbb{P}(C \oplus 1)$ corresponds to the "line" $0 \oplus 1 \hookrightarrow C \oplus 1$. As with any Proj, these projective cones come with universal line bundles $\mathcal{O}(1)$.

Suppose G acts on X and on $S_\bullet$, preserving the grading. This makes $C \to X$ a G-equivariant cone. The action extends to $C \oplus 1$, by letting G act trivially on the extra factor. We therefore obtain actions on $\mathbb{P}(C)$ and $\mathbb{P}(C \oplus 1)$, compatible with inclusions and projections, and making $\mathcal{O}(1)$ an equivariant line bundle.

Definition 17.2.1. Let $p \colon \mathbb{P}(C \oplus 1) \to X$ be the projection. The *equivariant Segre class* of C is the class

$$s^G(C) = p_* \left(\sum_{i \geq 0} c_1^G(\mathcal{O}(1))^i \frown [\mathbb{P}(C \oplus 1)]^G \right)$$

in $\prod \overline{H}_{2k}^G(X)$, where $\mathcal{O}(1)$ is the univeral line bundle on $\mathbb{P}(C \oplus 1)$.

When X has pure dimension n, one writes $s_k^G(C)$ for the component of $s^G(C)$ in $\overline{H}_{2n-2k}^G(X)$. In general, $s_k^G(C)$ is nonzero for arbitrarily large k.

Equivariant Segre classes may also be defined in terms of ordinary ones, as shown in the following exercise.

Exercise 17.2.2. Assume X is an algebraic variety of (pure) dimension n. Show that the classes

$$s_k(\mathbb{E} \times^G C) \in \overline{H}_{2n-2k+\dim_{\mathbb{R}} \mathbb{B}}(\mathbb{E} \times^G X)$$

are compatible with the Gysin pullbacks for different choices of $\mathbb{E} \times \mathbb{B}$, and therefore they determine an element of $\overline{H}_{2n-2k}^G(X)$. Show that this class is equal to the class $s_k^G(C)$ defined above.

Basic properties of equivariant Segre classes of cones follow from the corresponding ones for non-equivariant Segre classes. For example, suppose E is a G-equivariant vector bundle on a variety X, and $C \to X$ is an equivariant cone. Writing $C \oplus E$ for the cone $C \times_X E$, we have

$$s^G(C \oplus E) = c^G(E)^{-1} \frown s^G(C) \tag{17.4}$$

in $\prod \overline{H}_i^G(X)$. In particular,

$$s^G(E) = c^G(E)^{-1} \frown [X]^G$$

and

$$s^G(C \oplus 1) = s^G(C).$$

Example 17.2.3. Consider the case $X = \mathrm{pt}$. Let $C = V$ be a vector space, with a torus T acting by characters $\chi_1, \ldots, \chi_n$. Then

$$s^T(V) = \frac{1}{c^T(V)} = \frac{1}{(1 + \chi_1) \cdots (1 + \chi_n)} \tag{17.5}$$

in the completed ring $\prod H^k_G(\mathrm{pt}) = \prod \overline{H}^G_k(\mathrm{pt})$. Using the definition and a localization computation on $\mathbb{P}^n$, this becomes a nontrivial identity:

$$\sum_{k=1}^{n+1} \sum_{i \geq 0} \frac{(-\chi_k)^i}{\prod_{j \neq k}(\chi_j - \chi_k)} = \frac{1}{\prod_{k=1}^n (1 + \chi_k)}, \tag{17.6}$$

where $\chi_{n+1} = 0$.

When $X \subseteq Y$ is a G-invariant subvariety (or subscheme), with ideal sheaf $I \subseteq \mathcal{O}_Y$, the normal cone

$$C_X Y = \mathrm{Spec} \left(\bigoplus_{i \geq 0} I^i / I^{i+1} \right)$$

is naturally a G-equivariant cone on X. The *equivariant Segre class of X in Y* is defined as

$$s^G(X, Y) = s^G(C_X Y).$$

Basic properties of this class follow from the non-euqivariant case. For example, let $f \colon Y' \to Y$ be a G-equivariant morphism, inducing a fiber square

$$\begin{array}{ccc} X' & \longrightarrow & Y' \\ {\scriptstyle g}\downarrow & & \downarrow{\scriptstyle f} \\ X & \longrightarrow & Y, \end{array}$$

so $X' = f^{-1}X$. If f is proper and surjective, then

$$g_* s^G(X', Y') = d \cdot s^G(X, Y), \tag{17.7}$$

where d is the degree of $Y' \to Y$ (so $d = 0$ if $\dim Y' > \dim Y$). If f is a smooth morphism, then

$$g^* s^G(X, Y) = s^G(X', Y'). \tag{17.8}$$

(This holds more generally when f is flat.)

17.3 Localization

There are general localization theorems for the inclusion of the fixed point set. These are most useful for torus actions. We will focus on that case here, and prove a simple version which is sufficient for our purposes.

Let T be a torus acting on a variety (or pure-dimensional scheme) X. The fixed locus $X^T \subseteq X$ is closed, so the inclusion $\iota \colon X^T \hookrightarrow X$ induces a homomorphism

$$\iota_* \colon \overline{H}_*^T(X^T) \to \overline{H}_*^T(X).$$

Theorem 17.3.1. *Let $S \subset \Lambda$ be the multiplicative set generated by all nonzero characters in M. Then*

$$S^{-1}\iota_* \colon S^{-1}\overline{H}_*^T(X^T) \to S^{-1}\overline{H}_*^T(X)$$

is an isomorphism of $S^{-1}\Lambda$-modules.

Proof Applying the long exact sequence for the closed subset $X^T \subseteq X$, we have

$$\cdots \to \overline{H}_*^T(X^T) \to \overline{H}_*^T(X) \to \overline{H}_*^T(X \smallsetminus X^T) \to \overline{H}_{*-1}^T(X^T) \to \cdots,$$

so it is equivalent to show that $S^{-1}\overline{H}_*^T(X) = 0$ whenever $X^T = \emptyset$. To prove this, we can use induction on $\dim X$ and the long exact sequence again to reduce to the case where X is nonsingular. Since

$$S^{-1}\overline{H}_*^T(X) = S^{-1}H_G^{2\dim X - *}(X),$$

this case follows from the localization theorems we have already seen (Theorem 7.1.1). □

As before, in special situations one can be more specific about which characters are inverted. We will use a construction based on *specialization to the normal cone*. For G acting on X, with invariant subvariety Y, this leads to a specialization map

$$\sigma \colon \overline{H}_*^G(X) \to \overline{H}_*^G(C_Y X),$$

which is a homomorphism of Λ_G-modules, with

$$\sigma([X]^G) = [C_Y X]^G$$

and

$$\sigma([Y]^G) = [Y]^G,$$

where $[Y]^G \in \overline{H}_*^G(C_Y X)$ is the class of the zero section. The construction and basic properties are reviewed in Appendix B.

In our setting, we have a torus T acting on X. Let $p \in X^T$ be a fixed point, with corresponding maximal ideal $\mathfrak{m} \subseteq \mathcal{O}_{X,x}$, so the Zariski tangent space $T_p X = (\mathfrak{m}/\mathfrak{m}^2)^\vee$ is a representation of T, and the tangent cone $C_p X = \mathrm{Spec}\left(\bigoplus_{i \geq 0} \mathfrak{m}^i/\mathfrak{m}^{i+1}\right)$ is a closed T-invariant subscheme $C_p X \subseteq T_p X$.

A fixed point $p \in X^T$ is *nondegenerate* if the top Chern class $c_{top}^T(T_p X)$ is nonzero; equivalently, all weights for the torus action on $T_p X$ are nonzero. Any nondegenerate fixed point is isolated. (If p is contained in a positive-dimensional fixed subvariety of X, then $T_p X$ contains a copy of the trivial representation.) The converse is not true in general, although it does hold when X is nonsingular, as we saw in Lemma 5.1.5.

Example 17.3.2. Consider T acting on $\mathbb{P}^3$ via characters $0, 0, \chi, -\chi$, for some nonzero character χ. Let $X = \{x_2^2 - x_3 x_4\} \subseteq \mathbb{P}^3$, so X is T-invariant. The fixed point $p = [1, 0, 0, 0]$ is isolated in X, but degenerate, since T acts on $T_p X$ by characters $0, \chi, -\chi$.

Proposition 17.3.3. *If X^T consists of finitely many nondegenerate points, the homomorphism*

$$\iota_* \colon \overline{H}_*^T(X^T) \to \overline{H}_*^T(X)$$

is injective, and becomes an isomorphism after localizing at any multiplicative set $S \subseteq \Lambda$ which contains $c_{top}^T(T_p X)$ for all $p \in X^T$.

Proof The argument for injectivity is similar to the one we gave in Theorem 5.1.8. Writing $N = \#X^T$, consider the maps

$$\Lambda^{\oplus N} = \overline{H}_*^T(X^T) \xrightarrow{\iota_*} \overline{H}_*^T(X) \xrightarrow{\sigma} \bigoplus_{p \in X^T} \overline{H}_*^T(C_p X) \to \bigoplus_{p \in X^T} \overline{H}_*^T(T_p X) = \Lambda^{\oplus N}.$$

The composition $\sigma \circ \iota_*$ is diagonal. For $p \in X^T$, with $\dim T_p X = m$, the image of $[p]^T \in \overline{H}_0^T(X^T)$ in $\overline{H}_0^T(T_p X) = H_T^{2m}(\mathrm{pt})$ is the top Chern class $c_m^T(T_p X)$. Since each of these Chern classes is assumed to be nonzero, the composed map $\Lambda^{\oplus N} \to \Lambda^{\oplus N}$ is injective, and it follows that ι_* is, too.

To see $S^{-1}\iota_*$ is an isomorphism when S contains each $c_{top}^T(T_p X)$, one can again argue as in Theorem 5.1.8, since we already know $\mathrm{rk}\,\overline{H}_*^T X = \#X^T$ by Theorem 17.3.1. $\qquad\square$

Exercise 17.3.4. Let $T = \mathbb{C}^*$ act on $\mathbb{P}^1$ with character t, where t is coordinate on T, and let X be the nodal curve obtained by identifying the points 0 and ∞. (This was considered in Exercise 3.5.1 and Exercise 7.3.6.)

(i) Show that $\overline{H}_*^T(X)$ is a free Λ-module with generator $[X]^T$ in $\overline{H}_2^T(X)$.

(ii) Let $p \in X$ be the node. The tangent cone $C_p X = L_1 \cup L_2$ is a union of two lines; show that $\overline{H}_*^T(C_p X)$ is generated by the elements $[L_1]^T$ and $[L_2]^T$ over $\Lambda = \mathbb{Z}[t]$, with the relation $t \cdot ([L_1]^T - [L_2]^T) = 0$.

(iii) Show that the composition
$$\Lambda = \overline{H}_*^T(p) \xrightarrow{\iota_*} \overline{H}_*^T(X) \xrightarrow{\sigma} \overline{H}_*^T(C_p X) \to \overline{H}_*^T(T_p X) = \Lambda$$
sends 1 to $-t^2$.

17.4   Equivariant multiplicities

Some classes appearing in localization theorems provide a useful measure of singularities. In this section we will consider a nondegenerate fixed point $p \in X^T$. Let $n = \dim X$ and $m = \dim T_p X$, so $m \geq n$, with equality if and only if p is a nonsingular point.

There is a *local class* $\eta_p X \in \Lambda$, defined by
$$\eta_p X = [C_p X]^T$$

in $\overline{H}_{2n}^T(T_p X) = H_T^{2m-2n}(T_p X) = H_T^{2m-2n}(\mathrm{pt})$. From the definition, $\eta_p X = 1$ if and only if p is nonsingular. (The local class is defined for any fixed point, but it may be zero if the fixed point is degenerate.)

For any class $\alpha \in \prod_i H_T^i(\mathrm{pt})$, we will write $\{\alpha\}^k$ for the component of α lying in $\Lambda^{2k} = H_T^{2k}(\mathrm{pt})$.

Proposition 17.4.1. *Suppose X is a T-invariant subvariety of an N-dimensional variety V, of codimension $k = N - n$. Assume the point $p \in X \subseteq V$ is nonsingular in V. Writing $\iota_p \colon \{p\} \hookrightarrow V$ for the inclusion, we have*

$$\iota_p^*([X]^T) = c_{N-m}^T(T_pV/T_pX) \cdot \eta_p^T X = \{c^T(T_pV) \frown s^T(p, X)\}^k,$$

in $H_T^{2k}(p) = \overline{H}_{-2k}^T(p)$, where $m = \dim T_pX$.

Proof This comes from general intersection theory, where one computes a pullback by deformation to the normal cone. The class of $C_pX \subseteq T_pV$ in $H_T^*(T_pV) = \Lambda$ is

$$[C_pX]^T = c_{N-m}^T(T_pV/T_pX) \cdot \eta_p^T X,$$

using the self-intersection formula for $C_pX \subseteq T_pX \subseteq T_pV$. Now we apply the "basic construction" of intersection theory to the situation

$$
\begin{array}{ccc}
\mathbb{E} \times^T \{p\} & \longhookrightarrow & \mathbb{E} \times^T X \\
\| & & \downarrow \\
\mathbb{E} \times^T \{p\} & \longhookrightarrow & \mathbb{E} \times^T V,
\end{array}
$$

noting that the inclusion $\mathbb{E} \times^T \{p\} \hookrightarrow \mathbb{E} \times^T V$ is a regular embedding (since $\{p\} \hookrightarrow V$ is), and the normal cone to $\mathbb{E} \times^T \{p\}$ in $\mathbb{E} \times^T X$ is canonically identified with $\mathbb{E} \times^T C_pX$. $\square$

Remark 17.4.2. The same proposition holds more generally, for the inclusion $\iota \colon Y \hookrightarrow V$ of a nonsingular invariant subvariety. We have

$$\iota^*([X]^T) = [C_{X \cap Y} X]^T$$

in $H_T^{2k}(N_{Y/V}) = H_T^{2k}(Y)$. One can also work with any linear algebraic group G in place of the torus T, using the same arguments.

Now we come to the main notion of this section.

Definition 17.4.3. Let $p \in X^T$ be a nondegenerate fixed point. The *equivariant multiplicity* of p in X is the element

$$\varepsilon_p^T(X) = \frac{\eta_p^T(X)}{c_m^T(T_pX)}$$

of $S^{-1}\Lambda$, for any homogeneous multiplicative set $S \subseteq \Lambda$ containing $c_m^T(T_pX)$.

The main properties of equivariant multiplicities are summarized as follows.

Proposition 17.4.4. *Let $p \in X$ be a nondegenerate fixed point.*

(i) *The equivariant multiplicity $\varepsilon_p^T(X)$ is a homogeneous element of degree $-\dim X$ in $S^{-1}\Lambda$, and it lies in the subring $\Lambda[1/c_m^T$ $(T_pX)]$.*

(ii) *The point p is nonsingular if and only if $\varepsilon_p^T(X) = 1/c_m^T(T_pX)$.*

(iii) *When X is an invariant subvariety of codimension k in an N-dimensional vector space V, we have*

$$\varepsilon_p^T(X) = \frac{\{c^T(V) \frown s^T(p, X)\}^k}{c_N^T(V)}.$$

(iv) *For the inclusion $\{p\} \hookrightarrow X$, the fundamental class of X specializes as*

$$[X]^T \mapsto \varepsilon_p^T(X) \frown [p]^T,$$

under the composition

$$\overline{H}_*^T(X) \xrightarrow{\sigma} \overline{H}_*^T(C_pX) \to \overline{H}_*^T(T_pX) = H_T^{2m-*}(\mathrm{pt}).$$

(v) *Assume all fixed points of X are nondegenerate. We have*

$$[X]^T = \sum_{p \in X^T} \varepsilon_p^T(X) \frown [p]^T$$

under the localization isomorphism $S^{-1}\overline{H}_*^T(X) \cong S^{-1}\overline{H}_*^T(X^T)$.

(vi) *Let $f \colon X' \to X$ be an equivariant proper surjective morphism of degree d, and assume all fixed points of X and X' are nondegenerate. Then*

$$d \cdot \varepsilon_p^T(X) = \sum_{p' \in f^{-1}(p)^T} \varepsilon_{p'}^T(X').$$

Proof Most of these are very easy. Properties (i) and (ii) are immediate from the definition, and (iii) is a special case of Proposition 17.4.1.

To prove (iv), we may replace X by C_pX, since $\sigma([X]^T) = [C_pX]^T$ and $\varepsilon_p^T(X) = \varepsilon_p^T(C_pX)$. Now (iv) follows from (iii), using $V = T_pX$.

To prove (v), recall that the classes $[p]^T$ form a basis for $S^{-1}\overline{H}_*^T X$ over $S^{-1}\Lambda$, by the localization theorem. Now apply (iv), using the fact that $[p]^T$ maps to $[p]^T$ under the composition

$$S^{-1}\overline{H}_*^T(p) \to S^{-1}\overline{H}_*^T(X) \to S^{-1}\overline{H}_*^T(C_pX) \to S^{-1}\overline{H}_*^T(T_pX),$$

as in the proof of Proposition 17.3.3.

For (vi), we have $f_*[X']^T = d \cdot [X]^T$, and the diagram

$$
\begin{array}{ccc}
\overline{H}_*^T(X'^T) & \longrightarrow & \overline{H}_*^T(X') \\
\downarrow & & \downarrow f_* \\
\overline{H}_*^T(X^T) & \longrightarrow & \overline{H}_*^T(X)
\end{array}
$$

commutes. The statement follows from (v), applied to X and X'. $\qquad\square$

When X is equivariantly embedded in an N-dimensional nonsingular variety V, with $[X]^T \in H_T^{2N-2n}(V)$, one can characterize the equivariant multiplicity of $p \in X$ in terms of the restriction homomorphism, by

$$\varepsilon_p^T(X) = \frac{[X]^T|_p}{c_N^T(T_pV)}. \tag{17.9}$$

(Use $[X]^T|_p = \iota_p^*[X]^T = c_N^T(T_pV) \cdot \varepsilon_p^T(X)$, from Proposition 17.4.1.) This leads to a characterization of $\varepsilon_p^T(X)$ as the coefficient of $[p]^T$ in $[X]^T$.

Corollary 17.4.5. *For an invariant subvariety X of a nonsingular variety V, we have*

$$[X]^T = \sum_{p \in V^T} \varepsilon_p^T(X)\,[p]^T$$

in H_T^*V.

The local class η_pX is a polynomial in Λ, homogeneous of degree $m-n$. Similarly, when $X \subseteq V$, the restriction $[X]^T|_p$ is a polynomial of degree $N - n$. The former is intrinsic to X, while the latter depends on an embedding in V; however the two polynomials differ only by the factor $c_{N-m}^T(T_pV/T_pX)$, as we have seen.

In Chapter 18, we will use equivariant multiplicities to prove a nonsingularity criterion for Schubert varieties. However, it is not always easy to identify singular points from the shape of $\varepsilon_p^T(X)$.

Exercise 17.4.6. Let T act on $\mathbb{P}^2$ via distinct characters χ_1, χ_2, χ_3, with $\chi_3 = 2\chi_1 + \chi_2$. The cuspidal curve $X = \{x_1^2 x_2 - x_3^3 = 0\}$ is T-invariant. Show that

$$\varepsilon_{p_1}^T(X) = \frac{1}{\chi_3 - \chi_1} \quad \text{and} \quad \varepsilon_{p_2}^T(X) = \frac{1}{\chi_1 - \chi_2},$$

although $p_1 = [1, 0, 0]$ is nonsingular in X, and $p_2 = [0, 1, 0]$ is singular. Verify that

$$[X]^T = \varepsilon_{p_1}^T(X)\,[p_1]^T + \varepsilon_{p_2}^T(X)\,[p_2]^T$$

in $H_T^*(\mathbb{P}^2)$.

Notes

As noted in Chapter 2, equivariant Chow groups were defined by Edidin and Graham. The same authors also defined equivariant Borel–Moore groups (Edidin and Graham, 1998, §2.8). These groups were further studied by Brion (2000).

In Chapter 2 we also noted that the space $\mathbb{E} \times^G X$ need not exist as a scheme; the general theory of equivariant Chow groups requires algebraic spaces, as in (Edidin and Graham, 1998). On the other hand, for a complex variety X, $\mathbb{E} \times^G X$ is always a complex-analytic space, and the construction of Borel–Moore groups for such spaces presents no special difficulties (see Appendix A).

Variations on equivariant multiplicities appear under various names in several areas of mathematics. For T acting linearly on an N-dimensional vector space V, an equivariant coherent sheaf corresponds to a graded R-module F, where $R = \mathbb{C}[x_1, \ldots, x_N]$. When the isotypic components F_χ (where T acts by the character χ) are all finite-dimensional, the generating function

$$\sum_\chi (\dim_{\mathbb{C}} F_\chi)\, e^\chi$$

can be written as a rational function in variables $e^{\chi_1}, \ldots, e^{\chi_N}$, and one obtains a polynomial by extracting the leading term of the numerator. In representation theory, polynomials arising this way are known as *character polynomials, Joseph polynomials,* or *equivariant Hilbert polynomials* (Joseph, 1984; Borho et al., 1989; Chriss and Ginzburg, 1997). In commutative algebra, this is often called a *multidegree*, especially when $F = R/I$ is the coordinate ring of an invariant subvariety $X \subseteq V$; in this case, it agrees with what we have called the local class, $\eta_p X$ (Miller and Sturmfels, 2005, §8). These polynomials are cases of Rossmann's *equivariant multiplicity*, which is defined for an equivariant coherent sheaf on a complex manifold (Rossmann, 1989).

Our version more closely follows a construction by Brion (1997b, §4), who defined equivariant multiplicities in Chow groups. A version of Rossmann's multiplicity for Chow groups is described in (Nyenhuis, 1993, §5.2).

The "basic construction" used in Proposition 17.4.1 is standard in intersection theory (Fulton, 1998, §6).

Hints for Exercises

Exercise 17.2.2. If C is a cone on Y, and $Y' \to Y$ is a morphism, there is a pullback cone $C' \to Y'$. This comes with a morphism $f \colon \mathbb{P}(C') \to \mathbb{P}(C)$, and the universal bundles are related by $\mathcal{O}_{\mathbb{P}(C')}(1) = f^*\mathcal{O}_{\mathbb{P}(C)}(1)$. Apply this to the cones $(\mathbb{E} \times \mathbb{E}') \times^G C \to \mathbb{E} \times^G C$.

Exercise 17.4.6. Writing $\zeta = c_1^T(\mathcal{O}(1))$, we have $[X]^T = 3\zeta + 3\chi_3$, since X is defined by the vanishing of an equivariant section of $\mathcal{O}(3) \otimes \mathbb{C}_{3\chi_3}$, so $\iota_{p_i}^*[X]^T = -3\chi_i + 3\chi_3$.

18

Bott–Samelson Varieties and Schubert Varieties

Schubert varieties in G/P admit explicit equivariant desingularizations by Bott–Samelson varieties. These are certain towers of $\mathbb{P}^1$-bundles, and their cohomology rings are relatively easy to compute.

In this chapter, we use the Bott–Samelson desingularization to obtain a positive formula for restricting a Schubert class to a fixed point. This, in turn, leads to a criterion for a point of a Schubert variety to be nonsingular.

18.1 Definitions, fixed points, and tangent spaces

Let $G \supset B \supset T$ be as usual: G is a semisimple (or reductive) group, with Borel subgroup B and maximal torus T. For each simple root α, we have a minimal parabolic subgroup P_α, and the corresponding projection of flag varieties is a $\mathbb{P}^1$-bundle, $G/B \to G/P_\alpha$. These spaces occur frequently in this chapter, so we will write

$$X = G/B \quad \text{and} \quad X_\alpha = G/P_\alpha$$

from now on.

For any sequence of simple roots $\underline{\alpha} = (\alpha_1, \ldots, \alpha_d)$, we have a *big Bott–Samelson variety* $Z(\underline{\alpha}) = Z(\alpha_1, \ldots, \alpha_d)$, defined by

$$Z(\underline{\alpha}) = X \times_{X_{\alpha_1}} X \times_{X_{\alpha_2}} \cdots \times_{X_{\alpha_d}} X.$$

Since each projection $X \to X_{\alpha_i}$ is a $\mathbb{P}^1$-bundle, $Z(\underline{\alpha})$ is a tower of $\mathbb{P}^1$-bundles over X. In particular, it is a nonsingular projective variety of dimension $\dim X + d$. The group G acts diagonally on $Z(\underline{\alpha})$, equivariantly for each projection $pr_i \colon Z(\underline{\alpha}) \to X$. (We index these projections from left to right by $0 \le i \le d$.)

Example 18.1.1. For $G = SL_n$, so $X = Fl(\mathbb{C}^n)$, a Bott–Samelson variety can be described as a sequence of flags, with the ith differing from the $(i-1)$st only in position j, if $\alpha_i = t_j - t_{j+1}$. That is,

$$Z(\underline{\alpha}) = \left\{ (F_\bullet^{(0)}, \ldots, F_\bullet^{(d)}) \,\middle|\, \begin{array}{l} E_k^{(i)} = E_k^{(i-1)} \text{ for all } k \ne j, \\ \text{where } \alpha_i = t_j - t_{j+1} \end{array} \right\}.$$

When $n = 3$, these can be represented as configurations of points and lines in $\mathbb{P}^2$. For instance, suppose $\alpha = t_1 - t_2$ and $\beta = t_2 - t_3$. Then a general point of $Z(\alpha, \beta, \alpha, \beta)$ looks like a quintuple of flags:

So from left to right, consecutive flags differ by moving the point, then the line, then the point, and finally the line again.

The T-fixed points of $Z(\underline{\alpha})$ are easily described. An $\underline{\alpha}$-*chain* (or simply *chain*) of elements of W is a sequence

$$\underline{v} = (v_0, v_1, \ldots, v_d)$$

such that for each i, either $v_i = v_{i-1}$ or $v_i = v_{i-1} \cdot s_{\alpha_i}$.

Exercise 18.1.2. Show that the T-fixed points of $Z(\underline{\alpha})$ are the $2^d \cdot |W|$ points

$$Z(\underline{\alpha})^T = \left\{ p_{\underline{v}} = (p_{v_0}, p_{v_1}, \ldots, p_{v_d}) \right\},$$

where each $\underline{v}$ is an α-chain.

The *(small) Bott–Samelson variety* is the fiber $X(\underline{\alpha}) = pr_0^{-1}(p_e)$, that is,

$$X(\underline{\alpha}) = \{p_e\} \times_{X_{\alpha_1}} X \times_{X_{\alpha_2}} \cdots \times_{X_{\alpha_d}} X.$$

The projection $X(\alpha_1, \ldots, \alpha_d) \to X(\alpha_1, \ldots, \alpha_{d-1})$ is a $\mathbb{P}^1$-bundle, so $X(\underline{\alpha})$ is a nonsingular projective variety of dimension d. Since p_e is fixed by B, the Bott–Samelson variety $X(\underline{\alpha})$ comes with an action of B (but not G, in general).

The Bott–Samelson variety $X(\underline{\alpha})$ has 2^d T-fixed points $p_{\underline{v}}$, for chains $\underline{v} = (e, v_1, \ldots, v_d)$. We will index these in two ways: using the chain $\underline{v}$, and using the subset $I = \{i_1 < \cdots < i_\ell\} \subseteq \{1, \ldots, d\}$ defined by

$$I = \Big\{ i \,\big|\, v_i = v_{i-1} \cdot s_{\alpha_i} \Big\}.$$

We often use the notation interchangeably, writing $p_{\underline{v}} = p_I$. Sometimes we write $I = I^{\underline{v}}$ and $\underline{v} = \underline{v}^I$ to indicate the bijection between chains and subsets.

For each subset $I \subseteq \{1, \ldots, d\}$, there is a B-invariant subvariety $X(I) \subseteq X(\underline{\alpha})$, defined by

$$X(I) = \Big\{ (x_1, \ldots, x_d) \in X(\underline{\alpha}) \,\big|\, x_j = x_{j-1} \text{ for } j \notin I \Big\}.$$

In fact, this is canonically isomorphic to another Bott–Samelson variety. Each subset $I = \{i_1 < \cdots < i_\ell\}$ corresponds to a subword $\underline{\alpha}(I) = (\alpha_{i_1}, \ldots, \alpha_{i_\ell})$, and we have

$$X(I) \cong X(\underline{\alpha}(I)).$$

(Use a diagonal embedding of $X^{\ell+1}$ in X^{d+1}.) Containment among these subvarieties corresponds to containment of subsets:

$$X(J) \subseteq X(I) \quad \text{iff} \quad J \subseteq I.$$

For example, $X(\{1, \ldots, d\}) = X(\underline{\alpha})$, and $X(\emptyset)$ is the point $p_\emptyset$.

Each $X(I)$ is the closure of a locally closed set $X(I)^\circ$, consisting of the points where $x_i \neq x_{i-1}$ for $i \in I$. In fact, these are cells.

Lemma 18.1.3.   *We have $X(I)^\circ \cong \mathbb{A}^\ell$, where $\ell = \#I$.*

Proof   It suffices to consider $I = \{1, \ldots, d\}$. Here one has the $\mathbb{P}^1$-bundle $X(\alpha_1, \ldots, \alpha_d) \to X(\alpha_1, \ldots, \alpha_{d-1})$. The complement of the locus where $x_{d-1} = x_d$ is an $\mathbb{A}^1$-bundle over $X(\alpha_1, \ldots, \alpha_{d-1})$, so the claim follows by induction on d. $\qquad\square$

The subvarieties $X(I)$ therefore determine a cell decomposition of $X(\underline{\alpha})$, and their classes $x(I) = [X(I)]^T$ form a basis for $H_T^* X(\underline{\alpha})$, as I varies over subsets of $\{1, \ldots, d\}$. It also follows that

$$p_J \in X(I) \quad \text{iff} \quad J \subseteq I.$$

We will need a description of the tangent spaces.

Lemma 18.1.4. *Let $\underline{v} = (e, v_1, \ldots, v_d)$ be an $\underline{\alpha}$-chain. The torus weights on $T_{p_{\underline{v}}} X(\underline{\alpha})$ are $\{-v_1(\alpha_1), \ldots, -v_d(\alpha_d)\}$.*

More generally, for $K \subseteq I$, with corresponding chains $\underline{v}^K$ and $\underline{v}^I$, the weights on $T_{p_K} X(I)$ are $-v_i^K(\alpha_i)$ for $i \in I$.

Proof We will find the weights at any fixed point of the big Bott–Samelson variety . For a chain $\underline{v} = (v_0, v_1, \ldots, v_d)$, consider the point $p = p_{\underline{v}} \in Z(\underline{\alpha})$. The tangent space to $Z(\underline{\alpha})$ at p is the fiber product of vector spaces

$$T_{p_0} X \underset{T_{p_{[1]}} X_{\alpha_1}}{\times} T_{p_1} X \underset{T_{p_{[2]}} X_{\alpha_2}}{\times} \cdots \underset{T_{p_{[d]}} X_{\alpha_d}}{\times} T_{p_d} X,$$

where we have written $p_i = p_{v_i} \in X$ and $p_{[i]} = p_{[v_i]} \in X_{\alpha_i}$ to economize on subscripts. (Note $[v_i] = [v_{i-1}]$ for each i, since $\underline{v}$ is an $\underline{\alpha}$-chain.) We have seen descriptions of each of these spaces in Chapter 15. The weights are $v_0(R^-)$, from the first factor, together with weights $-v_i(\alpha_i)$ for $1 \leq i \leq d$, since $\mathfrak{g}_{-v_i(\alpha_i)}$ is the kernel of $T_{p_i} X \to T_{p_{[i]}} X_{\alpha_i}$.

When $v_0 = e$, the variety $X(\underline{\alpha})$ is the fiber over p_e in the first factor, so the weights $R^- = v_0(R^-)$ are omitted, proving the first claim. The second claim follows from the first, using $X(I) \cong X(\alpha_{i_1}, \ldots, \alpha_{i_\ell})$. $\qquad\square$

18.2 Desingularizations of Schubert varieties

Let $f \colon X(\underline{\alpha}) \to X$ be the projection onto the last factor; that is, f is the restriction of $pr_d \colon Z(\underline{\alpha}) \to X$. For each $I \subseteq \{1, \ldots, d\}$, with corresponding $\underline{\alpha}$-chain $\underline{v} = (e, v_1, \ldots, v_d)$, we have $f(p_I) = p_{v_d}$. The subset I corresponds to the subword $(\alpha_{i_1}, \ldots, \alpha_{i_\ell})$ of $\underline{\alpha}$, and

$$v_d = s_{\alpha_{i_1}} \cdots s_{\alpha_{i_\ell}}.$$

Since f is proper and B-equivariant, $f(X(I))$ contains the Schubert variety $X(v_d) \subseteq X$. However, if $(\alpha_{i_1}, \ldots, \alpha_{i_\ell})$ is not a reduced word for v_d, the image of f may be larger.

Lemma 18.2.1. *Let $\underline{\alpha} = (\alpha_1, \ldots, \alpha_d)$ be a sequence of simple roots. The set of products $s_{\alpha_{i_1}} \cdots s_{\alpha_{i_\ell}}$ over subwords contains a unique maximum element $w(\underline{\alpha}) \in W$ in Bruhat order, and*

$$f(X(\underline{\alpha})) = X(w(\underline{\alpha})).$$

We have $w(\underline{\alpha}) = s_{\alpha_1} \cdots s_{\alpha_d}$ if the word $\underline{\alpha}$ is reduced.

Proof Since $X(\underline{\alpha})$ is irreducible, the image of the B-equivariant morphism $f \colon X(\underline{\alpha}) \to X$ must be some Schubert variety $X(w)$. It follows that $w = w(\underline{\alpha})$ satisfies the asserted properties. □

In fact, the maximal element $w(\underline{\alpha})$ can be easily computed. Let "$*$" be the associative product on W defined by

$$w * s_\alpha = \begin{cases} ws_\alpha & \text{if } \ell(ws_\alpha) > \ell(w); \\ w & \text{otherwise.} \end{cases}$$

This product is called the *Demazure product*.

Exercise 18.2.2.   Show that $w(\underline{\alpha}) = s_{\alpha_1} * \cdots * s_{\alpha_d}$, i.e., it is the Demazure product of reflections from $\underline{\alpha}$.

Lemma 18.2.3.   *The map $f \colon X(\underline{\alpha}) \to X(w)$ is birational if and only if α is a reduced word for $w = w(\underline{\alpha})$.*

Proof If $\underline{\alpha}$ is not a reduced word, then $w(\underline{\alpha})$ is the product of reflections for a proper subword, so it has length $\ell(w(\underline{\alpha})) < d$. In this case, f cannot be birational by dimension.

If $\underline{\alpha}$ is reduced, then $w = w(\underline{\alpha}) = s_{\alpha_1} \cdots s_{\alpha_d}$, and $f(p_{\{1,\ldots,d\}}) = p_w$. The map $f \colon X(\underline{\alpha})^\circ \to X(w)^\circ$ is B-equivariant, and therefore also equivariant for the subgroup $U(w) = \dot{w}U\dot{w}^{-1} \cap U$. Since the map $u \mapsto u \cdot p_w$ defines an isomorphism $U(w) \xrightarrow{\sim} X(w)^\circ$, it follows that $f \colon X(\underline{\alpha})^\circ \to X(w)^\circ$ is an isomorphism. □

For a reduced word $\underline{\alpha}$, one can also establish the birationality of $f \colon X(\underline{\alpha}) \to X(w)$ by examining tangent weights. The tangent space to $X(\underline{\alpha})$ at $p = p_{\{1,\ldots,d\}}$ has weights

$$\alpha_1, \, s_{\alpha_1}(\alpha_2), \ldots, \, s_{\alpha_1} \cdots s_{\alpha_{d-1}}(\alpha_d),$$

using Lemma 18.1.4, for $v_i = s_{\alpha_1} \cdots s_{\alpha_i}$. These are precisely the weights on $T_{p_w} X(w)$ (see Lemma 15.2.2).

Given a Schubert variety $X(w) \subseteq G/B$, one obtains a B-equivariant desingularization $f \colon X(\underline{\alpha}) \to X(w)$ by choosing a reduced word for w. For a parabolic subgroup P, the projection $G/B \to G/P$ maps $X(w^{\min})$ birationally onto $X[w]$, so we obtain desingularizations of these varieties, too.

Corollary 18.2.4. *For a Schubert variety $X[w] \subseteq G/B$, and any re-duced word $\underline{\alpha}$ for $w^{\min}$, one obtains a desingularization $X(\underline{\alpha}) \to X[w]$ by composing f with the projection $G/B \to G/P$.* □

These statements have evident analogues for the subvarieties $X(I) \subseteq X(\underline{\alpha})$. If I is a subset, with subword $\underline{\alpha}(I)$, we will write $w(I) = w(\underline{\alpha}(I))$ for the corresponding Demazure product.

Corollary 18.2.5. *Let I be a subset, and let $\underline{v} = (v_1, \dots, v_d)$ be the corresponding chain. The following are equivalent:*

(i) The map $X(I) \to X(w(I))$ is birational.

(ii) $w(I) = v_d$.

(iii) $\ell(v_d) = \#I$.

(iv) The subword $\underline{\alpha}(I)$ is a reduced word for v_d. □

Example 18.2.6. Let $\underline{\alpha} = (\alpha, \alpha)$, for some simple root α. Then $X(\underline{\alpha})$ is isomorphic to $\mathbb{P}^1 \times \mathbb{P}^1$. The Demazure product is $s_\alpha * s_\alpha = s_\alpha$, and the map $f \colon X(\alpha, \alpha) \to X(s_\alpha)$ is identified with the second projection $\mathbb{P}^1 \times \mathbb{P}^1 \to \mathbb{P}^1$. The subvarieties $X(I) = X(\underline{v})$ are

$$X(\{1,2\}) = X(s_\alpha, e) = X(\underline{\alpha}),$$
$$X(\{1\}) = X(s_\alpha, s_\alpha) = \delta(\mathbb{P}^1) \text{ (the diagonal in } \mathbb{P}^1 \times \mathbb{P}^1),$$
$$X(\{2\}) = X(e, s_\alpha) = \{p_e\} \times \mathbb{P}^1, \text{ and}$$
$$X(\emptyset) = X(e, e) = \{(p_e, p_e)\}.$$

While $X(\underline{\alpha})$ always has finitely many fixed points, it often has infinitely many invariant curves – even when $\underline{\alpha}$ is a reduced word.

Exercise 18.2.7. The following are equivalent, for a sequence of simple roots $\underline{\alpha} = (\alpha_1, \dots, \alpha_d)$:

(a) $X(\underline{\alpha})$ has finitely many T-curves.

(b) The roots $\alpha_1, \dots, \alpha_d$ are distinct.

(c) $X(\underline{\alpha})$ is a toric variety for the quotient of T whose character lattice has basis $\alpha_1, \dots, \alpha_d$.

(d) The map $f \colon X(\underline{\alpha}) \to X(w)$ is an isomorphism.

(Use the description of weights on tangent spaces.)

Another construction of the Bott–Samelson variety $X(\underline{\alpha})$ is sometimes useful.

Proposition 18.2.8.   *For a word $\underline{\alpha} = (\alpha_1, \ldots, \alpha_d)$, there is an isomorphism*

$$P_{\alpha_1} \times^B P_{\alpha_2} \times^B \cdots \times^B P_{\alpha_d}/B \to X(\underline{\alpha}),$$

given by $[p_1, \ldots, p_d] \mapsto (eB, p_1 B, p_1 p_2 B, \ldots, p_1 \cdots p_d B)$. This is B-equivariant, where B acts via left multiplication on P_{α_1}. The subvarieties $X(I) \subseteq X(\underline{\alpha})$ are identified with

$$X(I) = \{[p_1, \ldots, p_d] \mid p_i B = eB \text{ for } i \notin I\},$$

and the point p_I corresponds to $[\varepsilon_1, \ldots, \varepsilon_d]$, where $\varepsilon_i = \dot{e}$ for $i \in I$, and $\varepsilon_j = \dot{s}_{\alpha_j}$ for $j \notin I$.

Exercise 18.2.9.   Prove the proposition.

Remark 18.2.10.   Bott–Samelson varieties appear in the geometric construction of divided difference operators described in §16.1. Let $\underline{\alpha}$ be a reduced word for w. The big Bott–Samelson variety $Z(\underline{\alpha})$ maps birationally to the double Schubert variety

$$Z(w) = \overline{G \cdot (p_e, p_w)} \subseteq X \times X$$

via the projection $pr_0 \times pr_d$. Using Proposition 16.1.2, the operator $\mathrm{D}_{w^{-1}}$ on $H_T^* X$ is identified with $pr_{d*} pr_0^*$.

On the other hand, these projections factor as iterated $\mathbb{P}^1$-bundles, and the diagram

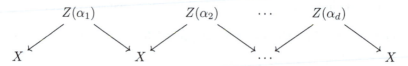

shows that $\mathrm{D}_{w^{-1}} = \mathrm{D}_{\alpha_\ell} \circ \cdots \circ \mathrm{D}_{\alpha_1}$ is independent of the choice of reduced word. One can also see this by restricting the diagram

to the fiber $pr_0^{-1}(p_e)$, obtaining

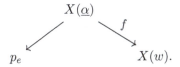

Since f is birational, we have

$$D_{w^{-1}}(x(e)) = pr_{d*}pr_0^*(x(e)) = f_*[X(\underline{\alpha})]^T = [X(w)]^T = x(w).$$

18.3   Poincaré duality and restriction to fixed points

We have seen that the classes $x(I) = [X(I)]^T$ form a Λ-module basis for $H_T^*X(\underline{\alpha})$. Next we will study their restrictions to fixed points, and determine the Poincaré dual basis.

Lemma 18.1.4 leads directly to a description of weights at the fixed points of $X(I) \subseteq X(\underline{\alpha})$. Suppose $K \subseteq I$, so $p_K \in X(I)$, and let $\underline{v}^K$ and $\underline{v}^I$ be the corresponding chains. The weights on $T_{p_K}X(I)$ are $-v_i^K(\alpha_i)$ for $i \in I$. This, in turn, gives a formula for restricting the classes $x(I) = [X(I)]^T$. For any $x \in H_T^*X(\underline{\alpha})$, its restriction to the fixed point p_I is denoted $x|_I$.

Corollary 18.3.1. *We have*

$$x(I)|_K = \begin{cases} \prod_{j \notin I} v_j^K(-\alpha_j) & \text{if } K \subseteq I; \\ 0 & \text{otherwise.} \end{cases}$$

Let $\{y(I)\}$ be the Poincaré dual basis to $\{x(I)\}$, meaning that $\rho_*(x(I) \cdot y(J)) = \delta_{I,J}$ in Λ, where $\rho: X(\underline{\alpha}) \to$ pt is the projection. As we saw in §4.6, such a basis always exists. It is natural to look for invariant subvarieties $Y(I)$ representing these Poincaré dual classes. However, no such algebraic subvarieties exist!

Example 18.3.2. Consider the variety $X(\alpha, \alpha) \cong \mathbb{P}^1 \times \mathbb{P}^1$ from Example 18.2.6. The basis $\{x(I)\}$ consists of the equivariant classes of

$$x(\emptyset) = [(p_e, p_e)]^T,$$
$$x(\{1\}) = [\delta(\mathbb{P}^1)]^T,$$
$$x(\{2\}) = [\{p_e\} \times \mathbb{P}^1]^T, \text{ and}$$
$$x(\{1, 2\}) = [\mathbb{P}^1 \times \mathbb{P}^1]^T.$$

Even non-equivariantly, the Poincaré dual basis cannot be represented by algebraic subvarieties: the class $y(\{2\})$ must have zero intersection with the diagonal class $x(\{1\})$, and no algebraic curve in $\mathbb{P}^1 \times \mathbb{P}^1$ can do this.

Another way of phrasing the conclusion of Example 18.3.2 is this: we seek a curve $Y(\{2\}) \subseteq \mathbb{P}^1 \times \mathbb{P}^1$ which consists of pairs (L, L') of lines in $\mathbb{C}^2$ such that $L \neq L'$ – but the complement of the diagonal is affine, so it contains no complete curves. In fact, this observation indicates a solution. Using the standard Hermitian metric on $\mathbb{C}^2$, we may consider pairs of perpendicular lines (L, L'); in terms of a coordinate z on $\mathbb{P}^1$, this is the set of pairs $(z, -1/\bar{z})$. This set is a non-algebraic submanifold $Y(\{2\}) \subseteq \mathbb{P}^1 \times \mathbb{P}^1$, which we orient by projecting onto the first factor. (Projection onto the second factor would give the opposite orientation, as the coordinate description shows.) Fixing the metric amounts to reducing GL_2 to the maximal compact subgroup $U(2)$, and identifying $\mathbb{P}^1 = GL_2/B$ with $U(2)/(T \cap U(2))$.

The general situation is similar: we construct (non-algebraic) submanifolds $Y(I) \subseteq X(\underline{\alpha})$ whose classes represent the Poincaré dual classes $y(I)$. Let $K \subseteq G$ be a maximal compact subgroup, with maximal compact torus $S = T \cap K$, so we have a diffeomorphism $K/S \cong G/B$, and the Weyl group $W = N_K(S)/S$ acts on the right. For $I \subseteq \{1, \ldots, d\}$, we define

$$Y(I) = \{(e, x_1, \ldots, x_d) \in X(\underline{\alpha}) \mid x_i = x_{i-1} \cdot s_{\alpha_i} \text{ for } i \in I\}.$$

This is a C^∞ submanifold, of real codimension $2 \cdot \#I$ in $X(\underline{\alpha})$, invariant for the action of the compact torus S. Containment among these submanifolds reverses containment of subsets:

$$Y(K) \subseteq Y(I) \quad \text{iff} \quad p_K \in Y(I) \quad \text{iff} \quad K \supseteq I.$$

Lemma 18.3.3. *Giving each $Y(I)$ an appropriate orientation (to be specified in the proof), the classes $y(I) = [Y(I)]^S$ form the Poincaré dual basis to $x(I)$.*

For $K \supset I$, with corresponding α-chains $\underline{v}^K$ and $\underline{v}^I$, the normal space to $Y(I) \subseteq X(\underline{\alpha})$ at the fixed point p_K has characters $-v_i^K(\alpha_i)$, for $i \in I$.

Proof To compute the tangent spaces of $Y(I)$, and to orient it, we work from the left, using induction on d. For $d = 1$, we have $Y(\{1\}) = \{\dot{s}_\alpha B\}$ (a point), and $Y(\emptyset) = X(\alpha) = \mathbb{P}^1$, so these are already oriented. Proceeding inductively, consider the projection $X(\alpha_1, \ldots, \alpha_d) \to X(\alpha_1, \ldots, \alpha_{d-1})$. If $d \in I$, this induces an isomorphism $Y(I) \to Y(I \smallsetminus \{d\})$. Otherwise, if $d \notin I$, it induces a $\mathbb{P}^1$-bundle, so there is a fiber square

$$
\begin{array}{ccc}
Y(I) & \lhook\joinrel\longrightarrow & X(\alpha_1, \ldots, \alpha_d) \\
\downarrow & & \downarrow \\
Y(\overline{I}) & \lhook\joinrel\longrightarrow & X(\alpha_1, \ldots, \alpha_{d-1}),
\end{array}
$$

where we have written $\overline{I} = I$ as a subset of $\{1, \ldots, d-1\}$. By the inductive assumption, we have an orientation of $Y(\overline{I})$. The canonical orientation of the $\mathbb{P}^1$ fiber then induces an orientation of $Y(I)$.

This construction also identifies the tangent spaces: assume $d \notin I$, and for $K \supseteq I$, write $p = p_K$ and $\overline{p}$ for the image of this point in $Y(\overline{I})$. The kernel of

$$T_p Y(I) \to T_{\overline{p}} Y(\overline{I})$$

is $\mathfrak{g}_\beta$, where $\beta = -v_d^K(\alpha_d)$.

It follows that $Y(I)$ meets $X(I)$ transversally in the point p_I. Indeed, we have weight decompositions of the tangent spaces as

$$T_{p_I} X(I) = \bigoplus_{i \notin I} \mathfrak{g}_{-v_i^I(\alpha_i)}$$

and

$$T_{p_I} Y(I) = \bigoplus_{i \in I} \mathfrak{g}_{-v_i^I(\alpha_i)}.$$

So these are complementary subspaces of $T_{p_I} X(\underline{\alpha})$. By considering fixed points, we see $X(I) \cap Y(J) = \emptyset$ unless $J \subseteq I$, and it follows that the classes $x(I)$ and $y(J)$ form Poincaré dual bases. $\qquad\square$

This description of tangent spaces proves a formula for restricting the classes $y(I)$.

Corollary 18.3.4.   *We have*

$$
y(I)|_K =
\begin{cases}
\prod_{i \in I} v_i^K(-\alpha_i) & \text{if } K \supseteq I; \\
0 & \text{otherwise.}
\end{cases}
$$

A more algebraic proof of Corollary 18.3.4 uses the localization formula. The dual classes $y(I)$ are uniquely determined by

$$
\sum_{p_K \in X(J)} \frac{y(I)|_K}{c_{top}^T(T_{p_K} X(J))} = \delta_{I,J}, \tag{18.1}
$$

for every subset $J \subseteq \{1, \dots, d\}$. We know $p_K \in X(J)$ if and only if $K \subseteq J$, and in this case $c_{top}^T(T_{p_K} X(J)) = \prod_{j \in J}(-v_j^K(\alpha_j))$. To prove the claimed formula for $y(I)|_K$, it remains to establish the identity

$$
\sum_{K: I \subseteq K \subseteq J} \frac{1}{\prod_{j \in J \smallsetminus I}(-v_j^K(\alpha_j))} = \delta_{I,J}. \tag{18.2}
$$

This is clear if $I = J$, or if $I \not\subseteq J$. When $I \subsetneq J$, the terms cancel in pairs, as follows. Suppose j is the largest index in $J \smallsetminus I$; then for each $K \not\ni j$, there is $K' = K \cup \{j\}$, and the corresponding terms cancel. (Indeed, $s_{\alpha_j}(\alpha_j) = -\alpha_j$, so $v_j^{K'}(\alpha_j) = -v_j^K(\alpha)$ and the other factors in the product are equal.)

Remark 18.3.5.   The identification $X = G/B = K/S$ leads to a third description of the Bott–Samelson varieties. Each $K_\alpha = K \cap P_\alpha$ is a maximal compact subgroup of the minimal parabolic P_α, and the evident map

$$
K_{\alpha_1} \times^S K_{\alpha_2} \times^S \cdots \times^S K_{\alpha_d}/S \to P_{\alpha_1} \times^B K_{\alpha_2} \times^B \cdots \times^B K_{\alpha_d}/B
$$

is a diffeomorphism. The submanifolds $Y(I) \subseteq X(\underline{\alpha})$ are easy to identify from this point of view:

$$
Y(I) = \Big\{ [k_1, \dots, k_d] \,\big|\, k_i S = \dot{s}_{\alpha_i} S \text{ for } i \in I \Big\}.
$$

For the corresponding projection $f \colon X(\underline{\alpha}) \to X$, one sees

$$
f(Y(\{1, \dots, k\})) = s_{\alpha_1} \cdots s_{\alpha_k} \cdot X(s_{\alpha_{k+1}} * \cdots * s_{\alpha_d})
$$

and

$$f(Y(\{k+1,\ldots,d\})) = X(s_{\alpha_1} * \cdots * s_{\alpha_k}) \cdot s_{\alpha_{k+1}} \cdots s_{\alpha_d},$$

where $w \cdot X(v)$ and $X(v) \cdot w$ denote the translations of Schubert varieties by the left and right W-actions.

18.4   A presentation for the cohomology ring

Multiplication in the basis $y(I)$ is particularly easy. To simplify the notation, we will write $p_i = p_{\{i\}}$, $p_{ij} = p_{\{i,j\}}$, $y_i = y(\{i\})$, and $y_{ij} = y(\{i,j\})$.

If $I \cap J = \emptyset$, then $Y(I)$ and $Y(J)$ meet transversally in $Y(I \cup J)$, so

$$y(I) \cdot y(J) = y(I \cup J) \quad \text{if } I \cap J = \emptyset. \tag{18.3}$$

In particular, $y_i \cdot y_j = y_{ij}$ if $i \neq j$, and $y(I) = y_{i_1} \cdots y_{i_\ell}$ if $I = \{i_1, \ldots, i_\ell\}$. To determine the structure of $H_T^* X(\underline{\alpha})$, it suffices to give a formula for y_i^2.

Proposition 18.4.1.   *We have*

$$y_i^2 = \sum_{j<i}(-\langle \alpha_i, \alpha_j^\vee \rangle) y_{ij} + \alpha_i y_i, \tag{18.4}$$

where $\langle \alpha, \beta^\vee \rangle$ is the pairing between roots and coroots.

Proof   By considering degrees and support, we have

$$y_i^2 = \sum_{j \neq i} c_{ij} y_{ij} + \lambda_i y_i, \tag{18.5}$$

for some $c_{ij} \in \mathbb{Z}$ and $\lambda_i \in M$. (Since $p_j \notin Y(\{i\})$ for $j \neq i$, we have $y_i|_{p_j} = 0$, so the classes y_j do not appear. Similarly, $p_\emptyset \notin Y(\{i\})$, so there is no "constant" term of degree 2 in Λ.) So we must determine these coefficients.

Using the restriction formula from Corollary 18.3.4, we have

$$y_i|_{p_i} = -v_i'(\alpha_i) = \alpha_i,$$

where the chain corresponding to $\{i\}$ is $\underline{v}' = (e, \ldots, e, s_{\alpha_i}, \ldots, s_{\alpha_i})$. Since $p_i \notin Y(\{i,j\})$ for $j \neq i$, restricting Equation (18.5) to this point gives

$$(\alpha_i)^2 = \lambda_i \alpha_i,$$

and it follows that $\lambda_i = \alpha_i$.

Similarly, we have

$$y_i|_{p_{ij}} = \begin{cases} \alpha_i & \text{if } i < j; \\ s_{\alpha_j}(\alpha_i) & \text{if } i > j. \end{cases}$$

(When $i < j$, the chain $\underline{v}'$ corresponding to $\{i,j\}$ has $v_i' = s_{\alpha_i}$, so $y_i|_{p_{ij}} = -s_{\alpha_i}(\alpha_i) = \alpha_i$. For $i > j$, the chain has $v_i' = s_{\alpha_j}s_{\alpha_i}$, so $y_i|_{p_{ij}} = -s_{\alpha_j}s_{\alpha_i}(\alpha_i) = s_{\alpha_j}(\alpha_i)$.) Likewise,

$$y_{ij}|_{p_{ij}} = \begin{cases} \alpha_i\, s_{\alpha_i}(\alpha_j) & \text{if } i < j; \\ \alpha_j\, s_{\alpha_j}(\alpha_i) & \text{if } i > j. \end{cases}$$

(For $i < j$, we have $v_j' = s_{\alpha_i}s_{\alpha_j}$, and $v_i' = s_{\alpha_i}$ as noted before, so Corollary 18.3.4 gives $y_{ij}|_{p_{ij}} = \alpha_i \cdot s_{\alpha_i}(\alpha_j)$. If $i > j$, swap the roles of i and j.)

By substituting $\lambda_i = \alpha_i$ and restricting (18.5) to p_{ij}, we obtain

$$\alpha_i^2 = c_{ij}\, \alpha_j\, s_{\alpha_j}(\alpha_i) + \alpha_i^2$$

for $i < j$, so $c_{ij} = 0$ in this case. Doing the same for $i > j$, we obtain

$$s_{\alpha_j}(\alpha_i)^2 = c_{ij}\, \alpha_j\, s_{\alpha_j}(\alpha_i) + \alpha_i\, s_{\alpha_j}(\alpha_i),$$

so $s_{\alpha_j}(\alpha_i) = c_{ij}\, \alpha_j + \alpha_i$. Since $s_{\alpha_j}(\alpha_i) = \alpha_i - \langle \alpha_i, \alpha_j^\vee \rangle \alpha_j$, the claim follows. □

As a consequence, we obtain a presentation for equivariant cohomology.

Corollary 18.4.2.   *The map $\eta_i \mapsto y_i$ defines an isomorphism*

$$H_T^* X(\underline{\alpha}) = \Lambda[\eta_1, \ldots, \eta_d]/\left(\eta_i^2 + \sum_{j<i} \langle \alpha_i, \alpha_j^\vee \rangle \eta_i\eta_j - \alpha_i\, \eta_i \right)_{1 \le i \le d}.$$

Similar formulas determine multiplication in the $x(I)$ basis for $H_T^* X(\underline{\alpha})$.

Exercise 18.4.3.   Writing $\beta_i = s_{\alpha_1} \cdots s_{\alpha_{i-1}}(\alpha_i)$, show that

$$x_i^2 = \sum_{j<i} (-\langle \beta_i, \beta_j^\vee \rangle)\, x_{ij} - \beta_i\, x_i,$$

where $x_i = x(\{1, \ldots, d\} \smallsetminus \{i\})$ and $x_{ij} = x(\{1, \ldots, d\} \smallsetminus \{i,j\})$.

The equivariant cohomology of G/B embeds in that of a Bott–Samelson variety. Let $(\alpha_1, \ldots, \alpha_N)$ be a reduced word for the longest element $w_\circ$, so $f\colon X(\underline{\alpha}) \to G/B$ is birational. From the projection formula, the composition $f_* \circ f^*$ is the identity.

Corollary 18.4.4. *Let*

$$R = \Lambda[f^* y(s_\alpha) : \alpha \in \Delta] \subseteq H_T^* X(\underline{\alpha})$$

be the subalgebra generated by pullbacks of divisor classes. The pullback f^* identifies $H_T^*(G/B)$ with the subalgebra of $H_T^* X(\underline{\alpha})$ consisting of elements x such that some integral multiple $c \cdot x$ lies in R.

Proof Using rational coefficients, we have seen that $H_T^*(G/B; \mathbb{Q})$ is generated over $\Lambda_\mathbb{Q} = H_T^*(\mathrm{pt}; \mathbb{Q})$ by the divisor classes $y(s_\alpha)$. (This follows from the Borel presentation given in Corollary 15.6.6. It also follows from Chevalley's formula, which we will see in Chapter 19.) Using the splitting $f_* \circ f^*$ and the fact that both $H_T^*(G/B)$ and $H_T^* X(\underline{\alpha})$ are free Λ-modules, it follows that

$$H_T^*(G/B) = H_T^*(X(\underline{\alpha})) \cap H_T^*(G/B; \mathbb{Q})$$

as submodules of $H_T^*(X(\underline{\alpha}); \mathbb{Q})$. $\qquad\square$

18.5 A restriction formula for Schubert varieties

A remarkable formula for the restrictions $y(w)|_v$ was discovered by Andersen–Jantzen–Soergel, and in a different context, by Billey.

Theorem 18.5.1 (Andersen–Jantzen–Soergel, Billey). *Fix a reduced word $(\alpha_1, \ldots, \alpha_d)$ for $v \in W$. For any $w \in W$,*

$$y(w)|_v = \sum \beta_{i_1} \cdots \beta_{i_\ell}, \tag{18.6}$$

the sum over all subsets $I = \{i_1 < \cdots < i_\ell\} \subseteq \{1, \ldots, d\}$ such that $\underline{\alpha}(I) = (\alpha_{i_1}, \ldots, \alpha_{i_\ell})$ is a reduced word for w.

Here $\beta_i = s_{\alpha_1} \cdots s_{\alpha_{i-1}}(\alpha_i)$, as in Lemma 15.1.6. By one of the many characterizations of Bruhat order there exists a subsequence $(\alpha_{i_1}, \ldots, \alpha_{i_\ell})$ as in the theorem if and only if $w \leq v$, i.e., whenever $p_v \in Y(w)$.

Considered as a formula for $y(w)|_v$, one appealing feature is that the right-hand side is positive: the roots β_i which appear are all in R^+, and it follows that $y(w)|_v$ is nonzero whenever $v \geq w$. Another remarkable consequence of the formula is that the polynomial on the right-hand side is independent of the choice of reduced word.

We will give two proofs of this theorem: one based on the geometry of Bott–Samelson varieties, and another using induction and some algebra. We need an easy lemma.

Lemma 18.5.2.   *For any word $\underline{\alpha} = (\alpha_1, \ldots, \alpha_d)$ and any $w \in W$, the pullback for $f \colon X(\underline{\alpha}) \to X$ is given by*

$$f^* y(w) = \sum y(I),$$

the sum over all subsets I such that $\#I = \ell(w)$ and the corresponding $\underline{\alpha}$-chain $\underline{v}$ has $v_d = w$.

Proof   Let $\langle a, b \rangle$ denote the usual pairing in cohomology, given by pushforward of $a \cdot b$ to a point. By the projection formula, we have $\langle f^* y(w), x(I) \rangle = \langle y(w), f_* x(I) \rangle$. Since $f_* x(I) = x(v_d)$ when $X(I) \to X(v_d)$ is birational, and $f_* x(I) = 0$ otherwise, the lemma follows from Corollary 18.2.5.   □

Remark 18.5.3.   Applying the lemma to divisor classes, we have $f^* y(s_\alpha) = \sum y_i$, the sum over $1 \leq i \leq d$ such that $\alpha_i = \alpha$. Combining this with Proposition 18.4.1 gives a method for computing in $H_T^*(G/B)$.

First proof of Theorem 18.5.1   Let $f \colon X(\underline{\alpha}) \to X$ is the projection, and let $\underline{v} = (v_1, \ldots, v_d)$ be the $\underline{\alpha}$-chain associated to $I = \{1, \ldots, d\}$, so $v_i = s_{\alpha_1} \cdots s_{\alpha_i}$, and in particular $v = v_d$. Then $f(p_{\underline{v}}) = p_v$, so $y(w)|_v = (f^* y(w))|_{p_{\underline{v}}}$. By Lemma 18.5.2, this is $\sum y(K)|_I$, the sum over all K such that $\#K = \ell(w)$ and the corresponding $\underline{\alpha}$-chain $\underline{v}^K$ has $v_d^K = w$. On the other hand, by Corollary 18.3.4, we have $y(K)|_I = \prod_{i \in K}(-v_i(\alpha_i))$. Since $-v_i(\alpha_i) = \beta_i$, the theorem is proved.   □

For the second proof, we use a variation on the functions ψ_v which we studied in Chapter 16. These were given by $\psi_v(w) = y(v)|_w$. Here we will use functions $\varphi_v \colon W \to \Lambda$, defined by

$$\varphi_v(w) = y(w)|_v = \psi_w(v).$$

Properties of these functions are immediate from the corresponding properties of ψ_w (Proposition 16.2.5). We only need an inductive formula.

Lemma 18.5.4. *We have*

$$\varphi_v(w) = \varphi_{vs_\alpha}(w) \qquad\qquad \text{if } \ell(ws_\alpha) > \ell(w); \qquad (18.7)$$

$$\varphi_v(w) = \varphi_{vs_\alpha}(w) - v(\alpha)\,\varphi_{vs_\alpha}(ws_\alpha) \qquad \text{if } \ell(ws_\alpha) < \ell(w). \qquad (18.8)$$

Proof Using the operators A_α from Proposition 16.2.5, we have

$$\psi_w(vs_\alpha) - \psi_w(v) = v(\alpha)\,(A_\alpha\psi_w)(v)$$

$$= \begin{cases} 0 & \text{if } \ell(ws_\alpha) > \ell(w); \\ v(\alpha)\,\psi_{ws_\alpha}(v) & \text{if } \ell(ws_\alpha) < \ell(w). \end{cases}$$

This immediately proves (18.7), as well as (18.8) with $\varphi_v(ws_\alpha)$ appearing on the right-hand side in place of $\varphi_{vs_\alpha}(ws_\alpha)$. But by (18.7), we have $\varphi_v(ws_\alpha) = \varphi_{vs_\alpha}(ws_\alpha)$ (since $\ell(ws_\alpha) > \ell(ws_\alpha \cdot s_\alpha)$). $\qquad\square$

Using the lemma, if we know the function φ_{vs_α}, for some α, then we know φ_v. For instance, we know

$$\varphi_e(w) = \begin{cases} 1 & \text{if } w = e; \\ 0 & \text{otherwise} \end{cases}$$

(since $p_e \notin Y(w)$ for $w \neq e$). This determines the rest!

Second proof of Theorem 18.5.1 We use induction on $\ell(v)$. For $\ell(v) = 0$, so $v = e$, this is the case observed above, so the theorem holds. In general, fix a reduced word for v as in the theorem. Set $f_v(w)$ to be the right-hand side of the formula (18.6), and let $\alpha = \alpha_d$. We assume the formula for φ_{vs_α} is known, using the reduced word $(\alpha_1, \ldots, \alpha_{d-1})$ for it.

If $\ell(ws_\alpha) > \ell(w)$, then no reduced word for w ends in α, and it follows that $f_v(w) = f_{vs_\alpha}(w)$. Since $\varphi_v(w) = \varphi_{vs_\alpha}(w)$ by Lemma 18.5.4, the formula holds in this case.

If $\ell(ws_\alpha) < \ell(w)$, then no reduced word for ws_α ends in α. Consider subsets $I = \{i_1 < \cdots < i_\ell\}$ corresponding to reduced words for w. For those I such that $i_\ell = d$, the sequence $(\alpha_{i_1}, \ldots, \alpha_{i_{\ell-1}})$ is a reduced word for ws_α, and $\beta_d = -v(\alpha) = (vs_\alpha)(\alpha)$. So the sum of such terms is

$$\sum_{I \text{ with } i_\ell = d} \beta_{i_1} \cdots \beta_{i_{\ell-1}} \beta_{i_\ell} = -v(\alpha)\, \varphi_{vs_\alpha}(ws_\alpha).$$

The other terms, where $i_\ell < d$, sum to $\varphi_{vs_\alpha}(w)$. Applying Lemma 18.5.4, the full sum is $\varphi_v(w)$, as required. $\square$

Example 18.5.5.   Theorem 18.5.1 includes a formula for the restrictions of divisor classes $y(s_\alpha)|_v$, as the sum of those β_i for which $\alpha_i = \alpha$. On the other hand, we saw $y(s_\alpha) = \varpi_\alpha - v(\varpi_\alpha)$ in Lemma 16.2.6. The latter is often simpler to use in this case. For example, with $G = SL_n$ and $\alpha = t_1 - t_2$, we have

$$\varpi_\alpha - v(\varpi_\alpha) = \alpha_1 + \cdots + \alpha_{v(1)-1}$$

for any permutation $v \in S_n$, without needing to find a reduced expression.

Exercise 18.5.6.   Check directly that the two formulas for $y(s_\alpha)|_v$ agree: show that

$$\varpi_\alpha - v(\varpi_\alpha) = \sum_{i:\alpha_i = \alpha} s_{\alpha_1} \cdots s_{\alpha_{i-1}}(\alpha_i)$$

for any simple root α, and any reduced word $(\alpha_1, \ldots, \alpha_d)$ for $v \in W$.

Example 18.5.7.   As noted above, Theorem 18.5.1 shows that $y(w)|_v$ is nonzero if and only if $p_v \in Y(w)$. This is a special property of the standard torus action on Schubert varieties. In general, for an invariant subvariety Y of a nonsingular variety V, with $[Y]^T \in H_T^* V$, one can have $[Y]^T|_p = 0$ for an isolated fixed point $p \in Y$.

For example, consider $V = \mathbb{P}^4$ with coordinates $x_1, \ldots, x_5$, and a torus T acting by characters $0, \chi_1, -\chi_1, \chi_2, -\chi_2$, where $\chi_1 \neq \chi_2$. Let Y be the hypersurface defined by $x_2 x_3 - x_4 x_5 = 0$, so $p = [1, 0, 0, 0, 0]$ is the singular point of Y. Writing $\zeta = c_1^T(\mathcal{O}(1))$, we have $[Y]^T = 2\zeta$ so $[Y]^T|_p = 0$.

Remark 18.5.8.   As we saw in Chapter 15, Equation (15.9), Schubert classes in G/P pull back to Schubert classes in G/B. Writing the projection as $\pi \colon G/B \to G/P$, we have $\pi^* y[w] = y(w^{\min})$. This is compatible with restriction to fixed points, and we have

$$y[w]|_{[v]} = y(w^{\min})|_v$$

for any coset representative $v \in [v]$. In particular, Theorem 18.5.1 includes a formula for restricting G/P Schubert classes.

18.6 Duality

In §16.4, we used an isomorphism $\Phi^w \colon G/B \xrightarrow{\sim} G/B^w$ to relate difference operators with the right W-action on G/B. The particular case where $w = w_\circ$, so $B^{w_\circ} = \dot{w}_\circ B \dot{w}_\circ^{-1} = B^-$, is especially useful for passing between formulas involving $y(w)$ and ones involving $x(w)$. Here we will state several such formulas; their proofs are all immediate from the functoriality of pullbacks.

To set up notation, let $\overline{X} = G/B^-$, with fixed points $\overline{p}_w = \dot{w}B^-$ and Schubert varieties

$$\overline{X}(w) = \overline{B^- \cdot \overline{p}_w} \quad \text{and} \quad \overline{Y}(w) = \overline{B \cdot \overline{p}_w}.$$

Let $\overline{x}(w)$ and $\overline{y}(w)$ be the corresponding Schubert classes in $H_T^* \overline{X}$. The entire discussion for Schubert classes in $\overline{X} = G/B^-$ is parallel to that of $X = G/B$, except that each root is replaced by its negative. For example,

$$\overline{y}(w)|_{\overline{p}_w} = \prod_{\beta \in w(R^-) \cap R^+} (-\beta) = (-1)^{\ell(w)} \, y(w)|_{p_w}.$$

Let $\tau \colon \Lambda \to \Lambda$ be the graded involution which is multiplication by $(-1)^r$ on $\mathrm{Sym}^r M$, so τ is induced by the involution of M taking each root to its negative. Then

$$\overline{y}(w)|_{\overline{p}_v} = \tau(y(w)|_{p_v}) \tag{18.9}$$

for every $w, v \in W$.

Write $\Phi = \Phi^{w_\circ}$ for the G-equivariant isomorphism $X \xrightarrow{\sim} \overline{X}$, so $\Phi(gB) = g\dot{w}_\circ B^-$. Since $\Phi(p_{ww_\circ}) = \overline{p}_w$, we see

$$\Phi(X(ww_\circ)) = \overline{Y}(w) \quad \text{and} \quad \Phi(Y(ww_\circ)) = \overline{X}(w).$$

So $\Phi^* \overline{y}(w) = x(ww_\circ)$ and $\Phi^* \overline{x}(w) = y(ww_\circ)$, and we have

$$x(w)|_{p_v} = \overline{y}(ww_\circ)|_{\overline{p}_{vw_\circ}}.$$

Combining this with (18.9), we obtain

$$x(w)|_{p_v} = \tau(y(ww_\circ)|_{p_{vw_\circ}}). \tag{18.10}$$

Next consider the automorphism $\tau_\circ = \tau_{w_\circ} \colon X \to X$, coming from the left action of W on G/B as in §16.5. The map $\tau_\circ$ is equivariant with respect to the automorphism $\sigma \colon g \mapsto \dot{w}_\circ g \dot{w}_\circ^{-1}$ of G. Restricting σ to the torus $T \subseteq G$, in turn, induces the algebra automorphism $w_\circ \colon \Lambda \to \Lambda$ given by $\lambda \mapsto w_\circ(\lambda)$ for $\lambda \in M$. Since $\tau_\circ$ maps $p_{w_\circ w}$ to p_w, we see $\tau_\circ(X(w_\circ w)) = Y(w)$ and therefore

$$x(w_\circ w)|_{p_{w_\circ v}} = w_\circ \cdot (y(w)|_{p_v}). \tag{18.11}$$

Like τ, the algebra automorphism $w_\circ$ sends a product of positive roots to a product of negative roots – but in general these are different automorphisms.

Finally, the isomorphism $\Phi \circ \tau_\circ \colon X \to \overline{X}$ is equivariant with respect to the automorphism σ, and takes $Y(w_\circ w w_\circ)$ to $\overline{Y}(w)$, so

$$y(w_\circ w w_\circ)|_{p_{w_\circ v w_\circ}} = w_\circ \cdot \tau(y(w)|_{p_v}). \tag{18.12}$$

These identities generalize ones we have seen for Schubert polynomials in type A. For instance, Equation (18.12) here corresponds to Chapter 11, §11.8, Equation (11.2).

18.7   A nonsingularity criterion

For $v \leq w$ in W, when is the Schubert variety $X(w)$ nonsingular at the fixed point $p_v \in X(w)$? We will see a criterion in terms of equivariant cohomology, due to Kumar.

We need some information about the tangent cone $C_{p_v} X(w)$. Let

$$V_v = \dot{v} U^- \dot{v}^{-1} \cdot p_v \subseteq X$$

be the T-invariant open affine neighborhood of p_v, and let

$$V(w)_v = X(w) \cap V_v$$

be the corresponding affine neighborhood in $X(w)$. We will write $V(w)_v = \operatorname{Spec} A$, and $\mathfrak{m} \subseteq A$ for the maximal ideal corresponding to $p_v \in V(w)_v$.

Lemma 18.7.1.   *For each $\beta \in v(R^-)$ such that $s_\beta v \leq w$, there is a function $f_\beta \in A$ which is an eigenfunction of weight β for the action of T. (That is, $f_\beta(t^{-1}x) = \beta(t) f(x)$ for all $t \in T$ and $x \in V(w)_v$.)*

Furthermore, the f_β generate an $\mathfrak{m}$-primary ideal in A. (That is, $f_\beta(p_v) = 0$ for each β, and p_v is their only common zero.)

From the description of invariant curves we saw in §15.4, the roots $\beta \in v(R^-)$ such that $s_\beta v \leq w$ are precisely the weights of the T-invariant curves in $X(w)$ through p_v.

We will state the nonsingularity criterion in terms of the equivariant multiplicities defined in Chapter 17.

Theorem 18.7.2. *For $v \leq w$, the point p_v is nonsingular in $X(w)$ if and only if*

$$\varepsilon^T_{p_v} X(w) = \prod_{\substack{\beta \in v(R^-) \\ s_\beta v \leq w}} \beta^{-1},$$

where $\varepsilon^T_v X(w)$ is the equivariant multiplicity of $X(w)$ at p_v.

Proof One direction is immediate. If $X(w)$ is nonsingular at p_v, the weights on $T_{p_v} X(w)$ coincide with the tangent weights to the T-invariant curves through p_v. (This is a general fact about nonsingular varieties with finitely many invariant curves; see Proposition 7.2.3.) Therefore

$$T_{p_v} X(w) = \bigoplus_{\substack{\beta \in v(R^-) \\ s_\beta v \leq w}} \mathfrak{g}_\beta.$$

By an elementary property of equivariant multiplicities, $\varepsilon^T_v X(w)$ is the inverse of the product of tangent weights (Proposition 17.4.4(ii)).

Conversely, assume the formula holds. Using the notation of Lemma 18.7.1, let $A' \subseteq A$ be the subring generated by the functions f_β. Since $\varepsilon^T_v X(w)$ has degree $-\dim X(w) = -\ell(w)$, there are $\ell(w)$ such f_β's. It follows that they form a system of parameters for A at $\mathfrak{m}$. So the subalgebra $A' \cong \mathbb{C}[\{f_\beta \mid \beta \in v(R^-), s_\beta v \leq w\}]$ is a polynomial ring, and A is a finitely generated module over A'.

Let $V = V(w)_v = \operatorname{Spec} A$ and $V' = \operatorname{Spec} A'$, and write $\pi \colon V \to V'$ for the corresponding equivariant map of affine varieties. Let $p' \in V'$ be the origin, and note that this is a nondegenerate fixed point, since the tangent weights β are all nonzero. Since the functions f_β are a system of parameters, we have $\pi^{-1}(p') = p_v$. It follows from another property of equivariant mulitplicities (Proposition 17.4.4(vi)) that

$$\varepsilon_{p_v}^T V = d \cdot \varepsilon_{p'}^T V',$$

where d is the degree of the finite map π; since equivariant multiplicities are local, we have $\varepsilon_v^T X(w) = \varepsilon_{p_v}^T V$. On the other hand, $p' \in V'$ is nonsingular, with tangent weights β, so as observed above we have

$$\varepsilon_{p'}^T V' = \prod_{\substack{\beta \in v(R^-) \\ s_\beta v \leq w}} \beta^{-1}.$$

It follows that $d = 1$, so $A = A'$ is a polynomial ring, and $V \cong \mathbb{A}^{\ell(w)}$. In particular, p_v is a nonsingular point. $\qquad \square$

The criterion may be rephrased in terms of restrictions of Schubert classes.

Corollary 18.7.3. *For $v \leq w$, the point p_v is nonsingular in $X(w)$ if and only if*

$$x(w)|_v = \prod_{\substack{\beta \in v(R^-) \cap R^- \\ s_\beta v \nleq w}} \beta.$$

Proof We have

$$x(w)|_v = c_N^T(T_{p_v} X) \cdot \varepsilon_v^T X(w)$$

$$= \left(\prod_{\beta \in v(R^-)} \beta \right) \cdot \varepsilon_v^T X(w),$$

using another characterization of equivariant multiplicities (§17.4, Equation (17.9)). Dividing both sides by $c_N^T(T_{p_v} X)$, the assertion follows from Theorem 18.7.2. (For any $\beta \in v(R^-) \cap R^-$, we have $s_\beta v < v \leq w$, so these weights cancel.) $\qquad \square$

Using the duality identities from the previous section, it is easy to deduce corresponding nonsingularity criteria for opposite Schubert varieties $Y(w)$. Using the notation of §18.6, the automorphism τ_o sends $p_{w_o v}$ to p_v and $X(w_o w)$ to $Y(w)$, so p_v is nonsingular in $Y(w)$ if and only if $p_{w_o v}$ is nonsingular in $X(w_o w)$. We obtain the following:

Corollary 18.7.4. *For $v \geq w$, the point p_v is nonsingular in $Y(w)$ if and only if*

$$y(w)|_v = \prod_{\substack{\beta \in v(R^-) \cap R^+ \\ s_\beta v \not\geq w}} \beta.$$

In this case, the tangent space $T_{p_v} Y(w)$ has weights $\beta \in v(R^-)$ such that $s_\beta v \geq w$.

(Applying Equation 18.11, it suffices to verify that

$$\{\beta \in v(R^-) \,|\, s_\beta v \not\geq w\} = w_\circ \left(\{\gamma \in w_\circ v(R^-) \,|\, s_\gamma w_\circ v \not\geq w_\circ w\}\right),$$

which is straightforward, using $w_\circ v \leq w_\circ w$ if and only if $v \geq w$.)

Combining this with the restriction formula of Theorem 18.5.1, we arrive at a combinatorial criterion for nonsingularity of $Y(w)$ at p_v.

Corollary 18.7.5. *Fix a reduced word $\underline{\alpha} = (\alpha_1, \ldots, \alpha_d)$ for v, and write $\beta_i = s_{\alpha_1} \cdots s_{\alpha_{i-1}}(\alpha_i)$. Then p_v is nonsingular in $Y(w)$ if and only if*

$$\sum \beta_{i_1} \cdots \beta_{i_\ell} = \prod_{\substack{\beta \in v(R^-) \cap R^+ \\ s_\beta v \not\geq w}} \beta,$$

where the sum on the left-hand side is over all $I \subseteq \{1, \ldots, d\}$ such that the corresponding subword $\underline{\alpha}(I)$ is a reduced word for w.

Exercise 18.7.6. If $\ell(v) = \ell(w) + 1$, show that $p_v \in Y(w)$ is nonsingular. Conclude that Schubert varieties are nonsingular in codimension one. (That is, the singular locus has codimension at least two.)

Exercise 18.7.7. For $G = SL_n$ and $\alpha = t_k - t_{k+1}$, so $s_\alpha = s_k$, show that the (opposite) Schubert variety $Y(s_k) \subseteq SL_n/B$ is singular at w if and only if $\#\{i \leq k \,|\, w(i) > k\} \geq 2$.

Exercise 18.7.8. Use $\mathfrak{S}_{2143} = (x_1 - y_1)(x_1 + x_2 + x_3 - y_1 - y_2 - y_3)$ to determine the singular locus of $Y(2143) = \Omega_{2143} \subseteq Fl(\mathbb{C}^4)$.

Remark 18.7.9. Using the Bott–Samelson resolution, the additivity property of equivariant multiplicities (Proposition 17.4.4(vi)) leads to another formula for $\varepsilon_v^T X(w)$. We have

$$\varepsilon_v^T X(w) = \sum_{\underline{v}} \left(\prod_{i=1}^\ell (-v_i(\alpha_i))\right)^{-1}, \tag{18.13}$$

where $\underline{\alpha} = (\alpha_1, \ldots, \alpha_\ell)$ is a fixed reduced word for w, and the sum is over all $\underline{\alpha}$-chains $\underline{v} = (e, v_1, \ldots, v_\ell)$ such that $v_\ell = v$. (These correspond to the fixed points $p_{\underline{v}} \in X(\underline{\alpha})$ mapping to p_v under the resolution $X(\underline{\alpha}) \to X(w)$, and the corresponding term is $\varepsilon_{\underline{v}}^T X(\underline{\alpha})$.) Clearing denominators, one obtains a formula for $x(w)|_v$ which is different from the one deduced from Billey's formula. In particular, note that the chains indexing terms of the sum need not correspond to *reduced* words for v.

Remark 18.7.10. As noted in Remark 18.5.8, knowing about Schubert varieties in G/B is enough to say something about Schubert varieties in G/P. The projection $\pi \colon G/B \to G/P$ makes $X(w^{\mathrm{max}}) \to X[w]$ and $Y(w^{\mathrm{min}}) \to Y[w]$ into fiber bundles, with nonsingular fiber P/B. So a point $p_{[v]} \in X[w]$ is nonsingular if and only if $p_v \in X(w^{\mathrm{max}})$ is nonsingular, for any coset representative $v \in [v]$; and similarly for $p_{[v]} \in Y[w]$. So Theorem 18.7.2 and Corollary 18.7.3 provide nonsingularity criteria for Schubert varieties in G/P.

Notes

Bott and Samelson (1955) gave a construction similar to the one indicated in Remark 18.3.5, and used it to study the cohomology of $G/B = K/S$. In particular, they prove a non-equivariant version of Corollary 18.4.4. The algebraic version which is more commonly used in Schubert calculus and representation theory was introduced by Demazure (1974) and Hansen (1974), and for this reason the varieties $X(\underline{\alpha})$ are sometimes called *Bott–Samelson–Demazure–Hansen* (or *BSDH*) varieties. The non-equivariant part of the formula for x_i^2 (Exercise 18.4.3) appears in (Demazure, 1974, §4.2).

Corollary 18.3.4 was proved by Willems (2004), using a localization argument similar to the second proof we gave. Our geometric argument, using the submanifolds $Y(I)$, appears to be new.

Theorem 18.5.1 appears as an exercise (without proof) in a book by Andersen, Jantzen, and Soergel (1994, p. 298). Billey (1999) discovered the formula independently, emphasizing the connection with Schubert calculus. Her proof proceeds by decreasing induction on w, with a separate argument that the polynomial is independent of the choice of reduced word. The result is sometimes known as the *AJSB formula*.

Example 18.5.7 is due to Brion (2000).

Among simple linear algebraic groups, the automorphisms τ and $w_\circ$ (from §18.6) are equal precisely in types B_n, C_n, D_{2n}, E_7, E_8, F_4, and G_2; see, e.g., (Humphreys, 1981, §31.6).

Theorem 18.7.2 is due to Kumar (1996, Theorem 5.5). A simplified argument was given by Brion (1997b, §6.5), and this is essentially the one we use. Lemma 18.7.1 follows from a result of Polo (1994, Prop. 2.2); see also (Kumar, 1996, Prop. 5.2) and its proof. A more detailed study of the tangent cones $C_{p_v}X(w)$ has been carried out by Carrell and Peterson; see, e.g., (Carrell, 1994). For a comprehensive account of the singular loci of Schubert varieties, see (Billey and Lakshmibai, 2000).

The formula (18.13) for $\varepsilon_v^T X(w)$ is due to Rossmann (1989, (3.8)).

Hints for Exercises

Exercise 18.2.2. Use the subword characterization of Bruhat order, and a greedy algorithm to see that $s_{\alpha_{i_1}} \cdots s_{\alpha_{i_\ell}} \le s_{\alpha_1} * \cdots * s_{\alpha_d}$ for any subword of $\underline{\alpha}$. See (Knutson and Miller, 2004, Lemma 3.4).

Exercise 18.2.7. Consider the point $p = p_{\{1,\dots,d\}} \in X(\underline{\alpha})$. Using terminology from §7.2, the tangent space $T_p X(\underline{\alpha})$ contains parallel weights whenever $\underline{\alpha}$ is a non-reduced word; in this case there are infinitely many T-curves through a neighborhood of p. Whenever the sequence $\underline{\alpha}$ has a repeated root, an instance of the variety considered in Example 18.2.6 occurs as a subvariety of $X(\underline{\alpha})$, and this has infinitely many T-curves.

To see that $X(\underline{\alpha})$ is toric when all roots are distinct, look at the tangent space to $p_\emptyset$: the characters form part of a basis for M, so there is a dense T-orbit. To see that f is an isomorphism in this case, keep track of fixed points.

Exercise 18.2.9. Use induction on d. The same argument shows that the analogous map

$$G \times^B P_{\alpha_1} \times^B \cdots \times^B P_{\alpha_d}/B \to Z(\underline{\alpha})$$

is an isomorphism.

Exercise 18.5.6. Argue inductively as in the second proof of Theorem 18.5.1. It is obvious for $v = e$. Suppose the equality is known for v, and β is a simple root such that $\ell(vs_\beta) = \ell(v) + 1$. If $\beta \ne \alpha$, the right-hand sides are clearly equal for v and vs_β; since $s_\beta(\varpi_\alpha) = \varpi_\alpha$ for $\beta \ne \alpha$, so are the left-hand sides. If $\beta = \alpha$, then the difference of the right-hand sides is $v(\alpha)$, and the difference of the left-hand sides is $v(\varpi_\alpha) - vs_\alpha(\varpi_\alpha) = v(\alpha)$.

Exercise 18.7.6. The claim about $p_v \in Y(w)$ being nonsingular follows easily from Billey's formula for $y(w)|_v$. Using B-equivariance, one sees that the nonsingular locus of $Y(w)$ contains the union of Schubert cells $Y(v)^\circ$ for $v \ge w$ and $\ell(v) \le \ell(w) + 1$. (The conclusion also follows from the general fact that Schubert are normal.)

Exercise 18.7.7. Use the formula for $y(s_\alpha)|_w$ in Exercise 10.7.2.

19

Structure Constants

A major problem in combinatorics and geometry is to determine the structure constants for multiplying Schubert classes in G/P. The ultimate goal is a positive combinatorial formula akin to the Littlewood–Richardson rule for Grassmannians, and in this sense the problem is open for most cases.

Here we will prove an equivariant version of Chevalley's rule for multiplying by a Schubert divisor. As we saw for Grassmannians and type A flag varieties, this recursively determines all equivariant structure constants. We conclude with two proofs of a theorem due to Graham, which specifies the sense in which equivariant structure constants are positive.

19.1 Chevalley's formula

We start with Schubert classes in G/B. Let α be a simple root, so $y(s_\alpha) \in H_T^*(G/B)$ is a divisor class. The equivariant Chevalley formula expresses $y(s_\alpha) \cdot y(v)$ as a sum of classes $y(w)$. Such a product takes place in $Y(s_\alpha) \cap Y(v)$, so the classes $y(w)$ which appear will have $w \gtrdot v$ (that is, $w > v$ with $\ell(w) = \ell(v) + 1$), together with $w = v$.

Using the characterization of Bruhat order from Lemma 15.2.1(9), the covering relation $w \gtrdot v$ may be described as follows.

Lemma 19.1.1. *We have $w \gtrdot v$ if and only if $\ell(w) = \ell(v) + 1$, and there is a positive root β, with $\gamma = v(\beta)$ also positive, such that $w = v s_\beta = s_\gamma v$.*

(If $(\alpha_1, \ldots, \alpha_\ell)$ is a reduced word for w, so that $(\alpha_1, \ldots, \widehat{\alpha_k}, \ldots, \alpha_\ell)$ is a reduced for v, then $\beta = s_{\alpha_\ell} \cdots s_{\alpha_{k+1}}(\alpha_k)$.)

Here is the formula.

Theorem 19.1.2 (Equivariant Chevalley formula). *For a simple root α and $v \in W$, we have*

$$y(s_\alpha) \cdot y(v) = (\varpi_\alpha - v(\varpi_\alpha))\, y(v) + \sum_{\substack{w = v\, s_\beta \\ \ell(w) = \ell(v)+1}} \langle \varpi_\alpha, \beta^\vee \rangle\, y(w) \qquad (*)$$

in $H_T^*(G/B)$.

As remarked in §15.1.2, using a W-invariant inner product $(\, , \,)$ one can write the coefficients as

$$\langle \varpi_\alpha, \beta^\vee \rangle = n_\beta^\alpha \cdot \frac{(\alpha, \alpha)}{(\beta, \beta)},$$

where n_β^α is the coefficient of α when β is written as a sum of simple roots.

Proof As noted above, for degree and support reasons, the terms appearing in $y(s_\alpha) \cdot y(v)$ must be those $y(w)$ for which $w \geq v$ and $\ell(w) \leq \ell(v) + 1$. By Lemma 19.1.1, these are the ones appearing on the right-hand side of $(*)$. For each such w, except v itself, $p_v \notin Y(w)$. So restricting both sides of $(*)$ to p_v, we obtain

$$y(s_\alpha)|_v \cdot y(v)|_v = \lambda \cdot y(v)|_v + 0$$

for some $\lambda \in M$. Since $y(v)|_v \neq 0$, it follows that $\lambda = y(s_\alpha)|_v$, and we have seen the formula for this restriction (Lemma 16.2.6). So we have the first term.

The other coefficients are classical, but it is easier to compute them equivariantly. By Poincaré duality, the coefficient of $y(w)$ is

$$\rho_* \left(y(s_\alpha) \cdot y(v) \cdot x(w) \right),$$

where $\rho\colon G/B \to \mathrm{pt}$ is the projection. For $\ell(w) = \ell(v) + 1$, the Schubert varieties $Y(v)$ and $X(w)$ meet transversally in the T-invariant curve

$$E = Y(v) \cap X(w) = \overline{G_\gamma \cdot p_v}$$

containing the fixed points p_v and $p_{s_\gamma v} = p_{vs_\beta}$. This follows from the fact that Schubert varieties are nonsingular in codimension one. (See Exercise 18.7.6.) Since the subspace $T_{p_v}X(v) \subseteq T_{p_v}X(w)$ has codimension one, and meets $T_{p_v}Y(v)$ transversally, it follows that $T_{p_v}X(w) \cap T_{p_v}Y(v)$ is one-dimensional; and similarly for the subspace $T_{p_w}X(w) \cap T_{p_w}Y(v)$.

From the characterization of invariant curves, E has character γ, with tangent weight γ on $T_{p_w}E$ and $-\gamma$ on $T_{p_v}E$. By transversality of the intersection, we have

$$y(v) \cdot x(w) = [E]^T.$$

Writing $f \colon E \hookrightarrow G/B$ for the inclusion, and $\eta \colon E \to \mathrm{pt}$, an application of the projection formula shows

$$\rho_* \left(y(s_\alpha) \cdot y(v) \cdot x(w) \right) = \eta_* \left(f^* y(s_\alpha) \right).$$

Now we use localization to compute this integral. The restriction of $f^* y(s_\alpha)$ to p_w is $y(s_\alpha)|_w$, and the restriction to p_v is $y(s_\alpha)|_v$, so using the formulas we know for these, we have

$$\begin{aligned}
\rho_* \left(y(s_\alpha) \cdot y(v) \cdot x(w) \right) &= \frac{y(s_\alpha)|_w}{\gamma} + \frac{y(s_\alpha)|_v}{-\gamma} \\
&= \frac{y(s_\alpha)|_w - y(s_\alpha)|_v}{\gamma} \\
&= \frac{v(\varpi_\alpha) - w(\varpi_\alpha)}{\gamma}.
\end{aligned}$$

Since $w = s_\gamma v$, we can write

$$\begin{aligned}
v(\varpi_\alpha) - w(\varpi_\alpha) &= v(\varpi_\alpha) - s_\gamma v(\varpi_\alpha) \\
&= \langle v(\varpi_\alpha), \gamma^\vee \rangle \, \gamma,
\end{aligned}$$

and it follows that the desired coefficient is $\langle v(\varpi_\alpha), \gamma^\vee \rangle$. Since the pairing is W-invariant and $\gamma = v(\beta)$, this is equal to $\langle \varpi_\alpha, \beta^\vee \rangle$, as claimed. $\qquad\square$

Exercise 19.1.3. Prove that the same coefficients give the corresponding formula for multiplication by B-invariant divisor classes $x(w_\circ s_\alpha)$:

$$x(w_\circ s_\alpha) \cdot x(w) = (x(w_\circ s_\alpha)|_w) \cdot x(w) + \sum_v \langle \varpi_\alpha, \beta^\vee \rangle \, x(v),$$

the sum over $v \le w$ with $\ell(v) = \ell(w) - 1$, where $v = w s_\beta$.

Exercise 19.1.4. Prove Chevalley's formula for G/P. For $\alpha \in \Delta \smallsetminus \Delta_P$ and $v = v^{\min}$ a minimal representative for $[v] \in W/W_P$,

$$y[s_\alpha] \cdot y[v] = (\varpi_\alpha - v(\varpi_\alpha))\, y[v] + \sum_{w = v^{\min} s_\beta} \langle \varpi_\alpha, \beta^\vee \rangle\, y[w],$$

the sum over $w \geq v^{\min}$ with $\ell(w) = \ell(v^{\min}) + 1$ and $[w] \neq [v]$. Note that the coefficients $(\varpi_\alpha - v(\varpi_\alpha))$ and $\langle \varpi_\alpha, \beta^\vee \rangle$ appearing in the formula are independent of the choice of coset representative.

Recall that the G-equivariant line bundle $\mathcal{L}_\lambda = G \times^B \mathbb{C}_\lambda$ has $c_1^T(\mathcal{L}_\lambda)|_w = w(\lambda)$. Since $y(s_\alpha)|_w = \varpi_\alpha - w(\varpi_\alpha)$, we see that

$$y(s_\alpha) = c_1^T(\mathcal{L}_{-\varpi_\alpha} \otimes \mathbb{C}_{\varpi_\alpha}) = c_1^T(\mathcal{L}_{-\varpi_\alpha}) + \varpi_\alpha.$$

There is a similar formula for multiplying by $c_1^T(\mathcal{L}_\lambda)$.

Exercise 19.1.5. For any weight $\lambda \in M$, show that

$$c_1^T(\mathcal{L}_\lambda) \cdot y(v) = v(\lambda) \cdot y(v) + \sum_{w = v\, s_\beta} \langle -\lambda, \beta^\vee \rangle\, y(w)$$

in $H_T^*(G/B)$, the sum over w such that $\ell(w) = \ell(v) + 1$, as before.

19.2   Characterization of structure constants

No formula as explicit as Theorem 19.1.2 is currently known for multiplying two general Schubert classes. However, the general coefficients are characterized by a recursion similar to what we have seen for $Gr(d, \mathbb{C}^n)$ and $Fl(\mathbb{C}^n)$.

For $u, v, w \in W$, let $c_{uv}^w \in \Lambda$ be defined by

$$y(u) \cdot y(v) = \sum_w c_{uv}^w\, y(w)$$

in $H_T^*(G/B)$. As before, special cases of these coefficients appear as restrictions of Schubert classes.

Exercise 19.2.1. Show that $c_{uv}^v = y(u)|_v$. In particular,

$$c_{uu}^u = y(u)|_u = \prod_{\beta \in u(R^-) \cap R^+} \beta.$$

Recall that $w \gtrdot v$ if and only if there is some positive root β such that $w = vs_\beta$ and $\ell(w) = \ell(v) + 1$. In this situation, it will be convenient to use the notation

$$c_\alpha(v, w) := \langle \varpi_\alpha, \beta^\vee \rangle$$

when stating the recursive characterization.

Theorem 19.2.2. *The polynomials c_{uv}^w, homogeneous of degree $\ell(u) + \ell(v) - \ell(w)$ in Λ, satisfy and are determined by the following properties, for all simple roots $\alpha \in \Delta$:*

$$c_{uu}^u = \prod_{\beta \in u(R^-) \cap R^+} \beta, \tag{i}$$

$$(y(s_\alpha)|_u - y(s_\alpha)|_v) \, c_{uv}^u = \sum_{v^+ \gtrdot v} c_\alpha(v, v^+) \, c_{uv^+}^u, \tag{ii}$$

and

$$(y(s_\alpha)|_w - y(s_\alpha)|_u) \, c_{uv}^w = \sum_{u^+ \gtrdot u} c_\alpha(u, u^+) \, c_{u^+v}^w$$
$$- \sum_{w^- \lessdot w} c_\alpha(w^-, w) \, c_{uv}^{w^-}, \tag{iii}$$

where the sums are over v^+ such that $\ell(v^+) = \ell(v) + 1$, u^+ such that $\ell(u^+) = \ell(u) + 1$, and w^- such that $\ell(w^-) = \ell(w) - 1$.

The proof goes as in Theorems 9.6.4 and 10.7.4, using the following observation.

Lemma 19.2.3. *For any $u \neq v$ in W, there is a simple root α such that $y(s_\alpha)|_u \neq y(s_\alpha)|_v$.*

Proof Using $y(s_\alpha)|_u = \varpi_\alpha - u(\varpi_\alpha)$, this follows from the fact that the fundamental weights ϖ_α form a basis for the vector space $M_\mathbb{R}$, and W acts faithfully on this space. $\square$

The involutions we used in §18.6 apply to give identities among structure constants for various bases. Recall that τ is the automorphism of Λ induced by $\lambda \mapsto -\lambda$ for $\lambda \in M$, and $w_\circ$ is the one induced by the usual W action, i.e., $\lambda \mapsto w_\circ(\lambda)$ for $\lambda \in M$.

Proposition 19.2.4. *With notation as in §18.6, we have*

$$\bar{y}(u) \cdot \bar{y}(v) = \sum_w \tau(c_{uv}^w)\, \bar{y}(w);$$

$$x(uw_\circ) \cdot x(vw_\circ) = \sum_w \tau(c_{uv}^w)\, x(ww_\circ);$$

$$x(w_\circ u) \cdot x(w_\circ v) = \sum_w w_\circ(c_{uv}^w)\, x(w_\circ w);$$

$$\bar{y}(w_\circ u w_\circ) \cdot \bar{y}(w_\circ v w_\circ) = \sum_w w_\circ\tau(c_{uv}^w)\, \bar{y}(w_\circ w w_\circ).$$

Note that $w_\circ\tau = \tau w_\circ$, and this preserves products of positive roots.

For general G/P, with cosets $[u], [v], [w] \in W/W_P$, we write

$$y[u] \cdot y[v] = \sum_{[w]} c_{[u][v]}^{[w]}\, y[w]$$

in $H_T^*(G/P)$. Since $H_T^*(G/P)$ embeds in $H_T^*(G/B)$, these coefficients occur among the c_{uv}^w, but they can be characterized directly.

Theorem 19.2.5. *For $[u], [v], [w] \in W/W_P$, fix minimal representatives $u = u^{\min}$, $v = v^{\min}$, and $w = w^{\min}$. The polynomials $c_{[u][v]}^{[w]}$, homogeneous of degree $\ell(u) + \ell(v) - \ell(w)$ in Λ, satisfy and are determined by the following properties, for all $\alpha \in \Delta \smallsetminus \Delta_P$:*

$$c_{[u][u]}^{[u]} = \prod_{\beta \in u(R^-) \cap R^+} \beta, \tag{i}$$

$$\left(y[s_\alpha]\big|_{[u]} - y[s_\alpha]\big|_{[v]}\right) c_{[u][v]}^{[u]} = \sum_{v^+ \gtrdot v} c_\alpha(v, v^+)\, c_{[u][v^+]}^{[u]}, \tag{ii}$$

and

$$\left(y(s_\alpha)\big|_w - y(s_\alpha)\big|_u\right) c_{[u][v]}^{[w]} = \sum_{u^+ \gtrdot u} c_\alpha(u, u^+)\, c_{[u^+][v]}^{[w]}$$

$$- \sum_{w^- \lessdot w} c_\alpha(w^-, w)\, c_{[u][v]}^{[w^-]}, \tag{iii}$$

where the sums are over v^+ such that $\ell(v^+) = \ell(v) + 1$, u^+ such that $\ell(u^+) = \ell(u) + 1$, and w^- such that $\ell(w^-) = \ell(w) - 1$.

The proof goes as usual, using the fact that for any distinct cosets $[u] \neq [v] \in W/W_P$, there is a simple root $\alpha \in \Delta \smallsetminus \Delta_P$ such that $u(\varpi_\alpha) \neq v(\varpi_\alpha)$. This is a consequence of a basic lemma:

Lemma 19.2.6.   *Let $M_P \subseteq M_{\mathbb{R}}$ be the subspace spanned by ϖ_α for simple roots $\alpha \in \Delta \setminus \Delta_P$. The isotropy group for the action of W on M_P is precisely the subgroup $W_P \subseteq W$.*

19.3   Positivity via transversality

When $\ell(w^{\min}) = \ell(u^{\min}) + \ell(v^{\min})$, the coefficients $c^{[w]}_{[u][v]}$ for multiplying Schubert classes in the ordinary cohomology ring $H^*(G/P)$ are non-negative integers. This is proved by an easy transversality argument, which we will review below. A subtler positivity property holds for the equivariant structure constants.

In fact, positivity holds for the class of any subvariety. Let us write the simple roots as $\Delta = \{\alpha_1, \ldots, \alpha_r\}$.

Theorem 19.3.1.   *Let $Z \subseteq G/P$ be a T-invariant subvariety, and write its class as*

$$[Z]^T = \sum_{[u]} c_{[w]} \, x[w]$$

in the Schubert basis $\{x[w]\}$. Then each $c_{[w]}$ lies in $\mathbb{Z}_{\geq 0}[\alpha_1, \ldots, \alpha_r]$.

Positivity of the structure constants with respect to the opposite Schubert basis (note the switch!) is a consequence:

Corollary 19.3.2 (Graham).   *Consider multiplication in $H^*_T(G/P)$ with respect to the basis $\{y[w]\}$. The corresponding structure constants $c^{[w]}_{[u][v]}$ are polynomials in the (positive) simple roots with nonnegative coefficients: $c^{[w]}_{[u][v]}$ lies in $\mathbb{Z}_{\geq 0}[\alpha_1, \ldots, \alpha_r]$.*

Before turning to the proof of the theorem, let us deduce the corollary. Consider the Richardson variety $Z = Y[v] \cap X[w]$. The coefficients $c_{[u]}$ in the expansion $[Z]^T = \sum c_{[u]} \, x[u]$ can be computed as

$$c_{[u]} = \rho_* \left(y[u] \cdot [Z]^T \right),$$

by Poincaré duality. The intersection $Y[v] \cap X[w]$ is proper and generically transverse, so $[Z]^T = y[v] \cdot x[w]$, and it follows that the RHS of the above equation is equal to $c^{[w]}_{[u][v]}$, using Poincaré duality again. By the theorem, these are positive: $c_{[u]} = c^{[w]}_{[u][v]}$ lies in $\mathbb{Z}_{\geq 0}[\alpha_1, \ldots, \alpha_r]$.   □

The theorem includes the statement that the coefficients $c_{[w]}$ are polynomials in roots: that is, they lie in the subring $\Lambda_{\mathrm{rt}} = \mathrm{Sym}^* M_{\mathrm{rt}}$ of $\Lambda = \mathrm{Sym}^* M$. For the structure constants, this can be deduced from the characterization in Theorem 19.2.5, but it is easy to see directly.

Lemma 19.3.3. *The coefficients $c_{[w]}$ lie in the subring $\Lambda_{\mathrm{rt}} \subseteq \Lambda$ of polynomials in roots.*

Proof The adjoint group $G^{\mathrm{ad}} = G/Z(G)$ has parabolic subgroup $P^{\mathrm{ad}} = P/Z(G)$ and maximal torus $T^{\mathrm{ad}} = T/Z(G)$. The map on character groups corresponding to the quotient map $T \to T^{\mathrm{ad}}$ is the inclusion of the root lattice $M_{\mathrm{rt}} \subseteq M$, so $\Lambda_{T^{\mathrm{ad}}} = \Lambda_{\mathrm{rt}}$. There is a canonical isomorphism $G/P \xrightarrow{\sim} G^{\mathrm{ad}}/P^{\mathrm{ad}}$, and the action of T factors through that of T^{ad}, so any T-invariant subvariety is also T^{ad}-invariant. The coefficients for expanding $[Z]^{T^{\mathrm{ad}}}$ in the Schubert basis of $H^*_{T^{\mathrm{ad}}}(G/P)$ lie in Λ_{rt}. The change-of-groups homomorphism embeds $H^*_{T^{\mathrm{ad}}}(G/P)$ as a subalgebra of $H^*_T(G/P)$, sends $[Z]^{T^{\mathrm{ad}}}$ to $[Z]^T$, and preserves the Schubert basis, so the claim follows. $\qquad\square$

We will see two proofs of the positivity theorem. The first uses a refinement of the transversality argument which proves non-equivariant positivity, this time applied to certain fiber bundles. Later we will see a second proof via degeneration, which leads to a more precise result.

As a warmup, let us quickly review the reason why the non-equivariant coefficients are nonnegative integers. The basic tool is a weak version of the Kleiman–Bertini transversality theorem. Recall that two subvarieties $Y, Z \subseteq X$ intersect *properly* if $Y \cap Z$ is either empty or pure-dimensional, with codimension (in X) equal to $\mathrm{codim}_X Y + \mathrm{codim}_X Z$.

Lemma 19.3.4. *Let Γ be a connected algebraic group acting transitively on an irreducible variety X, and let $Y, Z \subseteq X$ be irreducible subvarieties. There is a dense open subset $\Gamma^\circ \subseteq \Gamma$ such that gY meets Z properly for all $g \in \Gamma^\circ$.*

The proof is not difficult. The main point is to show that the action morphism $q \colon \Gamma \times Y \to X$ is flat.

We need a basic fact from intersection theory: If Y and Z are subvarieties of a nonsingular variety X which intersect properly, then

$$[Y] \cdot [Z] = \sum_V m_V [V]$$

in $H^* X$, where the sum is over irreducible components $V \subseteq Y \cap Z$, and the intersection multiplicities m_V are nonnegative integers. (See Proposition A.3.3 of Appendix A.)

Now let us calculate the coefficients $c_{[w]}$ in the non-equivariant case. Writing $\int \colon H^*(G/P) \to \mathbb{Z}$ for the pushforward to a point and using Poincaré duality, we have

$$c_{[w]} = \int y[w] \cdot [Z].$$

Applying Lemma 19.3.4 with $\Gamma = G$, $X = G/P$, and $Y = Y[w]$, we can find $g \in G$ so that gY meets Z properly. Since G is connected, we have $[gY] = [Y]$ for all $g \in G$, and then

$$c_{[w]} = \int y[w] \cdot [Z]$$

$$= \int [gY] \cdot [Z]$$

$$= \sum_V m_V \int [V],$$

where V ranges over components of $gY \cap Z$. The pushforward $\int [V]$ is equal to 1 if V is a point and 0 otherwise, and $m_V \geq 0$ always, so it follows that $c_{[w]} \geq 0$.

Our plan is to prove Theorem 19.3.1 by imitating the above argument for positivity of the non-equivariant coefficients. We will show that $c_{[w]}$ represents the class of a (possibly reducible) subvariety of an approximation space – that is, it is an *effective class*. For any weight the Kleiman–Bertini theorem directly is that the approximation spaces are not homogeneous. However, a small refinement suffices.

Lemma 19.3.5. *Let Γ be a connected algebraic group acting on a nonsingular irreducible variety X with finitely many orbits $X_1, \ldots, X_n$. Let $Y \subseteq X$ be an irreducible subvariety, and assume it meets each Γ-*

orbit properly. Given another irreducible subvariety Z, there is a dense open subset $\Gamma^\circ \subseteq \Gamma$ such that gY meets Z properly for all $g \in \Gamma^\circ$.

Proof Let us write $Y_i = Y \cap X_i$ and $Z_i = Z \cap X_i$ for the intersections with each orbit. If $c = \operatorname{codim}_X Y$, then also $c = \operatorname{codim}_{X_i} Y_i$ for each i, by the assumption that Y meets orbits properly. Let $d = \dim Z$ and $d_i = \dim Z_i$. For each i, we may apply Lemma 19.3.4 to find a dense open subset $\Gamma_i^\circ \subseteq \Gamma$ consisting of g such that $\dim(gY_i \cap Z_i) = d_i - c$ for all $g \in \Gamma_i^\circ$. We claim that $\Gamma^\circ = \bigcap_{i=1}^n \Gamma_i^\circ$ is the desired open subset.

Since X is nonsingular, for any $g \in \Gamma$, every nonempty component of $gY \cap Z$ has dimension at least equal to $d - c$. So it suffices to show that when $g \in \Gamma^\circ$, each component of this intersection has dimension at most $d - c$. Since Z is irreducible and $Z = \coprod_i Z_i$, there is a unique j such that $Z_j \subseteq Z$ is dense; in particular, $d_j = d = \dim Z$, and $d_i < d$ for all $i \neq j$. For each $g \in \Gamma^\circ$, we have $gY \cap Z = \coprod_i gY_i \cap Z_i$, and $\dim gY_i \cap Z_i = d_i - c \leq d - c$, with equality exactly for $i = j$. So $gY_j \cap Z_j \subseteq gY \cap Z$ is dense, and the claim follows. $\qquad\square$

Now we return to the situation of Theorem 19.3.1. First we must characterize effective classes. Any choice of basis $t_1, \ldots, t_r$ for M determines an isomorphism $\Lambda \cong \mathbb{Z}[t_1, \ldots, t_r]$. According to our standard conventions, the positive classes naturally determined by such a basis are polynomials with nonnegative coefficients in the negative variables $-t_1, \ldots, -t_r$. To see why, we employ the corresponding isomorphism $T \cong (\mathbb{C}^*)^r$ along with approximation spaces $\mathbb{E} = (\mathbb{C}^m \smallsetminus 0)^r$ and $\mathbb{B} = (\mathbb{P}^{m-1})^r$ (for $m \gg 0$). The variables are identified as $t_i = c_1(pr_i^* \mathcal{O}(-1))$, so $-t_i = c_1(pr_i^* \mathcal{O}(1))$ is a hyperplane class. Then a homogeneous polynomial in $\Lambda = \mathbb{Z}[t_1, \ldots, t_r]$ is effective if and only if it is a nonnegative combination of monomials in $-t_1, \ldots, -t_r$.

With Lemma 19.3.3 in mind, we will assume $G = G^{\mathrm{ad}}$ for the remainder of this section, so that $M_{\mathrm{rt}} = M$ and $\Lambda_{\mathrm{rt}} = \Lambda = \mathbb{Z}[\alpha_1, \ldots, \alpha_r]$. We fix the basis of negative simple roots $-\alpha_1, \ldots, -\alpha_r$ for M and use this to identify $T = (\mathbb{C}^*)^r$, so that the effective classes for the corresponding approximation space $\mathbb{B} = (\mathbb{P}^{m-1})^r$ are given by polynomials in $\mathbb{Z}_{\geq 0}[\alpha_1, \ldots, \alpha_r]$.

We wish to compute $c_{[w]}$ via an intersection on $\mathbb{E} \times^T G/P$, and show that it is an effective class. To apply Lemma 19.3.5, we need an appropriate group action on this approximation space.

First we set some notation. Let $\boldsymbol{X} = \mathbb{E} \times^T G/P$ be the approximation space, so $\boldsymbol{X} \to \mathbb{B}$ is a fiber bundle over $\mathbb{B} = (\mathbb{P}^{m-1})^r$ with fibers G/P. Similarly, let $\boldsymbol{Y}[w] = \mathbb{E} \times^T Y[w]$ and $\boldsymbol{X}[w] = \mathbb{E} \times^T X[w]$.

Consider T acting on G by conjugation, so $t \cdot g = tgt^{-1}$, and let $\boldsymbol{G} = \mathbb{E} \times^T G$. The projection $\boldsymbol{G} \to \mathbb{B}$ makes this a group scheme over $\mathbb{B}$, acting on $\boldsymbol{X}$ by

$$\boldsymbol{G} \times_{\mathbb{B}} \boldsymbol{X} \to \boldsymbol{X}, \qquad ([e, g], [e, x]) \mapsto [e, gx].$$

Writing $U \subseteq B \subseteq G$ as usual for the unipotent subgroup whose roots are R^+, we have a subgroup $\boldsymbol{U} = \mathbb{E} \times^T U$ of $\boldsymbol{G}$, also acting on $\boldsymbol{X}$.

As a variety with T-action, U is isomorphic to the vector space $\bigoplus_{\beta \in R^+} \mathbb{C}_\beta$, where $\mathbb{C}_\beta$ is the one-dimensional representation with character β. Writing $\beta = n_1 \alpha_1 + \cdots + n_r \alpha_r$ in terms of simple roots, it follows that

$$\boldsymbol{U} \cong \bigoplus_{\beta \in R^+} \mathcal{O}(\beta)$$

as varieties, where

$$\mathcal{O}(\beta) = pr_1^* \mathcal{O}(n_1) \otimes \cdots \otimes pr_r^* \mathcal{O}(n_r)$$

is a line bundle on $\mathbb{B} = (\mathbb{P}^{m-1})^r$.

The key observation is this: since β is a positive root, each $\mathcal{O}(\beta)$ is a globally generated line bundle, so $\boldsymbol{U}$ is generated by global sections.

Let $\Gamma_f = \Gamma(\mathbb{B}, \boldsymbol{U})$ be the space of global sections; this is a group by pointwise multiplication, and it is a finite-dimensional complex vector space, so it is connected. For any given $b \in \mathbb{B}$, there is an evaluation homomorphism $\Gamma_f \to U$, which we denote by $g \mapsto g_b$. This is surjective, since $\boldsymbol{U}$ is globally generated. The group Γ_f has a natural "fiberwise" action on $\boldsymbol{X}$, by $g \cdot [e, x] = [e, g_b \cdot x]$, where $e \in \mathbb{E}$ maps to $b \in \mathbb{B}$.

The group $(GL_m)^r$ acts transitively on $\mathbb{B} = (\mathbb{P}^{m-1})^r$, and this action lifts to the standard action on $\mathbb{E} = (\mathbb{C}^m \smallsetminus 0)^r$. This induces an action of $(GL_m)^r$ on $\boldsymbol{X}$, by $h \cdot [e, x] = [h \cdot e, x]$.

These two actions naturally intertwine: $(GL_m)^r$ acts on Γ_f by $(h \cdot g)_b = g_{h^{-1}b}$ for any $h \in (GL_m)^r$, $g \in \Gamma_f$, and $b \in \mathbb{B}$. Let

$$\Gamma = \Gamma_f \rtimes (GL_m)^r$$

be the corresponding semidirect product. Since $(GL_m)^r$ and Γ_f are connected algebraic groups, so is Γ. This group acts on $\boldsymbol{X}$ by

$$(g, h) \cdot [e, x] = [h \cdot e, g_{h \cdot b} \cdot x],$$

for $g \in \Gamma_f$, $h \in (GL_m)^r$, $x \in G/P$ and $e \in \mathbb{E}$ mapping to $b \in \mathbb{B}$.

Lemma 19.3.6. *The orbits for the action of Γ on the approximation space $\boldsymbol{X} = \mathbb{E} \times^T G/P$ are the approximation spaces $\boldsymbol{X}[w]^\circ = \mathbb{E} \times^T X[w]^\circ$ for Schubert cells.*

Proof The projection $\boldsymbol{X} \to \mathbb{B}$ is equivariant with respect to the projection $\Gamma \to (GL_m)^r$, and $(GL_m)^r$ acts transitively on $\mathbb{B} = (\mathbb{P}^{m-1})^r$. So to determine the orbits of Γ on $\boldsymbol{X}$, it suffices to determine the orbits of Γ_f on any fiber. For a point $b \in \mathbb{B}$, the fiberwise action is given by the evaluation homomorphism $\Gamma_f \to U$. Since this homomorphism is surjective, the orbits of Γ_f on a fiber are the same as those of U on G/P, and the claim follows. $\qquad\square$

In particular, there are finitely many orbits, and $\boldsymbol{Y}[w] = \mathbb{E} \times^T Y[w]$ intersects each of them properly. Now we can prove the positivity theorem, by applying Lemma 19.3.5 to the action of Γ on $\boldsymbol{X}$.

First proof of Theorem 19.3.1 Let $\rho \colon G/P \to \mathrm{pt}$ be the map to a point, with corresponding pushforward $\rho_* \colon H_T^*(G/P) \to \Lambda$. Using Poincaré duality, we know

$$c_{[w]} = \rho_*(y[w] \cdot [Z]^T).$$

There are finitely many coefficients $c_{[w]}$, so one can compute all of them using a single approximation space $\mathbb{E} = (\mathbb{C}^m \smallsetminus 0)^r$ for m sufficiently large. (In fact, $m > \ell(w_\circ^{\min}) = \dim G/P$ suffices, since this is the largest possible degree of $c_{[w]}$.) Fixing such $\mathbb{E} \to \mathbb{B}$, and writing $\boldsymbol{X} = \mathbb{E} \times^T G/P$ as above, let $\overline{\rho} \colon \boldsymbol{X} \to \mathbb{B}$ be the projection. Using the identification $H^i \mathbb{B} = \Lambda^i$ for $i < 2m - 1$, we have

$$c_{[w]} = \overline{\rho}_*([\boldsymbol{Y}[w]] \cdot [\boldsymbol{Z}])$$

where $\boldsymbol{Z} = \mathbb{E} \times^T Z$.

The hypotheses of Lemma 19.3.5 are satisfied with respect to Γ acting on $\boldsymbol{X}$, so there is a $g \in \Gamma$ such that $g\boldsymbol{Y}[w]$ meets $\boldsymbol{Z}$ properly. Since Γ is connected, it follows that

$$[\boldsymbol{Y}[w]] \cdot [\boldsymbol{Z}] = [g\boldsymbol{Y}[w]] \cdot [\boldsymbol{Z}] = [g\boldsymbol{Y}[w] \cap \boldsymbol{Z}].$$

Now $c_{[w]} = \overline{\rho}_* \left([g\boldsymbol{Y}[w] \cap \boldsymbol{Z}] \right)$ is the pushforward of an effective class. By basic properties of Gysin maps, pushforward preserves effective classes (see Appendix A, §A.6). It follows that $c_{[w]}$ is effective, proving the theorem. $\square$

Example 19.3.7. Consider $T = (\mathbb{C}^*)^n$ acting $\mathbb{C}^n$ by standard characters $t_1, \ldots, t_n$, and on the complete flag variety $Fl(\mathbb{C}^n)$ in the usual way. (This is not the adjoint torus!) The approximation space $\mathbb{E} \to \mathbb{B}$ is identified with $(\mathbb{C}^m \smallsetminus 0)^n \to (\mathbb{P}^{m-1})^n$ using the basis $-t_1 + t_2, \ldots, -t_{n-1} + t_n, -t_n$ for the character lattice of T to define the action on $\mathbb{E}$. Let

$$M_i = pr_i^* \mathcal{O}(1) \quad \text{and} \quad L_i = M_i \otimes \cdots \otimes M_n,$$

so $c_1(M_i) = t_i - t_{i+1}$ and $c_1(L_i) = t_i$. Forming the vector bundle $V = L_1 \oplus \cdots \oplus L_n$ on $\mathbb{B}$, we have

$$\text{End}(V) = \bigoplus_{i,j} L_i \otimes L_j^{-1}.$$

By construction, $\text{End}(V)$ has global sections in upper-triangular matrices; that is, $L_i \otimes L_j^{-1}$ is globally generated if and only if $i \leq j$. In this setting, the group Γ_f is formed by taking global sections of $L_i \otimes L_j^{-1}$ for $i < j$.

This construction also identifies $\mathbb{E} \times^T Fl(\mathbb{C}^n) = \mathbf{Fl}(V)$, and choosing opposite flags $\widetilde{E}_p = L_1 \oplus \cdots \oplus L_p$ and $E_p = L_{n+1-p} \oplus \cdots \oplus L_n$, we have $\boldsymbol{X}(w) = \Omega_{w_\circ w}(\widetilde{E}_\bullet)$ and $\boldsymbol{Y}(w) = \Omega_w(E_\bullet)$ inside $\mathbf{Fl}(V)$.

19.4 Positivity via degeneration

Here we will use a degeneration to prove the positivity theorem, closely following the argument originally given by Graham. We are concerned with the action of a connected solvable group B on a nonsingular variety X. Let T be the maximal torus. Our goal is to write the class of a T-invariant cycle in X as a sum of classes of B-invariant cycles, keeping control over the coefficients. To begin, we consider a basic example.

Example 19.4.1. Let B act on $\mathbb{P}^1$ via a homomorphism $B \to GL_2$, given by

$$b \mapsto \begin{bmatrix} \chi(b) & \varphi(b) \\ 0 & 1 \end{bmatrix},$$

where $\chi\colon B \to \mathbb{C}^*$ is a character, and $\varphi\colon B \to \mathbb{A}^1 = \mathbb{G}_a$ is a regular function satisfying the identity

$$\varphi(b_1 b_2) = \varphi(b_1) + \chi(b_1)\varphi(b_2)$$

for $b_1, b_2 \in B$. Note that $\varphi(t) = 0$ for $t \in T$, since T consists of semisimple elements of B, which must map to diagonalizable elements of GL_2. The torus T therefore acts on $\mathbb{C}^2$ by the characters χ and 0.

As we have seen (Example 2.6.2),

$$H_T^* \mathbb{P}^1 = \Lambda[\zeta]/(\zeta + \chi)\zeta,$$

where $\zeta = c_1^T(\mathcal{O}(1))$. The point $0 = [1, 0]$ is fixed by B, and $\infty = [0, 1]$ is fixed by T, and we have

$$[0]^T = \zeta \quad \text{and} \quad [\infty]^T = \zeta + \chi.$$

The relation

$$[\infty]^T = [0]^T + \chi \tag{19.1}$$

writes the class of the T-invariant cycle ∞ as a sum of the classes of the B-invariant cycles 0 and $\mathbb{P}^1$, with coefficients 1 and χ, respectively.

Next we consider a situation which will be used in the induction step of the main theorem. Let $U \subseteq B$ be the unipotent radical, so it is normal in B, with $B = U \cdot T$ and $U \cap T = \{e\}$. Let $U' \subseteq U$ be a normal subgroup of B, with $\dim(U/U') = 1$, so U/U' is isomorphic to the additive group $\mathbb{G}_a$. (Such a U' exists, by solvability.) The action of T on U by conjugation determines an action of T on $U/U' \cong \mathbb{A}^1$ by a character χ. Equivalently, if $\mathfrak{u} \supset \mathfrak{u}'$ are the corresponding Lie algebras, then χ is the weight of T acting on $\mathfrak{u}/\mathfrak{u}'$.

Let $\theta\colon U \to \mathbb{G}_a$ be any homomorphism with kernel U', determining another isomorphism $\bar{\theta}\colon U/U' \xrightarrow{\sim} \mathbb{G}_a$. This is unique up to multiplication by a nonzero scalar.

These two pieces of data, $U' \subset U$ and $\theta\colon U \to \mathbb{G}_a$, will determine a homomorphism $B \to GL_2$ and an action as in the above example. To see how, we record two lemmas.

Let $B' = U' \cdot T$, a (possibly non-normal) closed subgroup of B, with B acting on B/B' by left multiplication. Combining the isomorphisms $\bar{\theta}\colon U/U' \xrightarrow{\sim} \mathbb{G}_a$ and $U/U' \xrightarrow{\sim} B/B'$, we obtain an action of B on $\mathbb{G}_a$.

Lemma 19.4.2.   *The above action of B on $\mathbb{G}_a$ is given by*

$$b \cdot z = \theta(u) + \chi(t)z \qquad\qquad (19.2)$$

for $b = ut$ with $u \in U$, $t \in T$, and $z \in \mathbb{G}_a = \mathbb{C}$.

Proof Consider an element $vU' \in U/U'$, for some $v \in U$. Using the isomorphism $U/U' \cong B/B'$, the element $b = ut$ acts on U/U' by $b \cdot vU' = u \cdot (tvt^{-1})U'$. (To check this, write

$$b \cdot vB' = utvB' = u(tvt^{-1})tB' = u(tvt^{-1})B'$$

in B/B'.) Using the isomorphism $\bar{\theta} \colon U/U' \to \mathbb{G}_a$, sending vU' to $z = \theta(v)$, we see $b \cdot z = \theta(u(tvt^{-1})) = \theta(u) + \theta(tvt^{-1}) = \theta(u) + \chi(t)z$, as claimed.   □

Given a character $\chi \colon T \to \mathbb{C}^*$, let $\chi \colon B \to \mathbb{C}^*$ be the unique extension to a character of B, so $\chi(ut) = \chi(t)$.

Lemma 19.4.3.   *With θ and χ as above, define $\varphi \colon B \to \mathbb{C}$ by $\varphi(ut) = \theta(u)$. Then the zero scheme of φ is B', and $\varphi(b_1 b_2) = \varphi(b_1) + \chi(b_1)\varphi(b_2)$ for $b_1, b_2 \in B$. So the map*

$$b \mapsto \begin{bmatrix} \chi(b) & \varphi(b) \\ 0 & 1 \end{bmatrix}$$

is a homomorphism from B to GL_2.

Proof By our choice of $\theta \colon U \to \mathbb{G}_a$, its zero scheme is $U' \subset U$, and it follows that the zero scheme of φ is B'. Writing $b_i = u_i t_i$ for $i = 1, 2$, we compute

$$\begin{aligned}
\varphi(b_1 b_2) &= \varphi(u_1 t_1 u_2 t_1^{-1} \cdot t_1 t_2) \\
&= \theta(u_1 \cdot t_1 u_2 t_1^{-1}) \\
&= \theta(u_1) + \chi(t_1)\theta(u_2) \\
&= \varphi(b_1) + \chi(b_1)\varphi(b_2),
\end{aligned}$$

as claimed.   □

So B acts on $\mathbb{P}^1$, preserving $\mathbb{A}^1 = \{[z, 1] \mid z \in \mathbb{C}\} \cong B/B'$ and fixing the point $0 = [1, 0]$. Now suppose B acts on an n-dimensional nonsingular variety X, with subgroup $B' \subset B$ and character χ as above. The following proposition provides the main ingredient for an inductive proof of the positivity theorem.

Proposition 19.4.4. *Suppose $Y \subseteq X$ is a B'-invariant effective cycle of dimension k. There are canonically defined B-invariant effective cycles Z_1 and Z_2, with $\dim Z_1 = k$ and $\dim Z_2 = k+1$, such that*

$$[Y]^T = [Z_1]^T + \chi[Z_2]^T \tag{19.3}$$

in $H_T^{2n-2k} X$.

Proof It suffices to make the construction when Y is an irreducible subvariety. If Y is B-invariant, then $Z_1 = Y$ and $Z_2 = 0$ are the canonical choices. Otherwise, let W be the closure of $B \cdot Y$ in X; this is a subvariety of dimension $k+1$. Let

$$V = B \times^{B'} Y,$$

with the natural left action of B. This is a fiber bundle over $B/B' = \mathbb{A}^1$, with fiber Y. The mapping $V \to \mathbb{A}^1 \times X$, sending $[b, y]$ to $(\varphi(b), b \cdot y)$, is a B-equivariant closed embeddding, where B acts diagonally on $\mathbb{A}^1 \times X$.

Using the action of B on $\mathbb{P}^1$ which results from the two lemmas, we have a B-equivariant compactification $\mathbb{A}^1 \subset \mathbb{P}^1$. Let $\overline{V}$ be the closure of V in $\mathbb{P}^1 \times X$, so this is a B-invariant subvariety of dimension $k+1$ which maps onto $W \subseteq X$ by the second projection. Using the first projection $\mathbb{P}^1 \times X \to \mathbb{P}^1$ to pull back the relation (19.1), we obtain a relation

$$[\infty \times X]^T = [0 \times X]^T + \chi \tag{19.4}$$

in $H_T^2(\mathbb{P}^1 \times X)$.

We can intersect both sides of this relation with $[\overline{V}]^T$. On the left-hand side, $\infty \times X$ meets $\overline{V}$ scheme-theoretically in $\infty \times Y \subseteq V$. This is because the zero scheme of φ is B', which preserves Y.

On the other hand, $\overline{V} \to \mathbb{P}^1$ is dominant, so $\overline{V}$ meets $0 \times X$ properly in a positive k-dimensional cycle of the form $0 \times Z_1$. Since $\overline{V}$ and $0 \times X$ are B-invariant subvarieties of $\mathbb{P}^1 \times X$, so is their intersection, and therefore $Z_1 \subseteq X$ is a B-invariant k-cycle. It follows that

$$[\infty \times Y]^T = [0 \times Z_1]^T + \chi \cdot [\overline{V}]^T$$

in $H_T^{2n-2k+2}(\mathbb{P}^1 \times X)$. Pushing forward by the projection $\mathbb{P}^1 \times X \to X$ gives

$$[Y]^T = [Z_1]^T + \chi \cdot d[W]^T,$$

where d is the degree of the map $\overline{V} \to W$. So Z_1 and $Z_2 = dW$ are the required positive B-invariant cycles. $\qquad \square$

With the same setup, so B is a solvable group acting on a nonsingular variety X, suppose we have chosen a *normal chain of subgroups*

$$U = U_N \supset U_{N-1} \supset \cdots \supset U_1 \supset U_0,$$

meaning each U_i is normalized by T, with U_{i-1} normal in U_i and $\dim U_i/U_{i-1} = 1$. Let χ_i be the character of T on the Lie algebra $\mathfrak{u}_i/\mathfrak{u}_{i-1}$, and write $B_i = U_i \cdot T$.

Theorem 19.4.5. *Suppose Y is an effective B_0-invariant k-cycle in X. There are canonical effective B-invariant cycles Z_I for each subset $I \subseteq \{1, \dots, N\}$, with $\dim Z_I = k + \#I$, such that*

$$[Y]^T = \sum_I \left(\prod_{i \in I} \chi_i \right) [Z_I]^T \qquad (19.5)$$

in $H_T^{2n-2k} X$.

Proof We use induction on N, the base case $N = 0$ being trivial. So assume the result for $N-1$. For $J \subseteq \{1, \dots, N-1\}$, we have nonnegative cycles Z'_J, invariant for $B' = U_{N-1} \cdot T$, such that

$$[Y]^T = \sum_J \left(\prod_{j \in J} \chi_j \right) [Z'_J]^T.$$

Applying Proposition 19.4.4 to each Z'_J, we may write

$$[Z'_J]^T = [Z_J]^T + \chi_r [Z_{J \cup \{N\}}]^T$$

for some nonnegative B-invariant cycles Z_J and $Z_{J \cup \{N\}}$, and the result follows. $\qquad \square$

Let $\mathfrak{u} = \bigoplus \mathfrak{u}_\chi$ and $\mathfrak{u}_0 = \bigoplus (\mathfrak{u}_0)_\chi$ be the weight decompositions. In each monomial coefficient $\prod_{i \in I} \chi_i$ from the theorem, no χ occurs more often than $\dim(\mathfrak{u}/\mathfrak{u}_0)_\chi = \dim \mathfrak{u}_\chi - \dim(\mathfrak{u}_0)_\chi$. In particular, if each weight space $\mathfrak{u}_\chi$ is one-dimensional, these coefficients are square-free.

Remark. Given a subgroup $U' \subset U$ which is normalized by T, one can always find a normal chain of subgroups $U_\bullet$ with $U_0 = U'$ and $U_N = U$, and in general there are many ways to choose such a chain. (Assume $U' \subsetneq U$; otherwise there is nothing to choose.) It suffices to choose a corresponding chain of Lie algebras $\mathfrak{u}_\bullet$, which can be done as follows. Since $\mathfrak{u}$ is nilpotent, the normalizer $N_\mathfrak{u}(\mathfrak{u}_0) = \{X \in \mathfrak{u} \mid [X, \mathfrak{u}_0] \subset \mathfrak{u}_0\}$ properly contains $\mathfrak{u}_0$. So there is a one-dimensional subspace $\mathfrak{v}_1 \subset N_\mathfrak{u}(\mathfrak{u}_0)$ which is preserved by T and is not contained in $\mathfrak{u}_0$. Set $\mathfrak{u}_1 = \mathfrak{u}_0 \oplus \mathfrak{v}_1$, and continue inductively.

The normal chain $U_\bullet$ uniquely determines the positive cycles in Theorem 19.4.5, but different chains can produce different cycles.

For B acting on X, we will apply Theorem 19.4.5 to the action of $\widetilde{B} = (U \times U) \cdot T$ on $X \times X$, and deduce the positivity of structure constants. Choose any normal chain of subgroups $U_\bullet$ with U_0 being the diagonal $U \subseteq U \times U$ and $U_N = U \times U$. So $B_N = \widetilde{B}$, and $B_0 = B$ acting diagonally on X.

If Y is a B-invariant positive cycle in $X \cong \delta(X) \subset X \times X$, then Theorem 19.4.5 produces nonnegative $\widetilde{B}$-invariant cycles Z_I in $X \times X$ such that

$$[\delta(Y)]^T = \sum_I \left(\prod_{i \in I} \chi_i \right) [Z_I]^T \qquad (**)$$

in $H_T^*(X \times X)$, where $\delta \colon X \hookrightarrow X \times X$ is the diagonal embedding.

If U acts on X with finitely many orbits, we can be more precise about the cycles Z_I. In this case, $\widetilde{B} = (U \times U) \cdot T$ acts on $X \times X$ with finitely many orbits, and each $\widetilde{B}$-orbit is of the form $V^\circ \times W^\circ$, where V° and W° are B-orbits in X. (This is because U-orbits and B-orbits on X are the same, both being orbits of T-fixed points.) It follows that X is a finite union of $(U \times U)$-orbits $V^\circ \times W^\circ$, the orbits of $(T \times T)$-fixed points.

Let $N = \dim U$, and let $\chi_1, \dots, \chi_N$ be the weights of $\mathfrak{u}$, with repetition according to multiplicity.

Corollary 19.4.6. *Suppose U acts on X with finitely many orbits, and let Y be a B-invariant positive cycle. There are B-invariant subvarieties V_I and W_I of X, and nonnegative integers m_I, indexed by $I \subseteq \{1, \dots, N\}$, such that*

$$[\delta(Y)]^T = \sum_I m_I \left(\prod_{i \in I} \chi_i \right) [V_I \times W_I]^T$$

in $H_T^*(X \times X)$.

Proof Since the $\widetilde{B}$-invariant subvarieties of $X \times X$ are all of the form $V_I \times W_I$, each cycle Z_I appearing in $(**)$ may be written as $\sum m_I (V_I \times W_I)$. The weight spaces associated to a chain from U to $U \times U$ are the same as those occuring in U itself, since $\mathfrak{u} \oplus \mathfrak{u}/\delta(\mathfrak{u})$ is T-equivariantly isomorphic to $\mathfrak{u}$. $\qquad\square$

The case of Corollary 19.4.6 where B is a Borel subgroup of G and $X = G/P$ implies the positivity theorem. As noted above, the fact that weight spaces of $\mathfrak{u}$ are one-dimensional (with weights being the positive roots) implies that the coefficients are squarefree monomials.

Corollary 19.4.7. *For the diagonal embedding* $\delta \colon G/P \hookrightarrow G/P \times G/P$, *we have*

$$\delta_*(x[w]) = \sum_{[u],[v]} c_{[u][v]}^{[w]} \, x[u] \times x[v],$$

in $H_T^*(G/P \times G/P)$, *where each coefficient* $c_{[u][v]}^{[w]}$ *is a nonnegative sum of squarefree monomials in the positive roots.*

This gives another proof of Corollary 19.3.2, since the coefficients $c_{[u][v]}^{[w]}$ appearing in the corollary are the same as the structure constants for multiplying in the basis $\{y[w]\}$, by Poincaré duality (Proposition 3.7.2).

Notes

Chevalley (1994) proved the non-equivariant version of Theorem 19.1.2 for G/B (which is the main case). He also gave a direct argument that the intersection $Y(v) \cap X(w)$ is transverse when $\ell(w) = \ell(v) + 1$. The corresponding formula for G/P, as in Exercise 19.1.4, appears in (Fulton and Woodward, 2004).

Lemma 19.2.6 is proved in (Humphreys, 1990, §1.15).

The Kleiman–Bertini theorem (Lemma 19.3.4) was proved by Kleiman (1974), who gives a more refined statement. Further generalizations and refinements, akin to Lemma 19.3.5, were given by Speiser (1988).

The fact that U-orbits and B-orbits coincide when there are finitely many U-orbits is proved in (Graham, 2001, Lemma 3.3), where it is attributed to Brion.

Positivity of structure constants (Corollary 19.3.2) was conjectured by Billey and by Peterson; see (Billey, 1999, §8, Remark 3). The transversality argument given in §19.3 is based on (Anderson, 2007b; Anderson et al., 2011). Graham's argument via degeneration, as in §19.4, was built on similar constructions by Kumar and Nori (1998), who studied varieties with finitely many U-orbits. See also (Kumar, 2002, §11.4).

Lemma 19.3.4 is proved as in (Kleiman, 1974, Theorem 2(i)). Consider the diagram

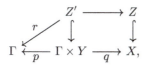

where $q(g, y) = g \cdot y$ is the action map, and p is the projection onto the first factor. One uses generic flatness together with homogeneity to show that q is flat. Then it follows Z' is pure-dimensional, with $\dim Z' = \dim Z + \dim \Gamma - \operatorname{codim}_X Y$. From the theorem on dimension of fibers, there is an open set $\Gamma^\circ \subset \Gamma$ such that either $r^{-1}(g)$ is empty for all g in Γ°, or else $r^{-1}(g)$ has dimension $\dim Z - \operatorname{codim}_X Y$ for all g in Γ°. And $r^{-1}(g) = gY \cap Z$ by definition.

Careful analysis of more explicit versions of the degeneration described in §19.4 have been used in the non-equivariant context to give geometric proofs of Pieri and Littlewood–Richardson rules in some cases (Sottile, 1997; Vakil, 2006). See (Coşkun and Vakil, 2009) for a survey of such results.

Hints for Exercises

Exercise 19.1.3. Use $x(w) = w_\circ \cdot y(w_\circ w)$ and $x(w)|_v = w_\circ \cdot (y(w_\circ w)|_{w_\circ v})$, as in §18.6. Note that the roles of v and w are swapped relative to Theorem 19.1.2.

Exercise 19.1.4. To compute in $H_T^*(G/P)$, one can use pullback by the projection $\pi \colon G/B \to G/P$ to embed the problem in $H_T^*(G/B)$. This gives $\pi^* y[w] = y(w^{\min})$ for any coset $[w] \in W/W_P$. For a simple root $\alpha \in \Delta \smallsetminus \Delta_P$, we have $\pi^* y[s_\alpha] = y(s_\alpha)$, and the desired formula follows from Theorem 19.1.2. (Terms with $\beta \in R_P$ do not appear, so $w = v^{\min} s_\beta$ is minimal in its coset. Indeed, $\langle \varpi_\alpha, \beta^\vee \rangle = 0$ for such β.)

Exercise 19.1.5. Either using the above formula for $y(s_\alpha)$ to reduce to Theorem 19.1.2, or argue more directly by imitating the proof of the theorem and computing the degree of $\mathcal{L}_\lambda|_E$.

Appendix A

Algebraic Topology

This appendix collects some basic facts from algebraic topology needed for our development of equivariant cohomology. Our basic references are the books by Dold (1980) and Spanier (1966) on general algebraic topology, together with Appendix B of (Fulton, 1997).

A.1 Homology and cohomology

We use $H_i X$ and $H^i X$ to denote the singular homology and cohomology of a space with coefficients in $\mathbb{Z}$ (except in §A.8, where we use Čech–Alexander–Spanier cohomology). Some basic properties are these:

- Homology is covariant for continuous maps, and cohomology is contravariant, so there are pushforward and pullback homomorphisms

$$f_* \colon H_i X \to H_i Y \quad \text{and} \quad f^* \colon H^i Y \to H^i X$$

  associated to a continuous map $f \colon X \to Y$. Both are invariant under homotopy: if $f, g \colon X \to Y$ are homotopic, then $f_* = g_*$ and $f^* = g^*$.

- $H^* X$ is a graded-commutative ring (with unit) under *cup product*, written $c \cdot d$. This means

$$c \cdot d = (-1)^{ij} d \cdot c$$

  when $c \in H^i X$ and $d \in H^j X$.

- H_*X is a graded module over H^*X via *cap product*, given by

$$H^i X \otimes H_j X \xrightarrow{\frown} H_{j-i}X.$$

There are relative groups $H^i(X, A)$ for a subspace $A \subseteq X$, with a long exact *sequence of pairs*

$$\cdots \to H^i(X, A) \to H^i(X, B) \to H^i(A, B) \to H^{i+1}(X, A) \to \cdots$$

when $B \subseteq A \subseteq X$. There is a similar sequence for relative homology groups $H_i(X, A)$.

If U and V are open sets in X, there is a long exact *Mayer–Vietoris sequence*

$$\cdots \to H^i(U \cup V) \to H^i U \oplus H^i V \to H^i(U \cap V) \to H^{i+1}(U \cup V) \to \cdots.$$

For subspaces $U \subseteq A \subseteq X$, there is also a canonical *excision isomorphism*

$$H^i(X, A) = H^i(X \smallsetminus U, A \smallsetminus U)$$

whenever the closure of U is contained in the interior of A. See (Spanier, 1966, Chapter 4, §6).

The *universal coefficient theorem* says there is a natural exact sequence

$$0 \to \mathrm{Ext}^1(H_{i-1}X, \mathbb{Z}) \to H^i X \to \mathrm{Hom}(H_i X, \mathbb{Z}) \to 0.$$

See (Spanier, 1966, Chapter 5, §5). The *Künneth theorem* gives natural short exact sequences

$$0 \to \bigoplus_k H_k X \otimes H_{i-k}Y \to H_i(X \times Y) \to \bigoplus_k \mathrm{Tor}_1(H_k X, H_{i-1-k}Y) \to 0.$$

Combined with the universal coefficient theorem, in cases where Ext^1 and Tor_1 vanish, one obtains an isomorphism

$$H^*(X \times Y) \cong H^*X \otimes H^*Y.$$

For example, this holds when $H_i X$ and $H_i Y$ are free, for all i.

There is a *tautness* property for nice enough spaces. Suppose X is locally contractible, paracompact, and Hausdorff, and $A \subseteq X$ is a closed subspace. Then there is an isomorphism

$$\varinjlim H^i(X, U) \xrightarrow{\sim} H^i(X, A),$$

the limit over all open neighborhoods U of A in X. This comes from the tautness property of Alexander–Spanier cohomology, together with the fact that this cohomology theory agrees with singular cohomology under the stated hypotheses (Spanier, 1966, Chapter 6, §6, §9).

The map $X \to$ pt determines a canonical pullback homomorphism $\mathbb{Z} = H^*(\mathrm{pt}) \to H^*X$, which is injective, being split by the pullback $H^*X \to H^*(\mathrm{pt})$ for the inclusion of any point. Often it is notationally convenient to consider the *reduced cohomology groups* $\widetilde{H}^*X := H^*X/H^*(\mathrm{pt})$. Of course, $\widetilde{H}^i(X) = H^iX$ for $i > 0$, and $\widetilde{H}^0X$ has rank one less than H^0X (so it is zero if X is path-connected).

For any space X, the *cohomology with compact support* is

$$H^i_c X = \varinjlim H^i(X, X \smallsetminus K),$$

the direct limit over all compact subspaces $K \subseteq X$. This is not homotopy invariant in general, and it is contravariant only for *proper* maps, i.e., maps so that the inverse image of a compact set is compact. When X is compact, however, $H^i_c X = H^i X$.

When M is an oriented n-manifold and $K \subseteq M$ is a compact subspace, there is an *orientation class* $\mu_K \in H_n(M, M \smallsetminus K)$. This is a local section of the *orientation sheaf* $U \mapsto H_n(M, M \smallsetminus U)$. (For us, a manifold is a Hausdorff space locally homeomorphic to $\mathbb{R}^n$; we follow conventions of [Dold, 1980, §VIII].) Capping with the orientation class defines maps

$$H^i(M, M \smallsetminus K) \to H_{n-i}M, \quad c \mapsto c \frown \mu_K,$$

inducing the *Poincaré isomorphism*

$$H^i_c M \xrightarrow{\sim} H_{n-i}M.$$

When M is compact, the Poincaré isomorphism is given by cap product with the *fundamental class* $\mu_M \in H_n M$. In this case, one has *Poincaré duality*, which says that the bilinear map $H^i M \times H^{n-i} M \to \mathbb{Z}$ given by

$$\langle c, d \rangle = \int_M (c \cdot d) := \rho_*((c \cdot d) \frown \mu_M)$$

is a perfect pairing. Here ρ_* is pushforward by $\rho \colon M \to$ pt.

One can interpret this as giving an isomorphism (modulo torsion)

$$H^i M \xrightarrow{\sim} \mathrm{Hom}(H^{n-i}M, \mathbb{Z}),$$

defined by $c \mapsto (d \mapsto \int c \cdot d)$, which factors into the Poincaré isomorphism $H^{n-i}M = H_i M$ and the *Kronecker isomorphism*, which identifies $H^i M = \mathrm{Hom}(H_i M, \mathbb{Z})$ (modulo torsion).

Suppose that $\{a_i\}$ and $\{b_i\}$ are bases for H^*M which are dual under this pairing, so $\langle a_i, b_j \rangle = \delta_{ij}$. We say $\{a_i\}$ and $\{b_i\}$ are *Poincaré dual bases* in this setting. For any class $c \in H^*X$, basic linear algebra gives

$$c = \sum_i \langle c, b_i \rangle \, a_i.$$

A.2 Borel–Moore homology

We will use a "non-compact" homology theory which pairs with H^* the way H_* pairs with H_c^*. In reference to (Borel and Moore, 1960; Borel and Haefliger, 1961), this is usually called *Borel–Moore homology* now, but it is closely related to constructions worked out by others; see especially (Massey, 1978). We will define these groups only for spaces X which can be embedded as a closed subspace in some oriented n-manifold M. (By the Whitney embedding theorem, one may take $M = \mathbb{R}^n$.) Our assumption that X be embeddable in an oriented manifold holds whenever X is a complex algebraic variety. Indeed, a variety can be covered by finitely many affine varieties, which embed by definition, and it follows that X itself embeds. See (Dold, 1980, §IV.8).

The definition is this:

$$\overline{H}_i X = H^{n-i}(M, M \smallsetminus X).$$

Here are some basic properties.

- The group $\overline{H}_i X$ is independent of the choice of embedding $X \hookrightarrow M$.
- Borel–Moore homology is covariant for proper maps, so there is a pushforward $f_* \colon \overline{H}_i X \to \overline{H}_i Y$ whenever $f \colon X \to Y$ is proper.
- There is a restriction homomorphism $\overline{H}_i X \to \overline{H}_i U$ for open subsets $U \subseteq X$. (This follows immediately from the inclusion of pairs $(M^\circ, M^\circ \smallsetminus U) \hookrightarrow (M, M \smallsetminus X)$, where $M^\circ \subseteq M$ is an open set so that $M^\circ \cap X = U$.)
- There is a long exact sequence

$$\cdots \to \overline{H}_i Z \to \overline{H}_i X \to \overline{H}_i U \to \overline{H}_{i-1} Z \to \cdots$$

  when $Z \subseteq X$ is closed and $U = X \smallsetminus Z$. (Use the cohomology exact sequence for $M \smallsetminus X \subseteq M \smallsetminus Z \subseteq M$.)
- If X is an oriented n-manifold, $H^i X \xrightarrow{\sim} \overline{H}_{n-i} X$. (Use $X = M$ and independence of choice in the definition.)

- If X is compact and locally contractible, then $\overline{H}_i X = H_i X$.
- If X embeds as a closed subspace of an n-manifold, then $\overline{H}_i X = 0$ for $i > n$.

Remark. The last property can fail for singular homology! For example, if X is a countable union of 2-spheres in $\mathbb{R}^3$, all tangent at the origin, then $H_i X \neq 0$ for all $i > 1$—in fact, these groups are uncountable. This computation is due to Barratt and Milnor (1962).

Example. We have

$$\overline{H}_i \mathbb{R}^n = \begin{cases} \mathbb{Z} & \text{if } i = n; \\ 0 & \text{otherwise.} \end{cases}$$

In particular, $\overline{H}_i X$ is not homotopy-invariant. (It is invariant for proper homotopies, i.e., if $F \colon X \times [0,1] \to Y$ is proper, then f_0 and f_1 induce the same homomorphisms $\overline{H}_i X \to \overline{H}_i Y$.)

A.3   Class of a subvariety

A closed subvariety of a nonsingular algebraic variety defines a canonical class in cohomology. We will need a straightforward generalization of this construction to include subvarieties of complex manifolds.

A *complex analytic space* (or *complex analytic variety*) has a sheaf of holomorphic functions, and is covered by open sets which are isomorphic to the zero sets of finitely many holomorphic functions in some domain in $\mathbb{C}^n$. The dimension of an analytic space may be defined as the maximal dimension of its local rings. Any analytic space has a unique decomposition into irreducible components, and its dimension is equal to the maximal dimension of irreducible components. The singular locus V_{sing} of an analytic space V is a closed analytic subspace; the smooth locus $V_{\text{sm}} = V \smallsetminus V_{\text{sing}}$ is a complex manifold. (The book by Grauert and Remmert (1984) is a good reference for details on analytic spaces.)

We will only consider finite-dimensional complex analytic spaces which can be embedded as closed subspaces of a Euclidean space.

Proposition A.3.1. *Let V be a complex analytic variety of dimension k. We have $\overline{H}_i V = 0$ for $i > 2k$, and $\overline{H}_{2k} V = \bigoplus \mathbb{Z}$, the sum over k-dimensional irreducible components.*

The proof is identical to the one given in (Fulton, 1997, §B.3) for algebraic varieties, using the fact that $\dim V_{\text{sing}} < \dim V$.

In particular, if V is irreducible, then it has a *fundamental class* $\eta_V \in \overline{H}_{2k}V$. This has the characterizing property that under the canonical isomorphism

$$0 = \overline{H}_{2k}V_{\text{sing}} \to \overline{H}_{2k}V \to \overline{H}_{2k}V_{\text{sm}} \to \overline{H}_{2k-1}V_{\text{sing}} = 0,$$

we have $\eta_V \mapsto \eta_{V_{\text{sm}}}$, where $\eta_{V_{\text{sm}}}$ is the fundamental class of the (oriented!) complex manifold V_{sm}.

If $V \subseteq X$ is a codimension-d closed irreducible analytic subvariety of a complex manifold, we obtain a class $[V] \in H^{2d}X$ as the image of η_V under

$$\overline{H}_{2k}V = H^{2d}(X, X \smallsetminus V) \to H^{2d}X.$$

A useful property of the fundamental class is that it is local. Suppose $X^\circ \subseteq X$ is an open set such that $V^\circ = V \cap X^\circ$ is connected. (For example, X° could be a small ball around $p \in V$.) Then the restriction $\overline{H}_{2k}V \to \overline{H}_{2k}V^\circ$ is an isomorphism, and it sends η_V to η_{V°.

We can describe how the class of a subvariety behaves under pullbacks and products. In §A.6 we will see how these classes behave under pushforward.

Let $f\colon X \to Y$ be a map of complex manifolds, with $W \subseteq Y$ a closed analytic subvariety of codimension d. Let $Y^\circ \subseteq Y$ be an open set so that $W^\circ = W \cap Y^\circ$ is smooth and connected, and $W^\circ \subseteq Y^\circ$ is defined by local holomorphic equations $h_1 = \cdots = h_d = 0$. Let $V = f^{-1}W \subseteq X$.

Proposition A.3.2. *If Y° can be chosen so that $V^\circ = V \cap f^{-1}Y^\circ$ is also connected, smooth, and defined by equations $h_1 \circ f = \cdots = h_d \circ f = 0$, then $f^*\eta_W = \eta_V$ in $H^{2d}(X, X \smallsetminus V)$. In particular, $f^*[W] = [f^{-1}W]$ in $H^{2d}X$.*

The proof is given in (Fulton, 1997, §B.3). A particularly useful case is when $f\colon X \to Y$ is a submersion of complex manifolds, or a smooth morphism of nonsingular varieties: then the hypotheses of the proposition are satisfied for any pure-dimensional closed analytic subspace $W \subseteq Y$.

Now consider a complex manifold X with closed analytic subvarieties V and W, of codimensions d and e, respectively. Under good conditions, we have $[V] \cdot [W] = [V \cap W]$ in $H^{2d+2e}X$. More precisely, we have:

Proposition A.3.3. *Suppose V and W intersect properly, meaning that each irreducible component $Z \subseteq V \cap W$ has codimension $d + e$. Then*

$$[V] \cdot [W] = \sum m_Z [Z]$$

for some nonnegative integers m_Z.

This can be deduced from the previous proposition, applied to the diagonal map $\Delta \colon X \to X \times X$. In fact, the product class $[V] \cdot [W]$ comes from $\eta_V \cdot \eta_W$, which in turn can be constructed from by a homomorphism

$$H^{2d}(X, X \smallsetminus V) \otimes H^{2e}(X, X \smallsetminus W) \to H^{2d+2e}(X \times X, (X \times X) \smallsetminus (V \times W))$$
$$\to H^{2d+2e}(X, X \smallsetminus (V \cap W)).$$

A similar argument appears in (Fulton, 1997, §B.3); see also (Fulton, 1998, §8).

The product $\eta_V \cdot \eta_W$ is supported on $V \cap W$ regardless of whether the intersection is proper. A particularly useful consequence is that $[V] \cdot [W] = 0$ whenever $V \cap W = \emptyset$.

A useful application of these ideas provides bases for cohomology, under assumptions that often hold. This is a kind of "cellular decomposition" lemma.

Proposition A.3.4. *Let $f \colon X \to Y$ be a morphism of complex manifolds, and suppose there is a filtration*

$$\emptyset \subset X_0 \subset X_1 \subset \cdots \subset X_m = X$$

by closed analytic subsets, such that for each $p \geq 0$,

$$X_p \smallsetminus X_{p-1} = \coprod_j U_{p,j}$$

is a disjoint union of (irreducible) complex manifolds. Assume that for some $N \leq \infty$, the homomorphisms $H^i Y \to H^i U_{p,j}$ are isomorphisms for all p, all j, and all $i < N$. Let $V_{p,j} = \overline{U}_{p,j} \subseteq X$, a closed irreducible analytic subvariety of codimension $d(p,j)$, defining a class $[V_{p,j}] \in H^{2d(p,j)}X$. Then for any $k < N$, any $c \in H^k X$ has a unique expression

$$c = \sum_{p,j} a_{p,j} [V_{p,j}] := \sum_{p,j} (f^* a_{p,j})[V_{p,j}]$$

for some $a_{p,j} \in H^{k-2d(p,j)} Y$.

Proof The hypotheses imply that the Borel–Moore long exact sequence for $X_{p-1} \subset X_p$ breaks into split short exact sequences

$$0 \to \overline{H}_i X_{p-1} \to \overline{H}_i X_p \to \overline{H}_i(X_p \smallsetminus X_{p-1}) \to 0$$

for all $i > 2 \dim X - N$, since the classes $[U_{p,j}] \in \overline{H}_*(X_p \smallsetminus X_{p-1})$ freely generate over $H^* Y$ in this range, and they come from $[V_{p,j}] \in \overline{H}_* X_p$. Applying induction on p proves the claim. $\square$

In the very special case where $Y = \mathrm{pt}$ and $U_{p,j} \cong \mathbb{C}^{n(p,j)}$ (so each $U_{p,j}$ is a *cell*), the proposition says that the classes of the closures of cells form a $\mathbb{Z}$-basis for $H^* X$. This is the usual cellular decomposition statement found in (Fulton, 1997, §B.3).

If $N = \infty$, the conclusion says that the classes $[V_{p,j}]$ form a (free) basis for $H^* X$ as an $H^* Y$-module. Many applications are a particular instance of this, where each $U_{p,j} \to Y$ is a locally trivial bundle whose fibers are affine spaces.

When X and Y are algebraic varieties and the filtration is by algebraic subsets, the same conclusion holds for Chow groups (by the same argument), so long as each $U_{p,j} \to Y$ is an affine bundle with affine-linear transition functions.

Exercise A.3.5. Suppose a connected Lie group G acts by holomorphic automorphisms on a complex manifold X. Show that $[gV] = [V]$ for any analytic subvariety $V \subseteq X$ and any $g \in G$.

Exercise A.3.6. Let $H \subseteq \mathbb{P}^{n-1}$ be a hyperplane, with class $h = [H]$ in $H^2 \mathbb{P}^{n-1}$. Show that $H^* \mathbb{P}^{n-1} = \mathbb{Z}[h]/(h^n)$.

A.4 Leray–Hirsch theorem

A continuous map $\rho \colon X \to Y$ is a *locally trivial fiber bundle*, with fiber F, if there is a covering of Y by open sets U so that $\rho^{-1} U \cong U \times F$, compatibly with the projection to U. The following is a basic fact about the cohomology of such bundles.

Theorem A.4.1. *Let $\rho \colon X \to Y$ be a locally trivial fiber bundle, with fiber $F = F_y = \rho^{-1}(y)$, and N a positive integer or ∞. Assume that for each $i < N$ there are finitely many elements $c_{ij} \in H^i X$ restricting to a*

$\mathbb{Z}$-*basis for* $H^i(F_y)$, *for all* $y \in Y$. *Then for all* $k < N$, *every* $c \in H^k X$ *can be written uniquely as*

$$c = \sum_{i,j} a_{ij} c_{ij}$$

for some $a_{ij} \in H^{k-i} Y$.

The case $N = \infty$ appears in Hatcher's book (Hatcher, 2002, Theorem 4D.1), and the general case is given as an exercise on the companion website. We will briefly sketch an argument.

Proof The basic case where the bundle is trivial, so $X = Y \times F$, follows from the Künneth theorem. When Y can be covered by finitely many trivializing open sets, one uses induction and the Mayer–Vietoris sequence to deduce the conclusion. (In particular, this suffices for compact Y.) A simple Mayer–Vietoris argument also gives the conclusion for an arbitrary *disjoint* union of open sets, when the result is known for each open individually.

The general case requires a bit more care, because the cohomology groups of an arbitrary union of open sets may not be the inverse limit of the cohomology groups of those opens. One may use CW-approximation to conclude the argument (as is done in Hatcher's book). Here is another way to conclude the proof, which works for all spaces of concern to us.

Assume Y is locally compact, Hausdorff, and σ-compact—so it is a union of nested compact sets K_n with $K_n \subseteq \text{int}(K_{n+1})$. Each "band" $K_n \smallsetminus \text{int}(K_{n-1})$ may be covered by an open set $U_n \subseteq \text{int}(K_{n+1}) \smallsetminus K_{n-2}$, which is a finite union of open sets for which the bundle is trivial. By the previous case, one knows the result over each of

$$\mathcal{U} = \coprod U_{3n}, \quad \mathcal{V} = \coprod U_{3n+1}, \quad \text{and} \quad \mathcal{W} = \coprod U_{3n+2},$$

as well as over their pairwise intersections. Since $Y = \mathcal{U} \cup \mathcal{V} \cup \mathcal{W}$, another Mayer–Vietoris argument proves the theorem in this case. $\square$

The analogous statement holds for fiber bundle pairs $(X, X') \to Y$, with the same proof. (That is, $X' \subseteq X$ is a subspace which is also a locally trivial fiber bundle over Y with fibers F'_y, and the hypothesis is that there are classes in $H^i(X, X')$ restricting to a basis for $H^i(F_y, F'_y)$.)

For instance, if $\pi\colon E \to Y$ is a (real) vector bundle of rank r, a *Thom class* is an element $\gamma_E \in H^r(E, E \smallsetminus 0)$ which restricts to a generator of $H^r(E_y, E_y \smallsetminus 0) \cong \mathbb{Z}$ for all $y \in Y$. (Such a class exists if and only if E is orientable.) A special case of the Leray–Hirsch theorem for pairs is the *Thom isomorphism*

$$H^i Y \xrightarrow{\sim} H^{i+r}(E, E \smallsetminus 0), \quad \text{by } c \mapsto \pi^* c \cdot \gamma_E,$$

for all $i \geq 0$.

Another special case is often useful:

Corollary A.4.2. *If $f\colon X \to Y$ is a locally trivial fiber bundle and $\widetilde{H}^i(f^{-1}(y)) = 0$ for all $i < N$, then $f^*\colon H^i X \to H^i Y$ is an isomorphism for all $i < N$.*

The following is proved in the same way as Theorem A.4.1.

Lemma A.4.3. *Let*

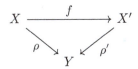

be a map of locally trivial fiber bundles, inducing maps of fibers $X_y \to X'_y$ for $y \in Y$. Assume the pullbacks $H^i X'_y \to H^i X_y$ are isomorphisms for all $y \in Y$ and all $i < N$. Then $f^*\colon H^i X' \to H^i X$ is an isomorphism for all $i < N$.

A.5 Chern classes

For this section, all spaces are at least paracompact and Hausdorff, so that partitions of unity exist. (Any complex analytic space or algebraic variety satisfies these conditions.)

For a complex vector bundle E on a space X, there are *Chern classes* $c_i(E) \in H^{2i} X$ satisfying the following formal properties.

(1) For any continuous map $f\colon X' \to X$, $f^* c_i(E) = c_i(f^* E)$.
(2) $c_i(E) = 0$ unless $0 \leq i \leq r := \mathrm{rk}(E)$, and $c_0(E) = 1$.
(3) For an exact sequence of vector bundles

$$0 \to E' \to E \to E'' \to 0,$$

there is the Whitney sum formula

$$c_k(E) = \sum_{i+j=k} c_i(E') \cdot c_j(E'').$$

Writing $c(E) = \sum c_i(E)u^i$ for the *total Chern class* (where u is a dummy variable), the Whitney formula is equivalent to $c(E) = c(E') \cdot c(E'')$.

There are also geometric properties, for which we assume X is a nonsingular algebraic variety or complex manifold. (Versions of these properties do hold more generally.)

(4) For line bundles L and M, we have $c_1(L \otimes M) = c_1(L) + c_1(M)$.

(5) If $s\colon X \to L$ is a nonzero section with zero locus $Z(s) \subseteq X$, then $[Z(s)] = c_1(L)$ in $H^2 X$.

(6) For the projective bundle of lines in E, $\pi\colon \mathbb{P}(E) \to X$, with tautological line bundle $\mathcal{O}(-1) \subseteq \pi^*E$ and its dual $\mathcal{O}(1)$, we have

$$H^*\mathbb{P}(E) = (H^*X)[\zeta]/(\zeta^r - c_1(E^*)\zeta^{r-1} + \cdots + (-1)^r c_r(E^*)),$$

where $\zeta = c_1(\mathcal{O}(1))$.

The *splitting principle* is useful when computing with Chern classes. Given any vector bundle $E \to X$, there is a map $f\colon X' \to X$ so that $f^*\colon H^*X \to H^*X'$ is injective, and $f^*E \cong L_1 \oplus \cdots \oplus L_r$ splits as a sum of line bundles on X'. (One can take X' to be the flag bundle $\mathbf{Fl}(E) \to X$, with tautological flag $\mathbb{S}_1 \subseteq \cdots \subseteq \mathbb{S}_r = f^*E$, and choose a Hermitian metric to split this filtration into its factors $L_i = \mathbb{S}_i/\mathbb{S}_{i-1}$.)

Exercise A.5.1.   Deduce the following properties from (1)–(6) above.

(7) If E splits as a direct sum $L_1 \oplus \cdots \oplus L_r$, with $x_i = c_1(L_i)$, then $c_i(E) = e_i(x_1, \ldots, x_r)$, the elementary symmetric polynomial in the x variables.

(8) $c_i(E^\vee) = (-1)^i c_i(E)$.

(9) If $s\colon X \to E$ is a section whose zero locus $Z(s) \subseteq X$ has pure codimension $r = \operatorname{rk} E$, then $[Z(s)] = c_r(E)$ in $H^{2r} X$.

(10) For a projective bundle $\pi\colon \mathbb{P}(E) \to X$ with tautological sequence $0 \to \mathcal{O}(-1) \to \pi^*E \to Q \to 0$, and any $\alpha \in H^*X$, we have

$$\pi_*(\pi^*\alpha \cdot c_j(Q)) = \begin{cases} \alpha & \text{if } j = r - 1; \\ 0 & \text{otherwise.} \end{cases}$$

There are also Chern classes for *virtual bundles* $E - F$, defined by the formula

$$c(E - F) = \frac{c(E)}{c(F)} = \frac{1 + c_1(E)u + c_2(E)u^2 + \cdots}{1 + c_1(F)u + c_2(F)u^2 + \cdots},$$

where one expands this rational function in power series about $u = 0$, so $c_k(E - F)$ is the coefficient of u^k in this expansion. This formal notation is often useful when dealing with degeneracy loci.

Exercise A.5.2. Prove the following properties of Chern classes for virtual bundles.

(11) For a vector bundle E of rank r, and a line bundle L, we have

$$c_r(E \otimes L^*) = c_r(E - L).$$

(In particular, if $s\colon L \to E$ is a *twisted* section whose zero locus $Z(s)$ has pure codimension r, then $[Z(s)] = c_r(E - L)$.)

(12) $(-c_1(L))^a \cdot c_b(E - L) = c_{a+b}(E - L)$ for any $a \geq 0$ and $b \geq r$.

A.6 Gysin homomorphisms

Borel–Moore homology can be used to define a *Gysin pushforward homomorphism* for a proper map $f\colon X \to Y$ of complex manifolds. Writing $d = \dim Y - \dim X$, the homomorphism

$$f_*\colon H^iX \to H^{i+2d}Y$$

is defined as the composition

$$H^iX = \overline{H}_{2\dim X - i}X \to \overline{H}_{2\dim X - i}Y = H^{i+2d}Y.$$

A basic instance is when the map $\rho\colon X \to Y$ is a smooth fiber bundle, with compact fibers. Here ρ_* may be identified with integration over the fiber. Another basic case is that of a closed embedding $\iota\colon X \hookrightarrow Y$. Here we have $\iota_*(1) = [X] \in H^{2d}Y$, from the definitions.

Some general properties of Gysin homomorphisms are these.

(1) (Functoriality) For proper maps $X \xrightarrow{f} Y \xrightarrow{g} Z$, the pushforward by the composition is the composition of pushforwards: $(g \circ f)_* = g_* f_*$.

(2) (Projection formula) For elements $b \in H^*Y$ and $a \in H^*X$, we have $f_*(f^*b \cdot a) = b \cdot f_* a$.

(3) (Naturality) For a fiber square

$$
\begin{array}{ccc}
X' & \xrightarrow{g'} & X \\
{\scriptstyle f'}\downarrow & & \downarrow{\scriptstyle f} \\
Y' & \xrightarrow{g} & Y,
\end{array}
$$

with f (and hence f') proper, and $d = \dim Y - \dim X = \dim Y' - \dim X'$, we have $g^* f_* = f'_* (g')^*$. (In the case where $X' = \emptyset$, here we use the convention that $\dim \emptyset$ can be any integer, so this equation holds and says the pushforward-pullback composition is the zero homomorphism.)

(4) (Self-intersection) If $\iota \colon X \hookrightarrow Y$ is a closed embedding with normal bundle N of rank d, then $\iota^* \iota_*(a) = c_d(N) \cdot a$ for any $a \in H^*X$.

(5) (Finite cover) If $f \colon X \to Y$ is proper, $V \subseteq X$ is an irreducible subvariety (possibly singular), and $W = f(V) \subseteq Y$, then

$$
f_*[V] = \begin{cases} \deg(V/W)[W] & \text{if } \dim W = \dim V; \\ 0 & \text{otherwise.} \end{cases}
$$

A.7   The complement of a variety in affine space

With the existence of the fundamental class for an algebraic variety (or algebraic set), one can give a sharp bound on the vanishing of its complement.

Proposition A.7.1.   *If $Z \subset \mathbb{C}^N$ is a Zariski-closed set, of codimension d, then $\pi_i(\mathbb{C}^N \setminus Z) = 0$ for $0 < i \leq 2d - 2$. This is always sharp: $\pi_{2d-1}(\mathbb{C}^N \setminus Z) \neq 0$ if Z is nonempty.*

Proof Identify $\mathbb{C}^n$ with $\mathbb{R}^{2n}$. For a smooth (C^∞) map $f : S^i \to \mathbb{C}^n \smallsetminus Z$, let $\mathscr{S} = \{p \in$ (real) line between Z and $f(S^i)\}$. (This is analogous to a secant variety in algebraic geometry.) Consider the number

$$\dim_{\mathbb{R}} \mathscr{S} = \dim_{\mathbb{R}} Z + \dim_{\mathbb{R}} f(S^i) + \dim_{\mathbb{R}} \mathbb{R}$$
$$\leq 2n - 2d + i + 1,$$

since smoothness of f implies $\dim f(S^i) \leq i$. The condition that this number be less than $2n$ is exactly that $i \leq 2d - 2$. For such i, then, $\mathscr{S} \subsetneq \mathbb{C}^n$; therefore we can find a point $p \notin \mathscr{S}$. Since p does not lie on any line joining Z and $f(S^i)$, the set of line segments between p and $f(S^i)$ lies in $\mathbb{C}^n \smallsetminus Z$. Use this to extend f to a map of the ball $\tilde{f} : D^{i+1} \to \mathbb{C}^n \smallsetminus Z$, thus showing that f is null-homotopic.

Since every continuous map between smooth manifolds is homotopic to a smooth map (Bott and Tu, 1995, pp. 213–214), the homotopy groups can be computed using only smooth maps. Thus $\pi_i(\mathbb{C}^n \smallsetminus Z) = 0$ for $i \leq 2d - 2$.

On the other hand, it follows from this, together with the Hurewicz isomorphism theorem, that $\pi_{2d-1}(\mathbb{C}^n \smallsetminus Z) = H_{2d-1}(\mathbb{C}^n \smallsetminus Z)$ and $H_{2d-2}(\mathbb{C}^n \smallsetminus Z) = 0$. By the universal coefficient theorem and the long exact sequence for the pair $(\mathbb{C}^n, \mathbb{C}^n \smallsetminus Z)$, we have

$$H_{2d-1}(\mathbb{C}^n \smallsetminus Z)^\vee \cong H^{2d-1}(\mathbb{C}^n \smallsetminus Z)$$
$$\cong H^{2d}(\mathbb{C}^n, \mathbb{C}^n \smallsetminus Z)$$
$$= \overline{H}_{2n-2d} Z,$$

and we know this top Borel–Moore homology group is nonzero. $\qquad\square$

Arguing as in the last paragraph, it also follows that the reduced cohomology $\tilde{H}^i(\mathbb{C}^n \smallsetminus Z)$ vanishes for $i \leq 2d - 2$.

We thank D. Speyer for supplying the first part of the proof.

A.8 Limits

In this section, we use Čech–Alexander–Spanier cohomology. As noted earlier, this agrees with singular cohomology for locally finite CW complexes—or more generally for locally contractible paracompact Hausdorff spaces—but it is different for some common examples of infinite-dimensional spaces. For instance, the theories may differ on the inverse limit of locally contractible spaces (which need not be locally contractible).

Consider a directed system $\{X_n\}$ of topological spaces, with direct limit $X = \varinjlim X_n$. There is always a canonical homomorphism $H^*X \to \varprojlim H^*X_n$, and a naive expectation is that this should be an isomorphism. Similarly, given an inverse system $\{X_n\}$ of spaces, with inverse limit $X = \varprojlim X_n$, there is a canonical homomorphism $\varinjlim H^*X_n \to H^*X$, and one naively expects this to be an isomorphism. We will identify some hypotheses which make these naive expectations true, and which are commonly satisfied in applications to algebraic geometry.

For direct systems, we consider CW complexes $\{X_n\}$, where for each $n \le n'$, the map $X_n \to X_{n'}$ is a closed embedding of complexes. The direct limit is the union

$$X = \varinjlim X_n = \bigcup X_n,$$

and is also a CW complex. Then for each i, there is a natural exact sequence

$$0 \to \varprojlim_n{}^1 H^{i-1}X_n \to H^iX \to \varprojlim_n H^iX_n \to 0,$$

where $\varprojlim^1$ is the derived functor of $\varprojlim$. In particular, if H^*X_n vanishes in odd degrees for all n, then there is a natural isomorphism $H^*X \cong \varprojlim H^*X_n$. See (Hatcher, 2002, Theorem 3F.8).

For inverse systems, we consider arbitrary (paracompact Hausdorff) spaces $\{X_n\}$, where for $n \le n'$, the map $X_{n'} \to X_n$ is surjective. It follows that the inverse limit $X = \varprojlim X_n$ maps surjectively to each X_n. Under these conditions, Čech–Alexander–Spanier cohomology satisfies the *continuity axiom*, which says

$$\varinjlim_n H^*X_n \to H^*X$$

is an isomorphism. When all X_n are compact Hausdorff, no surjectivity conditions are needed; see (Spanier, 1966, Chapter 6, Section 6). For non-compact spaces, see (Watanabe, 1987). Note that in general, the limit X need not be paracompact!

These isomorphisms are natural. Suppose we have a map of direct systems $\{f_n\colon X_n \to Y_n\}$, so that each diagram

$$\begin{array}{ccc} X_n & \longrightarrow & X_{n'} \\ f_n \downarrow & & \downarrow f_{n'} \\ Y_n & \longrightarrow & Y_{n'} \end{array}$$

commutes, inducing a map $f\colon X \to Y$ of direct limits. Then the induced homomorphism $\varprojlim H^*Y_n \to \varprojlim H^*X_n$ is the pullback homomorphism $f^*\colon H^*Y \to H^*X$. The same holds for inverse limits of spaces.

Under further conditions, one can define Gysin pushforwards. Suppose each $f_n\colon X_n \to Y_n$ is a proper map of complex manifolds, with $d = \dim Y_n - \dim X_n$ constant for all n. Furthermore, assume each square

$$\begin{array}{ccc} X_n & \xrightarrow{\alpha_{n,n'}} & X_{n'} \\ f_n \downarrow & & \downarrow f_{n'} \\ Y_n & \xrightarrow{\beta_{n,n'}} & Y_{n'} \end{array}$$

is a fiber square, so that $\beta_{n,n'}^*(f_{n'})_* = (f_n)_*\alpha_{n,n'}^*$ by naturality of Gysin homomorphisms. Then the pushforwards $(f_n)_*$ define a Gysin homomorphism

$$f_*\colon H^i X \to H^{i+2d}Y$$

on the cohomology of limit spaces.

One also has fundamental classes of subvarieties. First, we consider a direct system of complex manifolds $\{X_n\}$, with union X. Suppose $V_n \subseteq X_n$ is a direct system of closed subvarieties, such that $V_{n'} \cap X_n = V_n$ for all $n \leq n'$; suppose also that this intersection is transverse, so that $\alpha_{n,n'}^*[V_{n'}] = [V_n]$, and each $V_n \subseteq X_n$ has the same codimension, say d. Then the classes $[V_n] \in H^{2d}X_n$ define an element $([V_n])$ in $\varprojlim H^{2d}X_n$. We take this as a definition of $[V] \in H^{2d}X$, for the closed subspace $V = \varinjlim V_n \subseteq X$.

Next, we consider an inverse system $\{X_n\}$ of spaces, with limit X. For fixed n, suppose a closed subspace $V_n \subseteq X_n$ has a fundamental class $[V_n] \in H^{2d}X_n$. This determines a class in the limit, by the canonical homomorphism $H^{2d}X_n \to H^{2d}X$. For any $n' \geq n$, let $V_{n'} = \pi_{n',n}^{-1}V_n \subseteq X_{n'}$. If, for all $n' \geq n$, the maps $\pi_{n',n}\colon X_{n'} \to X_n$ are such that $\pi_{n',n}^*[V_n] = [V_{n'}]$ in $H^{2d}X_{n'}$, then the classes $[V_{n'}]$ all determine the same class in

$H^{2d}X$. We denote this class by $[V]$, where $V = \pi_n^{-1}V_n \subseteq X$. For example, this holds if all $\pi_{n',n}\colon X_{n'} \to X_n$ are smooth maps of complex manifolds (Proposition A.3.2).

Example A.8.1. Let $X_n = \mathbb{P}^{n-1} = \mathbb{P}(\mathbb{C}^n)$, with $X_n \hookrightarrow X_{n+1}$ given by the linear embedding of $\mathbb{C}^n$ as the span of the first n standard basis vectors in $\mathbb{C}^{n+1}$. Then $X = \bigcup X_n \cong \mathbb{P}^\infty$, so

$$
\begin{aligned}
H^*\mathbb{P}^\infty &= \varprojlim H^*\mathbb{P}^{n-1} \\
&= \varprojlim \mathbb{Z}[t]/(t^n) \\
&= \mathbb{Z}[t],
\end{aligned}
$$

recalling that this is the inverse limit in the category of graded rings.

Let $H_n \subseteq \mathbb{P}^{n-1}$ be the hyperplane spanned by the last $n-1$ standard basis vectors. Then $H = \bigcup H_n \subseteq \mathbb{P}^\infty$ is the subspace where the first coordinate is zero, and $H \cap \mathbb{P}^{n-1} = H_n$, transversely for all n. Thus we can identify $t = [H]$ in $H^*\mathbb{P}^\infty$.

Example A.8.2. Fix a basepoint $p \in \mathbb{P}^\infty$. Let $X_n = \prod_{k=1}^n \mathbb{P}^\infty$, embedded in X_{n+1} by $(p_1, \ldots, p_n) \mapsto (p_1, \ldots, p_n, p)$. The direct limit is the *restricted product* of projective spaces,

$$
X = \varinjlim X_n \cong \prod_{k \geq 1}' \mathbb{P}^\infty,
$$

whose points are countable tuples $(p_1, p_2, \ldots)$ such that $p_i = p$ is the basepoint for all but finitely many coordinates. The inverse limit of cohomology rings is

$$
H^*X = \varprojlim H^*X_n = \varprojlim \mathbb{Z}[t_1, \ldots, t_n] = \mathbb{Z}[\![t_1, t_2, \ldots]\!]_{\mathrm{gr}}.
$$

Here the notation $\mathbb{Z}[\![t]\!]_{\mathrm{gr}} = \mathbb{Z}[\![t_1, t_2, \ldots]\!]_{\mathrm{gr}}$ is used for the *graded formal series* ring. This ring consists of formal sums $\sum c_\alpha t^\alpha$, where each $t^\alpha = t_1^{\alpha_1} t_2^{\alpha_2} \cdots$ is a monomial in finitely many t-variables, and $c_\alpha \in \mathbb{Z}$; the sum may have infinitely many terms but the total degree must be bounded. (For example, $t_1 + t_2 + \cdots$ is an element of this ring.)

Note that H^2X has uncountable rank as a $\mathbb{Z}$-module, and it is not free. (It is isomorphic to the direct product of countably many copies of $\mathbb{Z}$. This is the dual of H_2X, which is isomorphic to the direct sum of countably many copies of $\mathbb{Z}$.)

Example A.8.3. Consider $X_n = \prod_{k=1}^n \mathbb{P}^\infty$ as in the previous example, but as an inverse system via the projection $X_n \to X_{n-1}$ on the first $n-1$ factors. Then

$$X = \varprojlim X_n = \prod_{k \geq 1} \mathbb{P}^\infty$$

is just the usual product of countably many projective spaces. Its cohomology ring is

$$H^*X = \varinjlim H^*X_n = \varinjlim \mathbb{Z}[t_1, \ldots, t_n] = \mathbb{Z}[t_1, t_2, \ldots],$$

the polynomial ring in countably many variables.

Fix i, and let $V(i)_n \subseteq X_n = \prod_{k=1}^n \mathbb{P}^\infty$ be the subspace where the first coordinate of the ith factor is zero; in the notation of Example A.8.1, this is

$$V(i)_n = \prod_{k=1}^{i-1} \mathbb{P}^\infty \times H \times \prod_{k=i+1}^n \mathbb{P}^\infty.$$

Then $[V(i)_n] = t_i$ in $H^*X_n = \mathbb{Z}[t_1, \ldots, t_n]$ for all n, so $[V(i)] = t_i$ in H^*X.

Comparing with the previous example, the embedding of $\prod'_{k \geq 1} \mathbb{P}^\infty$ in $\prod_{k \geq 1} \mathbb{P}^\infty$ induces an inclusion of rings $\mathbb{Z}[t_1, t_2, \ldots] \hookrightarrow \mathbb{Z}[\![t_1, t_2, \ldots]\!]_{\mathrm{gr}}$.

This example also provides an instance of the distinction between singular and Čech–Alexander–Spanier cohomology. Since the Čech–Alexander–Spanier cohomology H^2X is a free module of countable rank, it cannot be the (algebraic) dual of any $\mathbb{Z}$-module, violating the universal coefficient theorem for singular cohomology. Here X is paracompact and Hausdorff, but not locally contractible.

Appendix B

Specialization in Equivariant Borel–Moore Homology

Let Z be a complex variety or algebraic scheme, and let $\pi\colon Z \to \mathbb{A}^1$ be a mapping such that $\pi^{-1}(\mathbb{A}^1 \smallsetminus \{0\}) \to \mathbb{A}^1 \smallsetminus \{0\}$ is a locally trivial topological fibration. Write $Z_0 = \pi^{-1}(0)$ and $Z_1 = \pi^{-1}(1)$. There are *specialization homomorphisms*

$$\sigma\colon \overline{H}_k Z_1 \to \overline{H}_k Z_0.$$

Roughly speaking, specialization spreads out a k-cycle on Z_1 to a $(k+1)$-cycle on $\pi^{-1}((0,1]) \cong Z_1 \times (0,1]$, and takes the limiting k-cycle on $Z_0 = \pi^{-1}(0)$ obtained by taking the closure and restricting to the fiber over 0. The bivariant language of (Fulton and MacPherson, 1981) is a convenient tool for making this precise. Given a in $\overline{H}_k Z_1 = H^{-k}(Z_1 \to \{1\})$, there is a unique α in $H^{-k}(Z \to [0,1])$ which restricts to a at 1. The specialization $\sigma(a)$ is defined to be the restriction of α at 0, in $H^{-k}(Z_0 \to \{0\}) = \overline{H}_k(Z_0)$. Here the bivariant group $H^p(X \to Y)$ is defined as the singular cohomology group $H^{p+n}(Y \times \mathbb{R}^n, Y \times \mathbb{R}^n \smallsetminus X)$, for a closed embedding of X in some $Y \times \mathbb{R}^n$ compatible with the given mapping from X to Y. For details on this construction of specialization see (Fulton and MacPherson, 1981, §3.4).

We will use this in the setup of the *deformation to the normal cone* (Fulton, 1998, §5.1). Let X be a subvariety (or closed subscheme) of a variety (or scheme) Y. One constructs the space

$$M^\circ = M_X^\circ Y = \mathrm{Bl}_{X \times \{0\}}(Y \times \mathbb{A}^1) \smallsetminus \widetilde{Y},$$

where $\widetilde{Y}$ is the proper transform of $Y \times \{0\}$ in the blowup. This comes with an embedding $X \times \mathbb{A}^1 \hookrightarrow M^\circ$. The projection

$$\pi\colon M^{\circ} \to \mathbb{A}^1$$

is trivial away from 0, that is, $\pi^{-1}(\mathbb{A}^1 \smallsetminus \{0\}) \cong Y \times (\mathbb{A}^1 \smallsetminus \{0\})$, with $X \times (\mathbb{A}^1 \smallsetminus \{0\})$ embedded naturally. The special fiber is $\pi^{-1}(0)$ is the normal cone $C_X Y$ to X in Y, which comes with a zero section embedding of X in $C_X Y$ and a projection from $C_X Y$ to X. We obtain specialization maps

$$\sigma = \sigma_{X/Y}\colon \overline{H}_k Y \to \overline{H}_k C_X Y.$$

Lemma. *The specialization to the normal cone satisfies the following properties:*

(1) *Let $f\colon Y' \to Y$ be a smooth morphism of relative (complex) dimension d. Let $X' = f^{-1}(X)$, so f induces a morphism*

$$\overline{f}\colon C_{X'} Y' = C_X Y \times_X X' \to C_X Y,$$

also smooth of relative dimension d. Then the diagram

$$
\begin{array}{ccc}
\overline{H}_k Y & \xrightarrow{\ \sigma_{X/Y}\ } & \overline{H}_k C_X Y \\[2pt]
{\scriptstyle f^*}\downarrow & & \downarrow{\scriptstyle \overline{f}^*} \\[2pt]
\overline{H}_{k+2d} Y' & \xrightarrow{\ \sigma_{X'/Y'}\ } & \overline{H}_{k+2d} C_{X'} Y'
\end{array}
$$

commutes.

(2) *Let $c \in H^i Y$, $a \in \overline{H}_k Y$. Then*

$$\sigma_{X/Y}(c \frown a) = \varphi^*(c) \frown \sigma_{X/Y}(a)$$

in $\overline{H}_{k-i}(C_X Y)$, where φ is the composite $C_X Y \to X \to Y$.

(3) *If V is a closed subscheme of Y of pure dimension n, then $C_{X \cap V} V$ is a closed subscheme of $C_X Y$ of pure dimension n, and*

$$\sigma_{X/Y}([V]) = [C_{X \cap V} V]$$

in $\overline{H}_{2n} C_X Y$. In particular, if Y and X are pure-dimensional, then $\sigma_{X/Y}([Y]) = [C_X Y]$ and $\sigma_{X/Y}([X]) = [X]$.

Proof We use the basic notions of (Fulton and MacPherson, 1981).

(1) The morphism $M^{\circ}_{X'} Y' \to \mathbb{A}^1$ factors as

$$M^{\circ}_{X'} Y' \xrightarrow{F} M^{\circ}_X Y \xrightarrow{\pi} \mathbb{A}^1,$$

with F smooth of relative dimension d. Restricting over the interval $[0, 1]$, we have mappings

$$M' \to M \to [0, 1].$$

The first map has a canonical orientation class θ in $H^{-2d}(M' \to M)$, which is the restriction of the orientation class of the smooth mapping F. Let α in $H^{-k}(M \to [0, 1])$ restrict to a in $H^{-k}(Y \to \{1\}) = \overline{H}_k(Y)$ at 1. The product $\theta \cdot \alpha$ in $H^{-2d-k}(M' \to [0, 1])$ restricts to $f^*(a)$ at 1 and to $\overline{f}^*(\sigma_{X/Y}(a))$ at 0. This gives the required equation $\sigma_{X'/Y'}(f^*(a)) = \overline{f}^*(\sigma_{X/Y}(a))$.

(2) With the notation just set up, let $\widetilde{c} \in H^i(M)$ be the pullback of c by the projection $M \to Y \times [0, 1] \to Y$. Then $\widetilde{c} \cdot \alpha$ restricts to $c \frown a$ at 1 (since bivariant products and pullbacks commute), so it restricts to $\sigma_{X/Y}(c \frown a)$ at 0. But $\widetilde{c}$ restricts to $\varphi^*(c)$ and α restricts $\sigma_{X/Y}(a)$ at 0, so their product $\widetilde{c} \cdot \alpha$ restricts to $\varphi^*(c) \frown \sigma_{X/Y}(a)$ at 0, as required.

(3) This follows from the fact that $[M^{\circ}_{X \cap V} V]$ restricts to $[V]$ at 1 and to $[C_{X \cap V} V]$ at 0. $\qquad \square$

Using this lemma, we can define specialization in equivariant Borel–Moore homology. Let $X \subseteq Y$ be a G-invariant closed subvariety or subscheme. For any approximation $\mathbb{E} \to \mathbb{B}$ to $\mathbb{E}G \to \mathbb{B}G$, we have specialization maps

$$\overline{H}_{k+2b}(\mathbb{E} \times^G Y) \to \overline{H}_{k+2b}(\mathbb{E} \times^G C_X Y) \qquad \text{(B.1)}$$

where $b = \dim(\mathbb{B}) = \dim(\mathbb{E}) - \dim(G)$. If $\mathbb{E}' \to \mathbb{B}'$ is another approximation, the two specialization maps can be compared by using the product $\mathbb{E} \times \mathbb{E}' \to \mathbb{B} \times \mathbb{B}'$:

$$
\begin{array}{ccc}
\overline{H}_{k+2b}(\mathbb{E} \times^G Y) & \longrightarrow & \overline{H}_{k+2b}(\mathbb{E} \times^G C_X Y) \\
\downarrow & & \downarrow \\
\overline{H}_{k+2b''}(\mathbb{E} \times \mathbb{E}' \times^G Y) & \longrightarrow & \overline{H}_{k+2b''}(\mathbb{E} \times \mathbb{E}' \times^G C_X Y) \\
\uparrow & & \uparrow \\
\overline{H}_{k+2b'}(\mathbb{E}' \times^G Y) & \longrightarrow & \overline{H}_{k+2b'}(\mathbb{E}' \times^G C_X Y)
\end{array}
$$

where $b' = \dim(\mathbb{B}')$ and $b'' = \dim(\mathbb{E}) + \dim(\mathbb{E}') - \dim(G)$. This diagram commutes by (1) of the lemma. Hence we have well-defined *equivariant specialization maps*

$$\sigma^G_{X/Y} \colon \overline{H}^G_k Y \to \overline{H}^G_k C_X Y, \tag{B.2}$$

defined to be the maps of (B.1) for any approximation that is sufficiently contractible for the given k. These take $[V]^G$ to $[C_{X \cap V} V]^G$ for any pure-dimensional closed G-invariant subscheme V of Y, by (3). In particular, for Y and X pure-dimensional,

$$\sigma^G_{X/Y}([Y]^G) = [C_X Y]^G \quad \text{and} \quad \sigma^G_{X/Y}([X]^G) = [X]^G.$$

It follows from (2) that $\sigma^G_{X/Y}$ is a homomorphism of Λ_G-modules.

A similar specialization map is defined for equivariant Chow groups. In this case, the construction is more direct: one uses the setup of (Fulton, 1998, §5.2) applied to approximation spaces. Part (3) of the lemma shows that the specialization homomorphisms commute with the canonical maps from equivariant Chow groups to equivariant Borel–Moore homology groups.

Appendix C

Pfaffians and Q-Polynomials

The main goal of this appendix is to provide a characterization of certain symmetric functions, known as double Q-polynomials, which is analogous to the interpolation characterization of the double Schur polynomials (Chapter 9, §9.4). As preparation, we provide some background on Pfaffians and the classical (single) Q-polynomials.

C.1 Pfaffians

We quickly review some basic facts about Pfaffians. This is standard material; (Knuth, 1996) is a good source for refinements and historical context.

Let $M = (m_{ij})_{1 \leq i,j \leq 2r}$ be a skew-symmetric matrix of even size, with entries in a commutative ring. The determinant of such a matrix has a canonical square root, called the *Pfaffian*. It may be defined as

$$\mathrm{Pf}(M) := \sum_{\pi} (-1)^{\ell(\pi)} m_{\pi(1)\pi(2)} \cdots m_{\pi(2r-1)\pi(2r)},$$

the sum over all *matchings* $\pi \in S_{2r}$—that is, permutations such that $\pi(2i - 1) < \pi(2i)$ for all i, and $\pi(1) < \pi(3) < \cdots < \pi(2r - 1)$. For example, the Pfaffian of a 2×2 matrix is $\mathrm{Pf}(M) = m_{12}$, and for a 4×4 matrix, it is

$$\mathrm{Pf} \begin{pmatrix} \cdot & m_{12} & m_{13} & m_{14} \\ \cdot & \cdot & m_{23} & m_{24} \\ \cdot & \cdot & \cdot & m_{34} \\ \cdot & \cdot & \cdot & \cdot \end{pmatrix} = m_{12}m_{34} - m_{13}m_{24} + m_{14}m_{23}.$$

Just as for determinants, if the $2r \times 2r$ matrix M has rank less than $2r$, then $\text{Pf}(M) = 0$. Pfaffians generalize determinants, in the sense that if M is a block matrix of the form

$$M = \left(\begin{array}{c|c} 0 & A \\ \hline -A^\dagger & 0 \end{array} \right),$$

then $\text{Pf}(M) = \pm \det(A)$. (The sign is $(-1)^{\binom{r}{2}}$.)

A Laplace-type expansion formula is useful for computing recursively:

$$\text{Pf}(M) = \sum_{i=1}^{2r-1} (-1)^{i-1} m_{i,2r} \, \text{Pf}(M_{\widehat{i},\widehat{2r}}),$$

where $M_{\widehat{i},\widehat{j}}$ is the $(2r-2) \times (2r-2)$ submatrix obtained by removing the ith and jth rows and columns.

Like the determinant, the Pfaffian is multilinear in the rows and columns of a matrix, but to preserve skew-symmetry, one must modify rows and columns simultaneously. More precisely, let $M = (m_{ij})$, $M' = (m'_{ij})$, and $M'' = (m''_{ij})$ be skew-symmetric matrices of the same size. For some fixed k, suppose $m_{ik} = a\, m'_{ik} + b\, m''_{ik}$ for all i, and $m_{ij} = m'_{ij} = m''_{ij}$ for all i,j not equal to k. Then $\text{Pf}(M) = a\, \text{Pf}(M') + b\, \text{Pf}(M'')$.

In particular, the Pfaffian is unchanged by adding a multiple of one row to another, and simultaneously doing the same for corresponding columns. (Take the kth row and column of M'' to be equal to the ℓth row and column of M'.) For a skew-symmetric matrix M and any $2r \times 2r$ matrix A, we have

$$\text{Pf}(A^\dagger M A) = \det(A) \cdot \text{Pf}(M).$$

Finally, Schur's identity

$$\text{Pf}\left(\frac{x_i - x_j}{x_i + x_j} \right) = \prod_{i<j} \frac{x_i - x_j}{x_i + x_j}$$

plays a role analogous to that of the Vandermonde identity for determinants.

C.2   Schur Q-polynomials

Before discussing the double Q-polynomials, we review some properties of the "single" Q-polynomials which they generalize. These are classical symmetric functions defined by Schur in his study of projective representations of the symmetric group. More details and proofs may be found in Macdonald's book (Macdonald 1995, §III.8).

C.2.1　Pfaffian formula.

We consider formal variables $q_1, q_2, \ldots$, with $q_0 = 1$, often writing $q = 1 + q_1 + q_2 + \cdots$ for the corresponding generating series. Our Q-polynomials will be elements of the polynomial ring in these variables. We will define $Q_\lambda = Q_\lambda(q)$ for any sequence $\lambda = (\lambda_1, \ldots, \lambda_s)$ of nonnegative integers.

For $s = 1$, we set $Q_{(a)} = 1$.

For $s = 2$, we set

$$Q_{ab} = q_a\, q_b + 2 \sum_{i=1}^{b} (-1)^i q_{a+i}\, q_{b-i} \tag{C.1}$$

$$= q_a\, q_b - 2\, q_{a+1} q_{b-1} + 2\, q_{a+2} q_{b-2} - \cdots + (-1)^b 2\, q_{a+b}.$$

In particular, $Q_{(a,0)} = Q_{(a)} = q_a$.

For $s > 2$, we define a skew-symmetric matrix $M_\lambda(q)$ by specifying its entries m_{ij} for $i < j$, as follows. If s is even, then $M_\lambda(q)$ is the $s \times s$ skew-symmetric matrix with entries $m_{ij} = Q_{\lambda_i, \lambda_j}$. If s is odd, we append $\lambda_{s+1} = 0$ and set $M_\lambda(q) = M_{(\lambda_1, \ldots, \lambda_s, 0)}(q)$.

Definition.　The *Schur Q-polynomial* $Q_\lambda = Q_\lambda(q)$ is defined to be the Pfaffian

$$Q_\lambda(q) = \mathrm{Pf}(M_\lambda(q)).$$

In particular, $Q_{(a)} = Q_{a0} = q_a$, and $Q_{(a,b)} = Q_{ab}$.

The Laplace-type expansion formula gives an alternative way to compute Q_λ for a strict partition with an odd number of parts:

$$Q_\lambda = \sum_{i=1}^{s} (-1)^{i-1} q_{\lambda_i} \cdot Q_{(\lambda_1, \ldots, \widehat{\lambda_i}, \ldots, \lambda_s)}.$$

For example,

$$Q_{(a,b,c)} = q_a Q_{bc} - q_b Q_{ac} + q_c Q_{ab} \qquad \text{and}$$

$$Q_{(a,b,c,d)} = Q_{ab} Q_{cd} - Q_{ac} Q_{bd} + Q_{ad} Q_{bc}.$$

The second reduces to the first when $d = 0$.

C.2.2 Raising operators. Another way of formulating the definition of Q_λ will be useful when discussing double Q-polynomials. We consider formal, labelled, monomials $q(1)_{a_1} \cdots q(s)_{a_s}$, allowing negative indices a_i. For $i < j$, the *raising operator* R_{ij} acts on such labelled monomials by taking one away from the jth index and adding it to the ith index. That is,

$$R_{ij} \cdot (\cdots q_{a_i}(i) \cdots q_{a_j}(j) \cdots) = \cdots q_{a_i+1}(i) \cdots q_{a_j-1}(j) \cdots .$$

A monomial R in R_{ij}'s is also called a raising operator.

By taking linear combinations one obtains compact formulas. For example,

$$Q_{ab} = \frac{1 - R_{12}}{1 + R_{12}} \cdot q(1)_a \, q(2)_b,$$

as can be seen by expanding the fraction as

$$\frac{1 - R_{12}}{1 + R_{12}} = 1 - 2R_{12} + 2R_{12}^2 - 2R_{12}^3 + \cdots .$$

More generally, the Pfaffian formula for Q-polynomials takes the form

$$Q_\lambda = \left(\prod_{i<j} \frac{1 - R_{ij}}{1 + R_{ij}} \right) \cdot q(1)_{\lambda_1} \cdots q(s)_{\lambda_s},$$

regardless of the parity of s. To see this, first write $R_{ij} = T_i \cdot T_j^{-1}$, where T_i is the operator which adds 1 to the ith index. In the case where s is even, let A be the diagonal matrix with entries $q(1)_{\lambda_1}, \ldots, q(s)_{\lambda_s}$, and let M be the skew-symmetric matrix with entries

$$\frac{1 - R_{ij}}{1 + R_{ij}} = \frac{T_j - T_i}{T_j + T_i}.$$

So the matrix $A^\dagger M A$ has entries

$$Q_{\lambda_i,\lambda_j} = \frac{1 - R_{ij}}{1 + R_{ij}} \cdot q(i)_{\lambda_i} \, q(j)_{\lambda_j}.$$

Applying Schur's identity and using $\mathrm{Pf}(A^\dagger M A) = \mathrm{Pf}(M) \cdot \det(A)$, we obtain the asserted equality between the Pfaffian and raising-operator formulas for Q_λ. The case where s is odd can be reduced to the even case by appending $\lambda_{s+1} = 0$ as before, because in the expression

$$\left(\prod_{1 \le i < j \le s+1} \frac{1 - R_{ij}}{1 + R_{ij}} \right) \cdot q(1)_{\lambda_1} \cdots q(s)_{\lambda_s} \cdot q(s+1)_0,$$

no terms involving $s + 1$ remain after evaluating $q(s + 1)_0 = 1$ and $q(s + 1)_a = 0$ for $a < 0$.

See (Macdonald, 1995, §I.1, §I.3, §III.8) and (Anderson and Fulton, 2018, §A.1) for more details on raising operators and Pfaffians.

C.2.3 The ring of Q-polynomials.

We are particularly interested in the images of the Q-polynomials in a certain quotient ring.

Definition. Let

$$\Gamma = \mathbb{Z}[q_1, q_2, \ldots]/(Q_{pp})_{p>0},$$

where the relations

$$Q_{pp} = q_p^2 - 2\,q_{p+1}\,q_{p-1} + 2\,q_{p+2}\,q_{p-2} - \cdots$$

are as in formula (C.1) from §C.2.1.

For $n > 0$, let

$$\Gamma^{(n)} = \Gamma/(q_{n+1}, q_{n+2}, \ldots)$$
$$= \mathbb{Z}[q_1, \ldots, q_n]/(Q_{pp})_{1 \le p \le n}.$$

We will recycle notation, writing q and Q_λ for their images in Γ or $\Gamma^{(n)}$. The relations $Q_{pp} = 0$ imply $Q_{ab} = -Q_{ba}$ for all a, b, and more generally, that Q_λ is alternating in the indices λ. In particular, Q_λ is zero if λ contains repeated indices, so we may consider only the polynomials indexed by *strict partitions* $\lambda = (\lambda_1 > \cdots > \lambda_s > 0)$.

Just as the Schur polynomials s_λ form a basis for the ring of symmetric functions, the Q-polynomials are a basis for Γ.

Exercise C.2.1. Show that $\Gamma^{(n)}$ has a basis of squarefree monomials in $q_1, \ldots, q_n$. That is,

$$\Gamma^{(n)} = \bigoplus_\lambda \mathbb{Z} \cdot q_{\lambda_1} \cdots q_{\lambda_s},$$

where the sum ranges over strict partitions λ with no part larger than n. Conclude that $\Gamma^{(n)}$ is free of rank 2^n. Then deduce that Γ has a basis of squarefree monomials in $q_1, q_2, \ldots$; that is, of monomials $q_{\lambda_1} \cdots q_{\lambda_s}$ as λ ranges over all strict partitions.

Exercise C.2.2. Show that $\{Q_\lambda(q) \mid \lambda = (n \geq \lambda_1 > \cdots > \lambda_s \geq 0)\}$ form a basis for $\Gamma^{(n)}$. Similarly, show that $\{Q_\lambda(q) \mid \lambda = (\lambda_1 > \cdots > \lambda_s \geq 0)\}$ form a basis for Γ.

C.2.4 Symmetric functions and a tableau formula. The classical Schur Q-polynomials are obtained by further specializing the q variables. Let $x = (x_1, x_2, \ldots)$ be a set of variables, and consider the evaluation

$$q \mapsto q(x) := \prod_{i \geq 1} \frac{1 + x_i}{1 - x_i}.$$

Since $q(x) \cdot q(-x) = 1$, the relations $Q_{pp} = 0$ hold for $p > 0$, so this specialization descends to Γ. In fact, it defines an embedding of Γ into the ring of symmetric functions in x. For a strict partition λ, Schur's Q-function $Q_\lambda(x)$ is the image of Q_λ under this homomorphism.

The Q-polynomial $Q_\lambda(x) = Q_\lambda(x_1, \ldots, x_n)$ is defined similarly for a finite set of variables, by specializing

$$q \mapsto q(x) := \prod_{i=1}^{n} \frac{1 + x_i}{1 - x_i}.$$

There is a combinatorial tableau formula for $Q_\lambda(x)$, analogous to the one for Schur functions $s_\lambda(x)$ in terms of semistandard Young tableaux. This will be a sum over certain fillings of the *shifted diagram* of a strict partition: one writes represents λ by placing λ_i boxes in the ith row, as usual, but the ith row is indented by $i - 1$. For instance, $\lambda = (6, 4, 2, 1)$ has shifted diagram

A *shifted primed tableau* of shape λ is a filling $\mathcal{T}$ of the shifted diagram using the ordered entries $1' < 1 < 2' < 2 < \cdots$, so that rows and columns are weakly increasing; furthermore, the unprimed entries strictly increase down columns, and the primed entries strictly increase along rows. The corresponding monomial weight $x^{\mathcal{T}}$ records a factor x_i

for each i or i' appearing in $\mathcal{T}$. For example, with $\lambda = (6, 4, 2, 1)$, the tableau

has weight $x^{\mathcal{T}} = x_1^2 \, x_2^4 \, x_3^3 \, x_4^3 \, x_5$.

Then

$$Q_\lambda(x) = \sum_{\mathcal{T}} x^{\mathcal{T}},$$

the sum over all shifted primed tableaux of shape λ. See (Macdonald, 1995, §III.8), where this is deduced from a version of the Pieri rule for Q-polynomials. To obtain a formula for $Q_\lambda(x_1, \ldots, x_n)$, one simply restricts the entries to $1' < 1 < \cdots < n' < n$.

From the rule defining a shifted marked tableau, it is easy to see that the entries along the main (southwest) diagonal may be freely chosen between primed or unprimed labels. It follows that the symmetric function $Q_\lambda(x)$ is divisible by 2^s, whenever λ has s nonzero parts.

Furthermore, the entry in the ith position along the main diagonal must be greater than or equal to i'. From this one concludes

$$Q_\lambda(x_1, \ldots, x_n) = 0 \qquad \text{if } n < s, \tag{C.2}$$

that is, when the number of variables is less than the number of nonzero parts of λ.

Exercise C.2.3. Show that $Q_\rho(x) = 2^s s_\rho(x)$, where $\rho = (s, s - 1, \ldots, 1)$ is the staircase partition.

C.3   Double Q-polynomials and interpolation

The double Q-functions were introduced by Ivanov (2005), and their connection to the equivariant cohomology of the Lagrangian Grassmannian was established by Ikeda (2007); see also (Ikeda et al., 2011). These authors consider functions in two sets of variables x and y, which specialize to the Schur Q-function $Q_\lambda(x)$ when the y variables are 0.

We require a variation which is based on work of Kazarian (2000), and specializes to Ivanov's polynomial. Our $Q_\lambda(q|y)$ is a polynomial in two sets of variables, q and y, which specializes to the polynomials considered in the previous section when $y = 0$: $Q_\lambda(q|0) = Q_\lambda(q)$. Our main goal is to prove a theorem which characterizes these polynomials by their interpolation properties.

Here we will always assume the q variables satisfy the relations $Q_{pp} = 0$, so they belong to the ring Γ, and the double Q-polynomial lies in $\Gamma[y]$.

Definition. Fix a sequence $\lambda = (\lambda_1, \ldots, \lambda_s)$ of nonnegative integers. Let

$$q(k) = q \cdot (1 + y_1) \cdots (1 + y_{\lambda_k - 1})$$

for each $1 \le k \le s$, so

$$q(k)_a = q_a + e_1(y_1, \ldots, y_{\lambda_k - 1}) \, q_{a-1} + \cdots + e_a(y_1, \ldots, y_{\lambda_k - 1}),$$

where the e_i are elementary symmetric polynomials. The *double Q-polynomial* is

$$Q_\lambda(q|y) = \left(\prod_{1 \le i < j \le s} \frac{1 - R_{ij}}{1 + R_{ij}} \right) \cdot q(1)_{\lambda_1} \cdots q(s)_{\lambda_s}.$$

As with the single Q-polynomials, this raising operator formula is equivalent to a Pfaffian formula. We define

$$Q_{ab}(q|y) = q(1)_a \, q(2)_b + \sum_{i=1}^{b} (-1)^i q(1)_{a+i} \, q(2)_{b-i},$$

where $q(1) = q \cdot (1 + y_1) \cdots (1 + y_{a-1})$ and $q(2) = q \cdot (1 + y_1) \cdots (1 + y_{b-1})$. Let $M_\lambda(q|y)$ be the matrix with entries $m_{ij} = Q_{\lambda_i, \lambda_j}(q|y)$. Then

$$Q_\lambda(q|y) = \mathrm{Pf}(M_\lambda(q|y)).$$

where as before one augments λ by adding $\lambda_{s+1} = 0$ in case s is odd.

For example, if $s = 1$, so $\lambda = (a)$, we have

$$Q_{(a)}(q|y) = q(1)_a = q_a + e_1(y_1, \ldots, y_{a-1}) \, q_{a-1} + \cdots$$
$$+ e_{a-1}(y_1, \ldots, y_{a-1}) \, q_1.$$

(Note $e_a(y_1, \ldots, y_{a-1}) = 0$.) Similarly, we have $Q_{(a,b)}(q|y) = Q_{ab}(q|y)$.

Exercise C.3.1. Verify that the matrix $M_\lambda(q|y)$ is skew-symmetric; that is, $Q_{ab}(q|y) = -Q_{ba}(q|y)$ and $Q_{aa}(q|y) = 0$.

It follows from this exercise that $Q_\lambda(q|y)$ is alternating in the indices λ, so from now on we will consider only strict partitions $\lambda = (\lambda_1 > \cdots > \lambda_s > 0)$.

Exercise C.3.2. Show that the polynomials $Q_\lambda(q|y)$ form a basis for $\Gamma[y]$ over $\mathbb{Z}[y]$.

Given a set S of positive integers, we define a specialization homomorphism $\Gamma \to \mathbb{Z}[y_i \,|\, i \in S]$ by sending q to

$$q|_S = y^S := \prod_{i \in S} \frac{1 - y_i}{1 + y_i}.$$

Up to sign, when $S = \mathbb{Z}_{>0}$, this gives the standard embedding of Γ into the ring of symmetric functions in y. For $S = \{i_1, i_2, \ldots\}$, the image of the (single) Q-polynomial $Q_\lambda(q)$ under this specialization is the Schur Q-function $Q_\lambda(-y_{i_1}, -y_{i_2}, \ldots)$.

A strict partition $\mu = (\mu_1 > \cdots > \mu_r > 0)$ may be regarded as a set S, and we use the notation $q|_\mu = y^\mu$ to denote the corresponding specialization.

Here is the main theorem about the polynomials $Q_\lambda(q|y)$.

Theorem. *The double Q-polynomials satisfy and are characterized by two properties:*

$$Q_\lambda(y^\lambda|y) = \prod_{k=1}^{s} \prod_{\substack{\ell \leq \lambda_k \\ \ell \in \lambda}} (-y_\ell - y_{\lambda_k}) \cdot \prod_{\substack{\ell < \lambda_k \\ \ell \notin \lambda}} (y_\ell - y_{\lambda_k}) \qquad (*)$$

and

$$Q_\lambda(y^\mu|y) = 0 \qquad \text{if } \mu \not\supseteq \lambda. \qquad (**)$$

Proof To prove $(*)$, we need an easy fact (cf. Appendix A, §A.5 (12)). Suppose we have a series $c = 1 + c_1 + c_2 + \cdots$ which can be written as

$$c = \frac{1}{1 + z} \cdot (\text{a polynomial of degree } \leq d).$$

Then

$$(-z)^a \cdot c_p = c_{p+a} \qquad \text{for any } p \geq d. \qquad (C.3)$$

Now let

$$\widetilde{q}(k) = \frac{(1 + y_{\lambda_1}) \cdots (1 + y_{\lambda_{k-1}})}{(1 - y_{\lambda_1}) \cdots (1 - y_{\lambda_{k-1}})} \cdot q(k)|_\lambda$$

$$= \frac{1}{1 + y_{\lambda_k}} \cdot (1 \pm y_1) \cdots (1 \pm y_{\lambda_k - 1})(1 - y_{\lambda_k})$$

for each $1 \le k \le s$. (The factor is $1 + y_i$ if $i \in \lambda$, and $1 - y_i$ if $i \notin \lambda$, but this point is immaterial to the proof.) By (C.3), we have

$$(-y_{\lambda_k})^a \cdot \widetilde{q}(k)_p = \widetilde{q}(k)_{p+a}$$

for $p \ge \lambda_k$.

The claim in $(*)$ is that

$$Q_\lambda(y^\lambda|y) = \widetilde{q}(1)_{\lambda_1} \cdots \widetilde{q}(s)_{\lambda_s},$$

since one checks

$$\widetilde{q}(k)_{\lambda_k} = \prod_{\substack{\ell \le \lambda_k \\ \ell \in \lambda}} (-y_\ell - y_{\lambda_k}) \cdot \prod_{\substack{\ell < \lambda_k \\ \ell \notin \lambda}} (y_\ell - y_{\lambda_k}).$$

We will show

$$\left(\prod_{\substack{1 \le i < j \le s \\ r-1 < j}} \frac{1 - R_{ij}}{1 + R_{ij}} \right) \cdot \widetilde{q}(1)_{\lambda_1} \cdots \widetilde{q}(r-1)_{\lambda_{r-1}} \cdot q(r)_{\lambda_r}|_\lambda \cdots q(s)_{\lambda_s}|_\lambda$$

$$\text{(C.4)}$$

$$= \left(\prod_{\substack{1 \le i < j \le s \\ r < j}} \frac{1 - R_{ij}}{1 + R_{ij}} \right) \cdot \widetilde{q}(1)_{\lambda_1} \cdots \widetilde{q}(r)_{\lambda_r} \cdot q(r+1)_{\lambda_{r+1}}|_\lambda \cdots q(s)_{\lambda_s}|_\lambda$$

for each $1 \le r \le s$. This establishes the claim: at $r = 1$, the left-hand side of (C.4) is the left-hand side of $(*)$; at $r = s$, the right-hand side of (C.4) is the right-hand side of $(*)$.

We can write

$$\prod_{\substack{1 \le i < j \le s \\ r-1 < j}} \frac{1 - R_{ij}}{1 + R_{ij}} = \prod_{i=1}^{r-1} \frac{1 - R_{ir}}{1 + R_{ir}} \cdot \prod_{\substack{i < j \\ r < j}} \frac{1 - R_{ij}}{1 + R_{ij}}.$$

For $p_1 \geq \lambda_1, \ldots, p_r \geq \lambda_r$, we have

$$\tilde{q}(1)_{p_1} \cdots \tilde{q}(r)_{p_r} = \left(\prod_{i=1}^{r-1} \frac{1 - R_{ir}}{1 + R_{ir}} \right) \cdot \tilde{q}(1)_{p_1} \cdots \tilde{q}(r-1) \cdot q(r)_{p_r}|_\lambda,$$

using (C.3) together with

$$\tilde{q}(r)_p = \left(\prod_{i=1}^{r-1} \frac{1 + y_{\lambda_i}}{1 - y_{\lambda_i}} \cdot q(r)|_\lambda \right)_p$$

for all p. The identity (C.4) follows, so $(*)$ is proved.

For $(**)$, suppose $\mu \not\supseteq \lambda$, and let k be the smallest index such that $\mu_k < \lambda_k$, so $\mu_1 \geq \lambda_1, \ldots, \mu_{k-1} \geq \lambda_{k-1}$. For $\ell \leq k$, we have

$$q(\ell)|_\mu = \left(\frac{1 - y_{\mu_1}}{1 + y_{\mu_1}} \right) \cdots \left(\frac{1 - y_{\mu_{k-1}}}{1 + y_{\mu_{k-1}}} \right) \cdot (1 \pm y_1) \cdots (1 \pm y_{\lambda_\ell - 1}).$$

In particular, for $\ell \leq k$ and $p \geq \lambda_\ell$,

$$q(\ell)_p|_\mu = \overline{q}_p + e_1 \, \overline{q}_{p-1} + \cdots + e_{\lambda_\ell - 1} \, \overline{q}_{p - \lambda_\ell + 1},$$

where $\overline{q} = q|_{\{y_{\mu_1}, \ldots, y_{\mu_{k-1}}\}}$ and $e_i = e_i(\pm y_1, \ldots, \pm y_{\lambda_\ell - 1})$. (The main point here is that $p - \lambda_\ell + 1 > 0$.) For $\ell > k$, $q(\ell)_p|_\mu$ may be written as

$$q(\ell)_p|_\mu = \overline{q}_p + \alpha_1 \, \overline{q}_{p-1} + \cdots + \alpha_p,$$

where the α_i are some homogeneous polynomials in the y variables.

By expanding the $q(\ell)_p$ according to the above expressions, and using the multi-linearity of Pfaffians, we can write

$$Q_\lambda(y^\mu | y) = \left(\prod_{i<j} \frac{1 - R_{ij}}{1 + R_{ij}} \right) \cdot q(1)_{\lambda_1}|_\mu \cdots q(s)_{\lambda_s}|_\mu$$

$$= \sum_\nu \alpha_\nu \overline{Q}_\nu,$$

the sum taken over partitions ν with at least k nonzero parts, where $\overline{Q}_\nu = Q_\nu(-y_{\mu_1}, \ldots, -y_{\mu_{k-1}})$ is the (single) Schur Q-polynomial, and the coefficients α_ν are some homogeneous polynomials in y. Each $\overline{Q}_\nu = 0$ by (C.2) from §C.2.4, since it involves fewer variables than ν has parts. So $(**)$ is proved.

Finally, suppose $F(q|y)$ is a homogeneous polynomial of degree $|\lambda|$ which satisfies $(*)$ and $(**)$; we will show $F(q|y) = Q_\lambda(q|y)$. Since the double Q-polynomials form a basis for $\Gamma[y]$, we can write

$$F(q|y) = \sum_{\nu} \alpha_\nu \, Q_\nu(q|y)$$

for some homogeneous polynomials $\alpha_\nu \in \mathbb{Z}[y]$, with the sum over $|\nu| \leq |\lambda|$.

Using induction on the set of strict partitions ν partially ordered by containment, we see $\alpha_\nu = 0$ for $\nu \neq \lambda$ in this sum, by considering the specialization $F(y^\nu|y)$. (We assume by induction that $\alpha_\mu = 0$ for all $\mu \subsetneq \nu$. So the only term that survives is

$$F(y^\nu|y) = \alpha_\nu \, Q_\nu(y^\nu|y),$$

which equals zero since $\nu \not\supseteq \lambda$. By $(*)$, we know $Q_\nu(y^\nu) \neq 0$, and it follows that $\alpha_\nu = 0$. The base case $\nu = \emptyset$ is similar.) The same reasoning shows that $F(y^\lambda|y) = \alpha_\lambda \, Q_\lambda(y^\lambda|y)$, and since $F(q|y)$ satisfies $(*)$, it follows that $\alpha_\lambda = 1$. $\qquad\square$

Remark. Ivanov's polynomials $Q_\lambda(x|y)$ are recovered from our $Q_\lambda(q|y)$ by specializing $q = \prod \frac{1+x_i}{1-x_i}$. The evaluation $q \mapsto y^\mu$ can be written by further specializing these x variables as

$$x_i \mapsto \begin{cases} -y_i & \text{if } i \in \mu; \\ 0 & \text{if } i \notin \mu. \end{cases}$$

With this specialization, Ikeda uses the interpolation property and the localization theorem to show that $Q_\lambda(x|y)$ represents an equivariant Schubert class in the Lagrangian Grassmannian.

A different specialization of the q variables is more directly connected to the geometry of Lagrangian Grassmannians. Suppose the x and y variables satisfy relations

$$e_k(x_1^2, x_2^2, \ldots) = e_k(y_1^2, y_2^2, \ldots) \tag{C.5}$$

for all k, where e_k is the elementary symmetric function. Then we may set

$$q \mapsto \prod_{i \geq 1} \frac{1 - x_i}{1 + y_i}, \tag{C.6}$$

since $\left(\prod \frac{1-x_i}{1+y_i}\right) \cdot \left(\prod \frac{1+x_i}{1-y_i}\right) = 1$. Now the evaluation $q \mapsto y^\mu$ is obtained by specializing the x variables as

$$x_i \mapsto \begin{cases} y_i & \text{if } i \in \mu; \\ -y_i & \text{if } i \notin \mu. \end{cases}$$

Let us write $\widetilde{Q}_\lambda(x|y)$ for the image of $Q_\lambda(q|y)$ under the second specialization (C.6). The "single" polynomials $\widetilde{Q}_\lambda(x) = \widetilde{Q}_\lambda(x|0)$ were introduced by Pragacz and Ratajski in the study of the cohomology of the Lagrangian Grassmannian and degeneracy loci (Pragacz, 1988, 1991; Pragacz and Ratajski, 1997). The double polynomials $\widetilde{Q}_\lambda(x|y)$ are the natural generalizations to equivariant cohomology. These are distinct from the Schur and Ivanov Q-functions.

To explain the connection with geometry precisely, let V be a $2n$-dimensional symplectic vector space, with a torus T acting by characters $-y_n, \ldots, -y_1, y_1, \ldots, y_n$. Let $E \subseteq V$ be the isotropic subspace spanned by the last n basis vectors, and let $\mathbb{S} \subseteq V$ be the tautological subbundle on the Lagrangian Grassmannian. The Chern classes $c = c^T(V - \mathbb{S} - E)$ may be written as

$$c^T(V - \mathbb{S} - E) = c^T(\mathbb{S}^\vee - E) = \prod_{i=1}^{n} \frac{1 - x_i}{1 + y_i},$$

where x and y are equivariant Chern roots of $\mathbb{S}$ and E, respectively. The relations (C.5) follow from the identity $c^T(\mathbb{S} + \mathbb{S}^\vee) = c^T(E + E^\vee)$ (both sides being equal to $c^T(V)$). In this setting, Pragacz and Ratajski show that the cohomology class of a Schubert variety Ω_λ is equal to $\widetilde{Q}_\lambda(x)$. Theorem 14.2.2 shows that the equivariant class of Ω_λ is equal to $\widetilde{Q}_\lambda(x|y)$.

Hints for Exercises

Exercise C.2.1. Use the relations $Q_{pp} = 0$. See (Macdonald, 1995, §III.8, (8.6)).

Exercise C.2.2. From the raising operator formula, one has

$$Q_\lambda = q_\lambda + \sum_{\mu > \lambda} a_{\lambda\mu} q_\mu$$

for some integers $a_{\lambda\mu}$, where $>$ is the "dominance" partial order on partitions, generated by $R_{ij}\lambda > \lambda$. Now apply the previous exercise. See (Macdonald, 1995, §III.8, (8.9)).

Exercise C.2.3. Construct a weight-preserving bijection between (1) the set of primed tableaux on the shifted diagram of ρ, in which all entries along the main diagonal are unprimed; and (2) the set of semistandard tableaux on the usual diagram of ρ. See (Macdonald, 1995, §III.8, Ex. 3(b)) for another argument.

Exercise C.3.1. Use the relations $Q_{pp} = 0$, together with the fact that an elementary symmetric polynomial vanishes when the degree is greater than the number of variables.

Exercise C.3.2. One can write

$$Q_\lambda(q|y) = Q_\lambda(q) + \sum_{|\mu|<|\lambda|} \alpha_\mu Q_\mu(q)$$

for some polynomials α_μ in the y variables, by expanding the entries of the Pfaffian which defines $Q_\lambda(q|y)$ and using multi-linearity. Since the $Q_\lambda(q)$ form a basis for Γ over $\mathbb{Z}$, the claim follows.

Appendix D

Conventions for Schubert Varieties

The literature contains many conflicting conventions for describing and notating Schubert varieties, and especially in the equivariant situation, it is important to know exactly which set of conventions are in use. The following glossary is not meant to be a guide to the literature, but rather to the several sets of notation used in the book. We illustrate the translations with a running example.

Throughout, V is an n-dimensional vector space, often identified with $\mathbb{C}^n$ via a standard basis $\{e_1, \ldots, e_n\}$.

D.1 Grassmannians

Start with $Gr(d, V)$. Schubert varieties are indexed in three ways:

(1) by partitions λ fitting inside the $d \times (n - d)$ rectangle;
(2) by d-element subsets $I \subseteq \{1, \ldots, n\}$;
(3) by 01-sequences of length n, with d terms equal to 1.

The translation between the second two is easy: simply record the positions of the 1's in a 01-sequence to get a subset I. The first two indexings are related as

$$\lambda = (\lambda_1 \geq \cdots \geq \lambda_d \geq 0) \leftrightarrow I = \{i_1 < \cdots < i_d\},$$
$$\lambda_k = k - d - 1 + i_{d+1-k} \leftrightarrow i_k = k + \lambda_{d+1-k}.$$

Graphically, we usually represent λ by its Young diagram—a collection of boxes, with λ_k boxes in the kth row. As a subset of the $d \times (n - d)$

rectangle, the diagram is also determined by its southwest border—the path from the lower-left corner of the rectangle to the upper-right corner tracing the borders of the boxes. There are n steps in this path, and we get the corresponding d-subset I by recording the vertical (upward) steps.

Here is an example:

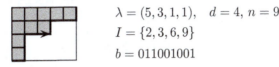

$$\lambda = (5, 3, 1, 1), \quad d = 4, \, n = 9$$
$$I = \{2, 3, 6, 9\}$$
$$b = 011001001$$

There is a duality which is important in this story. Given λ, the "dual" partition $\lambda^\vee$ is the complement of λ in the $d \times (n - d)$ rectangle, rotated 180 degrees. The dual subset $I^\vee$ and 01-sequence $b^\vee$ are obtained by reading the southwest border backwards. In formulas:

$$\lambda = (\lambda_1 \geq \lambda_2 \geq \cdots \geq \lambda_d \geq 0)$$
$$\lambda^\vee = (n - d - \lambda_d \geq n - d - \lambda_{d-1} \geq \cdots \geq n - d - \lambda_1 \geq 0)$$
$$\text{(that is, } \lambda_k^\vee = n - d - \lambda_{d+1-k}\text{);}$$

$$I = \{i_1 < i_2 < \cdots < i_d\}$$
$$I^\vee = \{n + 1 - i_d < n + 1 - i_{d-1} < \cdots < n + 1 - i_1\}$$
$$\text{(that is, } i_k^\vee = n + 1 - i_{d+1-k}\text{);}$$

$$b = b_1 b_2 \cdots b_n$$
$$b^\vee = b_n b_{n-1} \cdots b_1.$$

In pictures, continuing the above example:

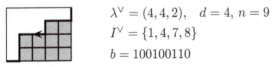

$$\lambda^\vee = (4, 4, 2), \quad d = 4, \, n = 9$$
$$I^\vee = \{1, 4, 7, 8\}$$
$$b = 100100110$$

The *size* of λ is $|\lambda| = \lambda_1 + \cdots + \lambda_d$. In terms of the subset I, this is equal to $\#\{j < i \mid i \in I, j \in \{1, \ldots, n\} \smallsetminus I\}$.

D.1.1 Degeneracy or incidence locus version.

The standard flag $E^\bullet$ has $E^q = \langle e_n, \ldots, e_{q+1} \rangle$. The opposite flag $\widetilde{E}^\bullet$ is defined by $\widetilde{E}^q = \langle e_1, \ldots, e_{n-q} \rangle$. The spaces E^q and $\widetilde{E}^q$ have codimension q in V. We write $E_q = E^{n-q}$ and $\widetilde{E}_q = \widetilde{E}^{n-q}$ for the corresponding q-dimensional subspaces.

The *Schubert variety* $\Omega_\lambda = \Omega_\lambda(E_\bullet)$ is

$$
\begin{aligned}
\Omega_\lambda &= \left\{ F \subset \mathbb{C}^n \;\middle|\; \dim(F \cap E^{d-k+\lambda_k}) \geq k, \text{ for } 1 \leq k \leq d \right\} \\
&= \left\{ F \subset \mathbb{C}^n \;\middle|\; \mathrm{rk}(F \to \mathbb{C}^n/E^{d-k+\lambda_k}) \leq d - k, \text{ for } 1 \leq k \leq d \right\}.
\end{aligned}
$$

$$\text{(D.1)}$$

It has codimension $|\lambda|$ in $Gr(d, \mathbb{C}^n)$. Its equivariant cohomology class is the double Schur polynomial $s_\lambda(x|y)$, where

$$x_1, \ldots, x_d \text{ are Chern roots of } \mathbb{S}^*$$

and $-y_1, \ldots, -y_n$ are characters of T acting on $\mathbb{C}^n$ (Chapter 9, §9.4).

This Schubert variety is the closure of the Schubert cell

$$
\begin{aligned}
\Omega_\lambda^\circ &= \left\{ F \;\middle|\; \begin{array}{c} \dim(F \cap E_s) = i \text{ for} \\ s \in [n - d + k - \lambda_k, n - d + k - \lambda_{k+1}], \; k = 0, \ldots, d \end{array} \right\} \\
&= \left\{ F \;\middle|\; \dim(F \cap E^{s-1}) = d - k \text{ for } s \in (i_k, i_{k+1}], 0 \leq k \leq d \right\},
\end{aligned}
$$

$$\text{(D.2)}$$

using the conventions $\lambda_0 = n - d$ and $\lambda_{d+1} = 0$, and $i_0 = 0$ and $i_{d+1} = n + 1$.

It is useful to parametrize the cells with matrices. Our convention—informed by group actions—is that spaces are *column-spans*, so a point in $Gr(d, \mathbb{C}^n)$ is represented by an $n \times d$ matrix. A point in $\Omega_\lambda^\circ = \Omega_I^\circ$ is uniquely represented by a matrix with "pivot 1's" forming an identity matrix on the rows $i_1, \ldots, i_d$, zeroes above these pivots, and arbitrary entries elsewhere.

For $d = 4$, $n = 9$, and $\lambda = (5, 3, 1, 1)$, so $I = \{2, 3, 6, 9\}$, the cell is this:

$$
\Omega_\lambda^\circ = \begin{bmatrix} 0 & 0 & 0 & 0 \\ 1 & 0 & 0 & 0 \\ 0 & 1 & 0 & 0 \\ * & * & 0 & 0 \\ * & * & 0 & 0 \\ 0 & 0 & 1 & 0 \\ * & * & * & 0 \\ * & * & * & 0 \\ 0 & 0 & 0 & 1 \end{bmatrix} \left. \begin{array}{c} 4 \\ 4 \\ 3 \\ 2 \\ \uparrow \; 2 \\ 2 \\ 1 \\ 1 \\ 1 \end{array} \right\} \dim(F \cap E_s).
$$

The numbers on the right record intersection dimensions with the fixed flag $E_\bullet$; note that jumps occur at rows labelled by I, which is also where the pivots are.

The opposite Schubert varieties $\widetilde{\Omega}_\lambda$ are defined similarly, but with respect to the flag $\widetilde{E}_\bullet$. Here is the opposite cell, for $\lambda = (5, 3, 1, 1)$ again (so $\lambda^\vee = (4, 4, 2)$):

$$\widetilde{\Omega}_{\lambda^\vee}^\circ = \left[\begin{matrix} * & * & * & * \\ 1 & 0 & 0 & 0 \\ 0 & 1 & 0 & 0 \\ 0 & 0 & * & * \\ 0 & 0 & * & * \\ 0 & 0 & 1 & 0 \\ 0 & 0 & 0 & * \\ 0 & 0 & 0 & * \\ 0 & 0 & 0 & 1 \end{matrix}\right] \left.\begin{matrix} 0 \\ 1 \\ 2 \\ 2 \\ \downarrow \ 2 \\ 3 \\ 3 \\ 3 \\ 4 \end{matrix}\right\} \dim(F \cap \widetilde{E}_s).$$

The basic fact is that

$$\Omega_\lambda \cap \widetilde{\Omega}_{\lambda^\vee} = p_\lambda,$$

and this intersection is transverse. The intersection point $p_\lambda = p_I$ is represented by the matrix with identity matrix on rows I, and zeroes elsewhere.

D.1.2 Orbit versions. The group GL_n acts on $Gr(d, V)$ via its action on $V = \mathbb{C}^n$. Consider the subgroups

$$P = \text{ block-upper triangular matrices of block sizes } d \text{ and } n - d,$$
$$B = \text{ upper-triangular matrices,}$$
$$B^- = \text{ lower-triangular matrices.}$$

The subspace $E = \widetilde{E}_d = \langle e_1, \ldots, e_d \rangle$ is stabilized by P, identifying the Grassmannian with GL_n/P (and this is the reason for using column spans). The flag $E_\bullet$ is stabilized by B^-, and $\widetilde{E}_\bullet$ is stabilized by B.

The "degeneracy locus" varieties satisfy:

$$\begin{array}{llll} \Omega_\lambda & \text{is:} & B^-\text{-invariant,} & \text{of codimension } |\lambda|; \\ \widetilde{\Omega}_\lambda & \text{is:} & B\text{-invariant,} & \text{of codimension } |\lambda|. \end{array}$$

In the context of representation theory, it is common to have Schubert varieties with *dimension* $|\lambda|$—so we define "orbit" Schubert varieties $X(\lambda) = X(I)$ and $Y(\lambda) = Y(I)$ satisfying

$$
\begin{array}{llll}
X(\lambda) & \text{is:} & B\text{-invariant,} & \text{of } \textit{dimension } |\lambda|; \\
Y(\lambda) & \text{is:} & B^-\text{-invariant,} & \text{of codimension } |\lambda|.
\end{array}
$$

They are closures of cells:

$$
X(I)^\circ = B \cdot p_I;
$$
$$
Y(I)^\circ = B^- \cdot p_I.
$$

A feature of these conventions is $X(\lambda) \cap Y(\lambda) = p_\lambda$, transversally.

For example, with $\lambda = (5, 3, 1, 1)$ and $I = \{2, 3, 6, 9\}$, recall that $p_\lambda = p_I$ is the subspace $\langle e_2, e_3, e_6, e_9 \rangle$. The cells are

$$
X(I)^\circ = \begin{bmatrix}
* & * & * & * \\
1 & 0 & 0 & 0 \\
0 & 1 & 0 & 0 \\
0 & 0 & * & * \\
0 & 0 & * & * \\
0 & 0 & 1 & 0 \\
0 & 0 & 0 & * \\
0 & 0 & 0 & * \\
0 & 0 & 0 & 1
\end{bmatrix}
\quad \text{and} \quad
Y(I)^\circ = \begin{bmatrix}
0 & 0 & 0 & 0 \\
1 & 0 & 0 & 0 \\
0 & 1 & 0 & 0 \\
* & * & 0 & 0 \\
* & * & 0 & 0 \\
0 & 0 & 1 & 0 \\
* & * & * & 0 \\
* & * & * & 0 \\
0 & 0 & 0 & 1
\end{bmatrix}.
$$

So one sees

$$
\tilde{\Omega}_\lambda = X(\lambda^\vee) \quad \text{and} \quad \Omega_\lambda = Y(\lambda). \tag{D.3}
$$

D.2   Flag varieties

Next consider $Fl(\mathbb{C}^n)$. The indexing set is S_n, the group of permutations of n letters. We write these in "one-line" notation, so $w = [w(1), w(2), \ldots, w(n)]$, sometimes omitting the brackets and commas. The longest permutation is $w_\circ = [n, n-1, \ldots, 1]$.

For $w \in S_n$, the corresponding permutation matrix A_w has 1's in positions $(w(i), i)$, and zeroes elsewhere. This is compatible with matrix multiplication: $A_{uv} = A_u A_v$. So it is just the standard representation of S_n on $\mathbb{C}^n$.

The *rank function* and *dimension function* are defined as

$$r_w(p,q) = \#\{i \le p \,|\, w(i) \le q\} \quad \text{and} \quad k_w(p,q) = \#\{i \le p \,|\, w(i) > q\}.$$

So $r_w(p,q)$ is the rank of the upper-left $q \times p$ submatrix of A_w, and $k_w(p,q)$ is the rank of the lower-left $(n-q) \times p$ submatrix.

The *length* of w is $\ell(w) = \#\{i < j \,|\, w(i) > w(j)\}$.

Example. The permutation $w = 2\ 3\ 6\ 9\ 1\ 4\ 5\ 7\ 8$ has length $\ell(w) = 10$. Its matrix A_w and rank function r_w are shown below.

$$
\begin{bmatrix}
0 & 0 & 0 & 0 & 1 & 0 & 0 & 0 & 0 \\
1 & 0 & 0 & 0 & 0 & 0 & 0 & 0 & 0 \\
0 & 1 & 0 & 0 & 0 & 0 & 0 & 0 & 0 \\
0 & 0 & 0 & 0 & 0 & 1 & 0 & 0 & 0 \\
0 & 0 & 0 & 0 & 0 & 0 & 1 & 0 & 0 \\
0 & 0 & 1 & 0 & 0 & 0 & 0 & 0 & 0 \\
0 & 0 & 0 & 0 & 0 & 0 & 0 & 1 & 0 \\
0 & 0 & 0 & 0 & 0 & 0 & 0 & 0 & 1 \\
0 & 0 & 0 & 1 & 0 & 0 & 0 & 0 & 0
\end{bmatrix}
\qquad
\overset{\xrightarrow{p}}{
\begin{bmatrix}
0 & 0 & 0 & 0 & 1 & 1 & 1 & 1 & 1 \\
1 & 1 & 1 & 1 & 2 & 2 & 2 & 2 & 2 \\
1 & 2 & 2 & 2 & 3 & 3 & 3 & 3 & 3 \\
1 & 2 & 2 & 2 & 3 & 4 & 4 & 4 & 4 \\
1 & 2 & 2 & 2 & 3 & 4 & 5 & 5 & 5 \\
1 & 2 & 3 & 3 & 4 & 5 & 6 & 6 & 6 \\
1 & 2 & 3 & 3 & 4 & 5 & 6 & 7 & 7 \\
1 & 2 & 3 & 3 & 4 & 5 & 6 & 7 & 8 \\
1 & 2 & 3 & 4 & 5 & 6 & 7 & 8 & 9
\end{bmatrix}} \;\downarrow q
$$

$$\qquad\quad A_w \qquad\qquad\qquad\qquad\qquad r_w$$

D.2.1 Degeneracy-locus version. The flag variety has universal bundles

$$\mathbb{S}_1 \hookrightarrow \mathbb{S}_2 \hookrightarrow \cdots \hookrightarrow \mathbb{S}_n = \mathbb{C}^n = \mathbb{Q}_n \twoheadrightarrow \cdots \twoheadrightarrow \mathbb{Q}_2 \twoheadrightarrow \mathbb{Q}_1.$$

For a fixed flag $E^\bullet$ as above, set

$$
\begin{aligned}
\Omega_w(E_\bullet) &= \left\{ F_\bullet \,\middle|\, \mathrm{rk}(F_p \to \mathbb{C}^n/E^q) \le r_w(p,q) \text{ for } 1 \le p,q \le n \right\} \\
&= \left\{ F_\bullet \,\middle|\, \mathrm{rk}((\mathbb{C}^n/E^q)^\vee \to F_p^\vee) \le r_w(p,q) \text{ for } 1 \le p,q \le n \right\} \\
&= \left\{ F_\bullet \,\middle|\, \dim(F_p \cap E^q) \ge k_w(p,q) \text{ for } 1 \le p,q \le n \right\}. \quad \text{(D.4)}
\end{aligned}
$$

(The last equality comes from $p - r_w(p,q) = k_w(p,q)$.) This has codimension $\ell(w)$. Its equivariant cohomology class is the double Schubert polynomial $\mathfrak{S}_w(x;y)$, where

$$x_i = -c_1^T(\mathbb{S}_i/\mathbb{S}_{i-1}) \quad \text{and} \quad y_i = -c_1^T(E^{i-1}/E^i)$$

(Chapter 10, §10.6).

The Schubert variety is the closure of a Schubert cell, which is defined by replacing inequalities with equalities:

$$\Omega_w^\circ(E_\bullet) = \Big\{ F_\bullet \mid \mathrm{rk}(F_p \to \mathbb{C}^n/E^q) = r_w(p,q) \text{ for } 1 \le p, q \le n \Big\}$$
$$= \Big\{ F_\bullet \mid \dim(F_p \cap E^q) = k_w(p,q) \text{ for } 1 \le p, q \le n \Big\}. \quad \text{(D.5)}$$

By taking column spans, a flag can be represented by an $n \times n$ matrix. The flag corresponding to the matrix A_w is the point p_w; it is

$$p_w = \langle e_{w(1)} \rangle \subset \langle e_{w(1)}, e_{w(2)} \rangle \subset \cdots \subset \langle e_{w(1)}, e_{w(2)}, \ldots, e_{w(n)} \rangle = \mathbb{C}^n.$$

In particular, the point p_w lies in the cell Ω_w°. Points in Ω_w° may therefore be represented by matrices whose rank functions are the same as that of A_w. More precisely, they have pivot 1's in positions $(w(i), i)$, zeroes below and to the right of the pivots, and free entries elsewhere.

Example. For $w = 2\,3\,6\,9\,1\,4\,5\,7\,8$, the Schubert cell Ω_w° is shown below, along with the rank function r_w.

$$\begin{bmatrix}
0 & 0 & 0 & 0 & 1 & 0 & 0 & 0 & 0 \\
1 & 0 & 0 & 0 & 0 & 0 & 0 & 0 & 0 \\
* & 1 & 0 & 0 & 0 & 0 & 0 & 0 & 0 \\
* & * & 0 & 0 & * & 1 & 0 & 0 & 0 \\
* & * & 0 & 0 & * & * & 1 & 0 & 0 \\
* & * & 1 & 0 & 0 & 0 & 0 & 0 & 0 \\
* & * & * & 0 & * & * & * & 1 & 0 \\
* & * & * & 0 & * & * & * & * & 1 \\
* & * & * & 1 & 0 & 0 & 0 & 0 & 0
\end{bmatrix}
\qquad
\begin{bmatrix}
0 & 0 & 0 & \mathbf{0} & 1 & 1 & 1 & 1 & 1 \\
1 & 1 & 1 & \mathbf{1} & 2 & 2 & 2 & 2 & 2 \\
1 & 2 & 2 & \mathbf{2} & 3 & 3 & 3 & 3 & 3 \\
1 & 2 & 2 & \mathbf{2} & 3 & 4 & 4 & 4 & 4 \\
1 & 2 & 2 & \mathbf{2} & 3 & 4 & 5 & 5 & 5 \\
1 & 2 & 3 & \mathbf{3} & 4 & 5 & 6 & 6 & 6 \\
1 & 2 & 3 & \mathbf{3} & 4 & 5 & 6 & 7 & 7 \\
1 & 2 & 3 & \mathbf{3} & 4 & 5 & 6 & 7 & 8 \\
1 & 2 & 3 & \mathbf{4} & 5 & 6 & 7 & 8 & 9
\end{bmatrix}$$

$$\underbrace{}_{\mathrm{rk}(F_p \to \mathbb{C}^n/E^q)}$$

It has dimension $26 = 36 - 10 = \dim Fl(\mathbb{C}^9) - \ell(w)$.

The conditions on F_4 (the bold column) are the same as those defining $\Omega_\lambda \subset Gr(4, \mathbb{C}^9)$, for $\lambda = (5, 3, 1, 1)$. In fact, these are the only relevant conditions: the other entries of the rank table are the smallest possible ones consistent with the fourth column.

In general, the relation between Schubert varieties in Gr and Fl is as follows. Given any partition λ in the $d \times (n-d)$ rectangle, there is a *Grassmannian permutation* $w(\lambda)$ defined as

$$w(\lambda) = i_1 \, i_2 \cdots i_d \, j_1 \, j_2 \cdots j_{n-d} \, ,$$

where $I = \{i_1 < \cdots < i_d\}$ is the subset corresponding to λ, and $J = \{j_1 < \cdots < j_{n-d}\} = \{1, \ldots, n\} \smallsetminus I$. Then

$$\pi^{-1}\Omega_\lambda = \Omega_{w(\lambda)},$$

where $\pi \colon Fl(\mathbb{C}^n) \to Gr(d, \mathbb{C}^n)$ is the projection.

As before, the opposite Schubert varieties are defined using the opposite flag, i.e., $\widetilde{\Omega}_w = \Omega_w(\widetilde{E}_\bullet)$. From the matrix point of view, it is easy to check that

$$\widetilde{\Omega}_w = w_\circ \cdot \Omega_w,$$

so $\widetilde{\Omega}_w$ contains the point $w_\circ \cdot p_w = p_{w_\circ w}$. We have

$$\Omega_w \cap \widetilde{\Omega}_{w_\circ w} = \{p_w\},$$

transversally.

D.2.2 Orbit versions. The upper-triangular Borel group B fixes $p_e = \widetilde{E}_\bullet$, and the flag variety is identified with GL_n/B. The "orbit Schubert varieties" are defined as before:

$$
\begin{array}{llll}
X(w) & \text{is:} & B\text{-invariant,} & \text{of dimension } \ell(w); \\
Y(w) & \text{is:} & B^-\text{-invariant,} & \text{of codimension } \ell(w).
\end{array}
$$

They are closures of orbits:

$$X(w)^\circ = B \cdot p_w,$$
$$Y(w)^\circ = B^- \cdot p_w,$$

and they intersect transversally in $X(w) \cap Y(w) = p_w$.

Comparing with the degeneracy locus versions, we have

$$\widetilde{\Omega}_w = X(w_\circ w) \qquad \text{and} \qquad \Omega_w = Y(w). \tag{D.6}$$

Note that $Y(w) = w_\circ \cdot X(w_\circ w)$.

D.3 General G/P

For a general homogeneous space G/P, fixed points and Schubert varieties are indexed by cosets W/W_P. Here W_P is the Weyl group of the parabolic P. Write $W^P_{\min}$ and $W^P_{\max}$ for the sets of minimal- and maximal-length coset representatives in W. Since minimal representatives are used more often, W^P means $W^P_{\min}$.

For any coset $[w] \in W/W_P$, write $w^{\min} \in W^P_{\min}$ and $w^{\max} \in W^P_{\max}$ for its minimal and maximal representatives. These are related by

$$w^{\max} = w^{\min} \cdot w^P_\circ,$$

where $w^P_\circ \in W_P$ is the longest element.

Write $p_{[w]} = wP/P$ for the fixed point in G/P corresponding to $w \in W^P$, and define

$$X[w]^\circ = B \cdot p_{[w]};$$
$$Y[w]^\circ = B^- \cdot p_{[w]}.$$

(One could take any $w \in W$, since $X[w] = X[w']$, etc., when $w \equiv w'$ in W/W_P. However, to get dimension counts right, one needs minimal representatives.) The closures of these cells satisfy

$$
\begin{aligned}
X[w] \quad &\text{is:} \quad B\text{-invariant,} \quad &&\text{of dimension } \ell(w); \\
Y[w] \quad &\text{is:} \quad B^-\text{-invariant,} \quad &&\text{of codimension } \ell(w);
\end{aligned}
$$

and $X[w] \cap Y[w] = \{p_{[w]}\}$, transversally.

In the case where $G/P = Gr(d, \mathbb{C}^n)$ is a Grassmannian, the notation corresponds as follows. For a partition λ in the $d \times (n-d)$ rectangle, the Grassmannian permutation $w(\lambda)$ is a minimal representative in W^P. Minimal representatives for the Poincaré dual classes are given by $w(\lambda^\vee) = w_\circ w(\lambda) w^P_\circ$.

Appendix E

Characteristic Classes and Equivariant Cohomology

The fiber bundles arising in the construction of equivariant cohomology—and their finite-dimensional approximations—are universal among all fiber bundles, in a sense to be made precise here. It follows that calculations in equivariant cohomology determine formulas for general fiber bundles.

We fix a category of (locally trivial) fiber bundles $\xi\colon \mathfrak{X} \to \mathbb{B}$, with fiber X and structure group G. Main examples include topological fiber bundles, Zariski-locally trivial bundles, or étale-locally trivial bundles. Any such bundle may be realized as $\mathfrak{X} = \mathbb{E} \times^G X$, for some principal G-bundle $\mathbb{E} \to \mathbb{B}$. See for example, (Husemoller 1975, Chapter 5).

We also fix a cohomology theory H^*, with the following key properties: it is a contravariant functor from spaces to (graded-commutative) rings, and pullback via an affine space bundle is injective. Main examples include singular cohomology and Chow cohomology.

By a *characteristic class (with values in H^i)* of fiber bundles with fiber X and structure group G, we mean an assignment α of a class in H^i to each fiber bundle, which satisfies the following compatibility. For every ξ there is $\alpha(\xi) \in H^i\mathfrak{X}$, and for every fiber square

$$
\begin{array}{ccc}
\mathfrak{X}' & \xrightarrow{\;\overline{f}\;} & \mathfrak{X} \\
{\scriptstyle \xi'}\downarrow & & \downarrow{\scriptstyle \xi} \\
\mathbb{B}' & \xrightarrow{\;f\;} & \mathbb{B},
\end{array}
$$

so $\xi' = f^*\xi$, we have $\alpha(f^*\xi) = \overline{f}^*\alpha(\xi)$. Characteristic classes with values in H^* form a graded-commutative ring under "pointwise" product: by definition, $(\alpha \cdot \beta)(\xi) = \alpha(\xi) \cdot \beta(\xi)$.

Theorem. *The ring of characteristic classes is naturally identified with the equivariant cohomology ring $H^*_G X$.*

Any characteristic class α certainly produces an equivariant class, by applying α to the bundle $\mathbb{E}_m G \times^G X \to \mathbb{B}_m G$. (Here the bundles $\mathbb{E}_m G \to \mathbb{B}_m G$ are finite-dimensional approximations to the classifying space.) Conversely, any characteristic class is induced from an equivariant class, in the following strong sense.

Proposition. *Given a fiber bundle ξ and an integer $i \geq 0$, there exist finite-dimensional approximation spaces $\mathbb{E}_m G \to \mathbb{B}_m G$ and fiber squares*

$$\begin{array}{ccccc}
\mathfrak{X} & \xleftarrow{\ \pi\ } & \mathfrak{X}' & \xrightarrow{\ \overline{\varphi}\ } & \mathbb{E}_m G \times^G X \\
{\scriptstyle \xi}\downarrow & & \downarrow{\scriptstyle \xi'} & & \downarrow{\scriptstyle \xi_\infty} \\
\mathbb{B} & \xleftarrow{\ \pi\ } & \mathbb{B}' & \xrightarrow{\ \varphi\ } & \mathbb{B}_m G
\end{array}$$

such that π^* is injective and $H^i_G X = H^i(\mathbb{E}_m G \times^G X)$.

Proof Write ξ as $\mathbb{E} \times^G X \to \mathbb{B}$ for some principal G-bundle $\mathbb{E} \to \mathbb{B}$. Then take approximations $\mathbb{B}_m G = (V \smallsetminus S)/G$ following Totaro, for some sufficiently high codimension closed subset S in a G-representation V. Setting $\mathbb{E}' = \mathbb{E} \times (V \smallsetminus S)$, the bundle $\xi' \colon \mathbb{E}' \times^G X \to \mathbb{E}'/G = \mathbb{B}'$ does the trick. $\qquad\square$

Proof of Theorem Given any characteristic class α with values in H^i, it follows from the proposition that the values $\alpha(\xi)$ on every fiber bundle ξ are determined by $\alpha(\xi_\infty) \in H^i_G X$. Indeed, one has $\pi^*\alpha(\xi) = \alpha(\xi') = \overline{\varphi}^*\alpha(\xi_\infty)$. To prove the theorem, it only remains to check compatibility with products. This follows by a similar argument: with notation as in the proof of the proposition, the pullback maps

$$\overline{\varphi}^* \colon H^*(\mathbb{E}_m G \times^G X) \to H^*(\mathfrak{X}')$$

are ring homomorphisms, and by choosing an appropriate approximation space, the product in $H^*(\mathbb{E}_m G \times^G X)$ agrees with that of $H^*_G X$ in any given degree. $\qquad\square$

The cases where G is GL_n, or the subgroup of upper-triangular matrices, are particularly relevant to the study of degeneracy loci.

Example 1. In the case $G = GL_n$, the Proposition takes on a concrete form. Let E be a rank n vector bundle on a variety $\mathbb{B}$. Then there is a variety $\mathbb{B}'$, together with morphisms $p\colon \mathbb{B}' \to \mathbb{B}$ and $f\colon \mathbb{B}' \to Gr(\mathbb{C}^m, n)$ (for $m \gg 0$), such that (1) $p^*\colon H^*\mathbb{B} \to H^*\mathbb{B}'$ is injective, and (2) $p^*E \cong f^*\mathbb{Q}$, where $\mathbb{Q}$ is the tautological quotient bundle on $\mathbb{B}_m GL_n = Gr(\mathbb{C}^m, n)$.

This follows from the Proposition, but here is an alternative argument. First suppose $\mathbb{B}$ is affine, with coordinate ring A. Then E corresponds to a finitely generated and locally free A-module M. Choosing m generators, one has $A^{\oplus m} \twoheadrightarrow M$; that is, a surjective homomorphism $\mathcal{O}_{\mathbb{B}}^{\oplus m} \twoheadrightarrow E$. By the universal property of the Grassmannian, this determines a morphism $f\colon \mathbb{B} \to Gr(\mathbb{C}^m, n)$ such that $f^*\mathbb{Q} = E$.

Next assume $\mathbb{B}$ is quasi-projective. There is an affine variety $\mathbb{B}'$, with a morphism $p\colon \mathbb{B}' \to \mathbb{B}$ which is locally trivial with affine space fibers. This is known as "Jouanolou's trick," and an exposition can be found in (Asok, 2009), along with some generalizations. (In our context, the construction is an easy and pleasant exercise.) So $p^*\colon H^*\mathbb{B} \to H^*\mathbb{B}'$ is injective, and we can apply the argument of the previous paragraph to p^*E on $\mathbb{B}'$.

For a general variety $\mathbb{B}$, one can reduce to the quasi-projective case by using a Chow envelope $\widetilde{\mathbb{B}} \to \mathbb{B}$; this means $\widetilde{\mathbb{B}}$ is quasi-projective and the corresponding pullback homomorphism is injective on cohomology. So one obtains maps $\mathbb{B}' \to \widetilde{\mathbb{B}} \to \mathbb{B}$ whose composition has the desired properties.

If one has a flag of quotient bundles $E_\bullet$, corresponding to the case where $G = B$ is a Borel subgroup of GL_n, the situation is entirely analogous: such a flag is pulled back from a universal flag, using $\mathbb{B}_m B = Fl(\mathbb{C}^m; n, \ldots, 1)$ and its tautological quotient flag $\mathbb{Q}_\bullet$ in place of the Grassmannian. Flags of subbundles are obtained by dualizing.

This argument is sketched in the final paragraphs of (Graham, 1997); as noted in the proof of the Proposition, the statements are implicit in (Totaro, 1999). See Chapter 11, §11.6 for applications.

Example 2. Let E and F be vector spaces of respective dimensions n and m, and let $X = \mathrm{Hom}(F, E)$, with the natural action by

$G = GL(E) \times GL(F)$. There is a characteristic class $\mathbb{D}_r$ for this situation, for each nonnegative integer r.

This is the class corresponding to $[D_r]^G \in H_G^* X$, where $D_r \subseteq X$ is the locus of homomorphisms of rank at most r. As a polynomial in $H_G^* X \cong \mathbb{Z}[a_1, \ldots, a_n, b_1, \ldots, b_m]$, $\mathbb{D}_r$ is computed by the Giambelli determinantal formula (Chapter 11, §11.1).

Example 3. With notation as in the previous example, fix complete flags

$$F_1 \subseteq \cdots \subseteq F_m = F \quad \text{and} \quad E = E_n \twoheadrightarrow \cdots \twoheadrightarrow E_1$$

of sub- and quotient spaces. Let $X = \mathrm{Hom}(F, E)$ as in the previous example, with a reduction of structure group from $G = GL(E) \times GL(F)$ to the Borel subgroup $B = B(E_\bullet) \times B(F_\bullet)$ which fixes the flags. In this situation, there is a characteristic class $\mathbb{D}_r$ for each $n \times m$ irreducible rank function r. (A rank function is a matrix of nonnegative integers $r = (r(p, q))_{1 \le p \le n, 1 \le q \le m}$. It is *irreducible* if it arises from a partial permutation matrix, i.e., $r(p, q)$ is the rank of the upper-left $p \times q$ submatrix of some 01-matrix with at most one 1 in each row and column.)

This class corresponds to $[D_r]^B \in H_B^* X$, where $D_r \subseteq X$ is the locus of homomorphisms $\varphi : F \to E$ such that the composite map $\varphi_{pq} : F_q \to E_p$ has rank at most $r(p, q)$. As a polynomial in $H_B^* X = \mathbb{Z}[x_1, \ldots, x_n, y_1, \ldots, y_m]$, $\mathbb{D}_r$ is equal to the Schubert polynomial $\mathfrak{S}_w(x; y)$, where $w = w(r)$ is the minimal permutation such that the upper-left $q \times p$ submatrix of the associated permutation matrix has rank $r(p, q)$ (Chapter 11, §11.2–11.4).

Example 4. Given a homomorphism $\varphi : \mathcal{F} \to \mathcal{E}$ of vector bundles on a variety Y, one has the degeneracy locus $D_r(\varphi) \subseteq Y$ of points $y \in Y$ where the corresponding linear map $\varphi_y : \mathcal{F}_y \to \mathcal{E}_y$ has rank at most r. Writing $\xi : \mathfrak{X} \to \mathbb{B}$ for the vector bundle $\mathrm{Hom}(\mathcal{F}, \mathcal{E}) \to Y$, the fibers are identified with $X = \mathrm{Hom}(F, E)$ as in the above examples, where $\dim E = \mathrm{rk}\, \mathcal{E}$ and $\dim F = \mathrm{rk}\, \mathcal{F}$. The homomorphism φ corresponds to a section of this bundle. Since ξ is a vector bundle, there is a canonical isomorphism $H^* \mathfrak{X} = H^* \mathbb{B}$. We use this to identify $\mathbb{D}_r$ with $\varphi^* \mathbb{D}_r$, for any section φ.

The homology class $\mathbb{D}_r(\xi) \frown [Y]$ is always supported on the degeneracy locus $D_r(\varphi)$. When Y is nonsingular, $\mathbb{D}_r(\xi) \frown [Y] = [D_r(\varphi)]$ if and only if D_r has expected codimension $(m-r)(n-r)$ in Y.

Conversely, if $\mathbb{D}_r(\xi) \frown [Y] \in H_*Y$ is not an effective class—that is, if it is not represented by an algebraic subvariety—then there is no homomorphism $\varphi \colon \mathcal{F} \to \mathcal{E}$ whose degeneracy locus $D_r(\varphi)$ has codimension $(m-r)(n-r)$. Effectivity of the class $\mathbb{D}_r(\xi) \frown [Y]$ can often be checked using the determinantal formula for $\mathbb{D}_r$.

References

Allday, C. and V. Puppe, 1993. *Cohomological Methods in Transformation Groups*, Cambridge University Press.

Allday, C., M. Franz, and V. Puppe, 2014. "Equivariant cohomology, syzygies and orbit structure," *Trans. Amer. Math. Soc.* **366**, 6567–6589.

Alper, J., J. Hall, and D. Rydh, 2020. "A Luna étale slice theorem for algebraic stacks," *Ann. Math.* **191**, no. 3, 675–738.

Andersen, H. H., J. C. Jantzen, and W. Soergel, 1994. "Representations of quantum groups at a pth root of unity and of semisimple groups in characteristic p: independence of p," *Astérisque* **220**.

Anderson, D. 2007a. "Double Schubert polynomials and double Schubert varieties," `www.people.math.osu.edu/anderson.2804/papers/geomschpolyn.pdf`.

——, 2007b. "Positivity in the cohomology of flag bundles (after Graham)," `arXiv:0711.0983`.

——, 2018. "Diagrams and essential sets for signed permutations," *Electron. J. Comb.* **25** (3), #P3.46.

Anderson, D. and W. Fulton, 2012. "Degeneracy loci, pfaffians, and vexillary signed permutations in types B, C, and D," `arXiv:1210.2066`.

——, 2018. "Chern class formulas for classical-type degeneracy loci," *Compositio Math.* **154**, 1746–1774.

——, 2021a. "Schubert polynomials in types A and C," preprint, `arXiv:2102.05731`.

——, 2021b. "Identities for Schur-type determinants and pfaffians," preprint, `arXiv:2103.16505`.

Anderson, D., S. Griffeth, and E. Miller, 2011. "Positivity and Kleiman transversality in equivariant K-theory of homogeneous spaces," *J. Eur. Math. Soc. (JEMS)* **13**, no. 1, 57–84.

Anderson, D. and A. Stapledon, 2013. "Arc spaces and equivariant cohomology," *Transform. Groups* **18**, no. 4, 931–969.

Arabia, A. 1986. "Cycles de Schubert et cohomologie équivariante de K/T," *Invent. Math.* **85**, no. 1, 39–52.

——, 1989. "Cohomologie T-équivariante de la variété de drapeaux d'un groupe de Kac–Moody," *Bull. Soc. Math. France* **117**, no. 2, 129–165.

Asok, A. 2009. "The Jouanolou-Thomason homotopy lemma," `www-bcf.usc.edu/~asok/notes/Jouanolou.pdf`.

Atiyah, M. F. 1974. *Elliptic Operators and Compact Groups*, Lecture Notes in Mathematics, Vol. 401, Springer-Verlag.

——, 1982. "Convexity and commuting Hamiltonians," *Bull. London Math. Soc.* **14**, no. 1, 1–15.

Atiyah M. F. 1984. and R. Bott, "The moment map and equivariant cohomology," *Topology* **23**, no. 1, 1–28.

Audin, M. 2004. *Torus Actions on Symplectic Manifolds*, Birkhäuser.

Bahri, A., M. Franz, and N. Ray, 2009. "The equivariant cohomology ring of weighted projective space," *Math. Proc. Cambridge Philos. Soc.* **146**, no. 2, 395–405.

Barratt, M. G. and J. Milnor, 1962. "An example of anomalous singular homology," *Proc. Amer. Math. Soc.* **13**, 293–297.

Behrend, K. 2004. "Cohomology of stacks," *Intersection Theory and Moduli*, 249–294, ICTP Lect. Notes, XIX, Abdus Salam Int. Cent. Theoret. Phys., Trieste.

Bergeron, N. and S. Billey, 1993. "RC-graphs and Schubert polynomials," *Experiment. Math.* **2**, no. 4, 257–269.

Berline, N. and M. Vergne, 1982. "Classes caractéristiques équivariantes, formule de localisation en cohomologie équivariante," *C. R. Acad. Sci. Paris* **295**, no. 9, 539–541.

Bernstein, I., I. Gelfand, and S. Gelfand, 1973. "Schubert cells and cohomology of the spaces G/P," *Russ. Math. Surveys* **28**, 1–26.

Bertram, A. 1987. "An existence theorem for Prym special divisors," *Invent. Math.* **90**, no. 3, 669–671.

Bhatnagar, G. 1999. "A short proof of an identity of Sylvester," *Int. J. Math. Math. Sci.* **22**, no. 2, 431–435.

Białynicki-Birula, A. 1973. 'Some theorems on actions of algebraic groups," *Ann. of Math. (2)* **98**, 480–497.

Bifet, E., C. De Concini, and C. Procesi, 1990. "Cohomology of regular embeddings," *Adv. Math.* **82**, no. 1, 1–34.

Billera, L., S. Billey, and V. Tewari, 2018. "Boolean product polynomials and Schur-positivity," *Sém. Lothar. Combin.* **80B**, Art. 91, 12 pp.

Billey, S. 1999. "Kostant polynomials and the cohomology ring for G/B," *Duke Math. J.* **96**, no. 1, 205–224.

Billey, S. and M. Haiman, 1995. "Schubert polynomials for the classical groups," *J. Amer. Math. Soc.* **8**, no. 2, 443–482.

Billey, S. and V. Lakshmibai, 2000. *Singular Loci of Schubert Varieties*, Birkhäuser.

Björner, A. and F. Brenti, 2005. *Combinatorics of Coxeter Groups*, Springer.

Bloch, S. and D. Gieseker, 1971. "The positivity of the Chern classes of an ample vector bundle," *Invent. Math.* **12**, 112–117.

Bogomolov, F. 1987. "The Brauer group of quotient spaces of linear representations," *Izv. Akad. Nauk SSSR Ser. Mat.* **51**, no. 3, 485–516, 688; English translation in *Math. USSR-Izv.* **30** (1988), no. 3, 455–485.

Borel, A. 1953. "Sur la cohomologie des espaces fibrés principaux et des espaces homogènes de groupes de Lie compacts," *Ann. Math.* **57**, 115–207.

——, 1960. *Seminar on Transformation Groups*, with contributions from G. Bredon, E. E. Floyd, D. Montgomery, and R. Palais, Annals of Mathematics Studies, No. 46, Princeton.

——, 1991. *Linear Algebraic Groups*, Springer.

Borel, A. and A. Haefliger, 1961. "La classe d'homologie fondamentale d'un espace analytique," *Bull. Soc. Math. France* **89**, 461–513.

Borel, A. and J. C. Moore, 1960. "Homology theory for locally compact spaces," *Michigan Math. J.* **7**, 137–159.

Borho, W., J.-L. Brylinski, and R. MacPherson, 1989. *Nilpotent Orbits, Primitive Ideals, and Characteristic Classes: A Geometric Perspective in Ring Theory*, Birkhäuser.

Borisov, L., L. Chen, and G. Smith, 2004. "The orbifold Chow ring of toric Deligne–Mumford stacks," *J. Amer. Math. Soc.* **18**, 193–215.

Bott, R. and H. Samelson, 1955. "The integral cohomology ring of G/T," *Proc. Nat. Acad. Sci.* **7**, 490–493.

Bott, R. and L. Tu, 1995. *Differential Forms in Algebraic Topology*, Springer.

Bourbaki, N. 1981. *Groupes et Algèbres de Lie, Chp. IV-VI*, Masson.

Braden, T., L. Chen, and F. Sottile, 2008. "The equivariant Chow rings of Quot schemes," *Pacific J. Math.* **238**, no. 2, 201–232.

Bredon, G. E. 1974. "The free part of a torus action and related numerical equalities," *Duke Math. J.* **41**, 843–854.

Brion, M. 1989. "Spherical varieties: An introduction," *Topological Methods in Algebraic Transformation Groups* (ed. H. Kraft, T. Petrie, and G. W. Schwarz), Progress in Mathematics **80**, Birkhäuser Boston, 11–26.

——, "Piecewise polynomial functions, convex polytopes and enumerative geometry," *Parameter Spaces (Warsaw, 1994)*, 25–44, Banach Center Publ. **36**, Polish Acad. Sci. Inst. Math., Warsaw.

——, 1997a. "The structure of the polytope algebra," *Tohoku Math. J.* **49**, no. 1, 1–32.

——, 1997b. "Equivariant Chow groups for torus actions," *Transform. Groups* **2**, no. 3, 225–267.

——, 1998. "Equivariant cohomology and equivariant intersection theory," notes by Alvaro Rittatore, in *Representation Theories and Algebraic Geometry (Montreal, 1997)*, 1–37, Kluwer.

——, 2000. "Poincaré duality and equivariant (co)homology," *Michigan Math. J.* **48**, 77–92.

——, 2015. "On linearization of line bundles," *J. Math. Sci. Univ. Tokyo* **22**, no. 1, 113–147.

Brion, M. and V. Lakshmibai, 2003. "A geometric approach to standard monomial theory," *Represent. Theory* **7**, 651–680.

Brion, M. and M. Vergne, 1997a. "An equivariant Riemann-Roch theorem for complete, simplicial toric varieties," *J. Reine Angew. Math.* **482**, 67–92.

——, 1997b. "On the localization theorem in equivariant cohomology," `arXiv:9711003`.

Brosnan, P. and T. Y. Chow, 2018. "Unit interval orders and the dot action on the cohomology of regular semisimple Hessenberg varieties," *Adv. Math.* **329**, 955–1001.

Brown, E. H. 1982. "The cohomology of BSO_n and BO_n with integer coefficients," *Proc. Amer. Math. Soc.* **85**, no. 2, 283–288.

Buch, A. 2015. "Mutations of puzzles and equivariant cohomology of two-step flag varieties," *Ann. Math.* **182**, no. 1, 173–220.

Buch, A., P.-E. Chaput, L. Mihalcea, and N. Perrin, 2018. "A Chevalley formula for the equivariant quantum K-theory of cominuscule varieties," *Algebr. Geom.* **5**, no. 5, 568–595.

Buch, A., A. Kresch, K. Purbhoo, and H. Tamvakis, 2016. "The puzzle conjecture for the cohomology of two-step flag manifolds," *J. Algebraic Combin.* **44**, no. 4, 973–1007.

Buch, A., A. Kresch, and H. Tamvakis, 2015. "A Giambelli formula for even orthogonal Grassmannians," *J. Reine Angew. Math.* **708**, 17–48.

——, 2017. "A Giambelli formula for isotropic Grassmannians," *Selecta Math.* **23**, no. 2, 869–914.

Buch, A. and R. Rimányi, 2004. "Specializations of Grothendieck polynomials," *C. R. Acad. Sci. Paris, Ser. I* **339**, 1–4.

Carrell, J. B. 1994. "The Bruhat graph of a Coxeter group, a conjecture of Deodhar, and rational smoothness of Schubert varieties," in *Algebraic Groups and Their Generalizations: Classical Methods* (American Mathematical Society, University Park, PA, 1991), 53–61, *Proc. Sympos. Pure Math.*, **56**, Part 1, Amer. Math. Soc.

Cartan, H. 1957. "Quotient d'un espace analytique par un groupe d'automorphismes," in *A Symposium in Honor of S. Lefschetz: Algebraic Geometry and Topology*, 90–102, Princeton University Press.

Cayley, A. 1849. "On the order of certain systems of algebraical equations," *Cambridge and Dublin Math. J.* **4**, 132–137.

Chang, T. and T. Skjelbred, 1974. "The topological Schur lemma and related results," *Ann. Math.* **100**, 307–321.

Chevalley, C. 1994. "Sur les décompositions cellulaires des espaces G/B," *Proc. Sympos. Pure Math.* **56** 1–23, Amer. Math. Soc., American Mathematical Society.

Chriss, N. and V. Ginzburg, 1997. *Representation Theory and Complex Geometry*, Birkhäuser.

Ciocan-Fontanine, I., B. Kim, and C. Sabbah, 2008. "The abelian/nonabelian correspondence and Frobenius manifolds," *Invent. Math.* **171**, no. 2, 301–343.

Coşkun, İ. 2009. "A Littlewood–Richardson rule for two-step flag varieties," *Invent. Math.* **176**, no. 2, 325–395.

Coşkun, İ. and R. Vakil, 2009. "Geometric positivity in the cohomology of homogeneous spaces and generalized Schubert calculus," *Algebraic geometry—Seattle 2005, Part 1, 77–124, Proc. Sympos. Pure Math.*, **80**, Amer. Math. Soc., American Mathematical Society.

Cox, D., J. Little, and H. Schenck, 2011. *Toric Varieties*, Amer. Math. Soc.

Danilov, V. 1978. "The geometry of toric varieties," *Russ. Math. Surveys* **33**, 97–154.

De Concini, C. and P. Pragacz, 1995. "On the class of Brill–Noether loci for Prym varieties," *Math. Ann.* **302**, no. 4, 687–697.

Demazure, M. 1974. "Désingularisation des variétés de Schubert généralisées," *Ann. Sci. École Norm. Sup. (4)* **7**, 53–88.

Deodhar, V. 1985. "On some geometric aspects of Bruhat orderings I: A finer decomposition of Bruhat cells," *Invent. Math.* **79**, no. 3, 499–511.

Dold, A. 1980. *Lectures on Algebraic Topology*, 2nd ed., Springer.

Duistermaat, J. and G. Heckman, 1982. "On the variation in the cohomology of the symplectic form of the reduced phase space," *Invent. Math.* **69**, no. 2, 259–268.

Edidin, D. and W. Graham, 1998. "Equivariant intersection theory," *Invent. Math.* **131**, 595–634.

Ehresmann, C. 1934. "Sur la topologie de certains espaces homogènes," *Ann. Math. (2)* **35**, no. 2, 396–443.

Ellingsrud, G. and S. A. Strømme, 1996. "Bott's formula and enumerative geometry," *J. Amer. Math. Soc.* **9**, no. 1, 175–193.

Evain, L. 2007. "The Chow ring of punctual Hilbert schemes on toric surfaces," *Transform. Groups* **12**, no. 2, 227–249.

Fehér, L. and R. Rimányi, 2002. "Classes of degeneracy loci for quivers—the Thom polynomial point of view," *Duke Math. J.* **114**, 193–213.

——, 2003. "Schur and Scubert polynomials as Thom polynomials—cohomology of moduli spaces," *Cent. European J. Math.* **4**, 418–434.

Feshbach, M. 1983. "The integral cohomology rings of the classifying spaces of $O(n)$ and $SO(n)$," *Indiana Univ. Math. J.* **32**, no. 4, 511–516.

Field, R. 2012. "The Chow ring of the classifying space $BSO(2n, \mathbb{C})$," *J. Algebra* **350**, 330–339.

Fomin, S. and C. Greene, 1998. "Noncommutative Schur functions and their applications," *Discrete Math.* **193**, 179–200.

Fomin, S. and A. N. Kirillov, 1996a. "The Yang-Baxter equation, symmetric functions, and Schubert polynomials" in *Proceedings of the 5th Conference on Formal Power Series and Algebraic Combinatorics* (Florence, 1993) and *Discrete Math.* **153**, no. 1–3, 123–143.

——, 1996b. "Combinatorial B_n-analogues of Schubert polynomials," *Trans. Amer. Math. Soc.* **348**, no. 9, 3591–3620.

Fomin, S. and R. P. Stanley, 1994. "Schubert polynomials and the nil-Coxeter algebra," *Adv. Math.* **103**, no. 2, 196–207.

Franz, M. and V. Puppe, 2007. "Exact cohomology sequences with integral coefficients for torus actions," *Transform. Groups* **12**, no. 1, 65–76.

——, 2011. "Exact sequences for equivariantly formal spaces," *C.R. Math. Acad. Sci. Soc. R. Can.* **33**, no. 1, 1–10.

Fulton, W. 1992. "Flags, Schubert polynomials, degeneracy loci, and determinantal formulas," *Duke Math. J.* **65**, no. 3, 381–420.

——, 1993. *Introduction to Toric Varieties*, Princeton University Press.

——, 1996a. "Schubert varieties in flag bundles for the classical groups," *Proceedings of the Hirzebruch 65 Conference on Algebraic Geometry (Ramat Gan, 1993)*, 241–262, Israel Math. Conf. Proc. **9**, Bar-Ilan University, Ramat Gan.

——, 1996b. "Determinantal formulas for orthogonal and symplectic degeneracy loci," *J. Differential Geom.* **43**, no. 2, 276–290.

——, 1997. *Young Tableaux*, Cambridge University Press.

——, 1998. *Intersection Theory*, 2nd ed., Springer.

Fulton, W. and R. Lazarsfeld, 1983. "Positive polynomials for ample vector bundles," *Ann. Math.* **118**, no. 1, 35–60.

Fulton, W. and R. MacPherson, 1978. "Defining algebraic intersections," *Algebraic geometry (Proc. Sympos., Univ. Tromsø, Tromsø, 1977)*, pp. 1–30, Lecture Notes in Math. 687, Springer, Berlin.

——, 1981. "Categorical framework for the study of singular spaces," *Mem. Amer. Math. Soc.* **31**, no. 243.

Fulton, W. and P. Pragacz, 1998. *Schubert Varieties and Degeneracy Loci*, Springer.

Fulton, W. and C. Woodward, 2004. "On the quantum product of Schubert classes," *J. Algebraic Geom.* **13**, 641–661.

Giambelli, G. 1904. "Ordine di una varietà più ampia di quella rappresentata coll'annullare tutti i minori di dato ordine estratti da una data matrice generica di forme," *Mem. R. Ist Lombardo* **11**, no. 3, 101–135.

——, 1906. "Sulle varietà rappresentate coll'annullare determinanti minori contenuti in un determinante simmetrico od emisimmetrico generico di forme," *Atti R. Acad. Torino)* **41**, 102–125.

Goresky, M., R. Kottwitz, and R. MacPherson, 1998. "Equivariant cohomology, Koszul duality, and the localization theorem," *Invent. Math.* **131**, no. 1, 25–83.

Graham, W. 1997. "The class of the diagonal in flag bundles," *J. Differential Geometry* **45**, 471–487.

——, 2001. "Positivity in equivariant Schubert calculus," *Duke Math. J.* **109**, 599–614.

Grauert, H. and R. Remmert, 1984. *Coherent Analytic Sheaves*, Springer-Verlag.

Grothendieck, A. 1957. "Sur quelques points d'algèbre homologique," *Tôhoku Math. J.* **9**, 119–221.

——, 1958. "Torsion homologique et sections rationelles," *Séminaire Claude Chevalley*, tome 3, exp. 5.

Grothendieck, A. and J. Dieudonné, 1971. *Eléments de Géométrie Algébrique I*, Springer.

Gu, X. 2021. "On the cohomology of the classifying spaces of projective unitary groups," *J. Topol. Anal.* **13**, no. 2, 535–573.

Guillemin, V. and S. Sternberg, 1982. "Convexity properties of the moment mapping," *Invent. Math.* **67**, no. 3, 491–513.

Guillemin, V. and C. Zara, 2001. "1-skeleta, Betti numbers, and equivariant cohomology," *Duke Math. J.* **107**, 283–349.

Hansen, H. C. 1974. "On cycles in flag manifolds," *Math. Scand.* **33** (1973), 269–274.

Harris, J. 1992. *Algebraic Geometry: A First Course*, Springer-Verlag.

Harris, J. and L. Tu, 1984. "On symmetric and skew-symmetric determinantal varieties," *Topology* **23**, 71–84.

Hatcher, A. 2002. *Algebraic Topology*, Cambridge University Press.

Helgason, S. 2001. *Differential Geometry, Lie Groups, and Symmetric Spaces*, Corrected reprint of the 1978 original, American Mathematical Society.

Higashitani, A., K. Kurimoto, and M. Masuda, 2022. "Cohomological rigidity for toric Fano manifolds of small dimensions or large Picard numbers," *Osaka J. Math.* **59**, no. 1, 177–215.

Holmann, H. 1960/1961. "Quotienten komplexer Räume," *Math. Ann.* **142**, 407–440.

Hsiang, W. Y. 1975. *Cohomology Theory of Topological Transformation Groups*, Springer-Verlag.

Humphreys, J. 1981. *Linear Algebraic Groups*, Springer.

——, 1990. *Reflection Groups and Coxeter Groups*, Cambridge University Press.

Husemoller, D. 1975. *Fibre Bundles*, Springer-Verlag.

Ikeda, T. 2007. "Schubert classes in the equivariant cohomology of the Lagrangian Grassmannian," *Adv. Math.* **215**, no. 1, 1–23.

Ikeda, T. and T. Matsumura, 2015. "Pfaffian sum formula for the symplectic Grassmannians," *Math. Z.* **280**, 269–306.

Ikeda, T., L. Mihalcea, and H. Naruse, 2011. "Double Schubert polynomials for the classical groups," *Adv. Math.* **226**, 840–886.

Ivanov, V. N. 2005. "Interpolation analogues of Schur Q-functions," *Zap. Nauchn. Sem. S.-Peterburg. Otdel. Mat. Inst. Steklov* **307** (2004), 99–119, 281–282; translation in *J. Math. Sci. (N. Y.)* **131**, 5495–5507.

Iversen, B. 1972. "A fixed point formula for action of tori on algebraic varieties," *Invent. Math.* **16**, 229–236.

Jacobson, N. 1985. *Basic Algebra I*, 2nd ed., W. H. Freeman.

Jantzen, J. C. 2003. *Representations of Algebraic Groups*, 2nd ed., American Mathematical Society.

Joseph, A. 1984. "On the variety of a highest weight module," *J. Algebra* **88**, no. 1, 238–278.

Józefiak, T., A. Lascoux, and P. Pragacz, 1981. "Classes of determinantal varieties associated with symmetric and skew-symmetric matrices," *Izv. Akad. Nauk SSSR Ser. Mat.* **45**, no. 3, 662–673.

Jurkiewicz, J. 1980. "Chow ring of projective nonsingular torus embedding," *Colloq. Math.* **43**, no. 2, 261–270.

Kazarian, M. 2000. "On Lagrange and symmetric degeneracy loci," Isaac Newton Institute for Mathematical Sciences Preprint Series.

Keel, S. 1992. "Intersection theory of moduli space of stable n-pointed curves of genus zero," *Trans. Amer. Math. Soc.* **330**, no. 2, 545–574.

Kempf, G. and D. Laksov, 1974. "The determinantal formula of Schubert calculus," *Acta Math.* **132**, 153–162.

Kleiman, S. 1974. "The transversality of a general translate," *Compositio Math.* **28**, 287–297.

——, 1980. "Chasles's enumerative theory of conics: a historical introduction," *Studies in Algebraic Geometry*, 117–138, MAA Stud. Math. 20, Math. Assoc. America, Washington, DC.

Kleiman, S. and D. Laksov, 1974. "Another proof of the existence of special divisors," *Acta Math.* **132**, 163–176.

Knuth, D. 1996. "Overlapping Pfaffians," *Electronic J. Combinatorics* **3**, no. 2, 13 pp.

Knutson, A. 2003. "A Schubert calculus recurrence from the noncomplex W-action on G/B," `arXiv:0306304`.

Knutson, A., T. Lam, and D. Speyer, 2013. "Positroid varieties: Juggling and geometry," *Compositio Math.* **149**, no. 10, 1710–1752.

Knutson, A. and E. Miller, 2004. "Subword complexes in Coxeter groups," *Adv. Math.* **184**, no. 1, 161–176.

——, 2005. "Gröbner geometry of Schubert polynomials," *Ann. Math.* **161** , no. 3, 1245–1318.

Knutson, A. and T. Tao, 2003. "Puzzles and (equivariant) cohomology of Grassmannians," *Duke Math. J.* **119**, no. 2, 221–260.

Knutson, A., A. Woo, and A. Yong, 2013. "Singularities of Richardson varieties," *Math. Res. Lett.* **20**, no. 2, 391–400.

Knutson, A. and P. Zinn-Justin, 2020. "Schubert puzzles and integrability I: invariant trilinear forms," preprint, `arXiv:1706.10019v6`.

Kontsevich, M. 1995. "Enumeration of rational curves via torus actions," *The Moduli Space of Curves (Texel Island, 1994)*, 335–368, Progr. Math., 129, Birkhäuser Boston, Boston, MA.

Kostant, B. and S. Kumar, 1986. "The nil Hecke ring and cohomology of G/P for a Kac–Moody group G," *Adv. Math.* **62**, 187–237.

Koszul, J.-L. 1953. "Sur certains groupes de transformations de Lie," *Géométrie différentielle. Colloques Internationaux du Centre National de la Recherche Scientifique, Strasbourg, 1953*, 137–141. Centre National de la Recherche Scientifique, Paris.

Kreiman, V. 2010. "Equivariant Littlewood–Richardson skew tableaux," *Trans. Amer. Math. Soc.* **362**, no. 5, 2589–2617.

Kresch, A. and H. Tamvakis, 2002. "Double Schubert polynomials and degeneracy loci for the classical groups," *Ann. Inst. Fourier (Grenoble)* **52**, no. 6, 1681–1727.

Kumar, S. 1996. "The nil Hecke ring and singularity of Schubert varieties," *Invent. Math.* **123**, no. 3, 471–506.

——, 2002. *Kac–Moody Groups, their Flag Varieties and Representation Theory*, Birkhäuser.

Kumar, S. and M. V. Nori, 1998. "Positivity of the cup product in cohomology of flag varieties associated to Kac–Moody groups," *Internat. Math. Res. Notices*, no. 14, 757–763.

Lam, T., S. Lee, and M. Shimozono, 2021. "Back stable Schubert calculus," *Compositio Math.* **157**, 883–962.

Lascoux, A. 1978. "Classes de Chern d'un produit tensoriel," *C. R. Acad. Sci. Paris* **286A**, 385–387.

Lascoux, A. and P. Pragacz, 1998. "Operator calculus for $\widetilde{Q}$-polynomials and Schubert polynomials," *Adv. Math.* **140**, 1–43.

Lascoux, A. and M.-P. Schützenberger, 1982. "Polynômes de Schubert," *C.R. Acad. Sci. Paris Sér. I Math.* **294**, 447–450.

——, 1985. "Interpolation de Newton à plusieurs variables," *Séminaire d'algèbre Paul Dubreil et Marie-Paule Malliavin, 36ème année (Paris, 1983-1984)*, 161–175, Lecture Notes in Math., 1146, Springer, Berlin.

Lazarsfeld, R. 2004. *Positivity in Algebraic Geometry*, Springer.

Li, L. 2009. "Chow motive of Fulton-MacPherson configuration spaces and wonderful compactifications," *Michigan Math. J.* **58**, no. 2, 565–598.

Macdonald, I. G. 1991. *Notes on Schubert Polynomials*, Publ. LACIM 6, Univ. de Québec à Montréal, Montréal.

——, 1992. "Schur functions: theme and variations," *Séminaire Lotharingien de Combinatoire (Saint-Nabor, 1992)*, 5–39, *Publ. Inst. Rech. Math. Av.* **498**, Strasbourg.

——, 1995. *Symmetric functions and Hall polynomials*, 2nd ed., Oxford University Press.

Massey, W. S. 1978. "How to give an exposition of the Čech–Alexander–Spanier type homology theory," *Amer. Math. Monthly* **85**, no. 2, 75–83.

Masuda, M. 2008. "Equivariant cohomology distinguishes toric manifolds," *Adv. Math.* **218**, no. 6, 2005–2012.

Metropolis, N., G. Nicoletti, and G.-C. Rota, 1981. "A new class of symmetric functions," *Mathematical Analysis and Applications, Part B*, 563–575, Adv. in Math. Suppl. Stud., 7b, Academic Press, New York-London.

Miller, E. and B. Sturmfels, 2005. *Combinatorial Commutative Algebra*, Springer-Verlag.

Milnor, J. W. 1956. "Construction of universal bundles II," *Ann. Math.* **63**, 430–436.

Milnor, J. W. and J. D. Stasheff. 1974. *Characteristic Classes*, Princeton University Press.

Miwa, T., M. Jimbo, and E. Date, 2000. *Solitons: Differential Equations, Symmetries and Infinite-Dimensional Algebras*, translated from the 1993 Japanese original by Miles Reid, Cambridge University Press.

Molev, A. 2009. "Littlewood–Richardson polynomials," *J. Algebra* **321**, no. 11, 3450–3468.

Molev, A. and B. Sagan, 1999. "A Littlewood–Richardson rule for factorial Schur functions," *Trans. Amer. Math. Soc.* **351**, no. 11, 4429–4443.

Monk, D. 1959. "The geometry of flag manifolds," *Proc. London Math. Soc.* **9**, 253–286.

Mumford, D., J. Fogarty, and F. Kirwan, 1994. *Geometric Invariant Theory*, third enlarged edition, Springer-Verlag.

Nimmo, J. J. C. 1990. "Hall-Littlewood symmetric functions and the BKP equation," *J. Phys. A* **23**, no. 5, 751–760.

Nyenhuis, M. 1993. *Equivariant Chow Groups and Multiplicites*, Ph. thesis, University of British Columbia.

Okounkov, A. 1996. "Quantum immanants and higher Capelli identities," *Transformation Groups* **1**, 99–126.

Okounkov, A. and G. Olshanski, 1997. "Shifted Schur functions," *Algebra i Analiz* **9**, no. 2, 73–146; translation in *St. Petersburg Math. J.* **9** (1998), no. 2, 239–300.

Pawlowski, B. 2021. "Universal graph Schubert varieties," *Transform. Groups* **26**, no. 4, 1417–1461.

Payne, S. 2006. "Equivariant Chow cohomology of toric varieties," *Math. Res. Lett.* **13**, no. 1, 29–41.

Polo, P. 1994. "On Zariski tangent spaces of Schubert varieties, and a proof of a conjecture of Deodhar," *Indag. Math.* **5**, no. 4, 483–493.

Pragacz, P. 1988. "Enumerative geometry of degeneracy loci," *Ann. Sci. École Norm. Sup.* **21**, no. 3, 413–454.

——, "Algebro-geometric applications of Schur S- and Q-polynomials," in *Topics in Invariant Theory, Séminaire d'Algèbre Dubreil-Malliavin 1989-1990* (M.-P. Malliavin ed.), Springer Lect. Notes in Math. **1478**, 130-191.

——, 1996. "Symmetric polynomials and divided differences in formulas of intersection theory," in *Parameter Spaces (Warsaw, 1994)*, 125–177, Banach Center Publ., 36, Polish Acad. Sci., Warsaw.

Pragacz, P. and J. Ratajski, 1997. "Formulas for Lagrangian and orthogonal degeneracy loci; $\widetilde{Q}$-polynomial approach," *Compositio Math.* **107**, no. 1, 11–87.

Pressley, A. and G. Segal, 1986. *Loop Groups*, Oxford University Press.

Quart, G. 1979. "Localization theorem in K-theory for singular varieties," *Acta Math.* **143**, no. 3–4, 213–217.

Quillen, D. 1971. "Elementary proofs of some results of cobordism theory using Steenrod operations," *Adv. Math.* **7**, 29–56.

Ramanathan, A. 1985. "Schubert varieties are arithmetically Cohen–Macaulay," *Invent. Math.* **80**, no. 2, 283–294.

Richardson, R. W. 1992. "Intersections of double cosets in algebraic groups," *Indag. Math. (N.S.)* **3**, no. 1, 69–77.

Rossmann, W. 1989. "Equivariant multiplicities on complex varieties," *Astérisque* **173–174**, no. 11, 313–330.

Rothe, H. A. 1800. "Über Permutationen, in Beziehung auf die Stellen ihrer Elemente. Anwendung der daraus abgeleiteten Satze auf das Eliminationsproblem," in Hindenburg, Carl, ed., *Sammlung Combinatorisch-Analytischer Abhandlungen*, 263–305, Bey G. Fleischer dem jüngern.

Salmon, G. 1852. *A Treatise on the Higher Plane Curves*, 1st ed., Hodges & Smith, Dublin.

Schenck, H. 2012. "Equivariant Chow cohomology of nonsimplicial toric varieties," *Trans. Amer. Math. Soc.* **364**, no. 8, 4041–4051.

Schur, I. 1911. "Über die Darstellung der symmetrischen und der alternierenden Gruppe durch gebrochene lineare Substitutionen," *J. Reine Angew. Math.* **139**, 155–250.

Schützenberger, M.-P. 1977. "La correspondance de Robinson," in *Combinatoire et représentation du groupe symétrique (Actes Table Ronde CNRS, Univ. Louis-Pasteur Strasbourg, Strasbourg, 1976)*, Springer Lecture Notes in Math. **579**, 59–113.

Sottile, F. 1997. "Pieri's formula via explicit rational equivalence," *Canad. J. Math.* **49**, no. 6, 1281–1298.

Spanier, E. 1966. *Algebraic Topology*, McGraw-Hill.

Speiser, R. 1988. "Transversality theorems for families of maps," *Algebraic Geometry (Sundance, UT, 1986)*, 235–252, Lecture Notes in Math. **1311**, Springer.

Springer, T. A. 1998. *Linear Algebraic Groups*, 2nd ed., Birkhäuser.

Stanley, R. P. 1984. "On the number of reduced decompositions of elements of Coxeter groups," *European J. Combin.* **5**, no. 4, 359–372.

——, 1996. *Combinatorics and Commutative Algebra*, 2nd ed., Birkhäuser.

——, 1999. *Enumerative Combinatorics*, Volume 2, with an appendix by S. Fomin, Cambridge University Press.

Steiner, J. 1848. "Elementare Lösung einer geometrischen Aufgabe, und über einige damit in Beziehung stehende Eigenschaften der Kegelschnitte," *J. Reine Angew. Math.* **37**, 161–192.

Stembridge, J. R. 1985. "A characterization of supersymmetric polynomials," *J. Algebra* **95**, 439–444.

Sumihiro, H. 1974. "Equivariant completion," *J. Math. Kyoto Univ.* **14**, 1–28.

——, 1975. "Equivariant completion II," *J. Math. Kyoto Univ.* **15**, 573–605.

Tamvakis, H., 2016a "Double eta polynomials and equivariant Giambelli formulas," *J. Lond. Math. Soc.* **94**, no. 1, 209–229.

——, 2016b. "Giambelli and degeneracy locus formulas for classical G/P spaces," *Mosc. Math. J.* **16**, no. 1, 125–177.

Tamvakis, H. and E. Wilson, 2016. "Double theta polynomials and equivariant Giambelli formulas," *Math. Proc. Cambridge Philos. Soc.* **160**, no. 2, 353–377.

Thomas, H. and A. Yong, 2009. "A combinatorial rule for (co)miniscule Schubert calculus," *Adv. Math.* **222**, no. 2, 596–620.

——, 2018. "Equivariant Schubert calculus and jeu de taquin," *Ann. Inst. Fourier (Grenoble)* **68**, no. 1, 275–318.

tom Dieck, T. 1987. *Transformation Groups*, de Gruyter, Berlin.

Totaro, B. 1999. "The Chow ring of a classifying space," *Proc. Sympos. Pure Math.* **67**, AMS, 249–281.

——, 2014. *Group Cohomology and Algebraic Cycles*, Cambridge University Press.

Tymoczko, J., 2008a. "Permutation actions on equivariant cohomology of flag varieties," *Toric topology*, 365–384, Contemp. Math. **460**, Amer. Math. Soc., Providence, RI.

——, 2008b. "Permutation representations on Schubert varieties," *Amer. J. Math.* **130**, 1171–1194.

——, 2008c. "Equivariant structure constants for ordinary and weighted projective space," arXiv:0806.3588.

Vakil, R. 2006. "A geometric Littlewood–Richardson rule, with an appendix with A. Knutson," *Ann. Math.* **164**, no. 2, 371–421.

Vistoli, A. 2007. "On the cohomology and the Chow ring of the classifying space of PGL_p," *J. Reine Angew. Math.* **610**, 181–227.

Watanabe, T. 1987. "The continuity axiom and the Čech homology," in *Geometric topology and Shape Theory (Dubrovnik, 1986)*, 221–239, Lecture Notes in Math. **1283**, Springer, Berlin.

Weber, A. 2003. "Formality of equivariant intersection cohomology of algebraic varieties," *Proc. Amer. Math. Soc.* **131**, no. 9, 2633–2638.

Willems, M. 2004. "Cohomologie et K-théorie équivariantes des variétés de Bott–Samelson et des variétés de drapeaux," *Bull. Soc. Math. France* **132**, 569–589.

Notation Index

Subject Index